Lacrimal Gland, Tear Film, and Dry Eye Syndromes 3

Basic Science and Clinical Relevance

Part B

ADVANCES IN EXPERIMENTAL MEDICINE AND BIOLOGY

Lacrimal Gland, Tear Film, and Dry Eye Syndromes 3

Basic Science and Clinical Relevance
Part B

Edited by

David A. Sullivan
Schepens Eye Research Institute and Harvard Medical School
Boston, Massachusetts, USA

Michael E. Stern
Allergan, Inc.
Irvine, California, USA

Kazuo Tsubota
Tokyo Dental College and Ichikawa General Hospital
Chiba, Japan

Darlene A. Dartt
Schepens Eye Research Institute and Harvard Medical School
Boston, Massachusetts, USA

Rose M. Sullivan
Schepens Eye Research Institute
Boston, Massachusetts, USA

and

B. Britt Bromberg
University of New Orleans
New Orleans, Louisiana, USA

Kluwer Academic / Plenum Publishers
New York, Boston, Dordrecht, London, Moscow

Library of Congress Cataloging-in-Publication Data

International Conference on the Lacrimal Gland, Tear Film, and Dry Eye Syndromes
(3rd: 2000: Maui, Hawaii)
    Lacrimal gland, tear film, and dry eye syndromes 3: basic science and clinical
relevance/edited by David A. Sullivan ... [et al.].
        p. ; cm. — (Advances in experimental medicine and biology; v. 506)
    "Proceedings of the 3rd International Conference on the Lacrimal Gland, Tear Film,
and Dry Eye Syndromes: Basic Science and Clinical Relevance, November 15–18, 2000,
Maui, Hawaii"—T.p. verso.
    Includes bibliographical references and index.
    ISBN 0-306-47282-1
     1. Dry eye syndromes—Congresses. 2. Lacrimal apparatus—Congresses. 3.
Tears—Congresses. I. Title: Lacrimal gland, tear film, and dry eye syndromes three. II.
Sullivan, David A. III. Title. IV. Series.
    [DNLM: 1. Lacrimal Apparatus—Congresses. 2. Dry Eye Syndromes—Congresses. 3.
Tears—physiology—Congresses. WW 208 I6125 2002]
RE216.D78 I55 2000
617.7′64—dc21

                          2002072992

Proceedings of the 3rd International Conference on the Lacrimal Gland, Tear Film and Dry Eye Syndromes:
Basic Science and Clinical Relevance, November 15–18, 2000; Maui, Hawaii

ISBN 0-306-47282-1

©2002 Kluwer Academic / Plenum Publishers, New York
233 Spring Street, New York, New York 10013

http://www.wkap.nl/

10 9 8 7 6 5 4 3 2 1

A C.I.P. record for this book is available from the Library of Congress

Immunity and Inflammation I: Influence of Cytokines, Chemokines, Adhesion Molecules and Other Immunoactive Agents

ROLE OF IMMUNITY AND INFLAMMATION IN CORNEAL AND OCULAR SURFACE DISEASE ASSOCIATED WITH DRY EYE

M. Reza Dana[1-3] and Pedram Hamrah[1]

[1]Laboratory of Immunology
Schepens Eye Research Institute and the
Department of Ophthalmology
Harvard Medical School
[2]Brigham and Women's Hospital
[3]Massachusetts Eye and Ear Infirmary
Boston, Massachusetts, USA

1. INTRODUCTION

Dry eye syndromes (DES) represent a common but highly heterogeneous group of ocular surface disorders (OSD) that affect millions of individuals, in particular women, in the United States alone.[1] While efforts to better classify disease categories[2,3] and determine relevant diagnostic tests for each disease subtype[2,4] continue, common features of DES are shared by the tear-deficient and non-tear-deficient (evaporative) forms of DES serving as common denominators of disease. These common denominators include ocular surface epitheliopathy, tear hyperosmolarity, an unstable preocular tear film and symptoms of dry eye associated with varying degrees of inflammation.[2] As such, an increasing number of investigators contend ocular irritation or discomfort, virtually invariable components of severe tear-deficient dry eye, are clinical correlates of ocular surface inflammation—which at the cellular and molecular levels are requisite factors for the pathogenesis of DES.[5-9] In this brief overview, we will first examine the factors involved in the generation of immune responsiveness to antigens in the ocular surface and anterior segment. We will then critically evaluate whether the data generated to date provide a causal link between the generation of adaptive immunity and OSD in DES, or if the well-described inflammatory response in the ocular surface could be primarily a consequence of the myriad pathological

Lacrimal Gland, Tear Film, and Dry Eye Syndromes 3
Edited by D. Sullivan *et al.*, Kluwer Academic/Plenum Publishers, 2002

729

phenomena that characterize all forms of DES. Finally, a hypothetical model is offered that will attempt to conceptually bridge the gap relating causality and empiric evidence in relation to immunoinflammatory responses in DES.

2. GENERATION OF ADAPTIVE IMMUNITY

"Adaptive immunity," whether (T) cell-mediated or   humoral (antibody-mediated), represents that arm of the immune response generated de novo or secondarily to specific antigen(s). As such, it is characterized by a delayed clonal response specific to certain epitopes. In contrast, "innate immunity" refers to that segment of the immune response whose receptors and effector elements are fixed in the genome, and hence the reaction is immediate and non-clonal.[10] The latter is typified by neutrophils and macrophages and their released proteins and peptides generated in response to local perturbations—whether traumatic (e.g., as occurs in wound healing), infectious or immune (involving new antigens or epitopes). Since "immune-mediated inflammation" is increasingly being cited as a primary process in the pathogenesis of both autoimmune (e.g., Sjøgren's) and non-autoimmune (e.g., primary sicca syndrome or evaporative DES secondary to meibomian gland dysfunction) dry eye, it is important to first provide a brief outline of the processes that mediate immune responses in tissues.

The first step in generation of inflammation is an inciting stimulus. This may be microbial, traumatic or due to introduction of novel antigens (e.g., as occurs in tissue allotransplantation). These stimuli may lead to the release of proinflammatory cytokines, nucleic acid fragments, heat shock proteins and various other mediators that, in the aggregate, signal the host that normal physiology and the microenvironment have been violated. In response to these signals, the second step in this cascade of events occurs when local (resident) tissue cells activate signal transduction pathways (e.g., NFκ-B) that augment or downmodulate expression of cellular cytokine genes and/or cytokine receptor genes that, in turn, dictate the response of these resident cells to paracrine signals in the microenvironment by other cells in proximity. These responses are not limited to classic immunoinflammatory mediators (e.g., interleukins and interferons) but also include other molecular classes such as growth factors, chemokines and adhesion factors[11] acting in a coordinated fashion to regulate the immune/inflammatory response in the tissue. Mild stimuli (e.g., a minor abrasive injury to an epithelialized tissue) may not generate a response beyond this step. However, generation of primary adaptive immune responses, mediated by stimulation of antigen-specific cells in lymphoid organs, requires recruitment and activation of bone marrow-derived cells. The latter is accomplished through the function of two classes of immunoactive agents—cell adhesion molecules (CAMs) and chemokines.

CAMs represent a heterogeneous group of molecules composed of members of the immunoglobulin superfamily (e.g., ICAM-1), selectins (e.g., E-selectin) and integrins that act in concert to activate leukocytes and enhance cell-cell interactions.[12] These cell-cell interactions are critical in increasing leukocyte adhesion to vascular endothelial cells

(VEC) and thereby allow them to "roll" on the VEC prior to integrin-mediated activation and finally transendothelial migration. Additionally, certain CAMs such as ICAM-1 are important as costimulatory factors that provide naïve T-cells with the requisite second (T-cell receptor-independent) signal for sensitization. In corneal immune and inflammatory diseases, the central role of ICAM-1 is now well established.[13,14]

Once the process of leukocyte activation and transendothelial migration is accomplished, leukocytes, whether antigen-presenting cells (APC) or effectors, require recruitment to the primary site of inflammation. If APC (e.g., dendritic cells or macrophages), they initiate the process of antigen uptake and processing prior to T-cell priming. However, whereas transendothelial migration from the intravascular compartment to the tissue compartment is a requisite step for leukocyte function in the periphery, it is insufficient since leukocytes do not have the capacity for independent homing. This directionality is provided to leukocytes by chemotactic cytokines, also known as chemokines, which provide a chemotactic gradient for leukocytes. To date, nearly 50 chemokine species (ligands) and nearly two dozen chemokine receptors have been characterized,[15] and chemokines are now recognized as important mediators of ocular surface and corneal disease including ocular allergy,[16] corneal graft rejection,[17,18] microbial keratitis[19,20] and possibly DES.[21] Examples of chemokines implicated in ocular immune disorders include RANTES, MIP-1alpha and IL-8, just to name a few. Hence, the coordinated and differential expression of select chemokines and CAMs leads to recruitment of APCs to sites of inflammation. These requisite steps in the generation of antigen-specific adaptive immunity are in turn related to upregulation in the activity of proinflammatory cytokines such as interleukin-1 (IL-1). For example, in the cornea and ocular surface the expression of ICAM-1 appears to be largely regulated by IL-1 expression.[13] Moreover, IL-1 in concert with tumor necrosis factor-alpha (TNF-α) leads to upregulation of select chemokines that in turn lead to activation and recruitment of ocular dendritic (e.g., Langerhans) cells.[22]

The role of "master molecules" such as IL-1 and TNF-α that are capable of (1) augmenting inflammation and innate immunity (e.g., IL-1, IL-6, TNF-α); (2) overexpressing factors involved in T helper cell differentiation, activation (e.g., IL-12, CD40) and proliferation (e.g., IL-2) and (3) APC maturation and recruitment is getting increasing attention in the corneal immunology literature,[22-25] because these molecules link innate and adaptive immune responses and may serve as therapeutic targets in a host of immune disorders.

3. OCULAR SURFACE INFLAMMATION IN DRY EYE SYNDROMES

A large body of evidence suggests clinically significant, and especially (but importantly not exclusively) tear-deficient, DES is associated with variable degrees of ocular surface inflammation. The hallmark of the inflamed and irritated eye is "red eye," characterized by vascular engorgement and variable degrees of matrix (stromal) edema. Nonspecific features of this tissue inflammation include extravasation of protein and fluid from "leaky" vessels with impaired barrier function at the level of their VEC, loss of epithelial barrier function and infiltration of leukocytes into the tissue matrix. At the molecular level, ample histological and flow cytometric data demonstrate enhanced expression of proinflammatory cytokine (e.g., IL-1, IL-6, IL-8, TNF-α) mRNA and protein by the ocular surface epithelium or tear film.[5,21,26–31] Conversely, treatment of DES, either clinically or in animal models of keratoconjunctivitis sicca, with anti-inflammatory agents (e.g., cyclosporin A) has been associated with improved disease endpoints either clinically[6,9,32–36] or at the level of the tissue.[5,26,37] In addition to these findings, DES-associated ocular surface epitheliopathy has been associated with a significantly enhanced level of class II major histocompatibility complex (MHC) antigen (HLA DR) expression by resident ocular surface epithelial cells.[5] Since presentation of processed (foreign or autoantigen) peptides in the context of class II MHC molecules is a classic pathway for priming of CD4-positive T-cells, it has been postulated but not proven that this phenomenon enables these epithelial cells to serve as "non-professional" (since they are not bone marrow-derived) APC,[5,38] as has been postulated for lacrimal gland epithelia.[39] The fact the expression of CD40 and its ligand (CD40L or CD154) is likewise significantly enhanced in OSD related to ocular sicca has similarly provided indirect support, but again no proof, these epithelial cells may have all the requisite phenotypic markers for stimulating a primary T-cell response.[5]

   In summary, the association between OSD in DES with immunoinflammatory markers on the one hand, and the subsequent downmodulation of these factors when dry eye is treated with anti-inflammatory/immunomodulatory treatments on the other, has provided indirect correlative evidence for the involvement of immunoactive agents in the pathophysiology of DES.

4. IS THE PATHOGENESIS OF DRY EYE SYNDROMES CAUSALLY RELATED TO IMMUNE MECHANISMS GENERATED IN THE OCULAR SURFACE?

   As noted above, the substantial empirical evidence linking DES with immunoinflammatory molecular and cellular markers in the ocular surface is sufficient to demonstrate a significant role for these mediators in the pathophysiology of OSD in dry eyes. The question, however, is not whether these mediators play a role, but rather whether this role is causal. Moreover, is the site of putative T-cell activation the ocular surface? If so, how would these molecular features fit with the paradigm of immunity described above, and with the various features of pathobiology described thus far in DES?

   We simply do not have adequate knowledge to answer any of these questions definitively. For example, the immune (HLA DR, CD40 overexpression) and inflammatory (IL-1, TNF-α, ICAM-1 overexpression) features of DES may similarly be applied to non-sicca-related ocular conditions such as atopic disease,[8] allograft rejection,[23,40] cicatrizing conjunctivitis,[41] ocular surface burns[42] and ocular graft-versus-host disease post-allogeneic bone marrow transplantation,[7] just to name a few. Hence, the overexpression of these factors is by no means specific to DES. In addition, reasons exist why some putative "immune" features of the OSD observed in DES may be non-specific consequences of inflammatory responses. For example, increased expression of HLA DR has been demonstrated in autoimmune (Sjøgren's) and non-autoimmune DES, and as has been argued by Baudouin and colleagues,[5] may well represent a nonspecific marker of inflammation, since proinflammatory cytokines such as interferon-gamma (IFN-γ) and TNF-α may lead to class II MHC overexpression.[31,43] Increased expression of CD40, a member of the TNF superfamily of receptors, which is expressed near-ubiquitously by nucleated cells, may likewise be secondary to inflammation[44] and not necessarily as a facet of APC-T-cell interaction. In fact, for the specific case of DES, central features of its pathology, namely alterations in the availability of lacrimal trophic factors and neural (and possibly endocrine) dysregulation, may amplify ocular surface inflammatory changes.[45,46]

We propose that while significant work remains in elucidating the molecular mechanisms of ocular surface inflammation in the pathophysiology of DES, it is unlikely a simple linear and highly reductionist cascade of events will emerge to explain all the pathological features of the different types of DES. However, ample scientific evidence exists from the ophthalmic and non-ophthalmic fields that can allow us to propose a model that could explain ocular surface inflammation as a cause and consequence of OSD in DES. In so doing, we hope to refrain from selective emphasis or de-emphasis of empiric evidence and provide investigators with a model system whose parts can be independently tested.

This model (Fig. 1) proposes that lacrimal gland disease, which itself has multifaceted dimensions (immune, apoptotic, neurendocrine), leads to OSD through various mechanisms—alterations in trophic factor availability, lacrimal insufficiency leading to secondary hyperosmolarity, etc. These features of the ocular surface pathology, potentially compounded by microabrasive effects of blinking in the dry eye state or other anatomic alterations, lead to upregulation of proinflammatory cytokines such as IL-1, IL-6 and TNF-α.[36] However, these cytokines are not simply "passive" markers of disease. Rather, these molecules are critical in amplifying the disease process in the local microenvironment of the cornea and conjunctiva. IL-1, for example, can lead to epithelial cell proliferation, keratinization and angiogenesis[47] and thereby possibly link OSD with lid margin disease—a commonly seen association. Additionally, IL-1 may lead to upregulation of matrix proteases,[48] including collagenases and therefore exacerbate stromal pathology as well as alter the paracrine effect of other cytokines on resident fibroblasts and epithelial cells. Inflammatory cytokines have also been implicated in the regulation of epithelial mucin expression[49] and may be relevant in the mucin alterations previously described in DES.[50,51] Ocular surface changes as a result of the cytokine microenvironment may lead to relative hypesthesia and changes in blink rate, a common feature of chronic ocular surface inflammatory disease, that may in turn exacerbate disease by altering the feedback to the lacrimal gland. This non-reductionist view of cytokine biology in the ocular surface relates disease parameters to an ocular-lacrimal gland "functional unit" as proposed by Stern and colleagues.[27, 52]

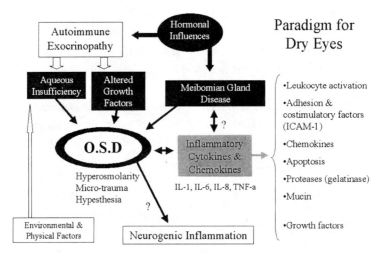

Figure 1. Proposed paradigm for dry eye disease.

5. FUTURE DIRECTIONS IN RESEARCH

Many critical questions should be answered in regard to the pathophysiology of OSD in DES. One critical area for further study is the activation and function of ocular surface APC in dry eye. Our laboratory has recently discovered that the dogma stating the cornea is devoid of any APC[53–60] is woefully inadequate. Interestingly, we found the normal cornea has a significant endowment of class II MHC⁻ but CD45+CD11c+ dendritic cells in the stroma and epithelium. These cells do not express B7 (CD80/CD86) costimulatory markers in the normal state and are hence "immature."[61] However, on exposure to inflammatory stimuli associated with inflammatory cytokine (IL-1, TNF-α) upregulation, they significantly alter their phenotype and express CD80, CD86 and class II MHC antigens (Fig. 2). We recently determined these cells are capable, at least in the transplant setting, of efficiently migrating to draining lymph nodes where they normally congregate in T-cell-rich areas in the parafollicular areas of the node.[62] Since the interaction of APC and T-cells in lymphoid organs, as opposed to peripherally, is far more efficient in inducing T-cell activation,[63] it is critical to determine how corneal and ocular surface APC function and trafficking is altered in DES, if indeed adaptive immune responses are germane to the disease mechanisms of OSD in DES.

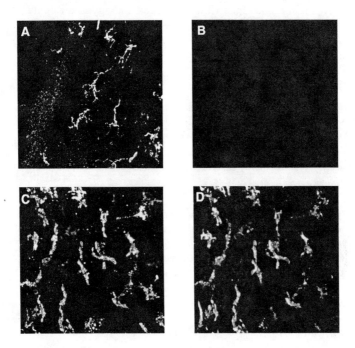

Figure 2. The center of the normal cornea is endowed with many CD11c+ dendritic cells (**A**). Double staining with CD80 (**B**) shows these cells do not express this maturation marker. In addition, these dendritic cells are MHC class II and CD86 negative. In the inflamed cornea, the number of CD11c+ dendritic cells increases (**C**). Moreover, significant change occurs in the phenotype of these cells and they upregulate CD80 (**D**), CD86 and MHC class II. Magnification (**A–D**) 400X.

Finally, it would be critical to determine if and to what extent the inflammatory changes observed in the ocular surface in severe DES are, at least in part, a protective and not solely deleterious mechanism. Accordingly, it is important to underscore the fact that immune responses are critical in host defense against pathogens.[10] The ocular surface epithelium relies on two main sources, the tear film and limbal vasculature, for provision of metabolites and growth factors. It may be provocative to hypothesize, in a disease state in which the normal contribution of the tear film to the ocular surface physiology is compromised, increased reliance on the vasculature (that also provides a conduit for immune and inflammatory effectors) may also facilitate the expression of inflammation.

In closing, much work remains to be done to elucidate the interface between immunity and OSD in DES. This work is important because it will likely help us better understand the disease mechanisms operative in DES and aid in better understanding of the role of inflammation and immunity in a host of diseases unrelated to dry eye.

REFERENCES

1. Schaumberg, DA, Sullivan, DA, and Dana, MR. Epidemiology of dry eye syndrome. *Adv Exp Med Biol.* 2001;in press:

2. Lemp, MA. Report of the national eye institute/industry workshop on clinical trials in dry eyes. *CLAO J.* 1995;21:221–232.

3. Fox, RI. Sjøgren's syndrome. *Curr Opin Rheumatol.* 1995;7:409–416.

4. Korb, DR. Survey of preferred tests for diagnosis of the tear film and dry eye. *Cornea.* 2000;19:483–486.

5. Brignole, F, Pisella, PJ, De Saint Jean, M, Goldschild, M, Goguel, A, and Baudouin, C. Flow cytometric analysis of inflammatory markers in KCS: 6-month treatment with topical cyclosporin A. *Invest Ophthalmol Vis Sci.* 2001;42:90–95.

6. Tsubota, K, Fujita, H, Tadano, K, et al. Improvement of lacrimal function by topical application of CyA in murine models of Sjøgren's syndrome. *Invest Ophthalmol Vis Sci.* 2001;42:101–110.

7. Ogawa, Y, Yamazaki, K, Kuwana, M, et al. A significant role of stromal fibroblasts in rapidly progressive dry eye in patients with chronic GVHD. *Invest Ophthalmol Vis Sci.* 2001;42:111–119.

8. Hingorani, M, Moodaley, L, Calder, VL, Buckley, RJ, and Lightman, S. A randomized, placebo-controlled trial of topical cyclosporin A in steroid-dependent atopic keratoconjunctivitis. *Ophthalmology.* 1998;105:1715–1720.

9. Stevenson, D, Tauber, J, and Reis, BL. Efficacy and safety of cyclosporin A ophthalmic emulsion in the treatment of moderate-to-severe dry eye disease: a dose-ranging, randomized trial. The cyclosporin A phase 2 study group. *Ophthalmology.* 2000;107:967–974.

10. Gura, T. Innate immunity. Ancient system gets new respect. *Science.* 2001;291:2068–2071.

11. Cotran, RS. Inflammation: Historical perspectives. In: Gallin, JI and Snyderman, R, eds. *Inflammation: Basic Principles and Clinical Correlates.* Third ed. Philadelphia: Lippincott Williams & Wilkins; 1999:5–10.

12. Springer, TA. Traffic signals for lymphocyte recirculation and leukocyte emigration: the multistep paradigm. *Cell.* 1994;76:301–314.

13. Zhu, SN and Dana, MR. Expression of cell adhesion molecules on limbal and neovascular endothelium in corneal inflammatory neovascularization. *Invest Ophthalmol Vis Sci.* 1999;40:1427–1434.

14. Zhu, SN, Yamada, J, Streilein, JW, and Dana, MR. ICAM-1 deficiency suppresses host allosensitization and rejection of MHC-disparate corneal transplants. *Transplantation.* 2000;69:1008–1013.

15. Mackay, CR. Chemokines: immunology's high impact factors. *Nat Immunol.* 2001;2:95–101.

16. Keane-Myers, AM, Miyazaki, D, Liu, G, Dekaris, I, Ono, S, and Dana, MR. Prevention of allergic eye disease by treatment with IL-1 receptor antagonist. *Invest Ophthalmol Vis Sci.* 1999;40:3041–3046.

17. Yamagami, S, Miyazaki, D, Ono, S, and Dana, MR. Differential chemokine gene expression after corneal transplantation. *Invest Ophthalmol Vis Sci.* 1999;40:2892–2897.

18. Hamrah, P, Liu, Y, Yamagami, S, Zhang, Q, Vora, S, and Dana, MR. Targeting the chemokine receptor CCR1 suppresses corneal alloimmunity and promotes allograft survival. *Clin Immunol.* 2001;(Suppl) *Cl Immunol* 2001; 99: 118.

19. Tumpey, TM, Cheng, H, Yan, XT, Oakes, JE, and Lausch, RN. Chemokine synthesis in the HSV-1-infected cornea and its suppression by interleukin-10. *J Leukoc Biol.* 1998;63:486–492.

20. Rudner, XL, Kernacki, KA, Barrett, RP, and Hazlett, LD. Prolonged elevation of IL-1 in pseudomonas aeruginosa ocular infection regulates macrophage-inflammatory protein-2 production, polymorphonuclear neutrophil persistence, and corneal perforation. *J Immunol.* 2000;164:6576–6582.

21. Pflugfelder, SC, Jones, D, Ji, Z, Afonso, A, and Monroy, D. Altered cytokine balance in the tear fluid and conjunctiva of patients with Sjøgren's syndrome keratoconjunctivitis sicca. *Curr Eye Res.* 1999;19:201–211.

22. Dekaris, I, Zhu, SN, and Dana, MR. TNF-α regulates corneal Langerhans cell migration. *J Immunol.* 1999;162:4235–4239.

23. Dana, MR, Yamada, J, and Streilein, JW. Topical interleukin 1 receptor antagonist promotes corneal transplant survival. *Transplantation.* 1997;63:1501–1507.

24. Yamada, J, Zhu, SN, Streilein, JW, and Dana, MR. Interleukin-1 receptor antagonist therapy and induction of anterior chamber-associated immune deviation-type tolerance after corneal transplantation. *Invest Ophthalmol Vis Sci.* 2000;41:4203–4208.

25. Qian, Y, Dekaris, I, Yamagami, S, and Dana, MR. Topical soluble tumor necrosis factor receptor type I suppresses ocular chemokine gene expression and rejection of allogeneic corneal transplants. *Arch Ophthalmol.* 2000;118:1666–1671.

26. Turner, K, Pflugfelder, SC, Ji, Z, Feuer, WJ, Stern, M, and Reis, BL. Interleukin-6 levels in the conjunctival epithelium of patients with dry eye disease treated with cyclosporine ophthalmic emulsion. *Cornea.* 2000;19:492–496.

27. Stern, ME, Beuerman, RW, Fox, RI, Gao, J, Mircheff, AK, and Pflugfelder, SC. The pathology of dry eye: The interaction between the ocular surface and lacrimal glands. *Cornea.* 1998;17:584–589.

28. Baudouin, C, Haouat, N, Brignole, F, Bayle, J, and Gastaud, P. Immunopathological findings in conjunctival cells using immunofluorescence staining of impression cytology specimens. *Br J Ophthalmol.* 1992;76:545–549.

29. Jones, DT, Monroy, D, Ji, Z, Atherton, SS, and Pflugfelder, SC. Sjøgren's syndrome: cytokine and Epstein-Barr viral gene expression within the conjunctival epithelium. *Invest Ophthalmol Vis Sci.* 1994;35:3493–3504.

30. Jones, DT, Monroy, D, Ji, Z, and Pflugfelder, SC. Alterations of ocular surface gene expression in Sjøgren's syndrome. *Adv Exp Med Biol.* 1998;438:533–536.

31. Tsubota, K, Fukagawa, K, Fujihara, T, et al. Regulation of human leukocyte antigen expression in human conjunctival epithelium. *Invest Ophthalmol Vis Sci.* 1999;40:28–34.

32. Tsubota, K, Saito, I, Ishimaru, N, and Hayashi, Y. Use of topical cyclosporin A in a primary Sjøgren's syndrome mouse model. *Invest Ophthalmol Vis Sci.* 1998;39:1551–1559.

33. Olivero, DK, Davidson, MG, English, RV, Nasisse, MP, Jamieson, VE, and Gerig, TM. Clinical evaluation of 1% cyclosporine for topical treatment of keratoconjunctivitis sicca in dogs. *J Am Vet Med Assoc.* 1991;199:1039–1042.

34. Laibovitz, RA, Solch, S, Andriano, K, O'Connell, M, and Silverman, MH. Pilot trial of cyclosporine 1% ophthalmic ointment in the treatment of keratoconjunctivitis sicca. *Cornea.* 1993;12:315–323.

35. Gunduz, K and Ozdemir, O. Topical cyclosporin treatment of keratoconjunctivitis sicca in secondary Sjøgren's syndrome. *Acta Ophthalmol (Copenh).* 1994;72:438–442.

36. Sall, K, Stevenson, OD, Mundorf, TK, and Reis, BL. Two multicenter, randomized studies of the efficacy and safety of cyclosporine ophthalmic emulsion in moderate to severe dry eye disease. Csa phase 3 study group. *Ophthalmology.* 2000;107:631–639.

37. Gao, J, Schwalb, TA, Addeo, JV, Ghosn, CR, and Stern, ME. The role of apoptosis in the pathogenesis of canine keratoconjunctivitis sicca: the effect of topical cyclosporin A therapy. *Cornea.* 1998;17:654–663.

38. Iwata, M, Kiritoshi, A, Roat, MI, Yagihashi, A, and Thoft, RA. Regulation of HLA class II antigen expression on cultured corneal epithelium by interferon-gamma. *Invest Ophthalmol Vis Sci.* 1992;33:2714–2721.

39. Mircheff, AK, Gierow, JP, and Wood, RL. Traffic of major histocompatibility complex class II molecules in rabbit lacrimal gland acinar cells. *Invest Ophthalmol Vis Sci.* 1994;35:3943–3951.

40. Zhu, S, Dekaris, I, Duncker, G, and Dana, MR. Early expression of proinflammatory cytokines interleukin-1 and tumor necrosis factor-alpha after corneal transplantation. *J Interferon Cytokine Res.* 1999;19:661–669.

41. Holland, EJ, Olsen, TW, Ketcham, JM, et al. Topical cyclosporin A in the treatment of anterior segment inflammatory disease. *Cornea.* 1993;12:413–419.

42. Sotozono, C, He, J, Matsumoto, Y, Kita, M, Imanishi, J, and Kinoshita, S. Cytokine expression in the alkali-burned cornea. *Curr Eye Res.* 1997;16:670–676.

43. De Saint Jean, M, Brignole, F, Feldmann, G, Goguel, A, and Baudouin, C. Interferon-gamma induces apoptosis and expression of inflammation-related proteins in Chang conjunctival cells. *Invest Ophthalmol Vis Sci.* 1999;40:2199–2212.

44. van Kooten, C and Banchereau, J. Functional role of CD40 and its ligand. *Int Arch Allergy Immunol.* 1997;113:393–399.

45. Damato, BE, Allan, D, Murray, SB, and Lee, WR. Senile atrophy of the human lacrimal gland: the contribution of chronic inflammatory disease. *Br J Ophthalmol.* 1984;68:674–680.

46. Williamson, J, Gibson, AA, Wilson, T, Forrester, JV, Whaley, K, and Dick, WC. Histology of the lacrimal gland in keratoconjunctivitis sicca. *Br J Ophthalmol.* 1973;57:852–858.

47. Dana, MR, Zhu, SN, and Yamada, J. Topical modulation of interleukin-1 activity in corneal neovascularization. *Cornea.* 1998;17:403–409.

48. Fini, ME, Strissel, KJ, Girard, MT, Mays, JW, and Rinehart, WB. Interleukin 1 alpha mediates collagenase synthesis stimulated by phorbol 12-myristate 13-acetate. *J Biol Chem.* 1994;269:11291–11298.

49. Yoon, JH, Kim, KS, Kim, HU, Linton, JA, and Lee, JG. Effects of TNF-alpha and IL-1 beta on mucin, lysozyme, IL-6 and IL-8 in passage-2 normal human nasal epithelial cells. *Acta Otolaryngol.* 1999;119:905–910.

50. Danjo, Y, Watanabe, H, Tisdale, AS, et al. Alteration of mucin in human conjunctival epithelia in dry eye. *Invest Ophthalmol Vis Sci.* 1998;39:2602–2609.

51. Kunert, KS, Tisdale, AS, Stern, ME, Smith, JA, and Gipson, IK. Analysis of topical cyclosporine treatment of patients with dry eye syndrome: effect on conjunctival lymphocytes. *Arch Ophthalmol.* 2000;118:1489–1496.

52. Pflugfelder, SC, Solomon, A, and Stern, ME. The diagnosis and management of dry eye: A twenty-five-year review. *Cornea.* 2000;19:644–649.

53. Gillette, TE, Chandler, JW, and Greiner, JV. Langerhans cells of the ocular surface. *Ophthalmology.* 1982;89:700–711.

54. Fujikawa, LS, Colvin, RB, Bhan, AK, Fuller, TC, and Foster, CS. Expression of HLA-A/B/C and -DR locus antigens on epithelial, stromal, and endothelial cells of the human cornea. *Cornea.* 1982;1:213–222.

55. Jager, MJ. Corneal Langerhans cells and ocular immunology. *Reg Immunol.* 1992;4:186–195.

56. Treseler, PA, Foulks, GN, and Sanfilippo, F. The expression of HLA antigens by cells in the human cornea. *Am J Ophthalmol.* 1984;98:763–772.

57. Streilein, JW, Toews, GB, and Bergstresser, PR. Corneal allografts fail to express Ia antigens. *Nature.* 1979;282:320–321.

58. Peeler, JS and Niederkorn, JY. Antigen presentation by Langerhans cells *in vivo*: donor-derived Ia⁺ Langerhans cells are required for induction of delayed-type hypersensitivity but not for cytotoxic T lymphocyte responses to alloantigens. *J Immunol.* 1986;136:4362–4371.

59. Pepose, JS, Gardner, KM, Nestor, MS, Foos, RY, and Pettit, TH. Detection of HLA class I and II antigens in rejected human corneal allografts. *Ophthalmology.* 1985;92:1480–1484.

60. Baudouin, C, Fredj-Reygrobellet, D, Gastaud, P, and Lapalus, P. HLA DR and DQ distribution in normal human ocular structures. *Curr Eye Res.* 1988;7:903–911.

61. Hamrah, P, Zhang, Q, and Dana, MR. The cornea, an immune privileged tissue, is endowed with significant numbers of resident MHC class II-negative dendritic cells [ARVO abstract]. *Invest Ophthalmol Vis Sci.* 2001;42(4):S470. Abstract nr 2535.

62. Liu, Y, Hamrah, P, Zhang, Q, Taylor, AW, and Dana, MR. Donor bone marrow-derived antigen-presenting cells (APC) traffic to draining lymph nodes after corneal transplantation [ARVO abstract]. *Invest Ophthalmol Vis Sci.* 2001;42(4):S470. Abstact nr 2536.

63. Banchereau, J and Steinman, RM. Dendritic cells and the control of immunity. *Nature.* 1998;392:245–252.

DRY EYE AND DELAYED TEAR CLEARANCE: "A CALL TO ARMS"

Stephen C. Pflugfelder,[1] Abraham Solomon,[2] Dilek Dursun,[2] and De-Quan Li[2]

[1]Ocular Surface Center
Cullen Eye Institute
Baylor College of Medicine
Houston, Texas, USA
[2]Department of Ophthalmology
Bascom Palmer Eye Institute
University of Miami School of Medicine
Miami, Florida, USA

1. INTRODUCTION

The ocular surface is constantly bombarded by insults, including allergans, ultraviolet light, bacterial infection and in some individuals mechanical trauma from a poorly lubricated ocular surface (keratoconjunctivitis sicca) and contact lenses. Mechanisms have evolved on the ocular surface to respond, including stimulated synthesis and release of cytokines, lipid mediators and matrix metalloproteinases (MMPs) by the epithelium and inflammatory cells on the ocular surface. Additionally, activation of latent cytokines and MMPs occurs in the tear fluid. While these mechanisms allow the ocular surface to react quickly, our results suggest they are susceptible to inappropriate activation. Regulation of these inflammatory processes requires adequate production and clearance of tears to dilute the concentration of these effectors, clear them from the ocular surface and deliver a fresh supply of their antagonists. The decreased production and clearance of tears that occur in dry eye disease result in excessive production and inappropriate activation of two important mediators of the ocular surface response mechanism, interleukin 1 (IL-1) and MMP-9. Excessive or uncontrolled activity of these factors on the ocular surface may be

Lacrimal Gland, Tear Film, and Dry Eye Syndromes 3
Edited by D. Sullivan *et al.*, Kluwer Academic/Plenum Publishers, 2002

739

responsible for some clinical manifestations of dry eye disease, including keratoconjunctivitis sicca (KCS) and corneal vascularization, opacification and ulceration.

2. THE CYTOKINE-MEDIATED INFLAMMATORY CYCLE IN DRY EYE DISEASE

IL-1 and MMP-9 have been implicated in the pathogenesis of many inflammatory conditions. IL-1 is a critical cytokine mediator of inflammation and overall immune response.[1] Its two forms, IL-1α and IL-1β, are synthesized as 35-kD precursor molecules that are proteolytically cleaved to 17-kD mature forms. Both precursor and mature forms of IL-1α are biologically active. In contrast, only the 17-kD form of IL-1β is active. Many ocular surface stresses that accompany dry eye disease increase IL-1 activity, including apoptosis, mechanical trauma, hyperosmolarity, free radicals and increased pro-inflammatory factors.[2-6] Hyperosmolarity activates stress-related mitogen-activated protein kinases (MAPKs) that increase IL-1 production by stabilizing its messenger RNA.[2,7,8] Since hyperosmolarity is a common feature of dry eye,[9] it could be a key-initiating factor in the ocular surface inflammatory cycle of dry eye disease. Once IL-1 binds to its signal transducing type 1 receptor, it induces production of other inflammatory mediators including MMPs, chemokines such as IL-8 and Gro-α and conversion of cell membrane phospholipids into prostaglandins.[2,10-13]

Our research indicates IL-1 activity increases in dry eye disease. We have evaluated the concentration of IL-1α and IL-1β in the tear fluid of patients with Sjögren's syndrome aqueous tear deficiency (SS-ATD) and meibomian gland disease (MGD) associated with ocular rosacea. Significantly greater concentrations of IL-1α were detected in the tear fluid of both dry eye conditions compared to normal control patients. In normal tear fluid, a high concentration of the precursor form of IL-1β (pro-IL-1β) and a low concentration of the mature form exist. In contrast, this shifts to predominantly mature IL-1β in the tear fluid of both of these dry eye conditions, resulting in a statistically lower IL-1β precursor/mature ratio (Table 1).

Table 1. IL-1 cytokine in tear fluid of normal and dry eye patients

	Normal (n = 10)	MGD (n = 19)	SS (n = 10)	p
IL-1α	43.1 ± 24	253.7 ± 90	443.3 ± 128.5	*<0.05 †<0.001
Pro-IL-1β	379.2 ± 73	54.6 ± 16	21.2 ± 10	*<0.01 †< 0.0001
Mature IL-1β	29.8 ± 10	187.7 ± 72	80.9 ± 22	*0.022
IL-1β pro/mature	19.1 ± 3	1.17 ± 0.7	0.35 ± 0.2	*†0.0001

* normal vs. MGD, † normal vs. SS; NS = not significant. Concentrations in pg/ml.

Accompanying the changes in the relative levels of IL-1 family cytokines in the tear fluid is increased immunoreactivity for IL-1α, precursor and mature IL-1β and IL-1 receptor antagonist in the conjunctival epithelium of patients with SS KCS compared to normal controls. As proof that reduction in aqueous tear production and clearance leads to increased concentration of IL-1 on the ocular surface, we evaluated the concentration of IL-1β in tear fluid samples obtained from normal mice and mice treated with transdermal scopolamine patches to induce aqueous tear deficiency. The IL-1β concentration was below the limit of detection in the tear fluid of normal mice but a mean IL-1β concentration of 180 ± 80 pg/ml was detected in the tear fluid of mice treated with scopolamine for 4 days ($P < 0.001$).

The severity of clinical keratoconjunctivitis sicca (KCS) measured by the severity of corneal fluorescein staining strongly correlated with the concentration of IL-1α in the tear fluid ($R^2 = 0.17$, $P = 0.02$) and the IL-1 pro/mature ratio in the tear fluid ($R^2 = 0.046$, $P = 0.001$). These findings suggest increased production of IL-1 by the ocular surface epithelium and conversion of pro-IL-1β in the tear fluid to its mature form.

3. CO-EXPRESSION OF MMP-9 IN THE TEAR FLUID OF DRY EYE PATIENTS

IL-1β is a more potent inducer of inflammatory processes than IL-1α.[14] Pro-IL-1β is cleaved to its mature active form intracellularly by interleukin 1β converting enzyme (ICE) and extracellularly by many proteases whose activity is increased in inflammatory conditions, including trypsin, plasmin, elastase, cathepsin G and MMP-9.[15-17] Among these, MMP-9 is one of the most rapid and effective activators of pro-IL-1β and has processed pro-IL-1β within minutes into a product that maintained biological activity for 72 h.[17]

MMP-9 (also know as gelatinase B and 92-kD gelatinase) is the principal MMP enzyme produced by the corneal epithelium.[18] Environmental factors may regulate its production. For example, phorbol esters and inflammatory cytokines such as IL-1β and TNF-α increase its production.[10] MMP-9 could be a key factor in the pathogenesis of sterile corneal ulceration and recurrent corneal epithelial erosion.[19,20] The level of MMP-9 is low or non-detectable in the tear fluid of normal eyes,[20] but the concentration and activity are significantly increased in the tear fluid of patients with MGD associated with ocular rosacea and SS KCS compared to normal controls. In agreement with previous reports, we observed 1 μg MMP-9 will completely convert recombinant pro-IL-1β to its mature form. Among 11 different growth factors and inflammatory cytokines the corneal epithelium is exposed to from the tear film and supporting stroma, we found IL-1β the most potent inducer of MMP-9 mRNA and protein by primary cultured human corneal epithelium. Taken together, these findings suggest IL-1 and MMP-9 may participate in an escalating cytokine-mediated inflammatory cycle on the ocular surface of dry eye (Fig. 1). In this scheme, dry eye and delayed tear clearance lead to increased IL-1 activity on the ocular surface. This, in turn, would stimulate MMP-9 production by the corneal epithelium. MMP-9 would activate latent IL-1β in the tear film, which would have adverse consequences on the ocular surface epithelium and alter ocular surface lacrimal gland interaction. The exact details of this mechanism remain to be confirmed in animal models.

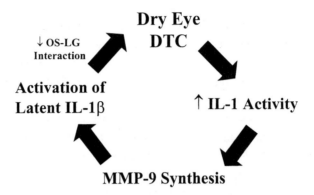

Figure 1. Cycle of cytokine-mediated inflammation on the ocular surface in dry eye disease. OS-LG, Ocular surface-Lacrimal gland.

4. SUMMARY

Dry eye may activate components of the ocular surface early warning system. Ocular surface stresses in dry eye, such as hyperosmolarity, could activate cellular stress pathways such as MAPKs. IL-1 and MMP-9 produced by the ocular surface epithelial cells may mediate initial events in the inflammatory cascade of dry eye through a dynamic interplay between them on the ocular surface.

REFERENCES

1.   Dinarello CA, Wolff SM. The role of interleukin-1 in disease. *N Engl J Med.*, 328: 106–113 (1993).
2.   Dinarello CA. Biologic basis for interleukin-1 in disease. *Blood*, 15: 2095–147 (1996).
3.   Lee RT, Briggs WH, Cheng GC, Rossiter HB, Libby P, Kupper T. Mechanical deformation promotes secretion of IL-1 alpha and IL-1 receptor antagonist. *J Immunol.*,159:5084–8 (1997).
4.   Shapiro L and Dinarello CA. Hyperosmotic stress as a stimulant for cytokine production. *Exp Cell Res.* 231:351–362 (1997).
5.   Chrousos GP. The hypothalmic-pituitary-adrenal axis and immune-mediated inflammation. *N Engl J Med.*, 332: 1351–1362 (1995).
6.   Navab M, Fogelman AM, Berliner JA, Territo MC, Demer LL, Frank JS, Watson AD, Edwards PA, Lusis AJ. Pathogenesis of atherosclerosis. *Am J Cardiol.*, 76:18C–23C (1995).
7.   Lee JC, Laydon JT, McDonnell PC, Gallagher TF, Kumar S, Green D, McNulty D, Blumenthal MJ, Heys JR, Landvatter SW, Strickler JE, McLaughlin MM, Slemens IR, Fisher SM, Livi GP, White JR, Adams JL, Young PR. A protein kinase involved in the regulation of inflammatory cytokine biosynthesis. *Nature*, 372:739–45 (1994).

8. Freshney NW, Rawlinson L, Guesdon F, Jones E, Cowley S, Hsuan J, Saklatvala J. Interleukin-1 activates a novel protein cascade that results in the phosphorylation of hsp27. *Cell*, 78:1039 (1994).

9. Farris RL. Tear osmolarity–a new gold standard? *Adv Exp Med Biol.*, 350:495–503 (1995).

10. Woessner, JF. Matrix metalloproteinases and their inhibitors in connective tissue remodeling. *FASEB J.*, 5: 2145–2154 (1991).

11. Miller MD, Krangel MS. Biology and biochemistry of the chemokines: a family of chemotactic and inflammatory cytokines. *Crit Rev Immunol.*, 12:17–46 (1992).

12. Gronich J, Konieczkowski M, Gelb MH, Nemenoff RA, Sodor JR. Interleukin-1α causes a rapid activation of cytosolic phospholipase A_2 by phosphorylation in rat mesangial cells. *J Clin Invest.*, 93:1224 (1994).

13. Farina M, Ribeiro ML, Ogando D, Gimeno M, Franchi AM. IL-1 alpha augments prostaglandin synthesis in pregnant rat uteri by a nitric oxide mediated mechanism. *Prostaglandins Leukot Essent Fatty Acids*, 62:243–7 (2000).

14. Geiger T, Towbin H, Consenti-Vargas A, Zingel O, Arnold J, Rordorf C, Glatt M, Vosbeck K. Neutralization of interleukin-1β activity in vivo with a monoclonal antibody alleviates collagen-induced arthritis in DBA/1 mice and prevents the associated acute phase response. *Clin Exp Rheumatol.*, 11:515 (1993).

15. Black RA, Kronheim SR, Sleath PR. Activation of interleukin-1β by a co-induced protease. *FEBS Lett.*, 247: 386–90 (1989).

16. Hazuda DJ, Strickler J, Simon P, Young PR. Structure-function mapping of interleukin-1 precursors. *J Biol Chem.*, 266:7081–7086 (1991).

17. Schonbeck V, Mach F, Libby P. Generation of biologically active IL-1β by matrix metalloproteinase, a novel caspase-1 independent pathway of IL-1β processing. *J Immunol.*, 161: 3340–3346 (1998).

18. Fini ME, Girard MT. Expression of collagenolytic/gelatinolytic metalloproteinases by normal cornea. *Invest Ophthalmol Vis Sci.*, 31:1779–88 (1990).

19. Matsubara M, Zieske JD, Fini ME. Mechanism of basement membrane dissolution preceding corneal ulceration. *Invest Ophthalmol Vis Sci.*, 32:3221–37 (1991).

20. Afonso A, Sobrin L, Monroy DC, Selzer M, Lokeshwar B, Pflugfelder SC. Tear fluid gelatinase B activity correlates with IL-1α concentration and fluorescein tear clearance. *Invest Ophthalmol Vis Sci.*, 40:2506–12 (1999).

INNATE IMMUNITY IN THE CORNEA: A PUTATIVE ROLE FOR KERATOCYTES IN THE CHEMOKINE RESPONSE TO VIRAL INFECTION OF THE HUMAN CORNEAL STROMA

Kanchana Natarajan,[1] James Chodosh,[1] and Ronald Kennedy[2]

[1]Department of Ophthalmology
Molecular Pathogenesis of Eye Infection Research Center
Dean A. McGee Eye Institute
[2]Department of Microbiology & Immunology
University of Oklahoma Health Sciences Center
Oklahoma City, Oklahoma, USA

1. INTRODUCTION

Keratocytes maintain the corneal stroma in an organized and transparent state, (Maurice, 1957; Muller et al., 1995) and modulate corneal responses to wounding (Fini, 1999). Less is known about the role of keratocytes in the inflammatory responses to stromal infection by pathogens. In this manuscript, we review evidence for the active participation of corneal cells not classically thought of as immunologically competent in the innate immune response to viral infection. We show that by the expression and secretion of chemokines, resident cells of the cornea, in particular the keratocytes, can respond quickly to an infectious insult and facilitate an inflammatory response to infection.

2. HUMAN KERATOCYTE BIOLOGY

The resident cells of the corneal stroma, the keratocytes, are highly active cells that communicate with one another by gap junctions (Spanakis et al., 1998; Watsky, 1995). Within the cornea, keratocytes are distinguishable as three distinct populations. Those just beneath Bowman's membrane (subepithelial keratocytes) form a particularly dense cellular network (Poole et al., 1993), contain twice as many mitochondria and considerably more

Lacrimal Gland, Tear Film, and Dry Eye Syndromes 3
Edited by D. Sullivan *et al.*, Kluwer Academic/Plenum Publishers, 2002

heterochromatin than keratocytes in the mid- and posterior regions (Muller et al., 1995) and are well innervated (Muller et al., 1996). When liberated from the excised cornea and grown in the absence of serum, these cells express keratan sulfate proteoglycan and replicate slowly. Exposure of keratocytes to serum in culture results in a change to a fibroblast phenotype (Fini, 1999; Fini and Girard, 1990), with increased assembly of f-actin into stress fibers, increased formation of focal adhesion complexes and greater expression of fibronectin, collagen and heparan sulfate, and is accompanied by more rapid growth and the capacity to degrade collagen (Mishima et al., 1998). Treatment of corneal fibroblasts with TGF-β (Jester et al., 1996) or wounding of keratocytes (Jester et al., 1995) induces a myofibroblast phenotype associated with expression of alpha-smooth muscle actin and biglycans, and results in an enhanced capacity to induce contraction of extracellular matrix (Jester et al., 1994; Masur et al., 1999). The transformation from keratocyte to myofibroblast involves a phosphotyrosine signal transduction pathway (Jester et al., 1999). These observations suggest keratocytes might play an important role in the immunopathologic responses to pathogens in the corneal stroma.

3. CHEMOKINE RESPONSES IN THE CORNEA

Chemokines are proteins that induce the migration and activation of cells such as neutrophils, macrophages and lymphocytes. Chemokine expression in the cornea (Table 1) has been demonstrated in vivo in animal models of corneal inflammation (Cole et al., 2000; Kernacki et al., 1998; Pearlman et al., 1997; Rudner et al., 2000; Su et al., 1996; Thomas et al., 1998; Tumpey et al., 1998; Yan et al., 1998) and human corneas removed at transplantation (Rosenbaum et al., 1995). Chemokine expression by all three major constitutive cell types of the cornea, the epithelial cells, keratocytes and endothelial cells, has been demonstrated in vitro in various experimental systems (Table 1).

Prevailing evidence suggests corneal cells participate in early stages of corneal inflammation by the secretion of chemokines. Corneal epithelial cells in vitro secrete interleukin-8 (IL-8) on exposure to secreted products of sensory nerves (Tran et al., 2000a; 2000b), indicating superficial cells in the cornea might facilitate the early innate response to an inflammatory stimulus. Corneal epithelial cells and keratocytes in vitro readily express neutrophil chemokines, such as IL-8 in response to treatment with the proinflammatory cytokines interleukin-1 (IL-1) and tumor necrosis factor–alpha (TNF-α). However, viral infection with adenovirus type 19 (Chodosh et al., 2000) or herpes simplex virus type 1 (Oakes et al., 1993) specifically stimulates the expression of neutrophil chemotactants, including IL-8 by human corneal fibroblasts and not corneal epithelial cells. This is consistent with the clinical observation that significant inflammation in the cornea following infection by a viral pathogen occurs coincident with breach of the stroma by the virus. These data indicate viral infection limited to the corneal epithelium may not be sufficient to induce significant inflammatory cell infiltration into the cornea.

Table 1. Evidence for chemokine expression in the cornea

Stimulus	Cell Type	Chemokine(s)	Author, Year
Adenovirus	HCF	GRO-alpha, IL-8	Chodosh, 2000
Bullous keratopathy	n.d.	IL-8	Rosenbaum, 1995
Calcitonin gene-related peptide	HCEC	IL-8	Tran, 2000b
Cytochalasin B	HCF	IL-8	West-Mays, 1997
E. coli endotoxin	n.d.	IL-8	Sobottka, 1997
Fibrin	Corneal endothelium	IL-8	Ramsby, 1994
Glycated human serum albumin	HCF	IL8, MCP-1	Bian, 1998
	HCEC	IL-8	Miyazaki, 1998
"	HCF	IL-8	Oakes, 1993
"	n.d.	IP-10, KC, MCP-1, MIP-1, MIP-2, RANTES	Su, 1996
"	n.d.	KC, MCP-1, MIP-1, MIP-2	Thomas, 1998
"	n.d.	MIP-2, MIP-1, MCP-1	Tumpey, 1998
"	n.d.	KC, MIP-2	Yan, 1998
Histamine	HCEC	IL-8	Sharif, 1998
IL-1, TNF	HCEC, HCF	IL-8	Cubitt, 1993
IL-1, TNF	HCEC, HCF	MCP-1, RANTES	Tran, 1996
IL-1, TNF	n.d.	IL-8	Elner, 1991
Onchocerca volvulus, IL-12	n.d.	Eotaxin, IP-10, MCP-1, MIP-1, RANTES	Pearlman, 1997
Pseudomonas aeruginosa	n.d.	KC	Cole, 2000
"	n.d.	Eotaxin, IP-10, MCP-1, MIP-1, MIP-2, RANTES	Kernacki, 1998
"	n.d.	MIP-2	Rudner, 2000
Substance P	HCEC	IL-8	Tran, 2000a
TNF	HCEC/HCF	IL-8, RANTES	Takano, 1999
TNF, IL-4, IL-13	HCF	Eotaxin	Kumagai, 2000
UV irradiation	HCF	IL-8	Kennedy, 1997

n.d.: whole cornea evaluated and cell type not differentiated; HCEC: human corneal epithelial cells; HCF: human corneal fibroblasts

4. CHEMOKINE RESPONSES TO VIRAL INFECTION

Eukaryotic cells produce chemokines early after infection by a broad array of viruses. A review of previously published studies reveals four principal mechanisms by which viral infection can induce host cell chemokine gene expression (Table 2): (1) viral binding to its host cell receptor may induce a signal transduction cascade, resulting in the activation of chemokine genes; (2) viral gene products may directly activate chemokine genes or increase the stability of chemokine mRNA; (3) viral replication may stimulate the expression of cellular IL-1 that, in turn, can activate chemokine gene expression, and (4) viral replication may result in the production of oxidative species such as H_2O_2 that can activate chemokine genes.

More than one mechanism may be active at different times during viral infection. In the case of infection of eukaryotic cells by adenoviruses, virus binding (Bruder and

Kovesdi, 1997), adenoviral E1A protein expression (Keicho et al., 1997) and IL-1 production in adenovirus infected cells (Schwarz et al., 1999) all can lead to IL-8 expression. It is of particular interest that if induction of an intracellular signaling cascade leads to chemokine gene expression prior to the onset of viral gene transcription (Bruder and Kovesdi, 1997), then antiviral therapy may not prevent the inflammation and secondary tissue damage associated with infection. Such a mechanism may occur in the pathogenesis of stromal infiltrates following group D adenovirus infection of the cornea (Chodosh et al., 2000). Host cell receptors have constitutive cellular functions aside from viral infection. From an evolutionary perspective, the induction of chemokine gene expression on viral binding to constitutive host cell ligands is intriguing. One might view the onset of chemokine expression on viral binding to a eukaryotic cell as a "cry for help" by the cell. However, chemokine expression may in some cases enhance viral replication (Murayama, 1998), suggesting subversion by the virus of the cell's response to infection.

5. SUMMARY

Existing evidence suggests that chemokine expression by virus-infected cells is a common response to viral infection. By such a mechanism, non-immunologic cells may participate in the generation of an early innate immune response to infection. In the absence of classic immunologic cells in the corneal stroma, keratocytes may play a similar role in the corneal responses to viral infection.

Table 2. Mechanisms of virus-induced chemokine expression

Virus	Chemokine(s)	Mechanism	Author, Year
Adenovirus	IL-8	Binding activates signaling pathway	Bruder, 1997
"	IL-8	Viral gene product (E1A)	Keicho, 1997
"	IL-8	Binding leads to IL-1 production	Schwarz, 1999
CMV	IL-8	Viral gene product	Murayama, 1998
EBV	IL-8	Viral gene product (LMP1)	Eliopoulos, 1999
"	IL-8, MIP-1	Binding	McSoll, 1997
HIV	MIP-1a, RANTES	Oxidant species (Nitric oxide)	Sherry, 2000
HSV-1	IL-8	Viral gene product	Oakes, 1993
"	IL-8, MCP-1, MIP-2, MIP-1, KC, lymphotactin	Viral gene product	Thomas, 1998
HTLV	IL-8	Viral gene product (Tax)	Mori, 1998
"	MIP-2, KC	IL-1	Yan, 1998
Measles Virus	MCP-1	Viral gene product (Tax)	Mori, 2000
Reovirus	RANTES	Viral gene product (Nucleocapsid protein)	Noe, 1999
Rhinovirus	IL-8	Oxidative species (H_2O_2)	Biagioli, 1999
"	IL-8	Binding	Hamamdzic, 1999
Rotavirus	IL-8	Binding	Johnston, 1998
"	MIP-1	Binding	Rollo, 1999
RSV	IL-8	IL-1, TNF	Arnold, 1994
"	IL-8	Both binding and viral gene product	Fiedler, 1995
"	IL-8	Viral gene product	Garofalo, 1996
"	RANTES	Viral gene product	Koga, 1999
"	IL-8	Oxidant species	Mastronarde, 1993
"	IL-8	Binding leads to IL-1 production	Patel, 1998

CMV: Cytomegalovirus; EBV: Epstein-Barr virus; HIV: human immunodeficiency virus; HSV-1: herpes simplex virus type 1; HTLV: human T-cell lymphotropic virus; RSV: respiratory syncytial virus

ACKNOWLEDGMENTS

This work was supported in part by Public Health Service grants EY00357 and EY12190, and by Research to Prevent Blindness.

REFERENCES

R. Arnold, B. Humbert, H. Werchau, H. Gallati, and W. Konig, Interleukin-8, interleukin-6, and soluble tumor necrosis factor receptor type I release from a human pulmonary epithelial cell line (A549) exposed to respiratory syncytial virus. *Immunology.* 82:126 (1994).

M.C. Biagoli, P. Kaul, I. Singh, and R.B. Turner, The role of oxidative stress in rhinovirus induced elaboration of IL-8 by respiratory epithelial cells. *Free Radic Biol Med.* 26:454 (1999).

Z.M. Bian, V.M. Elner, N.W. Lukacs, R.M. Strieter, S.L. Kunkel, and S.G. Elner, Glycated human serum albumin induces IL-8 and MCP-1 gene expression in human corneal keratocytes. *Curr Eye Res.* 17:65 (1998).

J.T. Bruder and I. Kovesdi , Adenovirus infection stimulates the Raf/MAPK signaling pathway and induces interleukin-8 expression. *J Virol.* 71:398 (1997).

J. Chodosh, R.A. Astley, M.G. Butler, R.C. Kennedy, Adenovirus keratitis: a role for interleukin-8. Invest *Ophthalmol Vis Sci.* 41:783 (2000).

N. Cole, S. Bao, A. Thakur, M. Willcox, and A.J. Husband, KC production in cornea in response to Pseudomonas aeruginosa challenge. *Immunol Cell Biol.* 78:1 (2000).

C.L. Cubitt, Q. Tang, C.A. Monteiro, R.N. Lausch, and J.E. Oakes, IL-8 gene expression in cultures of human corneal epithelial cells and keratocytes. *Invest Ophthalmol Vis Sci.* 34:3199 (1993).

A.G. Eliopoulos, N.J. Gallagher, S.M. Blake, C.W. Dawson, and L.S. Young, Activation of the p38 mitogen-activated protein kinase pathway by Epstein-Barr virus-encoded latent membrane protein 1 coregulates interleukin-6 and interleukin-8 production. *J Biol Chem.* 274:23 (1999).

V.M. Elner, R.M. Strieter, M.A. Pavilack, S.G. Elner, D.G. Remick, J.M. Danforth, and S.L. Kunkel, Human corneal interleukin-8. IL-1 and TNF-induced gene expression and secretion. *Am J Pathol.* 139:977 (1991).

M.A. Fiedler, K. Wernke-Dollries, and J.M. Stark, Respiratory syncytial virus increases IL-8 gene expression and protein release in A549 cells. *Am J Physiol.* 269:L865 (1995).

M.E. Fini, Keratocyte and fibroblast phenotypes in the repairing cornea. Prog Retin Eye Res.18:529 (1999).

M.E. Fini, and M.T. Girard, The pattern of metalloproteinase expression by corneal fibroblasts is altered by passage in cell culture. *J Cell Sci.* 97:373 (1990).

R. Garofalo, M. Sabry, M. Jamaluddin, R.K. Yu, A. Casola, P.L. Ogra, and A.R. Brasier, Transcriptional activation of the interleukin-8 gene by respiratory syncytial virus infection in alveolar epithelial cells: nuclear translocation of the RelA transcription factor as a mechanism producing airway mucosal inflammation. *J Virol.* 70:8773 (1996).

D. Hamamdzic, S. Altman-Hamamdzic, S.C. Bellum, T.J. Phillips-Dorsett, S.D. London, and L. London, Prolonged induction of IL-8 gene expression in a human fibroblast cell line infected with reovirus serotype 1 strain Lang. *Clin Immunol.* 91:25 (1999).

J.V. Jester, P.A. Barry, G.J. Lind, W.M. Petroll, R. Garana, and H.D. Cavanagh, Corneal keratocytes: In situ and *in vitro* organization of cytoskeletal contractile proteins. *Invest Ophthalmol Vis Sci.* 35:730 (1994).

J.V. Jester, W.M. Petroll, P.A. Barry, and H.D. Cavanagh, Expression of α-smooth muscle (α-SM) actin during corneal stromal wound healing. *Invest Ophthalmol Vis Sci.* 36:809 (1995).

J.V. Jester, P.A. Barry, H.D. Cavanagh, and W.M. Petroll, Induction of α-smooth muscle actin (α-SM) expression and myofibroblast transformation in cultured keratocytes. *Cornea.* 15:505 (1996).

J.V. Jester, J. Huang, P.A. Barry-Lane, W.W-y. Kao, W.M. Petroll, and H.D. Cavanagh, Transforming growth factor β-mediated corneal myofibroblast differentiation requires actin and fibronectin assembly. *Invest Ophthalmol Vis Sci.* 40:1959 (1999).

S.L. Johnston, A. Papi, P.J. Bates, J.G. Mastronarde, M.M. Monick, and G.W. Hunninghake, Low grade rhinovirus infection induces a prolonged release of IL-8 in pulmonary epithelium. *J Immunol.* 160:6172 (1998).

N. Keicho, W.M. Elliott, J.C. Hogg, and S. Hayashi, Adenovirus E1A upregulates interleukin-8 expression induced by endotoxin in pulmonary epithelial cells. *Am. J Physiol.* 272:L1046 (1997).

M. Kennedy, K.H. Kim, B. Harten, J. Brown, S. Planck, C. Meshul, H. Edelhauser, J.T. Rosenbaum, C.A. Armstrong, and J.C. Ansel, Ultraviolet radiation induces the production of multiple cytokines by human corneal cells. *Invest Ophthalmol Vis Sci.* 38:2483 (1997).

K.A. Kernacki, D.J. Goebel, M.S. Poosch, and L.D. Hazlett, Early cytokine and chemokine gene expression during Pseudomonas aeruginosa corneal infection in mice. *Infect Immun.* 66:376 (1998).

T. Koga, E. Sardina, R.M. Tidwell, M. Pelletyier, D.C. Look, and M.J. Holtzman, Virus-inducible expression of a host chemokine gene relies on replication-linked mRNA stabilization. *Proc Natl Acad Sci (USA).* 96:5680 (1999).

N. Kumagai, K. Fukuda, Y. Ishimura, and T. Nishida, Synergistic induction of eotaxin expression in human keratocytes by TNF-alpha and IL-4 or IL-13. *Invest Ophthalmol Vis Sci.* 41:1448 (2000).

J.G. Mastronarde, M.M. Monick, and G.W. Hunninghake, Oxidant tone regulates IL-8 production in epithelium infected with respiratory syncytial virus. *Am J Res Cell Mol Biol.* 13:237 (1995).

S.K. Masur, R.J. Conors, J.K-H. Cheung, and S. Antobi, Matrix adhesion characteristics of corneal myofibroblasts. *Invest Ophthalmol Vis Sci.* 40:904 (1999).

D.M. Maurice, The structure and transparency of the cornea. *J Physiol.* 136:263 (1957).

S.R. McSoll, C.J. Roberge, B. Larochelle, and J. Gosselin, EBV induces the production and release of IL-8 and macrophage inflammatory protein-1 alpha in human neutrophils. *J Immunol.* 159:6164 (1997).

H. Mishima, J. Okamoto, M. Nakamura, Y. Wada, and T. Otori, Collagenolytic activity of keratocytes cultured in a collagen matrix. *Jpn J Ophthalmol.* 42:79 (1998).

D. Miyazaki, Y. Inoue, K. Araki-Sasaki, Y. Shimomura, Y. Tano, and K. Hayashi, Neutrophil chemotaxis induced by corneal epithelial cells after herpes simplex virus type 1 infection. *Curr Eye Res.* 17:687 (1998).

N. Mori, N. Mukaida, D.W. Ballard, K. Matsushima, and N. Yamamoto, Human T-cell leukemia virus type I tax transactivates human interleukin 8 gene through acting concurrently on AP-1 and nuclear factor-kappaB-like sites. *Cancer Res.* 58:3993 (1998).

N. Mori, A. Ueda, S. Ikeda, Y. Yamasaki, Y. Yamada, M. Tomonaga, S. Morikawa, R. Geleziunas, T. Yoshimura, and N. Yamamoto, Human T-cell leukemia virus type I tax activates transcription of the human monocyte chemoattractant protein –1 gene through two nuclear factor-kappaB sites. *Cancer Res.* 60:4939 (2000).

L.J. Muller, L. Pels, and G.F.J.M. Vrensen, Novel aspects of the ultrastructural organization of human corneal keratocytes. *Invest Ophthalmol Vis Sci.* 36:2557 (1995).

L.J. Muller, L. Pels, and G.F.J.M. Vrensen, Ultrastructural organization of human corneal nerves. *Invest Ophthalmol Vis Sci.* 37:476 (1996).

T. Murayama, Interrelationship between human cytomegalovirus infection and chemokine. *Nippon Rinsho.* 56:69 (1998).

K.H. Noe, C. Cenciarelli, S.A. Moyer, P.A. Rota, and M.L. Shin, Requirements for measles virus induction of RANTES chemokine in human astrocytoma-derived U373 cells. *J Virol.* 73:3117 (1999).

J.E. Oakes, C.A. Monteiro, C.L. Cubitt, and R.N. Lausch, Induction of interleukin-8 gene expression is associated with herpes simplex virus infection of human corneal keratocytes but not human corneal epithelial cells. *J Virol.* 67:4777 (1993).

J.A. Patel, Z. Jiang, N. Nakajima, and M. Kunimoto, Autocrine regulation of interleukin-8 by interleukin-1alpha in respiratory syncytial virus-infected pulmonary epithelial cells *in vitro. Immunology.* 95:501 (1998).

E. Pearlman, J.H. Lass, D.S. Bardenstein, E. Diaconu, F.E. Jr. Hazlett, J. Albright, A.W. Higgins, and J.W. Kazura, IL-12 exacerbates helminth-mediated corneal pathology by augmenting inflammatory cell recruitment and chemokine expression. *J Immunol.* 158:827 (1997).

C.A. Poole, N. Brookes, and G.M. Clover, Keratocyte networks visualized in the living cornea using vital dyes. *J Cell Sci.* 106:685 (1993).

M.L. Ramsby, and D.L. Kreutzer, Fibrin induction of interleukin-8 expression in corneal endothelial cells *in vitro. Invest Ophthalmol Vis Sci.* 35:3980 (1994).

E.E. Rollo, K.P. Kumar, N.C. Reich, J. Cohen, J. Angel, H.B. Greenberg, R. Sheth, J. Anderson, B. Oh, S.J. Hempson, E.R. Mackow, and R.D. Shaw, The epithelial cell response to rotavirus infection. *J Immunol.* 163:4442 (1999).

J.T. Rosenbaum, S.T. Planck, X.N. Huang, L. Rich, and J.C. Ansel, Detection of mRNA for cytokines, interleukin-1 alpha and interleukin-8 in corneas from patients with pseudophakic bullous keratopathy. *Invest Ophthalmol Vis Sci.* 36:2151 (1995).

X.L. Rudner, K.A. Kernacki, R.P, Barrett, and L.D. Hazlett, Prolonged elevation of IL-1 in Pseudomonas aeruginosa ocular infection regulates macrophage-inflammatory protein-2 production, polymorphonuclear neutrophil persistence, and corneal perforation. *J Immunol.* 164:6576 (2000).

Y.A. Schwarz, R.S. Amin, J.M. Stark, B.C. Trapnell, and R.W. Wilmott, Interleukin-1 receptor antagonist inhibits interleukin-8 expression in A549 respiratory epithelial cells infected *in vitro* with a replication-deficient recombinant adenovirus vector. *Am. J Respir Cell Mol Biol.* 21:388 (1999).

N.A. Sharif, T.K. Wiernas, B.W. Griffin, and T.L. Davis, Pharmacology of [3H]-pyrilamine binding and of the histamine-induced inositol phosphates generation, intracellular Ca2+-mobilization and cytokine release from human corneal epithelial cells. *Br J Pharmacol.* 125:1336 (1998).

B. Sherry, H. Schmidtmayerova, G. Zybarth, L. Dubrovsky, T. Raabe, and M. Bukrinsky, Nitric oxide regulates MIP-1alpha expression in primary macrophages and T-lymphocytes: implications for anti-HIV response. *Mol Med.* 6:542 (2000).

A.C. Sobottka Ventura, K. Engelmann, C. Dahinden, and M. Bohnke, Endotoxins modulate the autocrine function of organ cultured donor corneas and increase the incidence of endothelial cell death. *Br J Ophthalmol.* 81:1093 (1997).

S.G. Spanakis, S. Petridou, and S.K. Masur, Functional gap junctions in corneal fibroblasts and myofibroblasts. *Invest Ophthalmol Vis Sci.* 39:1320 (1998).

Y.H. Su, X.T. Yan, J.E. Oakes, and R.N. Lausch, Protective antibody therapy is associated with reduced chemokine transcripts in herpes simplex virus type 1 corneal infection. *J Virol.* 70:1277 (1996).

Y. Takano, K. Fukugawa, S. Shimmura, K. Tsubota, Y. Oguchi, and H. Saito, IL-4 regulates chemokine production induced by TNF-alpha in keratocytes and corneal epithelial cells. *Br J Ophthalmol.* 83:1074 (1999).

J. Thomas, S. Kanangat, and B.T. Rouse, Herpes simplex virus replication-induced expression of chemokines and proinflammatory cytokines in the eye: implications of herpetic stromal keratitis. *J Interferon Cytokine Res.* 18:681 (1998).

M.T. Tran, M. Tellaetxe-Isusi, V. Elner, R.M. Strieter, R.N. Lausch, and J.E. Oakes, , Proinflammatory cytokines induce RANTES and MCP-1 synthesis in human corneal keratocytes but not in corneal epithelial cells. Beta-chemokine synthesis in corneal cells. *Invest Ophthalmol Vis Sci.* 37:987 (1996).

M.T. Tran, R.N. Lausch, and J.E. Oakes, Substance P differentially stimulates IL-8 synthesis in human corneal epithelial cells. *Invest Ophthalmol Vis Sci.* 41:3871 (2000)a.

M.T. Tran, M.H. Ritchie, R.N. Lausch, and J.E. Oakes, Calcitonin gene-related peptide induces IL-8 synthesis in human corneal epithelial cells. *J Immunol.* 162:4307 (2000)b.

T.M. Tumpey, H. Cheng, X.T. Yan, J.E. Oakes, and R.N. Lausch, Chemokine synthesis in the HSV-1-infected cornea and its suppression by interleukin-10. *J Leukoc Biol.* 63:486 (1998).

J.A. West-Mays, P.M. Sadow, T.W. Tobin, K.J. Strissel, C. Cintron, and M.E. Fini, Repair phenotype in corneal fibroblasts is controlled by an interleukin-1 alpha autocrine feedback loop. *Invest Ophthalmol Vis Sci.* 38:1367 (1997).

M.A. Watsky, Keratocyte gap junctional communication in normal and wounded rabbit corneas and human corneas. *Invest Ophthalmol Vis Sci.* 36:2568 (1995).

X.T. Yan, T.M. Tumpey, S.L. Kunkel, J.E. Oakes, and R.N. Lausch, Role of MIP-2 in neutrophil migration and tissue injury in the herpes simplex virus-1-infected cornea. *Invest Ophthalmol Vis Sci.* 39:1854 (1998).

THE ROLE OF ICAM-1 AS A SIGNAL PROTEIN FOR PREDISPOSITION OF OCULAR SURFACE INFLAMMATION

Michael E. Stern,[1] Jianping Gao,[1] Grant A. Morgan,[1] Douglas Brees,[1] Tammy A. Schwalb,[1] Michael Humphreys-Behr,[2] and Janine A. Smith[3]

[1]Allergan, Inc.
Irvine, California, USA
University of Florida
[2]Gainesville, Florida, USA
[3]National Eye Institute
Bethesda, Maryland, USA

1. INTRODUCTION

Recent research in our laboratory and others has demonstrated keratoconjunctivitis sicca (KCS, dry eye) is caused by an autoimmune inflammatory process affecting the lacrimal "functional unit".[1] This is composed of the lacrimal glands (main and accessory), ocular surface and interconnecting innervation. The histological hallmark of this disease, regardless if an individual is diagnosed with systemic autoimmune syndrome or is subject to a local autoimmune event, is an immune cell infiltration of the lacrimal glands (main and accessory) as well as the conjunctival substantia propria and epithelium. The infiltrating cells are typically CD4-positive T-helper cells. They target the tissues in response to a local epithelial antigen presentation resulting from a neurogenic inflammation of the lacrimal glands and ocular tissues. This inflammation is facilitated by an age-related (menopause) or treatment-related (anti-androgen therapy) loss of circulating androgens in the affected tissues.[2,3] The purpose of this study was to evaluate mouse models of autoimmunity to investigate initiation of the observed lacrimal gland and ocular surface inflammation and determine the role of ICAM-1 in the pathophysiology of dry eye.

2. MATERIALS AND METHODS

2.1. NOD and NOD.B10-H2^b Mouse Models

Lacrimal glands, conjunctiva, cornea and eyelids from male and female NOD, NOD.B10-H2^b and control C57BL/6 mice were excised and immersed in OCT and immediately frozen in liquid nitrogen. Frozen sections (5-8 μm) were cut and processed for immunohistochemistry with primary antibodies (BD Pharmingen, San Diego, CA) for T-cells (CD3+) and B-cells (CD45R/B220) as well as ICAM-1 and MHC II. Negative controls were performed by omitting the primary antibody, using a preimmune serum from the species in which the antibodies were raised and, when possible, using a tissue that does not express the marker of interest.

2.2. MRL/lpr Mouse Model

To study the pattern of ICAM-1 expression over time, female MRL/lpr mice (n = 16) and the age-matched C57BL/6 mice (n = 16) at 3 weeks of age were purchased from Jackson Laboratory. Animals (4 mice/group) were sacrificed at 1, 2, 3 and 4 months by CO_2 asphyxiation. Fresh frozen lacrimal glands and lids were embedded in OCT for immunofluorescent evaluation using purified anti-ICAM-1 or LFA-1 antibodies (Pharmingen). Plasma samples were collected for evaluation of soluble ICAM-1 (sICAM-1) concentration using mouse soluble ICAM-1 (CD54) ELISA kit (Endogen, Woburn, MA).

2.3. Flow Cytometry

Following CO_2 asphyxiation, lacrimal glands from 18-week-old (n = 6) MRL/lpr and MRL/lpr control (MRL/MpJ) mice were removed and placed in media. Acinar lobes were teased away from surrounding material using 18-gauge needles and placed into media containing soybean trypsin inhibitor (STI media). Lobes were cut into 1-mm cubes and subjected to three rounds of chelation and digestion with PBS + EDTA and STI media + collagenase / DNase 1 (10 min at 37°C). The supernatants from each digestion were combined, washed and resuspended in 200 μl PBS/BSA/azide at 1 x 10^6 cells/sample. Samples were then incubated for 30 min at 4°C with or without fluorescent or biotin-conjugated antibodies to either CD3 (pan T-cell marker), CD4 (helper T-cell subset), CD8 (cytotoxic T-cell subset), ICAM-1 (involved in activation and migration), CD25 (IL-2α receptor; upregulated during active proliferation) or LFA-1 (receptor for ICAM-1). Samples were then washed and resuspended in PBS/BSA/azide ± streptavidin-conjugated PerCp (all antibodies and secondary reagents from Pharmingen). Cells were incubated for a second time as described above before washing and resuspension in PBS + 1% formulin. Fluorescence-associated marker expression was then measured by flow cytometry (FACSCalibur, BD Biosciences, San Diego, CA).

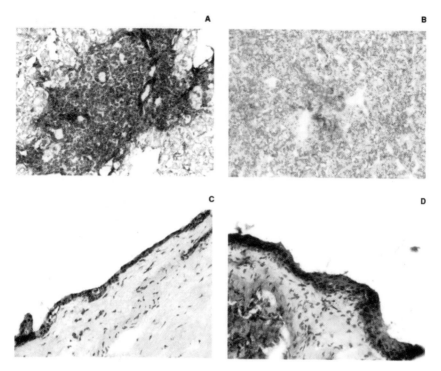

Figure 1. ICAM-1 expression was detected in the lacrimal gland of male NOD.B10-H2^b Mice (**A**), but not that of female NOD.B10-H2^b Mice (**B**).

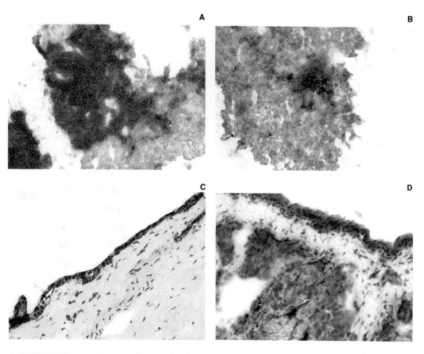

Figure 2. MHC II Expression was detected in the lacrimal gland of male NOD.B10-H2^b mice (**A**) and to a lesser extent, also in female NOD.B10-H2^b mice (**B**).

3. RESULTS

3.1. Characterization of Immune Cells in Ocular Surface Tissues from the Autoimmune Mice: NOD and a Congenic Variant, NOD.B10-H2^b

The lymphocytic infiltration of the lacrimal gland has been shown in NOD/J and NOD.B10-H2^b mice, and these results reflect the findings reported in the literature.[4] Large infiltrates were present in the lacrimal glands of male NOD/J and NOD.B10-H2^b versus female (Figs. 1 and 2). The infiltrates stained positively for CD3+ T-cells and CD45R/B220+ B-cells (data not shown). ICAM-1 staining was also present in the lacrimal glands of male NOD.B10-H2^b (Fig. 1A) and NOD/J (data not shown), but not in the glands of corresponding females (Fig. 1B). The ICAM-1 staining was localized to structures resembling high endothelial venules and not infiltrated cells. MHC II staining was observed primarily in the lacrimal glands of male and to a lesser extent female NOD.B10-H2^b mice (Figs. 2A and 2B, respectively). In contrast to the ICAM-1 staining, MHC II was localized to the infiltrated cells and not vessel structures.

In contrast to lacrimal glands, none of the ocular surface tissues (conjunctiva, cornea, lid) in the NOD.B10-H2^b or NOD/J mice had any signs of lymphocytic infiltration. Although no cellular infiltrates were observed, staining for ICAM-1 (Figs. 1C and 1D) and MHC II (Figs. 2C and 2D) was positive primarily in the epithelium of conjunctiva and periorbital lacrimal glands of male and to lesser extent female NOD.B10-H2^b mice. More experiments are necessary to determine if and what MHC II is presenting. It seems reasonable to determine if anti-androgens affect the ocular surface tissues expressing ICAM-1 and MHC II to facilitate cellular infiltration.

3.2. ICAM-1 Expression in the Ocular Tissues of the MRL-lpr Mouse Over Time

ICAM-1 immunoreactivities were detected in the lacrimal gland of MRL/lpr but not in the age-matched control C57BL/6 mice.[5] ICAM-1 positivity was found on the vascular endothelial cells, infiltrating lymphocytes and the lacrimal acinar epithelial cell membrane. The expression of ICAM-1 on different cell types depended on severity of the disease. The endothelial ICAM-1 expression was detected in young mice at 1 month. The lymphocytic ICAM-1 expression was evident at 2 months and increased at disease onset (3 months). No acinar epithelial cell ICAM-1 was detected until 3 months and markedly increased by 4 months. ICAM-1 was also detected in the conjunctival epithelial cells of MRL/lpr mice as early as 2 months. The expression level did not differ significantly over time.

3.3. Plasma sICAM-1 Concentrations in the MRL/lpr Mouse Over Time

The mean sICAM-1 levels were significantly higher in MRL/lpr mice than in the C57BL/6 controls at 2, 3 and 4 months (Fig. 3). In MRL/lpr mice, the mean sICAM-1 levels gradually and significantly increased over time, and peaked at 4 months. In the control C57BL/6 mice, however, the plasma sICAM-1 level did not increase significantly until 4 months.

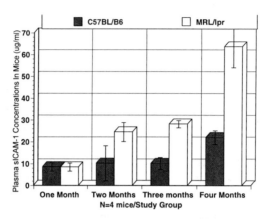

Figure 3. Plasma-soluble ICAM-1 concentrations in MRL/lpr and C57BL/6 mice.

3.4. Profile of Infiltrating Immune Cells in the Lacrimal Glands of MRL/lpr Mice

Lacrimal glands of MRL/lpr mice were heavily infiltrated with T-cells and to a lesser extent B-cells (B-cell data not presented) at 18 weeks in comparison with MRL/lpr controls (MRL/MpJ). Both CD4 and CD8 cells were observed, with little or no apparent shift toward either T-cell subset in comparison to the T-cell subset ratio of the control mice (Figs. 4 and 5). Using three-color analysis, we examined the expression of ICAM-1 (Fig. 4) and CD25 (Fig. 5) on the infiltrating T-cells. Fig. 4 clearly demonstrates both CD8 positive (middle row) and CD4 positive (right row) mostly expressed high levels of ICAM-1 on their surfaces. This indicates both subsets of T-cells had previously upregulated ICAM-1 before infiltration into the lacrimal gland. This is further supported by the fact that even in the control mice, where little infiltration is seen, ICAM-1 is also upregulated on the relatively small number of T-cells present in the lacrimal gland (lower panels, Fig. 4).

During active proliferation T-cells will upregulate the IL-2 receptor and secrete IL-2 in an autocrine and paracrine fashion. Fig. 4 illustrates that although some T-cells were in an activated status (high ICAM-1 expression; Fig. 4 and upregulated CD69; data not shown), they did not express the CD25 cell marker (IL-2 receptor; Fig. 5). Therefore, we conclude they are not actively proliferating in the lacrimal gland.

Consistent with the elevated levels of ICAM-1 expression in CD4 and CD8 cells, Fig. 6 demonstrates almost all CD3-positive T-cells present in the lacrimal gland of the MRL/lpr mice have upregulated the receptor for ICAM-1, LFA-1. The presence of two distinct populations of LFA-1 expressing T-cells (Fig. 6), may be of significance and further studies are underway to investigate this.

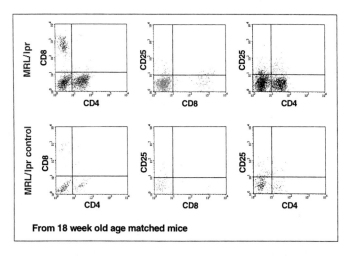

Figure 4. Lacrimal gland immune activation profile of MRL/lpr mice: ICAM-1. Flow cytometry immune profiles dot plots of lacrimal gland-derived cells taken from either MRL/lpr or control MRL/MpJ mice (n = 6 /group; duplicate samples of pooled cells). This illustrates the relative percentage of CD4+ and CD8+ T-cells (**left panels**) and their expression profile of ICAM-1 (ICAM-1 expressing CD8+ cells (**middle panels**); ICAM-1 expressing CD4+ cells (**right panels**)).

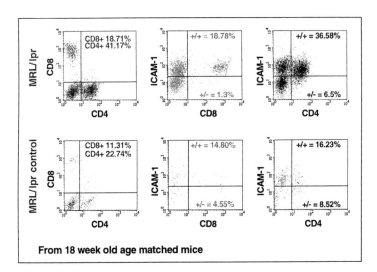

Figure 5. Lacrimal gland immune activation profile of MRL/lpr mice: CD25. Flow cytometry immune profiles dot plots of lacrimal gland-derived cells taken from either MRL/lpr or control MRL/MpJ mice (n = 6 / group; duplicate samples of pooled cells). This illustrates the relative percentage of CD4+ and CD8+ T-cells (**left panels**) and their expression profile of CD25 (CD25 expressing CD8+ cells (**middle panels**); CD25 expressing CD4+ cells (**right panels**)).

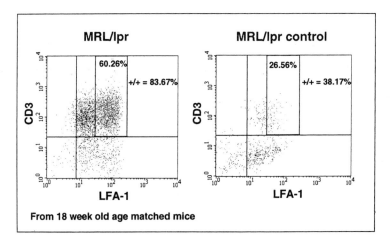

Figure 6. Lacrimal gland immune profile of MRL/lpr mice: LFA-1. Flow cytometry immune profiles dot plots of lacrimal gland-derived cells taken from either MRL/lpr or control MRL/lpr (MpJ) mice (n = 6 / group; duplicate samples of pooled cells). Presented is LFA-1 expression on CD3 T-cells. Notice the higher proportion of CD3+ cells expressing high levels of LFA-1 in the MRL/lpr mice (**left panel**) compared to the control (**right panel**).

4. DISCUSSION

This study demonstrates several facets of the development of the autoimmune response in the etiology and pathophysiology of lacrimal gland inflammation in two mouse models. The upregulation and presentation of ICAM-1 to the surface of lymphocytes as they diapedese from vessels and infiltrate the intralobular space in the lacrimal gland are an indication of an alteration in cell signaling and orientation. This accumulation of lymphocytes signals a pathological discontinuation of normal apoptotic cell death as seen in the conjunctiva of patients with KCS.[6,7] Factors in the gland itself or a regulatory event within the lymphocyte may induce the ICAM-1 expression seen here. Increased ICAM-1 expression was also present on the resident lacrimal and conjunctival epithelial cells, along with concurrent expression of MHC-II. The fact that neither model exhibited any noticeable ocular surface inflammation causes speculation on recent evidence surrounding the presence of "empty MHC-II".[8]

The increases in systemic sICAM-1 shown here may demonstrate a presdisposition toward the inflammatory state with aging. This may contribute to the presence of local inflammations as well. Further evaluation of immune cell populations in ocular inflammation may assist development of more specific therapeutic targets in controlling or curing these syndromes.

REFERENCES

1. M.E. Stern, R.W. Beuerman, R.I. Fox, J. Gao, A.K. Mircheff, S.C. Pflugfelder. The pathology of dry eye: the interaction between the ocular surface and lacrimal glands. *Cornea* 17(6): 584-589 (1998).
2. M Ono, FJ Rocha, DA Sullivan. Immunocytochemical location and hormonal control of androgen receptors in lacrimal tissues of the female MRL/Mp-lpr/lpr mouse model of Sjøgren's syndrome. *Exp Eye Res.* 61(6):659-66 (1995).
3. N. Mamalis, D. Harrison, G. Hiura, et al. Are dry eyes a sign of testosterone deficiency in women? The International Endocrine Society Meeting, San Francisco, CA, *The Programs and Abstract Book*, P3-379 (2): 849 (1996).
4. C.P. Robinson, S. Yamachika, D.I. Bounous, J. Brayer, R. Jonsson, R. Holmdahl, A.B. Peck, M.G. Humphreys-Beher. A novel NOD-derived murine model of primary Sjøgren's syndrome. *Arthritis Rheum.* 41(1):150-6 (1998).
5. J. Gao, G.A. Morgan, T.A. Schwalb, M.E. Stern. ICAM-1: it's role in the pathophysiology of immune activation in the MRL/lpr mouse. International Conference on the Lacrimal Gland, Tear Film and Dry Eye Syndromes: Basic Science and Clinical Relevance, Maui, HI. *Cornea* 19(6): S88, (2000).
6. J. Gao, T.A. Schwalb, J.V. Addeo, C.R. Ghosn, M.E. Stern. The role of apoptosis in the pathogenesis of canine keratoconjunctivitis sicca: the effect of topical Cyclosporin A therapy. *Cornea* 17(6):654-663 (1998).
7. M. E. Stern, J. Gao, T. A. Schwalb, M. Ngo, D. D. Tieu, C. Chan, B. L. Reis, S M. Whitcup, J. A. Smith. Sjøgren's and non-Sjøgren's keratoconjunctivitis sicca patients have comparable conjunctival T-lymphocyte sub-populations. (Submitted 2001).
8. L. Santambrogio, A.K. Sato, F.R. Fischer, M.E. Dorf, L.J. Stern. Abundant empty class II MHC molecules on the surface of immature dendritic cells. *Proc Natl Acad Sci. USA* 96(26):15050-5 (1999).

FLOW CYTOMETRIC ANALYSIS OF THE INFLAMMATORY MARKER HLA DR IN DRY EYE SYNDROME: RESULTS FROM 12 MONTHS OF RANDOMIZED TREATMENT WITH TOPICAL CYCLOSPORIN A

Christophe Baudouin,[1,2] Francois Brignole,[3] Pierre-Jean Pisella,[1] Magda De Saint Jean,[1] and Alain Goguel[3]

[1]Department of Ophthalmology
[3]Department of Immunohematology
Ambroise Paré Hospital
University of Paris V
[2]Department of Ophthalmology
 Quinze-Vingts National Ophthalmology Hospital
Paris, France

1. INTRODUCTION

Dry eye disease or keratoconjunctivitis sicca (KCS) is a common ocular surface disease that causes chronic ocular irritation and dry or gritty sensations of the ocular surface. Traditionally, KCS has been described as a secretion disorder of the lacrimal glands that leads to a deficiency in tear production. However, a growing body of evidence suggests KCS has a complex multifactorial etiology, which comprises tear dysfunction and a localized immune-mediated inflammatory response affecting the entire ocular surface (Stern et al., 1998). The inflammatory response observed in KCS patients with and without Sjögren's syndrome (SS) encompasses cellular and immunopathological abnormalities of the conjunctival epithelium including upregulation of inflammatory cytokines and molecular markers of the inflammatory pathway (Baudouin et al., 1992; Jones et al., 1994; 1998; Pflugfelder et al., 1990; 1999; Tsubota et al., 1999a; Smith et al., 2000). Mircheff and colleagues (1994) proposed that in KCS patients without SS, this inflammatory process results from altered membrane "trafficking" of the lacrimal acinar cells. These authors suggested that these changes arise due to expression of major histocompatibility

Lacrimal Gland, Tear Film, and Dry Eye Syndromes 3
Edited by D. Sullivan *et al.*, Kluwer Academic/Plenum Publishers, 2002

761

complex (MHC) molecules, which subsequently initiate an autoimmune response within the lacrimal gland. Furthermore, the results from more recent studies have demonstrated increased expression of human leukocyte antigen (HLA), the human MHC, in patients with KCS, further advocating a possible role for class II immune mediation in dry eye (Baudouin et al., 1997; Tsubota et al., 1999b; Brignole et al., 2000; Smith et al., 2000).

Despite the widespread prevalence of dry eye disease (Bjerrum, 1997; Hikichi et al., 1995; Schein et al., 1997), no therapeutic modality is available to treat this ocular condition. Conventional palliative therapy relies primarily on the use of tear substitutes solely designed to supplement natural ocular lubrication. While tear substitutes may alleviate the symptoms of the disease, the underlying immunological problem remains inadequately addressed. However, recent clinical trials have shown that topical treatment with the immunomodulatory agent cyclosporin A (CsA) may be effective against dry eye disease (Stevenson et al., 2000; Sall et al., 2000). The object of this study was to investigate the expression of the inflammatory marker HLA DR by conjunctival epithelial cells obtained by impression cytology from patients with KCS during long-term treatment with CsA.

2. MATERIALS AND METHODS

The study design and flow cytometric methods have been described in detail (Brignole et al., 2000). Briefly, a multicenter, double-masked, randomized, vehicle-controlled, parallel-group study of the safety and efficacy of CsA 0.05% and 0.1% ophthalmic emulsions used twice daily in patients with moderate to severe KCS was designed by Allergan Inc., Irvine, CA, USA. Following an initial 2-week run-in phase during which they received only unpreserved tear substitute (Refresh®, Allergan), patients who fulfilled the inclusion criteria were randomized to twice daily treatment with CsA 0.05%, CsA 0.1% or vehicle emulsion. From day 0 to month 6, patients were allowed to use Refresh® daily, as needed, in addition to the masked treatment. At month 6, patients in the vehicle treatment group were switched to 0.1% CsA.

2.1. Impression Cytology

At selected centers, as part of the whole study, impression cytology specimens were collected at day 0 from the worse eye, defined as the one showing the higher degree of corneal staining, or the lower Schirmer's test when both eyes had the same corneal staining scores. If the two criteria were equal in both eyes, the right eye was chosen for impression cytology. The same eye was used throughout the study. Patients providing samples for this study were recruited in 29 centers from four European countries. Immediately after collection, specimens were shipped to the Immunohematology Department, Ambroise Paré Hospital, Boulogne, France, for processing and analyses in a centralized procedure. Impression cytology specimens were obtained from patients under topical anesthesia (0.04% oxyburocaine), using 0.20-µm polyethersulfone filters (Supor®, Gelman Sciences,

MI, USA) applied on the superior and supero-temporal bulbar conjunctiva, according to previously published procedures (Baudouin et al., 1997; Brignole et al., 1998; 2000). Specimens were collected at least 15 min after instillation of the last staining eyedrop to avoid any interference with immunofluorescence analyses. After collection, filters were immediately placed into tubes containing 0.05% paraformaldehyde in 1.5 ml cold phosphate-buffered saline (PBS). Tubes were kept at or below 4°C before impression collection and sent within 2 days to the Department of Immunohematology, Ambroise Paré Hospital, in cold-conditioned containers. Cells were extracted by gentle agitation for 30 min and centrifuged (1600 rpm for 5 min). The cells were then counted in a Malassez's cell before processing for flow cytometry, according to previously validated methods (Baudouin et al., 1997; Brignole et al., 1998; 2000).

2.2. Flow Cytometry

For the detection of HLA DR antigens, a mouse immunoglobulin (IgG1) anti-HLA DR alpha chain antibody (clone TAL.1B5, 50 μg/ml, DAKO SA, Copenhagen, Denmark) was used in an indirect immunofluorescence protocol. Fluorescein isothiocyanate-conjugated goat anti-mouse IgG was used as the secondary antibody (DAKO). Non-immune mouse IgG1 (DAKO) was used as a negative isotypic control. The monoclonal HLA DR antibody was used at a 1:50 dilution in PBS containing 1% bovine serum albumin. After a 30-min incubation and washing with PBS, the secondary anti-mouse IgG at a 1:50 dilution was added for 30 min. After incubation, cells were centrifuged in PBS (1600 rpm for 5 min), resuspended in 100 μl PBS and analyzed on a flow cytometer (Becton Dickinson), according to previously validated methods (Baudouin et al, 1997; Brignole et al, 1998). The same flow cytometer was used throughout the study.

For each specimen, at least 1,000 cells were analyzed by flow cytometry, and specimens were discarded if less than 10,000 cells were collected. The percentages of positive cells were obtained from logarithmic cytograms of mean fluorescence intensities, by comparison with the negative isotypic control. Fluorescence intensities were further quantified by using calibrated fluorospheres to translate the mean fluorescence of each sample into standardized arbitrary fluorescence units (AUF). A calibration curve was established during each flow cytometric procedure by using four different beads (Immunobrite, Coulter, Hayleh, FL, USA) with standardized fluorescence intensities (Philip et al., 1994). The actual AUF value was obtained by subtracting the isotypic negative control from the total AUF calculated for the marker.

2.3. Statistical Analyses

For the percentage of positive cells and AUF analyses, a non-parametric method was used due to the high variability of the data. A Kruskal-Wallis test was used to compare differences in change from baseline among treatment groups. If the test for among-group differences was significant ($P < 0.05$), then all three pairwise comparisons were performed using a Wilcoxon rank sum test. Within-group changes from baseline were analyzed by the

C. Baudouin *et al.*

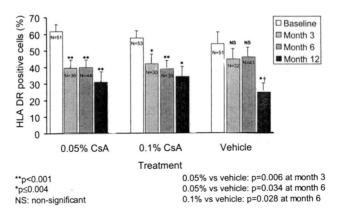

Figure 1. Changes in the mean percentage of HLA DR-positive conjunctival epithelial cells at baseline and following 12 months of treatment with cyclosporin emulsion or vehicle followed by cyclosporin emulsion. Error bars indicate standard errors. [†]6 months of vehicle followed by 6 months of 0.1% cyclosporin emulsion. Reproduced with kind permission from Brignole et al. © Association for Research in Vision and Ophthalmology 2001.

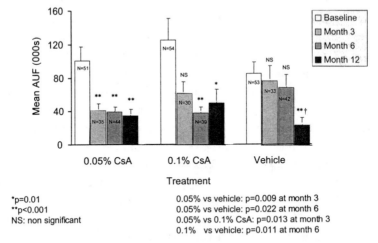

Figure 2. Changes in the mean levels of HLA DR expression by conjunctival cells at baseline and following 12 months of treatment with cyclosporin emulsion or vehicle followed by cyclosporin emulsion. Levels of HLA DR expressed as AUF. Error bars indicate standard errors. [†]6 months of vehicle followed by 6 months of 0.1% cyclosporin emulsion. Reproduced with kind permission from Brignole et al. © Association for Research in Vision and Ophthalmology 2001.

Wilcoxon signed rank test. The SAS computer package Release 6.12 for UNIX (SAS Institute) was used for computation and analysis.

3. RESULTS

At baseline, 169 patients aged 18–86 years (mean 57.1 years) yielded valid specimens. Of these patients, 86% were women and 41% had SS. Of these 169 samples, 158, 98, 125 and 68 allowed at least two analyses of HLA DR at baseline, 3, 6 and 12 months, respectively. No difference was found among the three treatment groups in mean age, sex ratio and coexisting medical conditions or systemic medications. Percentages of SS and non-SS patients did not differ among the three groups, and no differences in clinical data occurred between groups at baseline. Figs. 1 and 2 summarize mean values and standard errors for HLA DR-positive cells and AUF obtained at each time point, with statistical analyses of mean between- and within-group differences. At baseline, no statistical differences existed for HLA DR among the three groups.

For both CsA treatment groups, the percentage of HLA DR-positive conjunctival cells significantly decreased compared to baseline at months 3, 6 and 12. For patients receiving 0.05% CsA treatment, the percentage of HLA DR-positive conjunctival cells decreased from 61.67 ± 29.54% (mean ± SD) at baseline to 39.03 ± 31.36%, 39.45 ± 33.06% and 30.55 ± 29.40% at months 3, 6 and 12, respectively. In the 0.1% CsA treatment group, the percentage of HLA DR-positive conjunctival cells decreased from 57.53 ± 31.73% at baseline to 41.73 ± 33.57%, 38.59 ± 32.95% and 34.05 ± 28.78% at 3, 6 and 12 months, respectively. Patients receiving vehicle did not show any significant change from baseline (mean differences –5.09% and –6.68%, respectively, at months 3 and 6). However, after the vehicle group had been switched to CsA 0.1% at month 6, a significant change (–29.01%) was observed at month 12. In addition, a significant difference was found between 0.05% CsA and vehicle at month 3 (P = 0.006) and month 6 (P = 0.034), and between 0.1% CsA and vehicle at month 6 (P = 0.028).

For the 0.05% and 0.1% CsA groups the mean levels of expression of HLA DR decreased significantly at months 3, 6 and 12, and months 6 and 12, respectively, compared to baseline. Mean expression of HLA DR, expressed as AUF, generally mirrored the decrease observed with the mean percentage of positive HLA DR cells. In the 0.05% CsA treatment group, the mean expression of HLA DR fell from 101,004 ± 117,356 AUF (mean ± SD) at baseline to 41,557± 44,842 AUF, 39,748 ± 38,495 AUF and 34,697 ± 35,952 AUF at months 3, 6 and 12, respectively. For patients receiving 0.1% CsA, mean expression of HLA DR decreased from 124,883 ± 188,883 AUF at baseline to 61,949 ± 72,717 AUF, 38,492 ± 39,054 AUF and 50,521 ± 71,196 AUF at months 3, 6 and 12, respectively. Patients receiving vehicle did not show any significant difference from baseline (mean changes −13,140 and −16,387 AUF at months 3 and 6, respectively). However, as shown for the mean percentage of HLA DR-positive cells, after the vehicle group had been switched to 0.1% CsA at month 6, a significant change (-24,137 AUF) was observed at month 12. A significant difference was found between 0.05% CsA and vehicle at months 3 ($P = 0.009$) and 6 ($P = 0.022$), between 0.05% and 0.1% CsA at month 3 ($P = 0.013$) and between 0.1% CsA and vehicle ($P = 0.011$) at month 6.

4. DISCUSSION

Irrespective of the initial causes of dry eye, chronic desiccation of the ocular surface results in immunoregulatory dysfunction, which is characterized by widespread inflammation and destruction of the lacrimal gland and conjunctiva. Once the disease has developed, inflammation is the key mechanism of ocular surface injury as the cause and consequence of ocular damage. More than 80% of patients diagnosed with KCS exhibit immunological characteristics of conjunctival inflammation (Baudouin et al., 1992; 1997).

The results from the present study clearly demonstrate topical application of CsA is efficient in reducing the aberrant expression of the inflammatory marker HLA DR in conjunctival epithelial cells. Several studies have demonstrated the abnormal expression of HLA DR in conjunctival cells in KCS patients, including SS and non-SS patients (Baudouin et al., 1997; Tsubota et al., 1999a; Brignole et al., 2000; Pisella et al., 2000). HLA DR expression, in mean percentage of positive cells and mean levels of expression, was significantly reduced following 0.05% and 0.1% CsA treatment at 3, 6 and 12 months. The only exception was observed at month 3 in the 0.1% CsA treatment group, where the decrease in AUF was not significant. However, the levels of HLA DR expression in the eyes from KCS patients over 12 months of treatment with CsA demonstrate a consistent decrease in expression, with levels approaching those previously recorded for normal eyes (Brignole et al., 1998; 2000). Furthermore, both CsA emulsions were more effective than vehicle in reducing HLA DR at months 3 and 6 and month 6, respectively, with HLA DR remaining at low levels at month 12. A significant decrease was only observed in the vehicle group at month 12, after this treatment group had switched to 0.1% CsA treatment at month 6.

HLA DR is an important immune-related marker, normally expressed by immunocompetent cells, which is upregulated in epithelial cells in autoimmune and inflammatory disorders. However, in this study similar levels of HLA DR expression were observed for SS and non-SS patients with KCS at baseline and throughout the treatment period, which indicates MHC II expression is not a specific consequence of an autoimmune disease. In a recent study by Smith et al. (2000), the expression of class II HLA antigens and T-lymphocyte infiltration was characteristic of the immunopathological process of conjunctival inflammation. Also, upregulation of HLA DR expression in conjunctival epithelial cells may be induced by the cytokines interferon gamma (IFN-γ) and tumor necrosis factor alpha (TNF-α) (Tsubota et al., 1999h; De Saint Jean et al.,

1999). Furthermore, epithelial cells produce TNF-α (Jones et al., 1998). Consequently, it can be postulated that this combination of increased HLA DR antigen expression and elevated levels of inflammatory cytokines within the conjunctival epithelium may represent the focus for cytotoxic reactions between epithelial cells and infiltrating lymphocytes, which would also help to perpetuate the local immune response.

Whether or not the aberrant expression of HLA DR antigens by the conjunctival cells in KCS patients indicates a possible antigen-presenting role for these epithelial cells remains elusive. However, such properties have been ascribed to corneal epithelial and lacrimal acinar cells (Iwata et al., 1992; Mircheff et al., 1994). In addition, a similar autoantigen role has been proposed for α-fodrin, a cytoskeletal protein originally isolated from the salivary gland in a mouse model of SS. This molecule was subsequently shown to induce T-cell proliferation and expression of interleukin-2 and IFN-γ (Haneji et al., 1997). Although the precise role of MHC II antigen expression in the dry eye disease process remains speculative, it appears conjunctival epithelial cells represent an intrinsic component of the inflammatory process characteristic of KCS.

From the current results, it is clear that that long-term treatment with CsA is effective in reducing the expression of the inflammatory marker HLA DR in conjunctival epithelial cells. Additionally, these results corroborate those from previous studies by Kunert et al. (2000) and Turner et al. (2000), which demonstrated a significant reduction in lymphocyte activation, and interleukin-6 levels in the conjunctival epithelium after 6 months of treatment with 0.05% CsA emulsion. Furthermore, recent clinical trials with CsA have consistently demonstrated a significant clinical improvement in patients with KCS after CsA treatment (Tauber 1998; Stevenson et al., 2000; Sall et al., 2000). Current treatment of KCS involves artificial tears and punctal occlusion, which, at best, are palliative. However, CsA, through its mode of action as an immunomodulator, represents a new therapeutic modality that reduces ocular surface inflammation, the key to the immunopathology of dry eye disease. Consequently, CsA represents an important therapeutic advance in the management and treatment of moderate to severe KCS.

REFERENCES

C. Baudouin, N. Haouat, F. Brignole, J. Bayle, and P. Gastaud, Immunological findings in conjunctival cells using immunofluorescence staining of impression cytology specimens. *Br J Ophthalmol.* 76:545 (1992).

C. Baudouin, F. Brignole, F. Becquet, P.J. Pisella, and A. Goguel. Flow cytometry in impression cytology specimens: a new method for evaluation of conjunctival inflammation. *Invest Ophthalmol Vis Sci.* 8:1458 (1997).

K.B. Bjerrum. Keratoconjunctivitis sicca and primary Sjøgren's syndrome in a Danish population aged 30–60 years. *Acta Ophthalmol Scand.* 75:281 (1997).

F. Brignole, M. De Saint Jean, M. Goldschild, F. Becquet, A. Goguel, and C. Baudouin. Expression of Fas-Fas ligand antigens and apoptotic marker APO2.7 by the human conjunctival epithelium: positive correlation with class II HLA DR expression in inflammatory ocular surface disorders. *Exp Eye Res.* 67:687 (1998).

F. Brignole, P.J. Pisella, M. Goldschild, M. De Saint Jean, A. Goguel, and C. Baudouin, C. Flow cytometric analysis of inflammatory markers in conjunctival epithelial cells of patients with dry eyes. *Invest Ophthalmol Vis Sci.* 41:1356 (2000).

F. Brignole, P.J. Pisella, M. De Saint Jean, M. Goldschild, A. Goguel, and C. Baudouin. Flow cytometric analysis of inflammatory markers in KCS: 6-month treatment with topical cyclosporin A. *Invest Ophthalmol Vis Sci.* 42:90 (2001).

M. De Saint Jean, F. Brignole, G. Feldman, A. Goguel, and C. Baudouin. Interferon gamma induces apoptosis and expression of inflammation-related proteins in Chang conjunctival cells. *Invest Ophthalmol Vis Sci.* 40:2199 (1999).

N.Haneji, T. Nakamura, K. Takio, K. Yanagi, H. Higashiyama, I. Saito, S. Noji, H. Sugino, and Y. Hayashi. Identification of α-fodrin as a candidate autoantigen in primary Sjøgren's syndrome. *Science.* 276:604 (1997).

T. Hikichi, A. Yoshida, Y. Fukui, T. Hamano, M. Ri, K. Araki, K. Horimoto, E. Takamura, K. Kitagawa, M. Oyama, et al. Prevalence of dry eye in Japanese eye centers. *Graefe Arch Clin Exp Ophthalmol.* 233:555 (1995).

M. Iwata, A. Kiritoshi, M.I. Roat, A. Yagihashi, and R.A. Thoft. Regulation of HLA class II antigen expression on cultured corneal epithelium by interferon gamma. *Invest Ophthalmol Vis Sci.* 33:2714 (1992).

D.T. Jones, D. Monroy, Z. Ji, S.S. Atherton, and S.C. Pflugfelder. Sjøgren's syndrome: cytokine and Epstein-Barr viral gene expression within the conjunctival epithelium. *Invest Ophthlmol Vis Sci.* 35:3493 (1994).

D.T. Jones, D. Monroy, Z. Ji, and S.C. Pflugfelder. Alterations of ocular surface gene expression in Sjøgren's syndrome. *Adv Exp Med Biol.* 438:533 (1998).

K.S. Kunert, A.S. Tisdale, M.E. Stern, J.A. Smith, and I.K. Gipson. Analysis of topical cyclosporine treatment of patients with dry eye syndrome: effect on conjunctival lymphocytes. *Arch Ophthalmol.* 118:1489 (2000).

A.K. Mircheff, J.P. Gierow, and R.L. Wood. Traffic of major histocompatibility complex class II molecules in rabbit lacrimal gland acinar cells. *Invest Ophthalmol Vis Sci.* 35:3943 (1994).

S.C. Pflugfelder, A.J. Huang, W. Feuer, P.T. Chuchovski, I.C. Pereira, and S.C. Tseng. Conjunctival cytologic features of primary Sjøgren's syndrome. *Ophthalmology.* 97:985 (1990).

S.C. Pflugfelder, D. Jones, Z. Ji, A. Afonso, and D. Monroy. Altered cytokine balance in the tear fluid and conjunctiva of patients with Sjøgren's syndrome keratoconjunctivitis sicca. *Curr Eye Res.* 19:201 (1999).

P.J.M. Philip, C. Sartiaux, and the GEIL. Standardized multicentric quantimetry of differentiation antigens expression. The GEIL's approach in acute lymphoblastic leukemia. *Leukemia Lymphoma.* 13:45 (1994).

P.J. Pisella, F. Brignole, C. Debbasch, P.A. Lozato, C. Creuzot-Garcher, J. Bara, P. Saiag, J.M. Warnet, and C. Baudouin. Flow cytometric analysis of conjunctival epithelium in ocular rosacea and keratoconjunctivitis sicca. *Ophthalmology.* 107:1841 (2000).

K. Sall, O.D. Stevenson, T.K. Mundorf, B.L. Reis, and the CsA Phase 3 Study Group. Two multicenter, randomized studies of the efficacy and safety of cyclosporine ophthalmic emulsion in moderate to severe dry eye disease. *Ophthalmology.* 107:631 (2000).

O.D. Schein, B. Munoz, J.M. Tielsch, K. Bandeen-Roche, S. West. Prevalence of dry eye among the elderly. *Am J Ophthalmol.* 124:723 (1997).

J.A. Smith, M.E. Stern, J. Gao, T.A. Schwalb, D.C. Rupp, and S.M. Whitcup. Conjunctival inflammation in non-Sjøgren's syndrome and Sjøgren's syndrome keratoconjunctivitis sicca. *Invest Ophthalmol Vis Sci.* 41:S276 (2000).

M.E. Stern, R.W. Beurman, R.I. Fox, J. Gao, A.K. Mircheff, and S.C. Pflugfelder. The pathology of dry eye: the interaction between the ocular surface and lacrimal glands. *Cornea.* 17:584 (1998).

D. Stevenson, J. Tauber, and B.L. Reiss, B. Efficacy and safety of cyclosporin A ophthalmic emulsion in the treatment of moderate-to-severe dry eye disease: a dose-ranging, randomized trial. The Cyclosporin A Phase 2 Study Group. *Ophthalmology.* 107:967 (2000).

J. Tauber. A dose-ranging clinical trial to assess the safety and efficacy of cyclosporine ophthalmic emulsion in patients with keratoconjunctivitis sicca. *Adv Exp Med Biol.* 438:969 (1998).

K. Tsubota, T. Fujihara, K. Saito, and T. Takeuchi. Conjunctival epithelium expression of HLA-DR in dry eye patients. *Ophthalmologica.* 213:16 (1999a).

K. Tsubota, K. Fukagawa, T. Fujihara, S. Shimmura, I. Saito, K. Saito, and T. Takeuchi. Regulation of human leukocyte antigen expression in human conjunctival epithelium. *Invest Ophthalmol Vis Sci.* 40:28 (1999b).

K. Turner, S.C. Pflugfelder, Z. Ji, W.J. Feuer, M.E. Stern, and B.L. Reis. Interleukin-6 levels in the conjunctival epithelium of patients with dry eye disease treated with cyclosporine ophthalmic solution. *Cornea.* 19: 492 (2000).

PATHOGENESIS OF AUTOIMMUNE LACRIMAL GLAND DISEASE IN MRL/MPJ MICE

Douglas A. Jabs,[1,2] Robert A. Prendergast,[1] and Judith A. Whittum-Hudson[3]

Departments of [1]Ophthalmology and [2]Medicine
The Johns Hopkins University School of Medicine
Baltimore, Maryland, USA
[3]Department of Medicine
Wayne State University School of Medicine
Detroit, Michigan, USA

1. INTRODUCTION

MRL/MpJ mice spontaneously develop lacrimal and salivary gland inflammation and are a model for the human disorder Sjøgren's syndrome.[1-3] Two congenic substrains of MRL/MpJ mice exist, MRL/MpJ-+/+ (MRL/+) and MRL/MpJ-*lpr/lpr* (MRL/lpr). These substrains differ only at a single autosomal recessive gene locus, the *lpr* mutation. This mutation results in altered Fas protein, defective lymphocyte apoptosis, defective clonal deletion of autoreactive T-cells in peripheral lymphoid organs and accelerated autoimmune disease in MRL/lpr mice when compared to MRL/+ mice.[4,5] MRL/lpr mice typically die at 6 months of age, whereas MRL/+ mice often live to 2 years. Both substrains develop lacrimal gland inflammation, although the lacrimal gland disease develops earlier in MRL/lpr than MRL/+ mice, and at comparable ages, MRL/lpr mice have more severe and extensive disease.[3] The lacrimal gland lesions in both substrains are composed largely of T-cells (approximately 80%), the majority of which are CD4+ T-cells. Lesser numbers of CD8+ T-cells, B-cells and macrophages are present. In aged (18-month) MRL/+, mice B-cells accumulate in the lacrimal gland lesions.[2,3,6]

CD4+ helper T (Th)-cells differentiate via two pathways into Th1- or Th2-cells and have different effector mechanisms. Th1-cells produce interferon (IFN)-γ and tumor necrosis factor and are responsible primarily for cell-mediated immune responses. Th2-cells produce interleukin (IL)-4, IL-5 and IL-10 and provide help for B-cells in antibody

Lacrimal Gland, Tear Film, and Dry Eye Syndromes 3
Edited by D. Sullivan *et al.*, Kluwer Academic/Plenum Publishers, 2002

771

Table 1. Immunohistochemistry of inflammatory
lacrimal gland lesions in MRL/MpJ mice

Age (months)	IL-4	IFN-_	B7-1	B7-2
MRL/+				
2	30*	1	5	16
3	46	0	2	22
4	40	1	2	34
5	55	3	4	28
MRL/lpr				
2	30	5	10	26
3	67	1	10	20
4	30	3	2	38
5	40	3	6	20

*Median percent cells staining positive. Adapted from Jabs
DA, Lee B, Whittum-Hudson J, Prendergast RA. Th1
versus Th2 immune responses in autoimmune lacrimal
gland disease in MRL/MpJ mice. *Invest Ophthalmol Vis
Sci.* 2000;41:826-831. Used with permission.

production.[7,8] Cytokines are involved in directing immune responses towards a Th1 or Th2 type; IL-12 and IFN-γ production leads to Th1 responses, whereas IL-4 results in Th2 responses, and IL-10 inhibits Th1 responses.[9,10] B7 is a costimulatory molecule expressed on antigen-presenting cells and is required for effective stimulation of T-cells. Two major subtypes of B7 exist, B7-1 and B7-2, which stimulate Th1 and Th2 responses, respectively.[11,12] A series of experiments was undertaken to better understand the pathogenesis of autoimmune lacrimal gland disease in MRL/MpJ mice, in particular the roles of Th1- vs. Th2-cells in both substrains and the Fas-Fas ligand (FasL) system in accelerated lacrimal gland disease in MRL/lpr mice.

2. TH1 VS. TH2 CYTOKINES AND COSTIMULATORY MOLECULES IN THE LACRIMAL GLAND

MRL/MpJ mice of both substrains were obtained from the Jackson Laboratories (Bar Harbor, ME), kept under standard conditions, and groups of 5 mice were sacrificed at selected ages for evaluation of cytokine production and presence of costimulatory molecules. Lacrimal glands were removed, embedded in OCT (Miles, Elkhart, IN), frozen in liquid nitrogen, sectioned at 8 μm on a cryostat and stained with a panel of monoclonal antibodies (mAbs) and the avidin-biotin-peroxidase complex (ABC) technique. Monoclonal antibodies included anti-IL-4 (PharMingen, San Diego, CA), anti-IFN-γ (Biosource, Camarillo, CA), anti-B7-1 (PBL, New Brunswick, NJ) and anti-B7-2 (PharMingen, San Diego, CA). The staining was completed using ABC kits (Vector,

Burlingame, CA). The percentage of inflammatory cells staining positive for each mAb was counted with a micrometer disc mounted on a standard binocular microscope.[13]

Table 1 shows results of staining lacrimal gland sections for selected cytokines and B7.[13] Substantial staining for IL-4 was present in both substrains at all ages studied, whereas little staining for IFN-γ was detected. The proportion of cells staining positive for IL-4 was significantly greater than that for IFN-γ in both substrains; the mean difference between the proportion staining positive for IL-4 and IFN-γ was 42% for MRL/+ mice (P = 0.002) and 33% for MRL/lpr mice (P = 0.001). A significantly greater proportion of cells stained for B7-2 than for B7-1. The mean difference between the proportion of cells staining positive for B7-2 and B7-1 was 19% in MRL/+ mice (P = 0.006) and 15% in MRL/lpr mice (P = 0.0001).[13]

In preliminary experiments, competitive RT-PCR was used to quantify mRNA transcripts for the cytokines IL-4, IL-10, IL-12 and IFN-γ. Results were normalized to 1 pg HPRT.[14,15] IL-2 and IL-12 mRNA transcripts were below the limit of detection (10^{-3} fg/pg HPRT) in MRL/+ and MRL/lpr mice. IFN-γ mRNA transcripts were below the limit of detection in most samples. IL-4 mRNA transcripts were present in 100-fold to 1000-fold greater amounts than IFN-γ mRNA transcripts. IL-10 transcripts were detectable in MRL/+ and MRL/lpr mice. These results confirm the immunohistochemistry experiments and suggest the lacrimal gland lesions in MRL/+ and MRL/lpr mice are Th2 in nature.

3. FAS/FAS LIGAND-MEDIATED APOPTOSIS IN THE LACRIMAL GLAND

Frozen sections from the lacrimal glands of MRL/MpJ mice of both substrains were processed with TUNEL staining to determine the amount of apoptosis and for immunohistochemistry for Fas (Santa Cruz Biotechnology Inc., Santa Cruz, CA) and FasL (Boehringer Mannheim, Indianapolis, IN) with the ABC technique.[16] Groups of 9 B 11 mice were analyzed for lacrimal gland inflammation, apoptosis and Fas expression; 5 B 6 mice were used for FasL expression. The percentage of lacrimal gland replaced by inflammation was estimated within the micrometer disc. For TUNEL staining, the proportion of lymphocytes undergoing apoptosis was counted with a micrometer disc and the number of cells staining positive per unit area of inflammation calculated. For immunohistochemistry, a semi-quantitative scoring system for Fas staining on lymphocytes was used as follows: 0, no staining; 1+, < 25% cells positive; 2+, 25%–50% cells positive; 3+, 51%–75% cells positive; 4+, > 75% cells positive.[16]

MRL/lpr mice had a greater percentage of the lacrimal gland replaced by the inflammatory infiltrate than did MRL/+ mice (Table 2).[16] TUNEL staining demonstrated similar amounts of apoptosis per unit area of lacrimal gland inflammation in MRL/+ and MRL/lpr mice (Table 2).[16] Immunohistochemistry revealed Fas staining on infiltrating lymphocytes in MRL/+ but not MRL/lpr mice. Diffuse staining for FasL was present on epithelial structures in both substrains.[16]

Table 2. Apoptosis of lymphocytes in MRL/MpJ mouse lacrimal gland inflammation

	MRL/+	MRL/lpr	P value
Percent lacrimal gland replaced by inflammation	13.0 ± 3.0^a	30.3 ± 7.0	0.02
Apoptotic cells/unit area of inflammation	24.6 ± 6.0	23.8 ± 2.4	0.91
Fas expression	3+	0	
Fas ligand expression	+	+	

[a]Mean ± SEM. Adapted from Jabs DA, Lee B, Whittum-Hudson J, Prendergast RA. The role of Fas-Fas ligand-mediated apoptosis in autoimmune lacrimal gland disease in MRL/MpJ mice. *Invest Ophthalmol Vis Sci.* 2001;42:399–401. Used with permission.

4. DISCUSSION

As seen in minor salivary gland biopsy specimens from patients with Sjøgren's syndrome,[17,18] the lacrimal gland infiltrate in MRL/MpJ mice is composed largely of CD4+ T-cells with lesser numbers of CD8+ T-cells, B-cells and macrophages.[2,3] Our data suggest the large proportion of these cells is Th2-cells.[13] With immunocytohistochemistry for cytokines, a substantially greater proportion of cells stained for IL-4 than IFN-γ. Preliminary RT-PCR experiments confirmed these results, as mRNA transcripts for IL-4 were present in substantially greater amounts than those for INF-γ. On immunohistochemistry for accessory molecules, B7-2, which drives the immune response toward a Th2 response, was present in significantly greater amounts than B7-1, which drives the immune response toward a Th1 response.[13] Further support for a Th2 process was determined from the RT-PCR studies of other cytokines. IL-12, which drives the immune response toward a Th1 response, was not detected, whereas IL-10, which inhibits a Th1 response, was detected. Hence, the lacrimal gland lesions in MRL/MpJ mice of both substrains appear to be predominantly Th2 in nature. Although the evaluation of minor salivary gland biopsy specimens from patients with Sjøgren's syndrome has given variable results for cytokines detected,[19–21] one study reported IL-4 mRNA was detected by *in situ* hybridization in a greater proportion of the infiltrating mononuclear inflammatory cells than IFN-_ mRNA.[21] Further, only IL-4 mRNA-positive cells were detected in a statistically significant excess over control biopsy specimens. The authors of this report[21] concluded a Th2 process was present in the salivary gland inflammatory infiltrate in Sjøgren's syndrome, results similar to those seen in the lacrimal glands of MRL/MpJ mice.

The systemic autoimmune disease in MRL/lpr mice is related to the defective Fas protein and its immunologic consequences. Although both substrains have lacrimal gland disease, MRL/lpr mice have accelerated disease as indicated by the greater proportion of the lacrimal gland replaced by the inflammatory infiltrate at comparable ages.[16] Two possible mechanisms exist by which the defective apoptosis caused by the *lpr* mutation could accelerate lacrimal gland disease in MRL/MpJ mice: (1) defective apoptosis in the microenvironment of the lacrimal gland permits the infiltrating inflammatory cells to accumulate and expand; or (2) defective apoptosis in peripheral lymphoid tissues permits the accumulation of autoreactive lymphocytes, which then invade the lacrimal gland. Our results demonstrated similar amounts of apoptosis in the lacrimal gland in the two substrains.[16] These results suggest failure of apoptosis within the microenvironment of the lacrimal gland does not contribute to the accelerated disease in MRL/lpr mice, but rather an accumulation of autoreactive lymphocytes, which is a consequence of the defective apoptosis in peripheral lymphoid organs. This permits the expansion of a population of autoreactive T-cells that then invades the lacrimal gland in greater numbers in MRL/lpr than MRL/+ mice. Furthermore, MRL/lpr mice must use alternative pathways to the Fas/FasL system to initiate the apoptosis seen. In the human disease Sjøgren's syndrome, Fas is expressed on infiltrating lymphocytes in the glands, and FasL on acinar epithelial cells; however, apoptosis appears to be blocked. In conclusion, these results suggest the lacrimal gland lesions in both substrains of MRL/MpJ mice are largely Th2 in nature and accelerated disease in MRL/lpr mice occurs as a consequence of the failure of the deletion of autoreactive T-cells in peripheral lymphoid organs with subsequent invasion into the lacrimal gland.

REFERENCES

1.   D.A. Jabs, E.L. Alexander, and W.R. Green. Ocular inflammation in autoimmune MRL/Mp mice. *Invest Ophthalmol Vis Sci.* 26:1223 (1985).
2.   D.A. Jabs and R.A. Prendergast. Murine models of Sjøgren's syndrome: immunohistologic analysis of different strains. *Invest Ophthalmol Vis Sci.* 29:1437 (1988).
3.   D.A. Jabs, C. Enger, and R.A. Prendergast. Murine models of Sjøgren's syndrome: evolution of the lacrimal gland inflammatory lesions. *Invest Ophthalmol Vis Sci.* 32:371 (1991).
4.   R. Watanabe-Fukunaga, C.I. Brannan, N.G. Copeland et al. Lymphoproliferative disorder in mice explained by defects in Fas antigen that mediates apoptosis. *Nature.* 314 (1992).
5.   G.G. Singer and A.K. Abbas. The Fas antigen is involved in peripheral but not thymic deletion of T lymphocytes in T cell receptor transgenic mice. *Immunity.*1:365 (1994).
6.   D.A. Jabs and R.A. Prendergast. Reactive lymphocytes in lacrimal gland and renal vasculitic lesions of autoimmune MRL/lpr mice express L3T4. *J Exp Med.* 166:1198 (1982).
7.   T.R. Mossman, H.M. Cherwinski, M.W. Bond et al. Two types of murine helper T cell clone: I. definition according to profiles of lymphokine activities and secreted proteins. *J Immunol.* 136:2348 (1986).
8.   H.M. Cherwinski, J.H. Schumacher, K.D. Brown et al. Two types of mouse helper T cell clone. III. Further differences in lymphokine synthesis between Th1 and Th2 clones revealed by RNA hybridization, functionally monospecific bioassays, and monoclonal antibodies. *J Exp Med.* 166:1229 (1987).
9.   G. Trinchieri. Interleukin-12: a cytokine produced by antigen-presenting cells with immunoregulatory functions in the generation of T helper cells type 1 and cytotoxic lymphocytes. *Blood.* 84:4008 (1994).
10. D.J. Berg, M.W. Leach, R. Kuhn et al. Interleukin 10 but not interleukin 4 is a natural suppressant of cutaneous inflammatory responses. *J Exp Med.* 182:99 (1995).
11. V.K. Kuchroo, M.P. Das, J.A. Brown et al. B7-1 and B7-2 costimulatory molecules activate differentially the Th1/Th2 developmental pathways: application to autoimmune disease therapy. *Cell.* 80:707 (1995).

12. D.J. Lenschow, S.C. Ho, H. Satter et al. Differential effects of anti-B7-1 and anti-B7-2 monoclonal antibody treatment on the development of diabetes in the nonobese diabetic mouse. *J Exp Med.* 181:1145 (1995).

13. D.A. Jabs, B. Lee, J.A. Whittum-Hudson, and R.A. Prendergast. Th1 versus Th2 immune responses in autoimmune lacrimal gland disease in MRL/Mp mice. *Invest Ophthalmol Vis Sci.* 41:826 (2000).

14. K.M. Drescher, and J.A. Whittum-Hudson. Herpes simplex virus type 1 alters transcript levels of tumor necrosis factor-_ and interleukin-6 in retinal glial cells. *Invest Ophthalmol Vis Sci.* 37:2302 (1996).

15. S.L. Reiner, S. Zheng, D.B. Corry, and R.M. Locksley. Constructing polycompetitor cDNAs for quantitative PCR. *J Immunol Meth..* 165:37 (1993).

16. D.A. Jabs, B. Lee, J.A. Whittum-Hudson, and R.A. Prendergast. The role of Fas-Fas ligand-mediated apoptosis in autoimmune lacrimal gland disease in MRL/MpJ mice. *Invest Ophthalmol Vis Sci.* 42:399 (2001).

17. T.C. Adamson, R.I. Fox, D.M. Frisman, and F.V. Howell FV. Immunohistologic analysis of lymphoid infiltrates in primary Sjøgren's syndrome using monoclonal antibodies. *J Immunol.* 130:203 (1983).

18. R.I. Fox, S.A. Carstens, S. Fong et al. Use of monoclonal antibodies to analyze peripheral blood and salivary gland lymphocyte subsets in Sjøgren's syndrome. *Arthritis Rheum.* 25:419 (1982).

19. R.I. Fox, H.I. Kang, D. Ando, J. Abrams, E. Pisa. Cytokine mRNA expression in salivary gland biopsies of Sjøgren's syndrome. *J Immunol.* 152:5532 (1994).

20. Y. Ohyama, S. Nakamura, G. Matsuzaki et al. Cytokine messenger RNA expression in the labial salivary glands of patients with Sjøgren's syndrome. *Arthritis Rheum.* 39:1376 (1996).

21. K.E. Aziz, B. Markovic, P.J. McCluskey, D. Wakefield. A study of cytokines in minor salivary glands. In: Nussenblatt RB, Whitcup SM, Caspi RR, Gerry I, eds. *Advances in Ocular Immunology.* New York, NY: Elsevier (1994).

ICAM-1: ITS ROLE IN THE PATHOPHYSIOLOGY OF IMMUNE ACTIVATION IN THE MRL/LPR MOUSE

Jianping Gao, Grant Morgan, David D. Tieu, Tammy A. Schwalb, Mylinh Ngo, and Michael E. Stern

Department of Biological Science
Allergan
Irvine, California, USA

1. INTRODUCTION

Lymphocytic interaction with vascular endothelium and infiltration during inflammation requires the expression of adhesion molecules such as intercellular adhesion molecule-1 (ICAM-1). Expression of ICAM-1, as well as its receptor, lymphocyte function associated antigen-1 (LFA-1), plays an important role in lymphocyte signaling leading to lymphocytic extravasation and migration to the target tissues.[1] Progressive lymphocytic accumulation in the lacrimal gland and conjunctiva is characteristic of the ocular surface histopathology for dry eye syndrome. ICAM-1/LFA-1 is upregulated in lymphocytes and vascular endothelial cells in the salivary and lacrimal glands in patients with Sjøgren's syndrome.[2]

We have previously reported in humans and canines with dry eye that marked lymphocytic infiltration, with increased expression of various inflammatory and immune activation markers such as ICAM-1, LFA-1 and MHC class II antigens, was evident in the conjunctival as well as accessory lacrimal gland biopsies.[3] In the current report, we have undertaken a time course study to evaluate whether the expression of ICAM-1 locally and systemically correlates with disease progression and severity in MRL/lpr mice. Specifically, we have examined whether residential epithelial cells on the ocular surface are active components and can be induced to express ICAM-1 during immune-based inflammation. To determine if inhibition of ICAM-1/LFA-1 interactions and signaling could suppress the ocular inflammatory response, a subsequent study was carried out using anti-ICAM-1 and LFA-1 antibodies in this autoimmune mouse model.

Lacrimal Gland, Tear Film, and Dry Eye Syndromes 3
Edited by D. Sullivan *et al.*, Kluwer Academic/Plenum Publishers, 2002

777

2. MATERIALS AND METHODS

2.1. MRL/lpr Mouse Model

To study the pattern of ICAM-1 expression over time, female MRL/lpr mice (n = 16) and the age-matched C57 BL/6 mice (n = 16) at 3 weeks of age were purchased from Jackson Laboratory (Bar Harbour, ME). Animals (4 mice/group) were euthanized at 1, 2, 3 and 4 mo by CO_2 asphyxiation. Fresh frozen lacrimal glands and lids were embedded in OCT for immunofluorescent evaluation using purified anti-ICAM-1 or LFA-1 antibodies (PharMingen, San Diego, CA). Plasma samples were collected for evaluation of soluble ICAM-1 (sICAM-1) concentration using mouse sICAM-1 (CD54) ELISA kit (Endogen, Woburn, MA).

2.2. Antibody Treatment

To study the effectiveness of anti-ICAM-1/LFA-1 therapy, female MRL/lpr mice were randomly divided into four groups. At 3 weeks of age, each group (n = 4 mice), except controls, was intraperitoneally injected with purified monoclonal antibodies against ICAM-1, LFA-1 or ICAM-1 and LFA-1 (Pharmingen). The control animals (n = 12) were age-matched MRL/lpr mice without antibody treatment. Antibodies were given 150 _g/dose weekly until 8 weeks of age. Mice were sacrificed at week 17. Lacrimal glands, spleen and thymus were collected. The level of inflammation was evaluated by lacrimal histology. The areas of lymphocytic infiltration in the lacrimal gland with and without anti-ICAM-1/LFA-1 treatment were quantified. The image of the entire lacrimal section and areas of follicular infiltrates within the same section were traced on a computer screen using Image Pro Plus 4.0 software (Media Cybernetics, Silver Spring, MD). The ratio of the infiltrated area over the entire lacrimal gland area was determined for each sample. Differences in this ratio among test groups were analyzed by Student's *t*-test. Results were considered statistically significant if the one-tail P value was < 0.05.

2.3. Flow Cytometry

Isolated splenic and thymic tissues were disrupted using frosted glass slides and the resulting cell suspension (5 x 10^6/ml) was washed and resuspended in PBS/1% BSA/0.1% sodium azide. Aliquots (200 μl) were then added to microfuge tubes and incubated with or without fluorescent tagged anti-CD3, anti-CD4, anti-ICAM-1 or LFA-1 antibodies (1 mg/sample, Pharmingen) at 4°C in the dark for 30 min. The samples were then washed in PBS/BSA/azide and resuspended in 400 μl PBS/1% formalin. Subsequently, tubes were incubated at 4°C in the dark until analyzed by flow cytometry (FACSCalibur, Becton Dickinson) within 48 h.

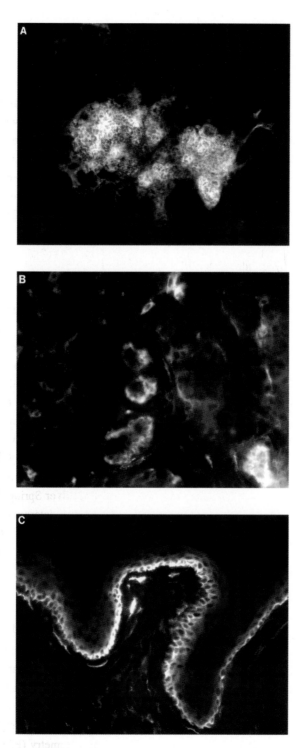

Figure 1. ICAM-1 immunoreactivity was detected in the lacrimal gland (**A, B**) and conjunctiva (**C**) of the MRL/lpr mouse at 4 mo of age. ICAM-1 was expressed by vascular endothelial cells (**A**), infiltrating lymphocytes (**A**), as well as epithelial cells (**B, C**).

3. RESULTS

3.1. ICAM-1 Expression in the Ocular Tissues of the MRL-lpr Mouse Over Time

ICAM-1 immunoreactivities were detected in the lacrimal gland of MRL/lpr (Figs. 1A, 1B) but not in the age-matched control C57 BL/6 mice (data not shown). ICAM-1 positivity was found on the vascular endothelial cells, infiltrating lymphocytes as well as the lacrimal acinar epithelial cell membrane. The expression of ICAM-1 on different cell types appeared to depend on the severity of the disease. Endothelial ICAM-1 expression was detected in young mice at 1 month. The lymphocytic ICAM-1 expression was evident at 2 mo and increased at disease onset (3 mo). No acinar epithelial cell ICAM-1 was detected until 3 mo and markedly increased by 4 months of age. ICAM-1 was also detected in the conjunctival epithelial cells of MRL/lpr mice at 2, 3 and 4 mo (Fig. 1C). The expression level, however, did not differ significantly over time.

For comparison, the cellular distribution and expression levels of LFA-1 over time were also investigated. LFA-1 was detected in the lacrimal tissues of MRL/lpr mice but not in the age-matched C57 BL/6 controls (data not shown). The positive signal was primarily found on the cell surface of infiltrating lymphocytes and vascular endothelial cells along blood vessels. Lymphocytic LFA-1 was evident from 2 mo in MRL/lpr mice and continued to increase as lymphocytic infiltration progressed. No significant LFA-1 immunoreactivity was found in the conjunctiva of MRL/lpr mice at any age studied.

3.2. Systemic sICAM-1 Concentrations in the MRL/lpr Mouse Over Time

Table 1 illustrates sICAM-1 levels in the plasma of MRL/lpr and C57BL/6 mice. Mean sICAM-1 levels were significantly higher in MRL/lpr mice than controls at 2, 3 and 4 mo. In MRL/lpr mice, the mean sICAM-1 levels gradually and significantly increased over time, with peak value of 63.65 µg/ml at 4 mo in MRL/lpr mice. In the control C57BL mice, however, the plasma sICAM-1 level did not increase significantly until 4 mo.

3.3. Anti-ICAM-1/LFA-1 Antibody Treatment in the MRL/lpr Mouse

In the preliminary study, a statistically significant ($P = 0.027$) decrease in the number of inflammatory infiltrates in the lacrimal gland was evident in the MRL/lpr mouse treated with purified monoclonal anti-ICAM-1/LFA-1 antibodies, compared to control mice (received no antibodies). The mean percentage change in the ratio of infiltrated area over the entire lacrimal section was –51.23%, compared to controls. Mice treated with anti-ICAM-1 or anti-LFA-1 antibody alone demonstrated a reduction in the level of inflammation in the lacrimal tissues to a certain degree, but the result was not statistically significant. Antibody administration had no significant effect on the expression of ICAM-1 or LFA-1 in spleen or thymus in the MRL/lpr mouse.

Table 1. Plasma-soluble ICAM-1 concentrations in MRL/lpr and C57BL/B6 mice

	1 Month	2 Months	3 Months	4 Months
C57BL/B6				
Mean ± SD	8.58 ± 1.89	10.21 ± 8.03	10.17 ± 2.75	22.01 ± 3.15
(n = 4, µg/ml)				
MRL/lpr				
Mean ± SD	8.51 ± 1.84	24.52 ± 4.35	28.13 ± 1.64	63.65 ± 9.61
(n = 4, µg/ml)				

4. DISCUSSION AND SUMMARY

This study has shown that during inflammation in the lacrimal gland of MRL/lpr mice, ICAM-1 was upregulated in the vascular endothelial cells and infiltrating lymphocytes as well as the acinar epithelial cells after the onset of disease. Epithelial cell ICAM-1 expression was also exhibited in the conjunctiva without apparent inflammation. The level of ICAM-1 expression, locally in the lacrimal gland and systemically in the plasma, increased over time and positively correlated with disease progression and severity in the Sjøgren's syndrome-like MRL/lpr mice. Treatment with anti-ICAM-1/LFA-1 antibody resulted in a significant decrease in lacrimal gland inflammation in MRL/lpr mice, but had no effect on ICAM-1 or LFA-1 expression in either spleen or thymus.

The presence of ICAM-1 in the epithelial cells in the lacrimal gland will allow direct interaction of epithelial cells with lymphocytes, resulting in lymphocyte homing and subsequent inflammatory damage such as epithelial cell apoptosis.[4] The upregulation of epithelial ICAM-1 expression on the ocular surface suggests that epithelial cells are active components and play an important role in the pathophysiology of immune activation and inflammation in MRL/lpr mice. ICAM-1 is a co-stimulator with MHC II in promoting antigen presentation during immune activation.[5] In the conjunctival epithelium of diseased MRL/lpr mice, the notion of increased ICAM-1 and MHC II expression in the absence of antigen and apparent inflammation suggests a predisposition may exist in the localized immune response whose initiation requires additional internal or external stimuli.[6,7]

It has been recently reported that the serum sICAM-1 level is elevated in patients with primary or secondary Sjøgren's syndrome.[8] It is not too surprising that sICAM-1 concentration increased in the MRL/lpr mouse with systemic autoimmunity in the present study. However, the positive correlation between the systemic level of sICAM-1 with the severity of lacrimal gland inflammation over time indicates upregulation in systemic sICAM-1 may contribute to local inflammation in the lacrimal gland of MRL/lpr mice. In fact, our recent finding suggests that lymphocytic accumulation in the lacrimal gland of the MRL/lpr mouse may be the result of progressive lymphocytic infiltration from the bloodstream instead of lymphocyte activation and proliferation locally in the lacrimal tissue (manuscript in preparation).

Finally, future study is necessary to determine the efficacy of anti-ICAM-1/LFA-1 therapy in suppressing lacrimal gland inflammation. The potential use of anti-ICAM-1/LFA-1 therapy in treating immune-based inflammatory diseases, such as dry eye, deserves further investigation.

REFERENCES

1. T.A. Springer. Adhesion receptors of the immune system. *Nature* 346:425 (1990).
2. I. Saito, K. Terauchi, M. Shimuta, S. Nishiimura, K.Yoshino, T. Takeuchi, K. Tsubota, N. Miyasaka. Expression of cell adhesion molecules in the salivary and lacrimal glands of Sjøgren's syndrome. *J Clin Lab Anal.* 7:180 (1993).
3. J.A. Smith, M.E. Stern, J. Gao, T.A. Schwalb, D.C. Rupp, S.M. Whitcup. Conjunctival Inflammation in non- Sjogren's Syndrome and Sjogren's Syndrome Keratoconjunctivitis sicca. ARVO Abstract; S276:1450 (2000).
4. J. Gao, T.A. Schwalb, J.V. Addeo, C.R. Ghosn, M.E. Stern. The role of apoptosis in the pathogenesis of canine keratoconjunctivitis sicca: the effect of topical cyclosporin A therapy. *Cornea* 17:654 (1998).
5. M. Croft, C. Dubey. Accessory molecule and costimulation requirements for CD4 T cell response. *Crit Rev Immunol.* 17(1):89 (1997).
6. M.E. Stern, J. Gao, G.A. Morgan, D.K. Brees, T.A Schwalb, M. Humphreys-Behr, J.A. Smith. The role of ICAM-1 as a signal protein for predisposition of ocular surface inflammation. *Cornea Suppl 2* 19:S126 (2000).
7. J.G. Lamphear, K.R. Stevens, R.R. Rich. Intercellular adhesion molecule-1 and leukocyte function-associated antigen-3 provide costimulation for superantigen-induced T lymphocyte proliferation in the absence of a specific presenting molecule. *J Immunol.* 160(2):615 (1998).
8. C. Andrys, J. Krejsek, H. Kralove et al. Serum soluble adhesion molecules (sICAM-1, sVCAM-1, sE-selectin) and neopterin in patients with Sjøgren's syndrome. *Acta Medica* 42:97 (1999).

PROINFLAMMATORY CYTOKINE INHIBITION OF LACRIMAL GLAND SECRETION

Claire Larkin Kublin, Robin R. Hodges, and Driss Zoukhri

Schepens Eye Research Institute
Department of Ophthalmology
Harvard Medical School
Boston, Massachusetts, USA

1. INTRODUCTION

A decrease in lacrimal and salivary gland secretion is a primary cause of dry eye and dry mouth. Sjögren's syndrome is the leading cause of the aqueous tear deficient type of dry eye.[1–3] It is an autoimmune disease, which occurs almost exclusively in females (>90), and involves an extensive lymphocytic infiltration of the lacrimal and salivary glands and destruction of epithelial cells.[4–5] No cure exists for this disease. Moreover, the exact etiology of Sjögren's syndrome is largely unknown, but may include factors of viral, endocrine, neural, genetic and environmental origin.[2–5]

The precise mechanism(s) responsible for the decreased tears and saliva secretion in Sjögren's syndrome is unknown.[2–5] The immune-mediated destruction of the epithelial cells due to the progressive lymphocytic infiltration of the lacrimal and salivary glands may be responsible for the decline in tear and saliva production leading to dry eye and dry mouth.[1–3–5] We have shown the innervation of the lacrimal and salivary glands of MRL/lpr mice, an animal model of human Sjögren's syndrome, was not altered with the onset and progression of the disease and acinar cells isolated from these animals were hyper-responsive to exogenous secretagogues.[6–7] Recently, we showed nerves of diseased lacrimal and salivary glands could not release their neurotransmitters in response to a depolarizing potassium solution.[8] In this paper we tested our hypothesis that increased production of proinflammatory cytokines, especially interleukin-1β (IL-1β), in diseased

Lacrimal Gland, Tear Film, and Dry Eye Syndromes 3
Edited by D. Sullivan *et al.*, Kluwer Academic/Plenum Publishers, 2002

783

animals leads to a blockade of neurotransmitter release and hence impaired lacrimal gland secretion.

2. METHODS

2.1. Measurement of Peroxidase Secretion

Lacrimal glands were removed from female BALB/c mice and cut into small lobules (~2 mm diameter). Lobules were placed in cell strainers and incubated at 37°C in Krebs-Ringer bicarbonate (KRB) buffer. The cell strainers containing lobules were transferred into fresh KRB solution every 20 min for a total of 60 min. The lobules were then incubated for 20 min in a total volume of 0.8 ml in normal KRB (spontaneous secretion) and depolarizing KRB (evoked secretion) solution in which the concentration of KCl was increased to 75 mM and that of NaCl was decreased to 55 mM to maintain isotonicity. Lacrimal gland lobules were further incubated for 20 min in 0.8 ml of normal KRB containing phenylephrine (an α_1-adrenergic agonist, 10^{-4} M). After incubation, the media were collected and centrifuged to remove debris. The lobules were homogenized in 10 mM Tris-HCl, pH 7.5. The amounts of peroxidase in the media and tissue homogenate were determined using a spectrofluorometric assay.

2.2. Effect of IL-1β, IL-1α or TNFα on Peroxidase Secretion

In one set of experiments, lacrimal gland lobules were incubated for 2 h in the absence or presence of recombinant human (rh)IL-1β, rhIL-1α (10 ng/ml each) or rhTNFα (50 ng/ml). Spontaneous, evoked and phenylephrine-induced peroxidase secretion was then measured. In another set of experiments, rhIL-1β (1 μg) or saline was injected (2 μl) into the lacrimal glands of anesthetized mice. To determine if the effects of IL-1β were specific and receptor-mediated, we used a monoclonal antibody against IL-1 receptor type 1 (clone 35F5, PharMingen). The lacrimal glands were removed 24 h post-injection, lobules were prepared and peroxidase secretion was measured.

2.3. Preparation of Lacrimal Gland Acini and Measurement of IL-1β Protein

Lacrimal glands were removed from 3-, 9- or 13-week-old BALB/c, MRL/+ and MRL/lpr mice. To prepare acini, lacrimal glands were minced and incubated in KRB buffer supplemented with 10 mM Hepes, 5.5 mM glucose, 0.5% BSA and collagenase (CLS III, 150 U/ml). To remove lymphocytes, acini were subjected to a Ficoll gradient of 2, 3 and 4%. Dispersed acini were allowed to recover for 30 min in fresh KRB-Hepes buffer containing 0.5% BSA after which they were homogenized in 0.3 ml of 10 mM Tris-HCl, pH 7.0. The amount of IL-1β in cell lysates was determined using an enzyme-linked immunosorbant assay (R&D Systems) and the amount of protein determined using the method of Bradford with BSA as a standard.

3. RESULTS AND DISCUSSION

In a recent report, we showed activation of inflamed lacrimal gland nerve endings with high KCl did not elicit acetylcholine release or peroxidase secretion.[8] We hypothesized elevated levels of proinflammatory cytokines in inflamed lacrimal glands inhibit neurotransmitter release and hence protein secretion. To test this hypothesis, lacrimal gland lobules were prepared from BALB/c mice and incubated in the presence or absence of rhIL-1α, rhIL-1β (10 ng/ml each) or rhTNFα (50 ng/ml). Peoxidase secretion in response to high KCl (75 mM) or an adrenergic agonist, phenylephrine (10^{-4} M), was then measured. KCl-induced peroxidase secretion was inhibited 62%, 66% and 53% by rhIL-1α, rhIL-1β and rhTNFα, respectively (Table 1). Phenylephrine-induced peroxidase secretion was also inhibited 62%, 66% and 36% by rhIL-1α, rhIL-1β and rhTNFα, respectively. These results show proinflammatory cytokines inhibit neurally induced lacrimal gland protein secretion.

To study the specificity of effect of inflammatory cytokines, rhIL-1β (1 μg) or saline was injected into lacrimal glands of anesthetized BALB/c mice. rhIL-1β inhibited KCl-induced peroxidase secretion by 72% and that induced by phenylephrine by 42% (Table 2). To determine if the effects of IL-1β were specific and receptor-mediated, we used a monoclonal antibody against murine IL-1R1. Co-injection of this antibody with the cytokine completely abolished the inhibitory effect of rhIL-1β(Table 2). These results show the inhibitory effect of rhIL-1β on lacrimal gland secretion is specific and mediated by IL-1R1.

We next sought to determine if the protein level of IL-1β were increased with age and disease in the lacrimal gland acinar cells of MRL/lpr mice. Acinar cells were isolated from control BALB/c and MRL/+ mice and diseased MRL/lpr mice; the amount of IL-1β was then measured by ELISA. A dramatic upregulation of IL1β protein occurred in lacrimal gland acinar cells isolated from 9- and 13-week-old MRL/lpr mice (Table 3). Compared to 3-week-old MRL/lpr mice, the amount of IL-1β protein was upregulated 15- and 21-fold in lacrimal gland acinar cells from 9- and 13-week-old MRL/lpr mice, respectively. The increase in the amount of IL-1β protein occurred in a time-dependent manner coinciding with the kinetics of the lymphocytic infiltration of the lacrimal gland in this murine model of human Sjögren's syndrome. These results show the amount of IL-1β protein increased in a time-dependent manner in MRL/lpr lacrimal gland acinar cells.

In summary, our results show the amount of IL-1β protein is elevated in diseased MRL/lpr mice and exogenous addition of this cytokine inhibits neurally mediated lacrimal gland secretion. Although we do not have direct evidence, the elevated levels of IL-1β in the MRL/lpr murine model of human Sjögren's syndrome may inhibit neurotransmitter release leading to insufficient lacrimal gland secretion. Further experiments are needed to test this hypothesis.

Table 1. Effect of inflammatory cytokines in vitro on lacrimal gland secretion

| | Peroxidase Secretion (% of total) | | | |
	Control	rhIL-1α	rhIL-1β	rhTNFα
Spontaneous	0.12 ± 0.02	0.11 ± 0.02	0.06 ± 0.01	0.15 ± 0.03
Evoked	1.78 ± 0.17*#	0.74 ± 0.12*	0.62 ± 0.15*	0.93 ± 0.17*
Phenylephrine	4.44 ± 0.58*	2.70 ± 0.38*	1.71 ± 0.28*	2.92 ± 0.33*

Values are means ± SEM, n = 3. *Significantly different from spontaneous secretion; #significantly different from evoked secretion in the presence of the cytokines ($P < 0.05$)

Table 2. Effect of rhIL-1β in vivo on lacrimal gland protein secretion

| | Peroxidase Secretion (% of total) | | | |
	Saline	rhIL-1β	Ab	Ab+rhIL-1β
Spontaneous	0.09 ± 0.03	0.07 ± 0.01	0.08 ± 0.02	0.07 ± 0.02
Evoked	0.45 ± 0.07*#	0.17 ± 0.05	0.65 ± 0.15*#	0.65 ± 0.03*#
Phenylephrine	1.13 ± 0.31*	0.68 ± 0.19*	1.45 ± 0.61*	1.25 ± 0.47*

Values are means ± SEM, n = 3. *Significantly different from spontaneous secretion; #significantly different from evoked secretion in the presence rhIL-1β ($P < 0.05$)

Table 3. Effect of age on the amount of IL-1β protein

| | Amount of IL-1β (pg/mg protein) | | |
Age (weeks)	BALB/c	MRL/+	MRL/lpr
3	58 ± 17	31 ± 11	65 ± 9
9	210 ± 49*	251 ± 20*	946 ± 245*#
13	131 ± 10	102 ± 21	1358 ± 338*#

Values are means ± SEM, n = 4–5. *Significantly different from 3-week-old animals; #significantly different from age-matched control animals ($P < 0.05$)

ACKNOWLEDGMENTS

The authors are grateful to Dr. Fara Sourie for her invaluable contribution to this work and Dr. Craig W. Reynolds (Biological Resources Branch, National Cancer Institute Preclinical Repository, Rockville, MD) for the generous gift of recombinant human cytokines. This research was supported by National Eye Institute grant EY12383.

REFERENCES

1. N. Talal, H.M. Moutsopoulos, and S.S. Kassan. *Sjögren's Syndrome. Clinical and Immunological Aspects,* Springer Verlag, Berlin (1987).
2. D.A. Sullivan, L.A. Wickham, K.L. Krenzer, E.M. Rocha, and I. Toda. Aqueous tear deficiency in Sjögren's syndrome: possible causes and potential treatment, in: *Oculodermal Diseases,* Pleyer U., and C. Hartmenn, eds., Aeolus Press, Buren (1997).
3. R.I. Fox, Pathogenesis of Sjögren's syndrome, *Rheum Dis Clin North Am.* 18:517 (1992).
4. R.I. Fox, J. Törnwall, and P. Michelson, Current issues in the diagnosis and treatment of Sjögren's syndrome, *Curr Opin Rheumatol.* 11:364 (1999).
5. D.A. Sullivan. Possible mechanisms involved in the reduced tear secretion in Sjögren's syndrome, in: *Sjögren's syndrome: state of the art,* Homma M., S. Sugai, T. Tojo, N. Miyasaka, and M. Akizuki, eds., Kugler publications, Amsterdam (1994).
6. D. Zoukhri, R.R. Hodges, and D.A. Dartt, Lacrimal gland innervation is not altered with the onset and progression of disease in a murine model of Sjögren's syndrome, *Clin Immunol Immunopathol.* 89:126 (1998).
7. D. Zoukhri, R.R. Hodges, I.M. Rawe, and D.A. Dartt, Ca^{2+} signaling by cholinergic and α_1-adrenergic agonists is up-regulated in lacrimal and submandibular glands in a murine model of Sjögren's syndrome, *Clin immunol Immunopathol.* 89:134 (1998).
8. D. Zoukhri, and C.L. Kublin, Impaired neurotransmitter release from lacrimal and salivary gland nerves of a murine model of Sjögren's syndrome, *Invest Ophthalmol Vis Sci.* 42:925 (2001).

EFFECT OF ANTI-INFLAMMATORY CYTOKINES ON THE ACTIVATION OF LYMPHOCYTES BY LACRIMAL GLAND ACINAR CELLS IN AN AUTOLOGOUS MIXED CELL REACTION

Melvin D. Trousdale,[1,2] Douglas Stevenson,[1] Zejin Zhu,[1]
Harvey R. Kaslow,[2] Joel E. Schechter,[2] Dwight W. Warren,[3]
Ana M. Azzarolo,[3] Thomas Ritter,[4] and Austin K. Mircheff[1,2]

[1]Doheny Eye Institute
[2]University of Southern California
Los Angeles, California, USA
[3]Florida Atlantic University
Boca Raton, Florida, USA
[4]Institute of Medical Immunology
Humboldt University
Berlin, Germany

1. INTRODUCTION

Lymphocytic infiltration of the lacrimal gland decreases the gland's ability to produce the fluid needed to keep the ocular surface healthy and comfortable. This immune-related lacrimal insufficiency is responsible for some severe forms of dry eye. Increasing anti-inflammatory cytokines in a dysfunctional lacrimal gland may suppress the inflammatory process and restore secretory function.

The ability of lacrimal gland acinar cells to stimulate lymphocyte proliferation has been demonstrated using a co-culture system consisting of purified lacrimal gland acinar cells (pLGAC) and peripheral blood lymphocytes from the same rabbit in a mixed-cell reaction (Kaslow et al., 1998; Guo et al., 2000a). The subsequent inoculation of these proliferating lymphocytes into the remaining lacrimal gland of the donor animal induces autoimmune dacryoadenitis (Guo et al, 2000b). This preliminary study evaluates four important immunoregulatory biologics that may interfere with the ability of acinar cells to induce lymphocyte proliferation in the autologous mixed-cell reaction. The biologics tested

Lacrimal Gland, Tear Film, and Dry Eye Syndromes 3
Edited by D. Sullivan *et al.*, Kluwer Academic/Plenum Publishers, 2002

789

included two anti-inflammatory cytokines (i.e., vIL-10 and TGF-β1), a chimeric construct capable of neutralizing the pro-inflammatory TNF-α (i.e., TNFRp55-Ig) and a multisite inhibitor of the inflammatory cascade (i.e., α-MSH).

2. METHODS

2.1. Purification and Culture of Lacrimal Gland Epithelial Cells and Lymphocytes

Inferior lacrimal gland acinar cells (pLGAC) and peripheral blood lymphocytes (PBL) were obtained from adult female New Zealand white rabbits and purified for in vitro studies using the methods described by Guo et al., 2000a; Schönthal et al., 2000.

2.2. Transduction and Autologous Mixed Cell Reaction

Autologous mixed cell reactions were performed in 96 well plates as previously described (Guo et al 2000a). Briefly, 100 µl of a purified suspension of pLGEC containing 10^6 cells/ml were seeded into each well. Then 10 µl of AdvIL-10 or AdTNFRp55-Ig, diluted to give the desired multiplicity of infection (MOI), was added to appropriate wells. After 48 hours in culture, the transduced and non-transduced pLGEC were gamma irradiated with 1,500 RAD to ensure blockage of cell proliferation while allowing gene expression. An equal number of PBL (i.e. 10^5/well) were added to the plate and the cells were co-cultured together for 5 days. After 5 days, co-cultured samples were pulsed with 1 µCi per well of ^{3}H-thymidine (New England Nuclear, Boston, MA), and 24 hours later the cells were collected using a Brandel model 290 PHD™ sample harvester (Gaithersburg, MD). ^{3}H-thymidine incorporation was determined in a LS 6000IC beta scintillation counter (Beckman Instruments, Inc. Fullerton, CA). Counts for each study group were determined from 6 replicate wells.

2.3. Recombinant adenovirus and Recombinant Proteins

The generation of recombinant adenoviruses (Ad) encoding for Epstein-Barr virus derived IL-10 (vIL-10) and TNF inhibitor has been described (Ritter et al., 2000 and Kollis et al., 1994). The TNF-inhibitor was a fusion protein formed by joining the human 55-kDa TNF receptor (TNFRp55) extracellular domain to a mouse IgG heavy chain (TNFRp55-Ig) (Kollis et al., 1994). The recombinant plasmid was cotransfected into the 293 cell line, a transformed, primary human embryonal kidney cell line purchased from American Type Culture Collection (ATCC CRL 1573),together with the large adenoviral plasmid pJM17 using Lipofectamin (Gibco/BRL, Eggenstein, Germany). Adenoviral genomes formed by homologous recombination between the pJM17 vector and the pACCMV vector contained the IL-10 or TNFRp55-Ig cDNA were efficiently packaged to form replication-deficient virus.

TGF-β1 (1–10 ng/ml, from R&D Systems) or α-MSH (10^{-7} M, from Novabiochem) was included in the acinar cell culture medium during the entire incubation (i.e., 48 hour mono-culture incubation period and 120 hour co-culture period with the lymphocytes) prior to the addition of ³H-thymidine.

3. RESULTS

Figs. 1 and 2 show results of studies of autologous mixed-cell reactions using irradiated pLGAC and purified lymphocytes. The co-culture of pLGAC cells and lymphocytes from the same animal resulted in increased lymphocyte proliferation, as indicated by increased [³H]-thymidine incorporation (Fig. 1). Approximately 9,000 cpm were detected in the co-culture samples, compared to background counts of 1,700 to 2,200 cpm for the controls (pLGAC or lymphocytes alone). Lymphocyte proliferation was significantly suppressed when acinar cells were first transduced with either vector Ad-vIL10 or Ad-TNFRp55-Ig prior to co-culture (i.e., ³H incorporation was reduced from 9,000 to less than 2,000 cpm with expression of vIL-10 or TNFRp55-Ig). Acinar cells had a stimulation index of 4.1 (i.e., 9,000 cpm ÷ 2,200 cpm = 4.1). The effect of our selected recombinant proteins was less impressive under the conditions employed. The presence of recombinant α-MSH (10^{-7} M) with cultured pLGAC before and during co-culture with lymphocytes had an insignificant effect on the stimulation of lymphocytes. ³H incorporation was reduced from 9,000 cpm in untreated co-cultures to 7,500 cpm for α-MSH-treated co-cultures.

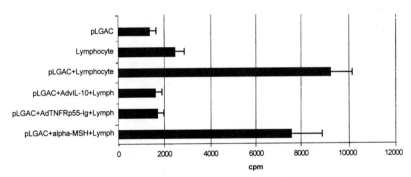

Figure 1. Effect of Ad-vIL10, Ad-TNFRp55-Ig and recombinant α-MSH on the activation of lymphocytes in an autologous mixed-cell reaction.

In a second experiment, co-cultured pLGAC and lymphocytes incorporated [³H]-thymidine with 13,000 cpm in the absence of TGF-β1 (10 ng/ml), giving a stimulation index of 3.4 (i.e., 13,000 cpm ÷ 3,800 cpm = 3.4), (Fig. 2). When pLGACs were incubated with recombinant TGF-β1 prior to and during co-culture with lymphocytes, the stimulatory activity was reduced by 50%. The cpm values for each study group were determined from six replicate samples.

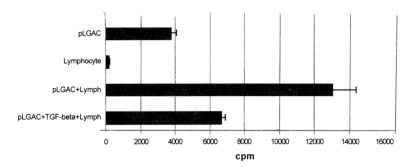

Figure 2. Effect of recombinant TGF-β1 on the activation of lymphocytes in an autologous mixed-cell reaction.

DISCUSSION

Gene-based therapy and delivery of recombinant proteins provide a means of regional and systemic administration of protein pharmaceuticals and may help treat the most severe cases of dry eye. Gene transfer was recently demonstrated successfully in a rat model after retrograde perfusion of the exocrine pancreas, liver and submandibular glands. Retrograde ductal administration of naked DNA resulted in the endocrine secretion of human insulin and growth hormone (Goldfine et al., 1997). Others have shown adenovirus-mediated transfer of human growth hormone gene leads to endocrine excretion with biological activity (He et al., 1998). Although IL-10 and TGF-β1 activities have been studied intensively in many in vivo models, investigations into the immunosuppressive potential have yielded conflicting results (Mathisen et al., 1997; Tsang et al., 1990). Evidence exists that IL-10 and TGF-β1 are capable of suppressing T-cell proliferation and cytokine release, both of which help control inflammation.

At least three species of IL-10 have been studied, including murine IL-10, human IL-10 and viral IL-10. The short half-life of each complicates their use in systemic treatments. We selected the viral IL-10 gene in an adenovirus vector for our studies for two reasons. First, we knew from previous studies that adenovirus vectors could successfully target lacrimal gland acinar cells (unpublished data); and second, earlier studies suggested viral IL-10 was superior to mouse and human IL-10 in suppressing Th1-mediated immune responses (Ma et al., 1998).

Kolls et al. (1994) reported a prolonged and effective blockade of tumor necrosis factor activity through adenovirus-mediated TNFRp55-Ig gene transfer in a mouse model.

Expression of this gene provides a fusion protein formed by joining the human 55-kDa TNF receptor extracellular domain to a mouse IgG heavy chain capable of binding TNF and engaging two of its three receptor-binding sites. More recently, Ritter et al. (2000) reported the neutralization of TNF-α with this same vector.

TGF-β1 inhibits proliferation of T-cells to polyclonal mitogens, and in mixed-cell leukocyte reactions it inhibits the maturation of cytolytic lymphocytes and activation of macrophages. In addition, TGF-β1 counteracts proinflammatory cytokine effects on polymorphonuclear leukocytes and endothelial cells and may be a signal for shutting off immune and inflammatory responses.

In summary, our in vitro mixed-cell reactions provide evidence that viral IL-10 and a TNF-inhibitor such as TNFRp55-Ig may interfere with stimulation of lymphocytes; therefore, both may be useful in controlling inflammation in the lacrimal gland. On the other hand, the recombinant proteins TGF-β1 and α-MSH had much less of an impact on the anti-stimulatory activity of acinar cells against lymphocytes in our in vitro system. The TGF-β1 effect was statistically significant while α-MSH was insignificant.

ACKNOWLEDGMENTS

Dr. Jay Kolls made the TNFRp55-Ig and Dr. Rene de Waal Malefyt provided the vIL-10 cDNA. The authors thank Ms. Susan Clarke for editorial assistance. The financial support was from NIH grants EY12689, EY 05801, EY 09405 and EY03040 and a Zumberge Fund grant.

REFERENCES

H.R. Kaslow, Z. Guo, D.W. Warren, R.L. Wood, and A.K. Mircheff, A method to study induction of autoimmunity in vitro: co culture of lacrimal cells and autologous immune system cells. Adv Exp Med Biol. 438:583 (1998).

Z. Guo, A.M. Azzarolo, J.E. Schechter, D.W. Warren, R.L. Wood, A.K. Mircheff and H.R. Kaslow. Lacrimal gland epithelial cells stimulate proliferation in autologous lymphocyte preparations. Exp Eye Res. 71:11 (2000).

Z. Guo, D. Song, A.M. Azzarolo, J.E. Schechter, D.W. Warren, R.L. Wood, A.K. Mircheff and H.R. Kaslow. Autologous lacrimal-lymphoid mixed-cell reactions induce dacryoadenitis in rabbits. Exp Eye Res. 71:23 (2000).

A.H. Schönthal, D.W. Warren, D. Stevenson, J.E. Schechter, A.M. Azzarolo, A.M. Mircheff, and M.D. Trousdale. Proliferation of lacrimal gland cells in primary culture. Stimulation by extracellular matrix, EGF, and DHT. Exp Eye Res. 70:639 (2000).

T. Ritter, G. Schröder, K. Risch, A. Vergopoulos, M.K. Shean, J. Kolls, J. Brock, M. Lehmann and H-D. Volk. Ischemia/reperfusion injury-mediated down-regulation of adenovirus-mediated gene expression in a rat heart transplantation model is inhibited by co-application of a TNFRp55-Ig chimeric construct. Gene Ther. 7: 1238 (2000).

J. Kolls, K. Peppel, M. Silva and B. Beutler. Prolonged and effective blockade of tumor necrosis factor activity through adenovirus-mediated gene transfer. Proc Natl Acad Sci U S A 91: 215 (1994).

I.D. Goldfine, M.S. German, H-C. Tseng, J. Wang, J.L. Bolaffi, J-W. Chen, D.C. Olson and S.S. Rothman. The endocrine secretion of human insulin and growth hormone by exocrine glands of the gastrointestinal tract. Nat Biotechnol. 15: 1378 (1997).

X. He, C.M. Goldsmith, Y. Marmary, R.B. Wellner, A.F. Parlow, L.K. Nieman and B.J. Baum. Systemic action of human growth hormone following adenovirus-mediated gene transfer to rat submandibular glands. Gene Ther. 5: 537 (1998).

gene transfer. J Immunol. 161: 1516 (1998).

P.M. Mathisen, M. Yu, J.M. Johnson, J.A. Drazba and V.K. Tuohy. Treatment of experimental autoimmune encephalomyelitis with genetically modified memory T cells. J Exp Med. 186: 159 (1997).

M.LS. Tsang, J.A. Weatherbee, M. Dietz, T. Kitamura, and R.C. Lucas.TGF-beta specifically inhibits the IL-4 dependent proliferation of multifactor-dependent murine T-helper and human hematopoietic cell lines. Lymphokine Res. 9:607 (1990).

Y. Ma, S. Thornton, L.E. Duwel, G.P. Boivin, E.H. Giannini, J.M. Leiden, J.A. Bluestone and R. Hirsch. Inhibition of collagen-induced arthritis in mice by viral IL-10

IL-2 IMMUNOREACTIVE PROTEINS IN LACRIMAL ACINAR CELLS

Yan Zhang,[1] Jiansong Xie,[1] Limin Qian,[1] Joel E. Schechter,[2] and Austin K. Mircheff[1]

[1]Department of Physiology & Biophysics
[2]Department of Cell & Neurobiology
Keck School of Medicine
University of Southern California
Los Angeles, California, USA

1. INTRODUCTION

Sjögren's syndrome is an autoimmune disease primarily involving the lacrimal and salivary glands. Recent studies on the initiation of autoimmune dacryoadenitis suggest that when acinar cells from rabbit lacrimal glands are induced to express MHC Class II molecules, the ability to present processed autoantigen peptides to CD4 T cells is also induced (Guo et al., 2000). However, it is likely that the outcome of MHC Class II molecule-mediated autoantigen presentation depends on the accessory signals present in the interstitial milieu. One potential accessory factor is interleukin 2 (IL-2), which is classically produced by Th1 CD4 cells.

The mature forms of human and rabbit IL-2 are 15 kDa single peptide chains of 133 amino acids, released from 17 kDa precursors by cleavage of hydrophobic secretory signal sequences containing the N-terminal 20 amino acids. IL-2 has several important roles in immunity: It is a crucial growth factor for all T cell populations, including suppressor and cytotoxic (CD8) T cells, as well as helper (CD4) T cells. In reacting with CD4 T cells it tends to promote immunity, while in reaction with CD8 T cells, it tends to depress immunity. It is also an important trophic factor for B cells. In addition, it can be produced by activated B cells (Benjamin et al., 1989), and it presumably performs an autocrine role or a paracrine role, sustaining T cell support for B cell activation. IL-2 has other functions aside from its classic immune functions, and it appears to be produced by other cell types

Lacrimal Gland, Tear Film, and Dry Eye Syndromes 3
Edited by D. Sullivan *et al.*, Kluwer Academic/Plenum Publishers, 2002

795

in addition to lymphocytes. Both IL-2 and IL-2 receptors have been found in neurons or glial cells in the central nervous system (Hanisch and Quirion, 1996), and in lung epithelial cells (Aoki et al., 1997). IL-2 enhances migration of intestinal epithelial cells into wounded areas in vitro (Dignass and Podolsky, 1996). There has been a preliminary report that acinar cells from rat lacrimal glands express IL-2 or an IL-2-like molecule (Stepkowski et al., 1993). In view of the important and diverse roles IL-2 plays in immunoregulation, it was of interest to investigate whether IL-2 immunoreactive proteins are also expressed by acinar cells of rabbit lacrimal glands.

2. MATERIALS AND METHODS

Female New Zealand white rabbits were obtained from Irish Farms (Norco, CA). Rabbits weighing between 1.8 and 2.2 kg were used for experiments requiring primary culture of lacrimal gland acinar cells. Sexually mature rabbits, weighing 4 kg, were used for lacrimal gland fluid collection and immunocytochemistry. Lacrimal gland fluid was collected (Azzarolo et al., 1997), and samples of lacrimal gland, cornea, conjunctiva, and other tissues were obtained. Acinar cells were isolated (Guo et al., 2000), and cell lysates were analyzed by differential sedimentation, density gradient centrifugation, and phase partitioning (Yang et al., 1999). IL-2 immunoreactivity was determined in cell lysates and isolated subcellular fractions by Western blotting and in frozen sections of lacrimal gland by immunocytochemistry using a mouse monoclonal antibody to human IL-2. Frozen sections of cornea and conjunctiva were also immunostained to examine the presence of IL-2 receptor (IL-2R) using a mouse anti-rabbit monoclonal antibody. Mouse anti-human interleukin-2 monoclonal antibody was obtained from Chemicon International (Temecula, CA) and Oncogene Research Products (Cambridge, MA). Recombinant human IL-2, used as positive control, was obtained from Oncogene Research Products (Cambridge, MA). Mouse anti-rabbit IL-2 Receptor α subunit monoclonal antibody was obtained from Serotec (Oxford, England). Goat anti-mouse IgG heavy and light chain antibody, labeled with peroxidase, was obtained from American Qualex (San Clemente, CA). Biotin-conjugated goat anti-mouse secondary antibody was obtained from Chemicon (Temecula, CA). Supersignal™ substrate for Western blot detection was obtained from Pierce (Rockford, IL).

3. RESULTS

IL-2 immunoreactive signals associated with 17 kDa and 15 kDa bands, corresponding to the molecular weights of precursor and mature forms of monomeric IL-2, were observed in lysates of lacrimal glands, spleen, brain, and peripheral blood lymphocytes. In contrast, there was little or no IL-2-like immunoreactivity at these molecular weights in muscle, kidney, heart, or salivary gland. The immunoreactivities in interstitial cells isolated from lacrimal glands were greater than the immunoreactivities observed in isolated acinar cell preparations, but interstitial cell contamination could account for no more than 10% of the total IL-2 immunoreactivity in the acinar cell preparation.

IL-2 immunoreactivities were also detected in supernatant cell culture media and in lacrimal gland fluid, and the concentration in lacrimal gland fluid was not changed when pilocarpine was injected to stimulate lacrimal gland fluid secretion. The precursor form was predominant in all tissue and cell samples analyzed, as well as in supernatant culture medium. In contrast, a mixture of the mature form and smaller forms, presumably proteolytic processing products of the mature form, was predominant in lacrimal gland fluid.

The precursor and mature forms were present in debris, soluble, and membrane fractions from acinar cell lysates. As described elsewhere in this volume (Mircheff et al.), subcellular fractionation analysis resolves a number of different endomembrane compartments. The precursor and mature IL-2 forms were associated with a variety of compartments, including Golgi complex (Gols), secretory vesicle membranes (svm), pre-lysosomes (preLys), basal-lateral membrane recycling endosome (blmre), high-density domains of the trans-Golgi network (hd-tgns) believed to mediate traffic to the secretory vesicles and to the pre-lysosomes, and low density domains of the trans-Golgi network (ld-tgns) believed to mediate traffic to the blm. Immunocytochemical staining revealed the presence of IL-2, predominantly in the apical cytoplasm of lacrimal acinar cells, but also in basal regions of the acinar cell cytoplasm.

Given the presence of IL-2 immunoreactive proteins in secretory vesicles and lacrimal gland fluid, it was of interest to test for the presence of IL-2 receptors in ocular surface tissues. IL-2R α subunit immunoreactivity in lysates of cornea was detected by immunoblotting. IL-2R α subunit immunoreactivity outlining corneal and conjunctival epithelial cells was demonstrated by immunocytochemistry.

4. DISCUSSION

The results obtained in this study demonstrate that lacrimal gland acinar cells express and secrete IL-2 immunoreactive proteins with apparent molecular weights of 17 kDa and 15 kDa, corresponding to precursor and mature forms of IL-2, as well as a range of more extensively processed forms of IL-2. These proteins were detected by a mouse monoclonal antibody to human IL-2, and also by a rabbit antibody to human IL-2. IL-2 is highly conserved, and rabbit IL-2 has an 85% amino acid homology to human IL-2 (Perkins et al., 1999). Since corresponding proteins were also detected in other tissues known to express IL-2, i.e., spleen, brain, and peripheral blood lymphocytes, while little or no signal was detected in muscle, kidney, heart, or salivary glands, it is likely that the IL-2 immunoreactive proteins in the lacrimal gland represent authentic IL-2. Immunocytochemical analysis of intact lacrimal glands indicated that IL-2 immunoreactivity was associated primarily with acinar and ductal epithelial cells, and secondarily with interstitial cells, confirming that the IL-2 immunoreactive proteins in acinar cell lysates and subcellular fractions were derived primarily from the acinar cells, rather than from contaminating lymphocytes.

IL-2 immunoreactivities in lacrimal gland fluid and in the isolated secretory vesicle membrane compartment were detected by immuoblotting, and IL-2 immunoreactivity in the acinar cell apical cytoplasm was detected by immunocytochemistry. These observations indicate that a portion of the precursor, mature, and additional proteolytically processed forms of IL-2 are directed into the regulated apical secretory pathway. It is possible to predict several roles for the IL-2 immunoreactive proteins secreted into the lacrimal gland fluid. In normal mice, the conjunctival epithelium contains a considerable number of lymphocytes, the majority of which are Thy-1+ cells. Topical application of IL-2 to mouse eyes increases the number of Lyt-1+ cells (Tagawa et al., 1985). Thus, the lacrimal IL-2 immunoreactive proteins may play an immunoregulatory role in the ocular surface tissues. Since IL-2 has significant influences on other epithelia, such as airway (Lesur et al., 1997) and intestinal epithelial cells (Dignass and Podolsky, 1996), the IL-2 immunoreactive proteins in lacrimal gland fluid may also play some role in modulating conjunctival and corneal function, and corneal epithelial wound repair. The observation that IL-2 receptor α subunits are expressed in the conjunctiva and corneal epithelium would seem to be consistent with this hypothesis.

The distribution of IL-2 immunoreactive proteins through the acinar cell endomembrane system is also consistent with the suggestion that acinar cells secrete a portion of the IL-2-like proteins that they produce into the interstitium via basal-lateral membrane recycling endosomes and the trans-Golgi network. Several different roles can be envisioned for IL-2 immunoreactive proteins in the lacrimal interstitium. Activated B cells express IL-2 receptors (Lipsky et al., 1988), and IL-2 has been found to be important in B cell differentiation (Jelinek and Lipsky, 1987). Thus, it is possible that the lacrimal IL-2 immunoreactive proteins promote secretion of dimeric IgA, an important constituent of lacrimal gland fluid. In the human lacrimal gland, CD8 T cells generally outnumber CD4 T cells by a ratio 2:1 (Weiczorek et al., 1988). Thus, lacrimal IL-2 immunoreactive proteins may support the function of a resident population of regulatory CD8 T cells that maintain peripheral tolerance to lacrimal gland autoantigens.

It is also possible that IL-2 plays a role in the pathogenesis of autoimmune disease. The lacrimal IL-2 immunoreactive proteins may stimulate positive accessory signals that promote activation and proliferation of CD4 cells recognizing lacrimal autoantigens. From this perspective, the critical event in initiation of autoimmune dacryoadenitis would seem to be the induction of MHC Class II molecule expression by acinar cells, which has been demonstrated by previous studies (Guo et al., 2000). Moreover, administration of IL-2 induces MHC Class II molecule expression by biliary epithelial cells (Himeno et al., 1992). Thus, it may be appropriate to consider scenarios in which some physiological or pathophysiological perturbation induces increased expression of the lacrimal IL-2 immunoreactive proteins, which in turn act both as autocrine factors, inducing MHC Class

II molecule expression by the acinar cells, and as paracrine factors, promoting activation of the CD4 cells that recognize autoantigen peptides presented by the acinar cell's MHC Class II molecules.

ACKNOWLEDGEMENTS

Supported by NIH grants EY 05801 and EY 10550 and a grant from Allergan.

REFERENCES

Y. Aoki, D. Qiu, A. Uyei, and P.N. Kao, Human airway epithelial cells express interleukin-2 in vitro, *Am. J. Physiol.* 272: L276 (1997).

A.M. Azzarolo, A.K. Mircheff, R.L. Kaswan, F.Z. Stanczyk, E. Gentschein, L. Becker, B. Nassir, and D.W. Warren, Androgen support of lacrimal gland function, *Endocrine.* 6: 39 (1997).

A.D. Dignass and D.K. Podolsky, Interleukin 2 modulates intestinal epithelial cell function in vitro, *Exp. Cell Res.* 225: 422 (1996).

D. Benjamin, D.P. Hartmann, L.S. Bazar, R.J. Jacobson, and M.S. Gilmore. Human B cell line can be triggered to secrete an interleukin 2-like molecule, *Cellular Immunology.* 121: 30 (1989).

Z. Guo, A.M. Azzarolo, J.E. Schechter, D.W. Warren, R.L. Wood, A.K. Mircheff, and H.R. Kaslow. Lacrimal gland epithelial cells stimulate proliferation in autologous lymphocyte preparations, *Exp. Eye Res.* 71: 11 (2000).

U.K. Hanisch and R. Quirion. Interleukin-2 as a neuroregulatory cytokine, *Brain Research Rev.* 21: 246 (1996).

H. Himeno, T. Saibara, S. Onishi, Y. Yamamoto, and H. Enzan, Administration of interleukin-2 induces major histocompatibility complex Class II expression on the biliary epithelial cells, possibly through endogenous interferon-γ production, *Hepatology.* 16: 409 (1992).

D.F. Jelinek and P.E. Lipsky, Regulation of human B lymphocyte activation, proliferation, and differentiation, *Advances in Immunology.* 40: 1 (1987).

O. Lesur, K. Arsalane, J. Berard, J.-P.M. Mukuna, A.J. de Brum-Fernandes, D. Lane, and M. Rola-Pleszczynski, Functional IL-2 receptor are expressed by rat lung type II epithelial cells, *Am. J. Physiol.* 273: L495 (1997).

P.E. Lipsky, S. Hirohata, D.F. Jelinek, L. McAnally, and J.B. Splawski, Regulation of human B lymphocyte responsiveness, *Scand J Rheumatology.* suppl.76: 229 (1988).

H.D. Perkins, B.H. van Leeuwen, C.M. Hardy, and P.J. Kerr, The complete cDNA sequences of IL-2, Il-4, IL-6, and IL-10 from the European rabbit (Oryctolagus cuniculus), *Cytokine* 12: 555 (1999).

S.M. Stepkowski, T. Li, and R.M. Franklin, Interleukin-2 like molecule produced by acinar cells in lacrimal gland regulates local immune response, *Invest. Ophthalmol. Vis. Sci.* 34: S1486 (1993).

Y. Tagawa, T. Takeuchi, M. Saito, T. Saga, and H. Matsuda, [Intraepithelial lymphocytes (IEL) in murine ocular surface epithelium and effects of topical application of OK-432 and Interleukin-2], *Acta Soc. Ophthalmol Japonicae* 89: 832-837 (1985).

R. Wieczorek, F.A. Jakobiec, E.H. Sacks, and D.M. Knowles, The immunoarchitecture of the normal human lacrimal gland. Relevancy for understanding pathologic conditions, *Ophthalmology.* 95: 100 (1988).

T. Yang, H. Zeng, H. Zhang, C.T. Okamoto, D.W. Warren, R.L. Wood, M. Bachmann, and A.K. Mircheff, Major histocompatibility complex Class II molecules, cathepsins, and La/SSb proteins in lacrimal acinar cell endomembranes, *Am. J. Physiol.* 277: C994 (1999).

IMMUNOPATHOGENESIS OF CONJUNCTIVAL HISTOPATHOLOGIC ALTERATION IN NON-SJÖGREN'S KERATOCONJUNCTIVITIS SICCA

Hongqing Q. Ye, Chi-Chao Chan, and Janine A. Smith

Laboratory of Immunology
National Eye Institute
National Institutes of Health
Bethesda, Maryland, USA

1. INTRODUCTION

Dry eye or keratoconjunctivitis sicca (KCS) is a common ocular surface disease characterized by quantitative or qualitative abnormality of the tear film, ocular surface damage and ocular discomfort. It is a major health problem affecting millions of people worldwide: 11% 30 to 60 years old and 14.6% over 65 (4.3 million).[1] Pflugfelder et al. have shown elevation of mRNA levels of inflammatory cytokines such as IL-1α, IL-6, IL-8 and TNF-α in conjunctival cytology specimens of Sjögren's syndrome with KCS (SS-KCS).[2,3] These findings suggest that the inflammatory response plays a role in pathogenesis of SS-KCS. Recent evidence suggests the inflammatory response may also play a role in pathogenesis of non-SS-related KCS (NS-KCS) that affects most patients with dry eye.[4] The conjunctival inflammatory status of NS-KCS has not been well studied. To investigate the role of inflammation in the pathogenesis of NS-KCS, we examined the expression of inflammatory cytokines in conjunctival biopsies from patients with NS-KCS and asymptomatic controls.

Lacrimal Gland, Tear Film, and Dry Eye Syndromes 3
Edited by D. Sullivan *et al.*, Kluwer Academic/Plenum Publishers, 2002

801

2. MATERIALS AND METHODS

2.1. Materials

Twelve bulbar conjunctival biopsies were obtained from six NS-KCS and six asymptomatic control patients diagnosed and treated at the National Eye Institute. Biopsies were freshly embedded in OCT and stored at -80°C prior to section. Eight-micron sections were processed for immunohistochemistry.

Antibodies used in the experiments were antihuman cytokine antibodies: IL-1α (1:100) IL-4 (1:25), IL-6 (1:25), IL-10 (1:25), IFN-γ (1:100)and TNF-α (1:50)(Biosource International, CA); mAb anti-human keratin 10 (1:100, DAKO, Denmark) and anti-human Ki67 (1:500, Novocastra Laboratories Ltd., UK). Secondary antibodies were donkey anti-mouse IgG and donkey anti-rabbit IgG, both conjugated with FITC or TRITC (1:100–200, Jackson Immunology Inc., PA).

2.2. Methods

Double- and triple-labeling immunofluorescent staining was performed as described.[5] Briefly, sections were blocked with 10% non-immune goat serum for 30 min before incubating with primary antibodies for 1 h. They were then washed with PBS three times for 5 min each and reacted with secondary antibodies for 30 min. After PBS wash, the sections were mounted with VectShield-DAPI medium. Finally the sections were examined under an Olympus BX 60 microscope equipped with an epi-fluorescence power supply and digital camera to record the multicolor images. Standard Hematoxylin & Eosin (H & E) and Periodic Acid-Schiff (PAS) staining were also used to study the morphological changes and goblet cell density.

3. RESULTS

PAS and H & E staining demonstrated the histopathologic characteristic of KCS: the squamous metaplasia in our NS-KCS samples and lack of these alterations in the control samples. Expression of the inflammatory cytokines IL-1α and IL-4 in the conjunctival epithelium of patients with NS-KCS was evident using immunofluorescent techniques (Fig. 1). No expression of IL-1α or IL-4 occurred in control conjunctival biopsies. In contrast, both cytokines were expressed in conjunctival epithelial cells in 4 of 6 NS-KCS samples. High levels of IL-1α and IL-4 expression existed in two NS-KCS cases within basal and suprabasal epithelial cells adjacent to areas of cell proliferation (Ki67+) and keratinization (CK10+) (Fig. 2).

T-cells infiltrated the epithelium and substantia propria of all NS-KCS patients (Fig. 1); only a few T-cells were present in control samples. IL-6, IL-10 (Fig. 3), INF-γ and TNF-α were not detected in epithelial cells of these KCS or control samples using immunofluorescent techniques.

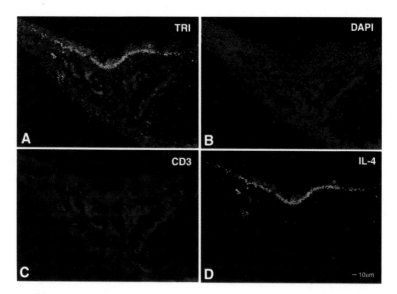

Figure 1. IL-4 expression and T-cell infiltration in NS-KCS. Triple-labeled immunostaining with IL-4 (Green), CD3 (Red) and DAPI-nucleus staining (Blue) was performed on frozen section from NS-KCS conjunctival biopsy. Top left superimposed image shows IL-4 is expressed by basal epithelial cells adjacent to the area of T-cell infiltration seen in the substantia propria.

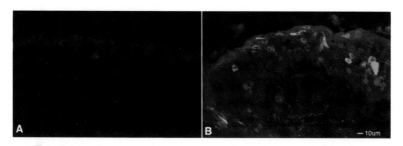

Figure 2. Keratinization and cell proliferation in NS-KCS. Double-labeled immunostaining with keratin 10 and Ki67 was performed on frozen sections from control and NS-KCS conjunctival biopsies. (A) Control sample. Only a few proliferating cells (Red, Ki67) exist, and no epithelial keratinization (Green, Keratin 10). (B) NS-KCS. Many proliferating basal epithelial cells exist with overlying expression of keratin 10 in the suprabasal and superficial epithelium.

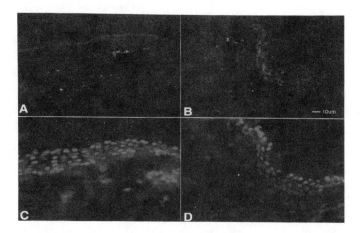

Figure 3. IL-10 expression in NS-KCS. Double-labeled immunostaining with IL-10 and DAPI was performed on frozen sections from control and NS-KCS conjunctival biopsies. **(A)** Minimal or no IL-10 staining in control sample. **(B)** Similar to **A**, few or no cells express IL-10 in NS-KCS at this stage. **(C)** Nuclear staining of control sample **A**. **(D)** Nuclear staining of NS-KCS sample **B**.

4. DISCUSSION

Cytokines are polypeptide factors that often work together in a coordinated fashion to achieve the range of biological effects collectively known as inflammation. Expression of pro-inflammatory cytokines is a hallmark of tissue inflammation. In this study, pro-inflammatory cytokines IL-1α and IL-4 were upregulated in conjunctival biopsies from patients with NS-KCS, indicative of ongoing ocular surface inflammation.

The localization of pro-inflammatory cytokines in NS-KCS conjunctiva was adjacent to areas of T-cell infiltration and conjunctival epithelial cell proliferation. Infiltrating T lymphocytes and resident ocular cells participate in the conjunctival inflammation seen in NS-KCS, suggesting inflammation plays a role in the histopathologic alterations seen in NS-KCS, similar to those found in SS-KCS.[2,3] Topical anti-inflammatory medications may be effective in treatment of this disease.

REFERENCES

1. O.D. Schein, B .Munoz, J.M. Tielsch, K .Bandeen-Roche, S.West. Prevalence of dry eye among the elderly. *Am J Ophthalmol.* 124:723 (1997).
2. D. Jones, D. Monroy et al. Sjøgren's syndrome: cytokine and Epstein-Barr viral gene expression within the conjunctival epithelium. *Invest Ophthalmol Vis Sci.* 35(9):3493 (1994).
3. S.C. Pflugfelder, D. Jones et al. Altered cytokine balance in the tear fluid and conjunctiva of patients with Sjøgren's syndrome keratoconjunctivitis sicca. *Curr Eye Res.* 19(3):201(1999).
4. M.E. Stern, R. Beuerman et al. The ocular surface in dry eye: a therapeutic target. *Socientas Ophthalmol Europaea* 11:911 (1997).
5. H. Ye and D.T. Azar. Expression of gelatinases A and B, and TIMPs 1 and 2 during corneal wound healing. *Invest Ophthalmol Vis Sci.* 39:913 (1998).

TOPICAL CYCLOSPORINE A (2%) EYEDROPS IN THE THERAPY OF ATOPIC KERATOCONJUNCTIVITIS AND KERATOCONJUNCTIVITIS VERNALIS

I. Tomida,[1] J. Bräuning,[1] T. Schlote,[1] P. E. Heide[2] and M. Zierhut[1]

[1]Department of Ophthalmology I
University Tuebingen
Tuebingen, Germany
[2]Pharmacy of the University Tuebingen

1. INTRODUCTION

The therapy for atopic and vernal keratoconjunctivitis (AKC and VKC) is often not satisfactory. Conventional therapeutic regimens consist of anti-allergic substances. Topically used immunosupressives such as Cyclosporine A (CsA) eyedrops are still part of clinical trials. CsA is an antibiotic produced by the fungus *Tolypocladium inflatum* Gams, and consists of a cyclic peptide composed of 11 aminoacids. Cyclosporin influences T-cells, restraining the synthesis of IL-2, IL-3 and γ-interferon and therefore reducing the humoral and cellular defense.[1-4]

VKC and AKC are some of the most serious ocular allergic diseases. Both forms show subtarsal conjunctival papillae, sometimes giant papillae. Trantas dots, representing whitish prominent regions at the limbus and consisting of eosinophils, are present during the active stage of the diseases. Dry eye syndrome may result, sometimes facilitating superinfections. Subjective symptoms such as itching, pain or epiphora can seriously affect the quality of life of some patients. A type-I allergic reaction mediated by mast cells and IgE initiates VKC and AKC. The basic cause for both may be a dysfunction of T-cells (type IV reaction) resulting in a predominance of Th2-cells and overproduction of IgE. In this way, the clinical presentation of these diseases can be very similar. This is the reason

Lacrimal Gland, Tear Film, and Dry Eye Syndromes 3
Edited by D. Sullivan *et al.*, Kluwer Academic/Plenum Publishers, 2002

for a more pragmatic differentiation of the two forms we apply; in case of a classic atopy (e.g., neurodermitis, bronchial asthma), we diagnose the disease as AKC, otherwise VKC.

Up to now, one of the therapeutic obstacles was the necessity for topical or oral corticosteroids in severe cases, in spite of well-known possible negative side effects. CsA is effective in dry eye diseases,[5] and the purpose of this study was to investigate the effect of topical CsA as an additive treatment for AKC and VKC, which are nearly always associated with dry eye.

2. METHODS

The data of 26 patients with severe AKC (age: 13.0 years ± 15.4; M:F = 19:7) and that of 12 with VKC (age: 10.5 years ± 8.25; M:F = 9:3) were analyzed after a follow-up period between 3 months to more than 2 years. The patients were treated with topical CsA 2% eyedrops, 2–3 times daily in addition to their basic anti-allergic therapy and lubricants. The eyedrops consisted of CsA 2% as a micro-emulsion dissolved in neutral oil. Patients had been interviewed regarding their tolerance of these eyedrops. Previous therapy already included intensive topical therapy containing antihistamines and mast cell stabilizing agents, in some cases (n = 6) supplementary topical corticosteroids.

The symptoms were assigned scores, depending on type (itching, tear secretion, foreign body sensation, spasm of eyelid, pain, pannus vasculosus, subtarsal papillae and Trantas dots), their intensity or frequency of recurrence (first exam/persisting: 3; reduced: 2; resolved: 1). Total scores up to 24 points were compared between the two diagnostic groups before and at the end of the follow-up period.

The objective clinical signs of the diseases before and under topical CsA 2% therapy were separately evaluated. These signs were subtarsal papillae, chronic conjunctivitis, pannus vasculosus, superficial keratitis, Trantas dots and corneal ulcer. The visual acuity at the beginning of this treatment was also compared to that at the end of the follow-up period.

3. RESULTS

The general assessment of the therapeutic effects of the CsA 2% eyedrops revealed an evident reduction of clinical signs and subjective symptoms for both AKC and VKC patients. The general median score for the symptoms listed above was reduced in the AKC group from 16.05 (± 4.04) at the first examination to 9.08 (± 2.94) at the last examination. In the VKC group the reduction was 13.58 (± 3.42) to 8.10 (± 2.47) (Fig. 1).

Before starting therapy with CsA 2% eyedrops, almost all patients revealed marked subjective symptoms related to dry eye syndrome associated with these two diseases. During therapy, all patients registered a reduction in all symptoms (Figs. 2 and 3).

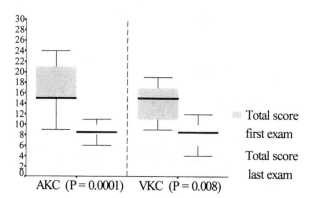

Figure 1. Total general mean scores for subjective and objective symptoms before and at the end of the follow-up period under topical CsA 2% therapy (P > 0.05; Wilcoxon-Test).

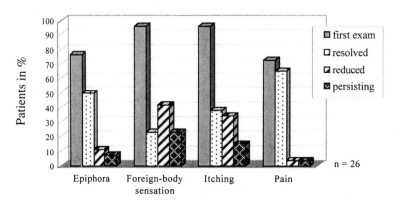

Figure 2. Subjective symptoms before and during therapy with topical CsA 2% in AKC patients.

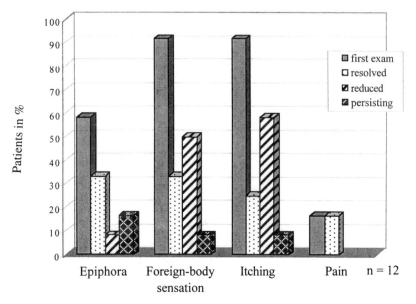

Figure 3. Subjective symptoms before and during therapy with topical CsA 2% in VKC patients.

Of the AKC patients, 96% suffered from marked foreign-body sensation. With CsA 2% eyedrops, 42.3% reported a reduced sensation, 23% symptom unchanged and 23% no such symptom. Of VKC patients, 91.6% had a marked foreign-body sensation. With therapy, 58.3% reported a reduced sensation, 8.3% symptom unchanged and 25% no such symptom.

Before application of CsA 2% eyedrops, 96.1% of AKC patients complained of itching. At the end of the follow-up period, 34.6% reported a marked reduction, 15.4% persisting itching and 38.4% no itching at all. In the VKC group, 91.6% complained of itching. With topical CsA 2% treatment, 58.3% reported diminished itching, 8.3% persistent itching and 25% no itching at the end of the follow-up period.

Before therapy 73% of AKC patients reported ocular pain; 65.4% felt no pain at all at the end of the follow-up period. Only 16.6% of VKC patients complained about ocular pain. None of these patients had this symptom at the end of the follow-up period.

Prior to topical therapy, 76% of AKC patients and 58.3% of VKC patients complained of epiphora. At the end of the follow-up period, 50% of these AKC patients had no such complaints and 33% of previously affected VKC patients reported no epiphora.

Topical application of CsA 2% also positively influenced visual acuity in both diagnostic groups. In the AKC group it increased from a medium of 0.65 ($\pm$ 0.51) at the first examination to 0.93 ($\pm$ 0.11) at the last; in the VKC group it improved from 0.72 ($\pm$ 0.56) to 0.98 ($\pm$ 0.58) (Fig. 4).

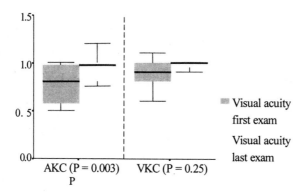

Figure 4. Effect of topical CsA 2% on visual acuity (P > 0.05; Wilcoxon-Test) in VKC and AKC groups.

Table 1. Objective clinical symptoms in AKC patients before and after topical CsA 2% therapy

	AKC (first exam) n = 26	AKC (last exam) n = 26
Keratitis superficialis	88.4%	30.4% resolved 43.4% reduced
Chronic conjunctivitis	92.3%	24.8% resolved 45.8% reduced
Pannus vasculosus	76.9%	39.9% reduced 55% persistent
Subtarsal papillae	96.1%	3.9% resolved 72% reduced
Blepharospasm	34.6%	88.7% resolved 10.9% reduced
Trantas dots	34.6%	44.2% resolved 44.2% reduced

CsA 2% eyedrops reduced subjective and clinical symptoms partly related to dry eye in AKC and VKC patients (Tables 1 and 2).

Table 2. Objective clinical symptoms in VKC patients before and after topical CsA 2% therapy

	VKC (first exam) n = 12	VKC (last exam) n = 12
Keratitis superficialis	50%	50% resolved 33.2% reduced
Chronic conjunctivitis	91.6%	27.3% resolved 54.6% reduced
Pannus vasculosus	91.6%	63.6% reduced 27.3% persistent
Subtarsal papillae	100%	8.3% resolved 66.6% reduced
Blepharospasm	25%	66.4% resolved 33.2% reduced
Trantas dots	25%	66.4% resolved 33.2% reduced

4. DISCUSSION

Our study indicates additional topical application of CsA can be effective in the therapy of severe AKC and VKC associated with dry eye symptoms, resulting in marked improvement for most affected patients. Probably because of the small groups, a statistical significance concerning the positive effect of CsA 2% eyedrops on the various subjective and objective symptoms in AKC and VKC could not be found here.

Additional therapy using corticosteroids was not necessary in any of our patients. On the other hand, the severity of the symptoms did not allow interruption of their previous topical anti-allergic therapy, so CsA 2% eyedrops were given as a completion of actual therapy. Few studies conducted with placebo groups suggest a monotherapy using CsA eyedrops in AKC or VCK is possible.[3,4]

In an experimental model of ocular allergy, Whitcup et al.[4] demonstrated a comparable effect between prednisolone acetate and CsA on the eosinophils, neutrophils and lymphocytes. In 1990, Secchi et al.[6] showed in two different studies (one open study for 3 months and one double-blind study for only 2 weeks) a significant effect of CsA 2% eyedrops in children with VKC. But a reliable evaluation of the effect of CsA eyedrops in this disorder is only possible after a longer follow-up period. In a prospective, double-blind placebo-controlled study of 12 steroid-dependent patients with AKC, Hingorani et al.[7] investigated the effect of CsA 2% eyedrops dissolved in maize oil. In their patients they demonstrated a significant steroid-sparing effect. The conjunctival biopsy of 8 patients revealed a reduction of HLA-DR expression, normalization of the CD4/CD8 ratio and reduction of B-cells. The expression of IL-2 and IFN-α was also reduced.

While previous showed a positive effect of topical CsA, they were of short duration (a maximum of 3 months). VKC especially, and in some cases AKC, showed seasonal variations. Another point of criticism is the galenic formula of the CsA eyedrops. Concerning subjective tolerance, in the study of Hingorani et al.[7] all 12 patients complained of a burning sensation; 9 even indicated a marked burning sensation. These side effects, especially in children, will lead to reduced compliance, and in many cases patients will cease therapy.

Until now, CsA eyedrops have been prepared in different ways. All galenic formulas led to more or less severe burning and redness. When prepared from the commercially available liquid Sandimmun Optoral®, the CsA eyedrops are extremely difficult to tolerate of the high alcohol content (1 – 2%) that causes severe burning and corneal toxicity. Since 1997, we have used a micro-emulsion of CsA in neutral oil for the topical treatment of AKC and VKC. Our patients, including the children, tolerate this solution quite well. In spite of only occasional "light burning" sensations in the beginning, none of the patients ended therapy.

4. CONCLUSION

CsA 2% eyedrops can successfully be used in the treatment of AKC and VKC, reducing subjective symptoms such as itching and foreign-body sensation that otherwise would stimulate continuous rubbing of the eyes. Besides reducing dry eye symptoms in ocular allergy, CsA 2% eyedrops may lessen the T-cell-mediated reaction in allergic diseases. While CsA might complement topical corticosteroids, in our study the patients had no need for them. It may be important to use CsA 2% eyedrops over a much longer period than might seem necessary due to seasonal variation of allergic diseases.

REFERENCES

1. M.W. Belin, C.S. Bouchard, and T.M. Phillips. Update on topical cyclosporine A. Cornea 9:184–195 (1990).
2. H.S. Dua, V.K. Jindal, J.A.P. Gomes, W.A. Amoaku, L.A. Donoso, P.R. Laibson, and K. Mahlberg. The effect of topical cyclosporine on conjunctiva-associated lymphoid tissue (CALT). Eye 10:433–438 (1996).
3. M. Hingorani, V.L. Calder, R.J. Buckley, and S. Lightman. The immunomodulatory effect of topical cyclosporine A in atopic keratoconjunctivitis. Invest Ophthalmol Vis Sci 40:392–399 (1999).
4. S.M. Whitcup, C.-C. Chan, D.A. Luyo, P. Bo, and Q. Li. Topical cyclosporine inhibits mast cell-mediated conjunctivitis. Invest Ophthalmol Vis Sci 37:2686–2693 (1996).
5. K. Sall, O.D. Stevenson, T.K. Mundorf, B.L. Reis. Two multicenter, randomized studies of the efficacy and safety of cyclosporine ophthalmic emulsion in moderate to severe dry eye disease. Ophthalmology 107:631–639 (2000).

6.   A.G. Secchi, M.S. Tognon, and A. Leonardi. Topical use of cyclosporine in the treatment of vernal keratoconjunctivitis. Am J Ophthalmol 110:641–645 (1990).

7.   M. Hingorani, L. Moodaley, V.L. Calder, R.J. Buckley, S. Lightman. A randomized, placebo-controlled trial of topical cyclosporine A in steroid-dependent atopic keratoconjunctivitis. Ophthalmology 105:1715–1720 (1998).

Immunity and Inflammation II: Role of Apoptosis, Antibodies and Mucosal Associated Lymphoid Tissue

APOPTOSIS: THE EYES HAVE IT

Bennett D. Elzey[1] and Thomas A. Ferguson[1,2]

[1]Department of Ophthalmology and Visual Sciences and
Department of Pathology
[2]Washington University School of Medicine
St. Louis, Missouri, USA

1. INTRODUCTION

Inflammatory immune responses are necessary for host protection against environmental pathogens. Critical to a successful immune response is the clonal expansion of specific lymphocytes. However, after infection, homeostasis must be returned to the immune system to prevent the development of tissue damage that could lead to autoimmunity. CD95-CD95L-mediated apoptosis is a central component of this process.[1] Indeed, mice deficient in functional CD95 or CD95L (*lpr* and *gld*, respectively) suffer from massive accumulation of lymphocytes and loss of self-tolerance.[2] In addition to reducing lymphocyte numbers, induction of apoptosis is associated with regulation or prevention of inflammatory responses to normal cellular antigens.[3]

Apoptosis should not stimulate inflammation. As normal cell turnover occurs via programmed cell death (apoptosis), it is undesirable to invoke an inflammatory response against normal cellular antigens. Although inflammation is usually a positive response to foreign antigens, the body contains several sites in which it can have deleterious consequences. Inflammation usually results in significant damage to surrounding tissue; therefore, in the brain, eyes, ovaries or testes it would be a serious threat to the life, sight and reproduction of the host. These sites have specifically designed attributes that inhibit inflammatory responses, or the immune privilege[4] phenomenon, first described in the late nineteenth century.

CD95L plays a pivotal role in ocular immune privilege.[5] When the eyes of CD95L-defective mice were virally infected, uncontrolled and damaging inflammation resulted. For years, immune privilege was largely attributed to physical barriers and antigen

Lacrimal Gland, Tear Film, and Dry Eye Syndromes 3
Edited by D. Sullivan *et al.*, Kluwer Academic/Plenum Publishers, 2002

815

sequestration from the immune system, but it is actually the culmination of several different factors as well as a physical barrier. These include, in addition to CD95L expression, locally produced immunosuppressive cytokines[6,7] and neuropeptides,[8] limited expression of major histocompatibility complex (MHC) molecules and strategically located antigen-presenting cells (APCs).[9] Antigen first introduced into an immune privileged site also results in antigen-specific immune deviation.[10–17] Using the eye as a model, CD95L-induced apoptosis of infiltrating lymphocytes is required for this effect.[18,19] So CD95L helps inhibit inflammation in an immune-privileged site and contributes to the regulation of subsequent responses.

2. CD95L AND OCULAR IMMUNE PRIVILEGE

CD95L's role in ocular immune privilege was recently established.[20] Mice naturally deficient in CD95L expression (*gld*) suffered from damaging inflammation in response to HSV injection into the eye, whereas wild-type (WT) mice did not. CD95L was constitutively expressed in strategic areas of the eye where it functions to destroy lymphocytes as they enter the organ. Bone marrow chimeras demonstrated it was ocular- and not hematopoietic-derived CD95L responsible for the death of the infiltrating cells. This function of CD95L predicts corneal allografts should survive quite well, and studies using the murine model confirmed this prediction.[21] More important, in humans, CD95L expression patterns are identical to that observed in mice and highly functional in their ability to kill CD95+ cells. Clinical data from patients receiving corneal transplants also implicate CD95L's role in ocular immune privilege. Without any immunosuppression, 1- and 5-year graft survival rates are 80% and 50–60%, respectively.[22]

3. IMMUNE TOLERANCE

It is important for the eye to control the first wave of infiltrating lymphocytes, but what about a sustained maturing immune response that could deliver enough activated lymphocytes to overrun the eye? Apoptosis in the eye controls the primary immune response and subsequent ones. Anterior chamber-associated immune deviation (ACAID) describes the phenomenon in which antigen first introduced into the anterior chamber of the eye induces a systemic antigen-specific immune deviation from a Th1 response to a Th2,[18] and is associated with activation of non-destructive antibody responses and attenuation of antigen-specific delayed-type hypersensitivity (DTH) responses. This loss of DTH does not result from a lack of response, but the presence of regulatory CD4+ and CD8+ T-cells. Although the mechanism for ACAID induction is not firmly established, the spleen plays a central role. ACAID cannot be induced in splenectomized mice.[23]

Injection of *lpr* splenocytes (CD95-defective) into the anterior chamber (AC) of WT mice does not result in ACAID, or WT splenocytes into *gld* (CD95L-defective) AC.[24] This indicates CD95L-mediated apoptosis is necessary for ACAID induction. However, if *lpr* cells are first irradiated and then injected into WT, ACAID is induced. Likewise, if WT cells are first irradiated, injection into *gld* AC also results in ACAID induction. Thus, the presence of apoptotic cells in the eye results in ACAID. These studies demonstrate that an important function of the CD95-CD95L system is to induce apoptosis to control initial lymphocytic AC infiltration and induce systemic tolerance to prevent subsequent inflammatory reactions. The importance of apoptosis in the eye was underscored by data showing removal of eyes containing apoptotic cells within the first 3 days following viral injection resulted in normal anti-viral immunity.[19] Therefore, sufficient viral particles leave the eye to induce an immune response, but the regulatory response as a result of apoptosis in the immune-privileged site imposes tolerance on any subsequent response.

How dead cells influence the immune response is not completely clear, but numerous reports suggest they can be tolerogenic. How this might be accomplished came from discovery that fas-induced apoptotic cells make IL-10 prior to death.[18] IL-10 is an immunosuppressive cytokine that can downregulate costimulatory molecules and prevent APC maturation.[25–28] Since apoptotic cells are rapidly targeted for phagocytosis by APC, one hypothesis is that apoptotic cells make IL-10 to induce ACAID, and the effects of IL-10 are mediated through APCs. IL-10$^{-/-}$ mice demonstrate this is indeed the case.[18] However, AC injections of haptenated IL-10$^{-/-}$ T-cells could not prevent DTH responses in WT mice, demonstrating it was not host-derived, but apoptotic cell-derived, IL-10 responsible for tolerance induction. Furthermore, peritoneal exudate cells (PEC) fed apoptotic IL-10^{+} cells were able to induce Th2 cell differentiation in vitro, whereas PEC fed IL-10$^{-/-}$ or non-apoptotic WT cells induced Th1 cell differentiation.

4. REGULATION OF APOPTOSIS IN THE EYE

Activation is required to render T-cells susceptible to CD95L-mediated death. However, both activated and inactivated T-cells are efficiently killed in the eye. This led to the investigation of other factors that could render T-cells sensitive to ocular CD95L. We determined that following AC injection of haptenated T-cells, ocular TNFα levels rapidly rise for the first 3 h before returning to baseline levels by 9 h.[29] Neutralization of TNFα by co-injection of anti-TNFα alone prevents apoptosis. However, it is not TNFα that directly kills the cells; it primes them for CD95L-mediated death exclusively through the TNFα p75 receptor (TNFR2). In vitro studies confirm TNFα and CD95L can cooperate to induce apoptosis. Here, WT lymphocytes incubated with either TNFα or anti-CD95 antibody do not undergo significant death; however, when the cells are preincubated with TNFα, a significant increase in death is observed following CD95 cross linking. Agonistic antibody to TNFR2, but not TNFR1, mimics the effect of TNFα, confirming the role for the p75 receptor. Studies with the ACAID model confirm cells from TNFR2 knockout mice cannot induce tolerance, whereas those from TNFR1 KO perform like WT. These studies suggest complicity between the inflammatory response and apoptosis in the control of potentially damaging immune and inflammatory responses in the eye.

5. CONCLUSION

The available evidence supports that CD95L is required for suppression of inflammatory response in the eye. It is also clear, however, that ocular immune privilege is the result of several cooperative factors including locally produced immunosuppressive cytokines and neuropeptides, limited expression of MHC molecules and tolerogenic APCs. Since *gld* mice do not express functional CD95L, one might expect them to suffer from increased lymphocytic infiltration or spontaneous pathological immune-mediated conditions. However, this is rarely the case, indicating CD95L prevents inflammation perhaps when the other barriers have been overcome.

Although the eye is carefully guarded, it can still suffer inflammatory reactions. Cells activated to retinal antigens can infiltrate the eye and cause blinding damage,[30] and HSV-1 can induce corneal keratitis in the presence of constitutive CD95L levels.[31] How the inflammatory cells have overcome the immune-privileged status of the eye is uncertain, although activated cells are resistant to CD95L-mediated death via high expression of the anti-apoptotic protein FLIP.[32] Additionally, activated CD8 T-cells are resistant to CD95L-mediated death, dying rather through a TNFα-mediated pathway.[33] Presumably, ocular CD95L induces apoptosis in lymphocytes infiltrating the site nonspecifically. It is unknown how the eye handles highly activated cells that enter specifically and encounter their cognate antigen. The eye may go to great lengths to avoid the generation of highly activated cells because it does not efficiently control them. On the other hand, activated T-cells can induce upregulation of CD95L on certain tissues that results in their deletion. The same may or may not be true in the eye.

Once viewed as a passive process, we now know ocular immune privilege is a very active regulatory process. To expand our current knowledge of immune privilege for clinical purposes regarding inflammatory eye diseases and transplantation, the complex regulation and interplay of the anti-inflammatory factors in the eye must be more clearly elucidated. Understanding what maintains the status quo in the absence of inflammation and how the eye deals with activated cells is crucial.

REFERENCES

1. S. Nagata and P. Golstein. Fas, the death factor. *Science*. 267:1449 (1995).
2. S. Nagata and T. Suda. Fas and Fas ligand. *lpr* and *gld* mutations. *Immunol Today*. 16:39 (1995).
3. J. Savill and V. Fadok. Corpse clearance defines the meaning of cell death. *Nature*. 407:784 (2000).
4. J.C.Van Dooremaal. Die Entwickelung der in fremden Grund versetzten lebenden Gewebe. *Albrecht von Graefe's Arch Ophthalmol*. 19:359 (1873).
5. T.S. Griffith, T. Brunner, S.M. Fletcher, D.R .Green, and T.A. Ferguson. Fas ligand-induced apoptosis as a mechanism of immune privilege. *Science*. 270:1189 (1995).
6. P. Hooper, N.S. Bora, H.J. Kaplan, and T.A. Ferguson. Inhibition of lymphocyte proliferation by resident ocular cells. *Curr Eye Res*. 10:363 (1991).

7.   G.A. Wilbanks and J.W. Streilein. Fluids from immune privileged sites endow macrophages with the capacity to induce antigen-specific immune deviation via a mechanism involving transforming growth factor-beta. *Eur J Immunol*. 22:1031 (1992).

8.   T.A. Ferguson, S. Fletcher, J.M. Herndon, and T.S. Griffith. Neuropeptides modulate immune deviation induced via the anterior chamber of the eye. *J Immunol*. 155:1746 (1995).

9.   T.L. Knisely, T.M. Anderson, M.E. Sherwood, T.J. Flotte, D.M. Albert, and R.D. Granstein. Morphologic and ultrastructural examination of I-A+ cells in the murine iris. *Invest Ophthalmol Vis Sci*. 32:2423 (1991).

10.   T.A. Ferguson, J.M. Herndon, and P. Dube. The immune response and the eye. IV. A role for tumor necrosis factor in anterior chamber associated immune deviation. *Invest Ophthal Vis Sci*. 35:2643 (1994).

11.   T. Ferguson, J.C. Waldrep, and H.J. Kaplan. The immune response and the eye. II. The nature of T suppressor cell induction in anterior chamber associated immune deviation (ACAID). *J Immunol*. 139:352 (1987).

12.   H.J.Kaplan and J.W. Streilein. Immune response to immunization via the anterior chamber of the eye: I. F1-lymphocyte induced immune deviation. *J Immunol*. 118:809 (1977).

13.   Y. Sonoda and J.W. Streilein. Impaired cell mediated immunity (ACAID) in mice bearing healthy orthotopic corneal allografts. *J Immunol*. 150:1727 (1993).

14.   C-K. Joo, J.S. Pepose, K.A. Laycock, and P.M. Stuart. Allogeneic graft rejection in a murine model of orthotopic corneal transplantation. In: Nussenblat t RB, SM Whitcup, RR Caspi, I Gery, editors. *Advances in ocular immunology*. Elsevier Science B. V.;115 (1994).

15.   J. Whittum-Hudson, M. Farazdaghi, and R.A. Prendergast. A role for T lymphocytes in preventing experimental herpes simplex type 1 retinitis. *Invest Ophthal Vis Sci*. 26:1524 (1985).

16.   R.P. Wetzig, S. Foster, and M.I. Greene. Ocular immune responses. I. Priming of A/J mice in the anterior with azobenzenearsonate-derivatized cells induces second-order-like suppressor T-cells. *J Immunol*. 128:1753 (1982).

17.   J.Y. Niederkorn and J.W. Streilein. Alloantigens placed into the anterior chamber of the eye induce specific suppression of delayed type hypersensitivity but normal cytotoxic T lymphocyte responses. *J Immunol*. 131:2587 (1983).

18.   Y. Gao, J.M. Herndon, H. Zhang, T.S. Griffith, and T.A. Ferguson. Antiinflammatory effects of CD95 ligand (FasL)-induced apoptosis. *J Exp Med*. 188:887 (1998).

19.   T.S. Griffith, X. Yu, J.M. Herndon, D.R. Green, and T.A. Ferguson. CD95-induced apoptosis of lymphocytes in an immune privileged site induces immunological tolerance. *Immunity*. 5:7 (1996).

20.   T.S. Griffith, T. Brunner, S.M. Fletcher, D.R. Green, and T.A. Ferguson. Fas ligand-induced apoptosis as a mechanism of immune privilege. *Science*. 270:1189 (1995).

21.   P.M. Stuart, T.S. Griffith, N. Usui, J.S. Pepose, X. Yu, and T.A. Ferguson. CD95 ligand (FasL)-induced apoptosis is necessary for corneal allograft survival. *J Clin Invest*. 99:396 (1997).

22.   S.E. Brady, J.C. Rapuano, J.J. Arentsen, E.J. Cohen, and P.R. Laibson. Clinical indications and procedures associated with penetrating keratoplasty. *Am J Ophthalmol*. 108:118 (1989).

23.   J.W. Streilein and J.Y. Niederkorn. Induction of anterior chamber associated immune deviation requires an intact, functional spleen. *J Exp Med*. 153:1058 (1981).

24.   T.A. Ferguson and T.S. Griffith. A vision of cell death: insights into immune privilege. *Immunological Reviews*. 156:167 (1997).

25.   C. Buelens, V. Verhasselt, D. De Groote, K. Thielemans, M. Goldman, and F. Willems. Human dendritic cell responses to lipopolysaccharide and CD40 ligation are differentially regulated by interleukin-10. *Eur J Immunol*. 27:1848 (1997).

26.   K. Steinbrink, M. Wolfl, H. Jonuleit, J. Knop, and A.H. Enk. Induction of tolerance by IL-10-treated dendritic cells. *J Immunol*. 159:4772 (1997).

27.   D.F. Fiorentino, A. Zlotnik, P. Vieira, T.R. Mosmann, M. Howard, K.W. Moore, and A. O'Garra. IL-10 acts on the antigen-presenting cell to inhibit cytokine production by Th1 cells. *J Immunol.* 146:3444 (1991).

28.   F. Willems, A. Marchant, J.P. Delville, C. Gerard, A. Delvaux, T. Velu, M. de Boer, and M. Goldman. Interleukin-10 inhibits B7 and intercellular adhesion molecule-1 expression on human monocytes. *Eur J Immunol.* 24:1007 (1994).

29.   B.D. Elzey, T.S. Griffith, J.M. Herndon, R. Barreiro, J. Tschopp, and T.A. Ferguson. Regulation of Fas ligand-induced apoptosis by TNF. Submitted. (2001).

30.   H. Sanui, T.M. Redmond, S. Kotake, B. Wiggert, L.H. Hu, H. Margalit, J.A. Berzofsky, G.J. Chader, and I. Gery. Identification of an immunodominant and highly immunopathogenic determinant in the retinal interphotoreceptor retinoid-binding protein (IRBP). *J Exp Med.* 169:1947 (1989).

31.   R.L. Hendricks. Immunopathogenesis of viral ocular infections. *Chem Immunol.* 73:120 (1999)

S. Kirchhoff, W.W. Müller, A. Krueger, I. Schmitz, and P.H. Krammer.. TCR-mediated up-regulation of c-FLIPshort correlates with resistance toward CD95-mediated apoptosis by blocking death-inducing signaling complex activity. *J Immunol.* 165:6293 (2000)

L. Zheng, G. Fisher, R.E. Miller, J. Peschon, D.H. Lynch, and M.J. Lenardo. Induction of apoptosis in mature T cells by tumour necrosis factor. *Nature.* 377:348 (1995)

APOPTOSIS IN THE CORNEA IN RESPONSE TO EPITHELIAL INJURY: SIGNIFICANCE TO WOUND HEALING AND DRY EYE

Steven E. Wilson,[1] Rahul R. Mohan,[1] JongWook Hong,[1,2] JongSoo Lee,[1,3] Rosan Choi,[1] Janice J. Liu,[1] and Rajiv R. Mohan[1]

[1]The Department of Ophthalmology
University of Washington School of Medicine
Seattle, Washington, USA
[2]The Department of Ophthalmology
Korea University
Seoul, Korea
[3]Department of Ophthalmology
Research Institute of Medical Science
College of Medicine
Pusan National University
Pusan, Korea

1. INTRODUCTION

Chronic ocular surface injury is a hallmark of dry eye disease. The extent of objective surface abnormality is highly variable between patients, and roughly correlates with the severity of the disease. These surface changes extend from mild conjunctival rose bengal staining to diffuse conjunctival and corneal rose bengal and fluorescein staining. Previous studies have demonstrated that corneal surface injury triggers keratocyte apoptosis and a subsequent wound healing response in the stroma. However, the majority of patients with mild to moderate surface abnormalities from dry eye have no evidence of stromal haze, unstable refractive error, or other changes that would be expected if significant ongoing stromal wound healing were triggered by the surface changes. Superficial injury to the corneal or conjunctival epithelium probably does not induce stromal wound healing because of the intact barrier function of the epithelium and lack of penetration of pro-apoptotic cytokines into the corneal stroma and subconjunctival tissues. New data suggest

Lacrimal Gland, Tear Film, and Dry Eye Syndromes 3
Edited by D. Sullivan *et al.*, Kluwer Academic/Plenum Publishers, 2002

821

cytokine-mediated epithelial to stroma to immune cell communications that attract inflammatory cells to the site of epithelial injury. These interactions could provide an important mechanism for inflammatory cell infiltration into the cornea and conjunctiva in eye disease. Similarly, these mechanisms could have relevance to sterile contact lens-related infiltrates, subepithelial infiltrates associated with adenoviral infection, and other conditions in which epithelial injury is accompanied by inflammation.

2. DRY EYE AND OCULAR SURFACE INJURY

Dry eye disease is associated with ocular surface injury. Typically, this injury is characterized by rose bengal staining of the conjunctival or corneal epithelium. Despite this ocular surface injury, there is no evidence that eyes with mild to moderate dry eye disease have an associated stromal wound healing response. This is true despite the association between ocular surface injury, apoptosis, and the stromal wound healing cascade.[1,2] Thus, stromal scarring and unstable refraction have not been associated with mild to moderate dry eye disease. There is, however, an iatrogenic condition that simulates the ocular surface changes associated with mild to moderate dry eye that may provide some insight.

About 1 to 2% of patients who undergo laser in situ keratomeliusis (LASIK) develop a dry eye-like condition associated with punctate epithelial erosions (PEE) and rose bengal staining of the flap (Fig. 1A). This condition develops within a few days to weeks of LASIK and typically persists for 6 to 8 months following surgery. In a recent study,[3] eyes with this condition had similar levels of tear production (measured with the Schirmer's test without anesthesia) to eyes that had LASIK without corneal epithelial rose bengal staining and PEE. Tear production was measured at the same time points following surgery in both groups. Importantly, none of the eyes in either group had symptoms or signs of dry eye prior to LASIK surgery. Since the corneal surface changes typically resolve in these eyes after 6 to 8 months (the same time course required for the corneal nerves to complete reinnervation of the flap)[4,5] we believe that this condition is actually LASIK-induced neurotrophic epitheliopathy (LNE).[3] Comparison between eyes with LNE and eyes that do not develop the condition revealed that there was no significant difference in refractive correction at 6 months after surgery, suggesting that this ocular surface abnormality, characterized by punctate epithelial erosions and rose bengal staining, is not associated with increased wound healing. None of these eyes develop subepithelial haze or stromal scaring.

About 5% of rabbits that undergo LASIK also develop LNE-like epithelial changes on the flap at 1 to 3 months after surgery (Mohan and Wilson, unpublished data, 2000). These eyes do not have keratocyte apoptosis detected with the TUNEL assay at 1 month or 3 months after LASIK when the surface abnormalities are present (Fig. 1B, 1C). Thus, ocular surface abnormalities are not necessarily associated with keratocyte apoptosis or stromal wound healing.

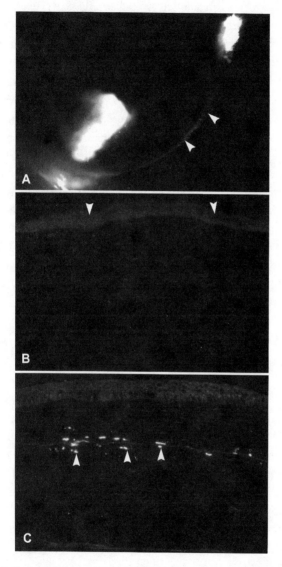

Figure 1. (A) Punctate epithelial erosions (PEE, arrow) characteristic of LASIK-induced neurotrophic epitheliopathy (LNE) at 3 months after surgery. Surface abnormalities resolved in this eye at 6 mo after surgery. (B) TUNEL staining to detect keratocyte apoptosis in a rabbit cornea that had PEE at the time of enucleation at 3 mo after LASIK. Arrowhead denotes the characteristic "blurred" epithelial surface when corneas with PEE are stained with the TUNEL assay with propidium iodide background staining. There was no detectable keratocyte apoptosis, although keratocytes are present (propidium iodide background stain-compare with **C**. Magnification 200X. (C) An example of a rabbit cornea at 24 h after LASIK. The epithelial surface is sharp in this cornea without PEE. TUNEL-positive keratocytes are noted at the interface (arrowheads). Magnification 200X.

3. OCULAR SURFACE INJURY AND KERATOCYTE APOPTOSIS

What determines if a particular ocular surface injury triggers keratocyte apoptosis? Corneal surface injuries that have been associated with keratocyte apoptosis are produced by viral infections, scrape injury, and surgical procedures.[1,6,7] All of these represent the more severe surface injuries associated with fluorescein staining of the epithelium, and thus, injuries sufficient to expose basement membrane. It is our hypothesis that ocular surface injury must be severe enough to break down the epithelial barrier function before cytokines released from the corneal epithelium can penetrate to the stroma and bind receptors to trigger keratocyte apoptosis and stromal wound healing. Epithelial barrier function is attributable to tight junctions between the epithelial cells.[8-10] However, this does not mean that the cytokines released from surface epithelial cells into the tear film do not have a role in the surface disease associated with mild to moderate dry eye. To the contrary, it seems likely that elevated levels of proinflammatory cytokines do have a role in the pathophysiology of ocular surface injury.[11,12] In some patients, this surface injury becomes sufficiently severe that deeper epithelial injury leads to stromal involvement that is characteristic of more severe dry eye disease. Severe dry eye patients frequently have corneal scarring and may note a refractive instability that is generally associated with the chronic disease.

4. CYTOKINE INVOLVEMENT IN WOUND HEALING AND INFLAMMATORY CELL INFILTRATION

Studies have demonstrated that increased levels of IL-1 and other pro-inflammatory and pro-apoptotic cytokines can be detected in the tear film of patients with dry eyes.[11,12] Most of these cytokines are likely released from superficial epithelial cells lining the ocular surface. Some of these cytokines may pass into the corneal or conjunctival stroma when there is sufficient injury to break down the epithelial barrier function or directly damage basal epithelial cells. Once these cytokines bind the underlying keratocyte or conjunctival fibroblast cells, many responses may be triggered. IL-1 is probably the best-characterized example. The multitude of effects IL-1 induces in keratocytes suggests that this cytokine acts as a master regulator of the corneal response to injury. Thus, IL-1 can modulate apoptosis,[1] induce negative chemotaxis,[13] upregulate expression of collagenases and metalloproteinases,[14-16] or upregulate the production of cytokines such as hepatocyte growth factor that modulate epithelial healing.[17] The specific responses triggered in a particular cell likely depends on the concentration of cytokine, the milieu of the cell, and other unknown factors.

Recently, we sought to use gene array technology to evaluate the overall effects of IL-1 on cytokine or receptor expression by corneal fibroblasts. We wanted to determine if IL-1 modulated the expression of any other cytokines or receptors in keratocytes. Interestingly, we found that one of the cytokines that is upregulated in corneal fibroblasts in response to IL-1 or tumor necrosis factor (TNF) alpha is monocyte chemotactic and activating factor (MCAF)[18] (also Hong, Liu, and Wilson, unpublished data, 2000). This upregulation of MCAF in response to IL-1 or TNF alpha stimulation was confirmed by RNAse protection assay (mRNA) and western blotting (protein). MCAF has chemotactic effects on monocyte and macrophage cells. This observation has led us to formulate a model to explain factors attracting inflammatory cells into corneal stroma or subconjunctival connective tissue in response to epithelial injury. This hypothesis holds that: 1) Significant injury to the epithelium (mechanical or other) results in the release of IL-1 and TNF alpha from the epithelial cells. 2) If the injury to the epithelium is sufficient to break down the epithelial barrier function, then IL-1 and TNF alpha penetrate into the stroma or subconjunctival tissues. Some types of injury might directly trigger release of IL-1 or TNF alpha from the basal epithelial cells. 3) IL-1 or TNF alpha that penetrates into the subepithelial tissues binds to receptors expressed by keratocytes or conjunctival fibroblasts, and triggers MCAF production. 4) MCAF attracts and activates macrophages. Work is in progress to characterize the effects of MCAF. We also hope to identify other cytokines released by keratocytes or conjunctival fibroblasts that attract T cells and polymorphonuclear cells in response to epithelial injury or other pathophysiological triggers.

5. CONCLUSIONS

Mild to moderate dry eye associated with rose bengal staining is not associated with a chronic wound healing response, refractive instability, or corneal scarring. As the disease becomes severe (characterized by fluorescein staining of the ocular surface) corneal scarring and other complications associated with a chronic wound healing response are common. Cytokine-mediated systems involving epithelial to keratocyte/conjunctival fibroblast to immune cell interactions may have an important role in attracting and activating cells associated with inflammation characteristic of dry eye disease.

REFERENCES

1. S.E. Wilson, Y-G. He, J. Weng, Q. Li, M. Vital, and E. L. Chwang. Epithelial injury induces keratocyte apoptosis: Hypothesized role for the interleukin-1 system in the modulation of corneal tissue organization. *Exp. Eye. Res.* 62:325 (1996).
2. S.E. Wilson. Keratocyte apoptosis in refractive surgery: Everett Kinsey Lecture. *CLAO Journal* 24:181 (1998).
3. S.E. Wilson. LASIK "Dry Eye" and LASIK-induced Neurotrophic Epitheliopathy (LNE). *Ophthalmology*, in press.
4. J.J. Perez-Santonja, H.F. Sakla, C. Cardona, E. Chipont, and J.L. Alio. Corneal sensitivity after photorefractive keratectomy and laser in situ keratomileusis for low myopia. *Am J Ophthalmol* 127:497 (1999).

5.   T.U. Linna, M.H. Vesaluoma, J.J. Perez-Santonja, W.M. Petroll, J.L. Alio, and T.M. Tervo. Effect of myopic LASIK on corneal sensitivity and morphology of subbasal nerves. *Invest Ophthalmol Vis Sci* 41:393 (2000).

6.   S.E. Wilson, L. Pedroza, R. Beuerman, and J.M. Hill. Herpes simplex virus type-1 infection of corneal epithelial cells induces apoptosis of the underlying keratocytes. *Exp Eye Res* 64:775 (1997).

7.   M.C Helena, F. Baerveldt, W-J. Kim, and S.E. Wilson. Keratocyte apoptosis after corneal surgery. *Invest Ophthalmol Vis Sci* 39:276 (1998).

8.   A.M. Tonjum. Permeability of rabbit corneal epithelium to horseradish peroxidase after the influence of benzalkonium chloride. *Acta Ophthalmol (Copenh).* 53:335 (1975).

9.   N.L. Burstein and S.D. Klyce. Electrophysiologic and morphologic effects of ophthalmic preparations on rabbit cornea epithelium. *Invest Ophthalmol Vis Sci* 16:899 (1977).

10.   B.J. McLaughlin, R.B. Caldwell, Y. Sasaki, and T.O. Wood. Freeze-fracture quantitative comparison of rabbit corneal epithelial and endothelial membranes. *Curr Eye Res* 4:951 (1985).

11.   S.C. Pflugfelder, D. Jones, Z. Ji, A. Afonso, and D. Monroy. Altered cytokine balance in the tear fluid and conjunctiva of patients with Sjogren's syndrome keratoconjunctivitis sicca. *Curr Eye Res* 19:201 (1999).

12.   S.C. Pflugfelder. Tear fluid influence on the ocular surface. *Adv Exp Med Biol.* 438:611 (1998).

13.   W-J. Kim, R.R. Mohan, R.R. Mohan, and S.E. Wilson. Effect of PDGF, IL-1 alpha, and BMP2/4 on corneal fibroblast chemotaxis: expression of the platelet-derived growth factor system in the cornea. *Invest Ophthalmol Vis Sci* 40:1364 (1999).

14.   K.J. Strissel, W.B. Rinehart, and M.E. Fini. Regulation of paracrine cytokine balance controlling collagenase synthesis by corneal cells. *Invest. Ophthalmol. Vis. Sci.* 38:546 (1997).

15.   K.J. Strissel, M.T. Girard, J.A. West-Mays, W.B. Rinehart, J.R. Cook, C.E. Brinckerhoff, and M.E. Fini. Role of serum amyloid A as an intermediate in the IL-1 and PMA-stimulated signaling pathways regulating expression of rabbit fibroblast collagenase. *Exp. Cell Res.* 15:237 (1997).

16.   J.A. West-Mays, K.J. Strissel, P.M. Sadow, and M.E. Fini. Competence for collagenase gene expression by tissue fibroblasts requires activation of an interleukin 1 alpha autocrine loop. *Proc. Natl. Acad. Sci. (U S A).* 92:6768 (1995).

17.   J. Weng, R.R. Mohan, Q. Li, and S.E. Wilson. IL-1 upregulates keratinocyte growth factor and hepato-cyte growth factor mRNA and protein production by cultured stromal fibroblast cells: Interleukin-1 beta expression in the cornea. *Cornea* 16:465 (1997).

18.   J.W. Hong, J.J. Liu, and S.E. Wilson. Effects of IL-1 alpha and TNF alpha on the expression of other cytokines and their receptors in human corneal fibroblast cells. *Invest Ophthalmol Vis Sci* 41:S265 (2000).

MECHANISMS OF APOPTOSIS IN HUMAN CORNEAL EPITHELIAL CELLS

Fiona Stapleton,[1] Jai-Min Kim,[2] Jason Kasses,[1] and
Mark D. P. Willcox[1]

[1]Cooperative Research Centre for Eye Research and Technology
The University of New South Wales
Sydney, Australia
[2]Dong Kang College of Ophthalmic Optics
Kwang-Ju, Korea

1. INTRODUCTION

Apoptosis describes programmed cell death, which is important in normal homeostasis, tissue development and wound healing (DeLong, 1998), in addition to pathological states such as autoimmune disease and neural degenerative diseases. Apoptotic cells are differentiated from necrotic cells by specific morphological and biochemical characteristics (Kerr et al., 1972). In the early stages these characteristics include condensation of chromatin, reduction in nuclear size, shrinkage of total cell volume, compaction of cytoplasmic organelles and dilation of endoplasmic reticulum. In the later stages, there is budding and separation of the nucleus and cytoplasm into membrane bound apoptotic bodies of various sizes which can be rapidly phagocytosed by neighbouring cells and phagocytes recruited to the area (Alison and Sarraf, 1992). This rapid system occurs without provoking an inflammatory response in the tissue. In contrast, necrosis is a random degenerative process, affecting groups of cells, where cells lose their ability to maintain cell volume, membrane integrity is lost and both the cytoplasm and mitochondria become swollen. There is no vesicle formation and the end stage is total cell lysis, with release of cellular contents. This process is often associated with inflammation.

Apoptosis occurs in ocular surface tissues in vitro and in vivo via multiple mechanisms. Viable cells are shed from the corneal epithelium by a process of classical apoptosis as part of normal homeostasis (Ren and Wilson, 1996, Estil et al., 2000). During

Lacrimal Gland, Tear Film, and Dry Eye Syndromes 3
Edited by D. Sullivan *et al.*, Kluwer Academic/Plenum Publishers, 2002

827

epithelial regeneration following injury, intraepithelial apoptosis appears to offset increased cellular proliferation to prevent epithelial hyperplasia (Glaso et al., 1993). In the corneal stroma, keratocyte apoptosis following corneal injury occurs through IL-1α and β released from damaged corneal epithelial cell (Wilson et al., 1996a) and potentially through the FAS/FAS ligand system (Wilson et al., 1996b). FAS antigen is a cell surface protein which is a member of the tumour necrosis factor (TNF)/nerve growth factor (NGF) receptor family. The binding of the membrane-spanning FAS receptor to FAS ligand, a member of the TNF cytokine family, on the extracellular side of the plasma membrane, mediates apoptosis via activation of caspases (Schultze-Ostoff and Peter, 1999). Herpes simplex infection of the corneal epithelium also induces apoptosis of underlying anterior stromal keratocytes (Wilson et al., 1997). Stromal fibroblasts also undergo TNFα mediated apoptosis, in the presence of NF-κB inhibition (Mohan et al., 2000). Interferon gamma causes both apoptosis and upregulation of inflammatory markers in conjunctival cells in vitro (De Saint Jean et al, 1999, 2000), which has been suggested as a pathogenic mechanism underlying conjunctival damage in dry eye disease.

The aim of the present study was to determine mechanisms of corneal epithelial cell apoptosis in vitro following exposure to anti-FAS and anti-FAS ligand antibody and during infection with *Mycoplasma* sp.

2. MATERIALS AND METHODS

2.1. Cell Line and Culture Techniques

SV40-immortalised human corneal epithelial (HCE) cells (Araki-Sasaki et al., 1995) were grown to confluency in minimum essential medium: Hams F-12 (1:1) supplemented with 5% fetal bovine serum, plus 100 U/ml penicillin, 100 μg/ml streptomycin, 0.5% dimethyl sulphotide, 1 μg/ml gentamicin, 10 ng/ml epidermal growth factor, 5 μg/ml insulin and 0.1 μg/ml cholera toxin. Cells were maintained in 10 cm^2 tissue culture flasks (Flacon, Becton Dickinson, Franklin Lakes, NJ) at 37°C in humidified 5% CO_2.

2.2. Procedures

The cell monolayers were washed and treated with anti-human FAS (N-18) goat polyclonal IgG (50, 200, 500 and 1000 ng/ml) (Santa Cruz Biotechnology, Inc., California) or anti-human FAS ligand (C-20) goat polyclonal IgG (500 ng/ml) (Santa Cruz Biotechnology, Inc., California) for 2 and 4 days. Control cells were maintained without a change of medium for the same exposure time. Following incubation both floating and attached cells were harvested by centrifugation and trypsinisation/centrifugation respectively. Experiments were repeated at least twice.

2.3. Effect of *Mycoplasma*

The original cell line was found to be contaminated by *Mycoplasma* sp. *Mycoplasma* removal agent (MRA, a 4-oxo-quinolone-3 carboxylic acid derivative, ICN Biomedicals Australasia, NSW, Australia), was used to eliminate the bacterium from the cell line. MRA (0.5 μl tissue culture medium) was added to the cell line and incubated for 1 week. The cell line underwent multiple passages in media not containing MRA and cells were grown to 50.–.80% confluency on coverslips and stained using the Hoescht stain provided in the kit to ensure *Mycoplasma* removal. Apoptosis experiments were performed before and after *Mycoplasma* removal.

2.4. Morphology

HCE cells were fixed in 70% ethanol and treated with 2 μg/ml Hoechst 33342 (Sigma Co. USA). Hoechst 33342 is taken up by all nuclei, and apoptotic cells were identified by chromatin aggregation, nuclear shrinkage and the formation of apoptotic bodies (Fig. 1). For quantitative analysis, more than 500 cells per slide were counted in 400X fields with fluorescent microscopy and the percentage of apoptotic cells was determined.

2.5. Annexin V-FITC and Propidium Iodide Staining

HCE cells were stained using the Annexin_V-FLUOS apoptotic detection kit (Roche Diagnostics, Mannheim, Germany). Briefly, cells were washed with phosphate buffered saline (PBS) and then incubated in the Annexin V-FLUOS, propidium iodide (PI) mixture incubation buffer. After 10 min, cells were washed with PBS and observed with fluorescence and light microscopy. Apoptotic cells were fluorescent and late apoptotic and necrotic cells were stained with PI. Cells in early apoptosis were stained with Annexin V which binds to phosphatidylserine on the cell membrane. Nonviable cells appeared red due to PI staining. The viable (non-apoptotic) cells were visualised by light microscopy and the percentage of dead cells estimated. Mean cell death data are presented as test–control (unchallenged cells).

3. RESULTS

3.1. Morphological Analysis

Mean background levels of apoptosis were 7.0 ± 1.1% (2 days) and 5.8 ± 0.4% (4 days) for the *Mycoplasma* contaminated cell line and 6.7 ± 1.5% (2 days) and 6.6 ± 1.9% (4 days) for the *Mycoplasma* free cell line. Thus, there were no significant differences in apoptosis between the cell lines ($P > 0.05$), nor was there any effect of time ($P > 0.05$). When compared to controls, anti-FAS antibody induced a time and concentration dependent increase in signs of late apoptosis. *Mycoplasma* contaminated cells were more

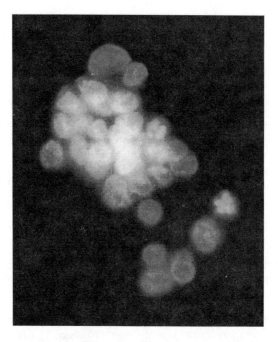

Figure 1. Fluorescent microscopy *Mycoplasma* free HCE cells treated with 500 ng/ml anti-human FAS for 2 days and stained with Hoechst 33342. Arrow shows an apoptotic cell characterised by chromatin condensation and nuclear fragmentation.

susceptible to anti-FAS antibody induced apoptosis and necrosis compared to *Mycoplasma* free cells (Figs. 2 and 3). Anti-FAS ligand antibody did not cause significant apoptosis above background in either the *Mycoplasma* contaminated or the free cell line at either time point.

3.2. ANNEXIN V-FITC AND PROPIDIUM IODIDE STAINING RESULTS

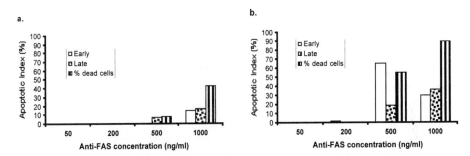

Figure 2. Apoptotic percentages for *Mycoplasma* free cells challenged with anti-FAS antibody for **(a)** 2 and **(b)** 4 days respectively. Early apoptotic percentages refer to Annexin V positive cells; Late apoptotic percentages refer to cells showing positive morphological changes on Hoescht staining; Dead cell percentages refer to cells showing positive PI nuclear staining and include apoptotic cells that have undergone secondary necrosis.

When compared to controls, cells treated with anti-FAS antibody showed a time- and concentration-dependent increase in signs of late apoptosis. As before, *Mycoplasma* contaminated cells were more susceptible to anti-FAS antibody induced apoptosis and necrosis compared to *Mycoplasma* free cells (Figs. 2 and 3). High concentrations of anti-FAS antibody (500 ng/ml and above in the *Mycoplasma* free cell line and 200 ng/ml and above in the *Mycoplasma* contaminated cell line) caused considerable cell death, particularly at the 4 day time point.

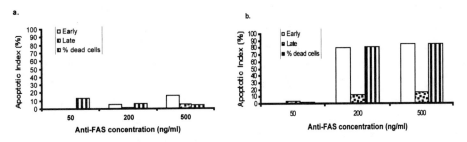

Figure 3. Apoptotic percentages for *Mycoplasma* contaminated cells challenged with anti-FAS antibody for **(a)** 2 and **(b)** 4 days respectively. See legend for Fig. 2 for specificity.

4. DISCUSSION

This study has demonstrated time- and concentration-dependent apoptosis in HCE cells treated with anti-FAS antibody, but not anti-FAS ligand antibody. Classically, the FAS apoptotic signal is mediated through binding of soluble FAS ligand to a membrane-bound receptor, causing apoptotic signal transduction. Several downstream signalling pathways have been proposed (Schulze-Osthoff and Peter, 1999). FAS receptors are expressed on corneal epithelial cells (Wilson et al., 1996), and cell death due to FAS activation in tissue culture has been demonstrated. The present data demonstrating apoptosis in cells challenged with anti-FAS antibody would suggest that FAS receptor is indeed expressed on HCE in culture. There is cross-linking of the FAS receptor by the anti-FAS antibody and initiation of the subsequent apoptosis pathways. Increased susceptibility to apoptosis in *Mycoplasma* contaminated cells via this pathway might suggest that there is increased expression of FAS receptor or a reduction in inhibitory mechanisms in contaminated cells.

A less well-characterised mechanism of FAS-induced apoptosis is via FAS ligand expression on the cell surface . Cytotoxic T cells express FAS ligand and bind to the membrane bound FAS receptor on host cells in autoimmune disease (Schulze-Osthoff and Peter, 1999). In the eye, FAS ligand is expressed by only a proportion of normal conjunctival cells, whereas FAS is expressed more consistently. Both FAS and FAS ligand mRNA are expressed in primary cultures of corneal epithelial cells (Wilson et al., 1996). In addition, injured ex-vivo corneal epithelial cells produce soluble FAS ligand (Mohan et al., 1997).

Intracellular contamination by *Mycoplasma*, frequently reported in immortalised cell lines (Coronato et al., 1994), appears to upregulate FAS-induced apoptosis and necrosis of HCE. Such contamination can affect a wide range of cell functions, including cell metabolism. Growth kinetics, however, are not invariably altered (Zhang et al., 2000), and growth curves for *Mycoplasma* contaminated and *Mycoplasma* free cells were similar in this study (data not shown). Additionally, background levels of apoptosis were unaffected by *Mycoplasma* contamination in this study, unlike hybridoma and lymphoma cell lines where *Mycoplasma* contamination induced higher levels of background apoptosis compared to uninfected cells (Sokolova et al., 1998). Differences in the host cell type and pathogenic traits of the infecting organism may account for these findings. *Mycoplasma* can activate cell death in immortalised cell lines, possibly through upregulation of certain cytokines (Rawadi et al., 1996). Some cell types appear to show increased sensitivity to various inducers of apoptosis through a range of signalling mechanisms (Sokolova et al., 2000) including endonuclease production by these intracellular organisms.

In summary, anti-FAS antibody induces apoptosis in HCE in a time- and concentration-dependent mechanism. Cell lines contaminated with *Mycoplasma* have an increased susceptibility to FAS induced apoptosis. Further study is required to determine the extent to which immortalised HCE cells are representative of the ocular surface in vivo.

ACKNOWLEDGEMENT

This work was supported by the Australian Federal Government through the Cooperative Research Centres programme. The authors would like to thank Vicky Vallas for removal of *Mycoplasma* from the cell line and David Miles for Fig. 1.

REFERENCES

M.R. Alison, and C.E. Sarraf. Apoptosis: a gene-directed programme of cell death, *J Royal College Physicians* 26:25–35 (1992).

K. Araki-Sasaki, Y. Ohashi, T. Sasabe, K. Hayashi, H. Watanabe, Y. Tano, and H. Handa. An SV40-immortalized human corneal epithelial cell line and its characterization, *Invest Ophthalmol Vis Sci* 36:614–621 (1995).

S. Coronato, D. Vullo, and C.E. Coto. A simple method to eliminate *Mycoplasma* from cell cultures. *J Virol Meth* 46:85–94 (1994).

M. De Saint Jean, F. Brignole, G. Feldmann, A. Goguel, and C. Baudouin. Interferon-γ induces apoptosis and expression of inflammation-related proteins in Chang conjunctival cells. *Invest Ophthalmol Vis Sci*, 40: 2199–2212 (1999).

M. De Saint Jean, C. Debbasch, M. Rahmani, F. Brignole, G. Feldmann, J-M. Warnet, and C. Baudouin. Fas- and Interferon γ-induced apoptosis in Chang conjunctival cells: Further investigations, *Invest Ophthalmol Vis Sci*, 41: 2531–2543 (2000).

M.J. DeLong. Apoptosis: a modulator of cellular homeostasis and disease states, *Ann NY Acad Sci*, 842:82–90 (1998).

S. Estil, E.J. Primo, and G. Wilson. Apoptosis in shed human corneal cells. *Invest Ophthalmol Vis Sci*, 41:3360–3364 (2000).

M. Glasφ, K.U. Sandvig, and E. Haaskjold. Apoptosis in the rat corneal epithelium during regeneration, *APMIS*, 101:914–922 (1993).

J.F.R. Kerr, A.K. Wyllie, and A.R. Currie. Apoptosis: A basic biological phenomenon with wide ranging implications in tissue kinetics. *Br J Cancer*, 26:25–35 (1972).

R.R. Mohan, Q. Liang, W-J. Kim, M. Helena, F. Baerveldt, and S.E. Wilson. Apoptosis in the cornea: further characterisation of Fas-Fas ligand system. *Exp Eye Res*, 65:575–589 (1997).

R.R. Mohan, R. Mohan, W-J. Kim, and S.E. Wilson. Modulation of TNF-α-induced apoptosis in corneal fibroblasts by transcription factor NF-κB. *Invest Ophthalmol Vis Sci*, 41: 1327–1336 (2000).

G. Rawadi, S. Roman-Roman, M. Castedo, M. V. Dutilleul, S. Susin, P. Marchetti, M. Geuskens, and G. Kroemer. Effects of *Mycoplasma* fermentans on the myelomonocytic lineage. Different molecular entities with cytokine-inducing and cytocidal potential. *J Immunol*, 156:670–678 (1996).

H. Ren, and G. Wilson. Apoptosis in the corneal epithelium. *Invest Ophthalmol Vis Sci*, 37:1017–1025 (1996).

K. Schulze-Osthoff, and M.E. Peter. Death Receptors, Chapter 1 in *Signalling Pathways in Apoptosis*, D.Watters, and M. Lavin, eds., Harwood Academic Publishing, QLD (1999).

I.A. Sokolova, A.T.M. Vaughan, and N. Khodarev. *Mycoplasma* infection can sensitise host cells to apoptosis through contribution of apoptotic-like endonucleases. *Immunol Cell Biol*, 76:526–534 (1998).

S.E. Wilson, Q. Li, J. Weng, P.A. Barry–Lane, J.V. Jester, Q. Liang, and R.J. Wordinger. The Fas-Fas Ligand system and other modulators of apoptosis in the Cornea. *Invest Ophthalmol Vis Sci*, 37:1582–1592 (1996a).

S.E. Wilson, Y.G. He, J. Weng, Q. Li, M. Vital, and W.L. Chwang. Epithelial injury induces apoptosis of the underlying keratocytes: Hypothesised role for the interleukin-1 system in the modulation of corneal tissue organisation, *Exp Eye Res*, 62:325–338 (1996b).

S.E. Wilson, L. Pedrosa, R. Beuerman, J.M. Hill. Herpes simplex virus type-1 infection of corneal epithelial cells induces apoptosis of the underlying keratocytes. *Exp Eye Res*, 64:775–779 (1997).

S. Zhang, D.J. Wear, and S. Lo. Mycoplasmal infections alter gene expression in cultured human prostatic and cervical epithelial cells. *FEMS Immunol Med Microbiol*, 27:43–50 (2000).

A FUNCTIONAL UNIT FOR OCULAR SURFACE IMMUNE DEFENSE FORMED BY THE LACRIMAL GLAND, CONJUNCTIVA AND LACRIMAL DRAINAGE SYSTEM

Erich Knop and Nadja Knop

Department for Cell Biology in Anatomy
Hannover Medical School
Hannover, Germany

1. INTRODUCTION

Mucosal organs represent a special moist compartment of the body's surface that is equipped with a diverse array of defense mechanisms to avoid microbial colonization. Besides an innate defense, the relevance of lymphoid cells that form a "mucosa-associated lymphoid tissue" (MALT) in these organs is increasingly recognized as an important factor for the preservation of mucosal integrity. The mucosal lymphoid tissue of the different organs together constitutes a "common mucosal immune system".[1] The components of this system interact through migrating lymphoid cells that are primed for antigens in follicular sites and later populate the same or similar tissues as effector cells.

The morphology and function of the mucosal immune system is distinct in that it consists, besides follicular "organized" lymphoid tissue (O-MALT), mainly of a "diffuse" lymphoid tissue (D-MALT) composed of lymphocytes and plasma cells. The latter exclude antigens from the mucosa mainly by the production of secretory IgA (sIgA) (Fig. 1A – C), and these tissues hence together constitute the so-called "secretory immune system".[2] Another important mucosal immune function is the generation of tolerance[3] which is of interest because these tissues are exposed to a variety of non-pathogenic ubiquitous antigens. Secretory immunity and mucosal tolerance are both important to avoid potentially destructive inflammatory reactions that endanger the tissue integrity. This is especially important for the delicate ocular tissues.

MALT represents an accepted component in organs like the intestine, respiratory system or genital tract. However, its presence at the normal human ocular surface is not fully recognized as yet, and the supply of the ocular surface with protective immunoglobulins is usually attributed to the lacrimal gland.[4] Our investigations focus on the presence and organization of lymphoid tissue at the ocular surface and appendage, it probable interaction and its undefined role up to date in ocular surface health and disease.

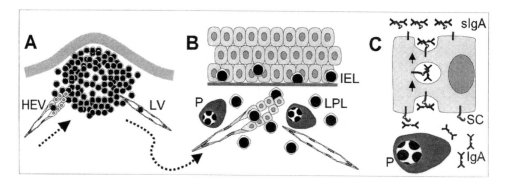

Figure 1. Mucosa-associated lymphoid tissue (MALT) consists of two types: An organized type (**A**) is composed of roundish lymphoid follicles with ordinary vessels and high endothelial venules (HEV) for cell immigration and lymph vessels (LV) for emigration. A diffuse type (**B**) is formed by plasma cells (P), primed at follicular sites, and lymphocytes in the lamina propria (LPL) together with intraepithelial lymphocytes (IEL) all immigrated after recirculation. One of the main functions (**C**) of the diffuse lymphoid tissue is the production of dimeric IgA by the plasma cells and its transcytosis by the epithelial molecule secretory component (SC) to build up a protective coat of secretory IgA (sIgA) at the mucosa.

2. DIFFUSE LYMPHIOD TISSUE AND THE SECRETORY IMMUNE SYSTEM

In the normal human conjunctiva[5,6] and lacrimal drainage system[7] there is an associated lymphoid tissue termed CALT[5] and LDALT[7] respectively. It consists, besides follicles, mainly of a diffuse type similar to other mucosal organs[8] that is composed of lymphocytes and plasma cells, spread diffusely in the subepithelial lamina propria, and of intraepithelial lymphocytes (IEL) in the basal epithelial layers. T-lymphocytes prevail (Fig. 2), IEL are mostly CD8-positive suppressor/cytotoxic cells, and lymphocytes expressing the human mucosa lymphocyte antigen (HML-1) regularly occur in CALT[9] and in LDALT[7] (Fig. 2D). This characterizes them as mucosa-specific and the respective tissues as a part of the MALT system. Most of the plasma cells stain positive for IgA (Fig. 3A) and a minority for IgM. In the overlying epithelium of the conjunctiva and lacrimal drainage system, the IgA- and IgM-transporter molecule poly-Ig-receptor,[3,4] the extracellular part of which is termed secretory component (SC),[1,2] is strongly expressed (Fig. 1C,3B), and IgA is present here as deposits or diffuse surface staining. Our findings also show that a diffuse lymphoid tissue with plasma cells and strong expression of SC in the epithelium is continuous from the acinar tissue of the lacrimal gland, along the excretory lacrimal ducts into the conjunctiva (Fig. 4) and from there, further along the lacrimal drainage system. This physical continuity of lymphoid tissue is another indication to assume that what is present in the lacrimal gland can also be present at the ocular surface.

The presence of plasma cells, however, was controversial in the human conjunctiva[4,5] and lacrimal drainage system. They were reported dating back to the nineteenth century although their normality was under discussion then and later.[10] In a histological investigation,[11] the consistent presence of plasma cells was shown. Their number in the conjunctiva accounted for about two thirds of that in the lacrimal gland. However, in immunofluorescence studies by the same group, IgA could not be located in normal conjunctival plasma cells and SC was not consistently found in the epithelium. Although the absence of IgA was confirmed in biopsies from only few presumably normal controls, some doubt persisted regarding the presence of plasma cells at the ocular surface.

In our histological, electron microscopical and immunohistochemical studies on a large number of complete and normal human tissues, numerous plasma cells were consistently identified in the conjunctiva[5] (except bulbar) and lacrimal drainage system.[7] IgA was found in the plasma cells and in deposits in the epithelium, together with a strong expression of SC (Fig. 3) similar to the lacrimal gland. The identity of IgA staining was verified by preadsorption of the antibody with commercial IgA. Differences to previous studies are probably related to different antibodies, staining techniques, tissue preparation or biopsy locations used.

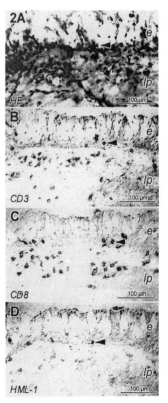

Further indication for an immunological ability of the conjunctiva may be derived from embryological development. The lacrimal gland arises from an epithelial bud of the conjunctiva, similar to the accessory glands that remain in closer topographical contact with the ocular surface. Both can hence, systematically, be considered as part of the gland-associated lymphoid tissue of the conjunctiva. The presence of plasma cells in the glandular tissues of the eye is commonly accepted.[4] Our findings therefore indicate that the lacrimal gland represents not the only source for immunoglobulins at the ocular surface and that the conjunctiva and lacrimal drainage system are able to contribute to ocular surface immune defense by production of secretory IgA. This may be required early, because after birth the lacrimal gland is not fully functional but the ocular surface needs immunological protection. The observation that the normal conjunctiva and drainage system, unlike the

Figure 2. T-cells represent the majority of lymphocytes in the diffuse lymphoid tissue of the lamina propria (lp) and epithelium (e). (A) They stain positive for CD3 (B), CD8 (C) and HML-1 (D); IEL are indicated by arrowheads. Sections from the human lacrimal sac; staining and

lacrimal gland, can have follicles, known to be most frequent before puberty,[12] may indicate that during an active learning phase, that later declines or transforms into a steady state, these tissues may also have the function of antigen uptake and plasma cell priming to provide the lacrimal gland with appropriate plasma cells.

3. FOLLICLES ARE A NORMAL COMPONENT OF CALT AND LDALT

Organized lymphoid follicles were reported in different amounts in the normal human conjunctiva. However, there are natural factors that influence their number because this declines with increasing age[12] and it also depends on the investigated location[5] which is important if the prevailing studies that use biopsies or incomplete conjunctival tissue are considered. In conjunctival wholemounts of an elderly human population follicles were shown to occur in 60% of cases, to be most frequent in the tarso-orbital conjunctiva and to have a bilateral symmetry[5]. This underlines their normal character even though they may not be found consistently in every single specimen. Such follicles similarly occur in the lacrimal drainage system,[7] have a complementary composition of B- and T-cells (Fig. 5) and represent the afferent limb of mucosal immunity where antigens are transferred into the tissue for antigen presentation to lymphocytes and their subsequent activation and pro-liferation.[8] S pecialized epithelial cells (M-cells) that allow antigen transfer through the intact epithelium into the lymphoid tissue are described for follicles of the intestine.[13] There is indication for cells with a similar morphology also in the conjunctiva, where they are also located at the apices of follicles with an epithelium free of SC and IgA to allow attachment of antigens.[5] Ultra-structural characteristics o f M-cells containing lymphocytes in intraepithelial pockets were observed in the lacrimal drainage system.[7] A germinal center (that contains CD68-positive macrophages–see another contribution to this book) and typical zones around (Fig. 5A) in some of these follicles verifies that antigen had indeed been presented here and indicates that eye-associated lymphoid tissue can perform an afferent immune function.

High endothelial venules (HEV) are consistently present at the ocular surface.[14] They have an ultrastructure like those found in other lymphoid tissues[15] and express cell adhesion molecules.[16] This shows that the eye-associated lymphoid tissue is an integral component of the MALT system and shares receptor molecules with other mucosal sites, connecting it to a regulated traffic of lymphocytes known as homing.[17] Thereby, ocular tissues are able, for example, to receive and distribute primed plasma cell precursors, a role which is mainly attributed to intestinal lymphoid tissue,[3,8] and may thus contribute to mucosal immunity by local priming of B-cells for antigens that are first taken up via the ocular tissues. In this way, the follicles of the conjunctiva and lacrimal drainage system are able to perform an afferent immune function. This is a strong indication that the human ocular tissues may thus serve, within the system of the eye-associated lymphoid tissue, for the population of the lacrimal gland with plasma cells that produce antibodies with specificities relevant for ocular surface needs. Data on recirculation of such cells in ocular

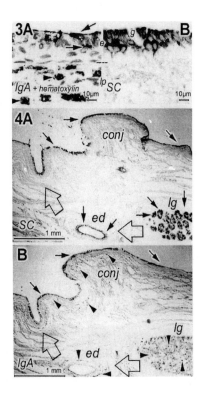

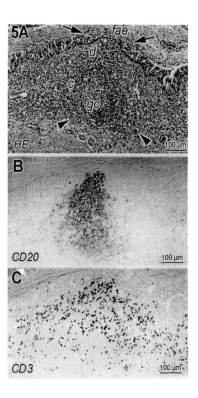

Figure 3. The normal human conjunctiva contains IgA-positive plasma cells (**A**, arrowheads) in the lamina propria (lp). IgA is seen as deposits (arrows) or diffuse staining in the epithelium and its transporter molecule SC (**B**) is in the upper layers of the epithelium (e), except in goblet cells (g). **Figure 4.** SC (**A**, arrows) occurs continuously in epithelial cells from the acini of the palpebral part of the lacrimal gland (lg), along one of its excretory ducts (ed, open arrows) into the conjunctiva (conj). (**B**) IgA is present along the same path in plasma cells (arrowheads) in the lacrimal gland, in the diffuse lymphoid tissue along its duct and in the conjunctiva (arrows).

Figure 5. A secondary follicle (**A**) in the human lacrimal sac shows a bright germinal center (gc), dark lymphocyte corona (c), dome-like zone (d) and flat follicle associated epithelium (fae) at the apex (arrows) and numerous parafollicular vessels (arrowheads). It is composed of central B - (**B**) and peripheral T-cells (**C**).

tissues is scarce compared to other organs and therefore requires further studies. The differential importance of the conjunctiva and the lacrimal gland for ocular surface immune defense is not clear as yet, however our findings give substantial indication for a reappraisal of the conjunctiva.

4. LYMPHOID CELLS CAN NOT GENERALLY BE ADDRESSED AS "INFLAMMATORY"

A problem complicating the view on lymphoid cells at the ocular surface is the fact that lymphocytes as well as plasma cells, have a long tradition[10] of being referred to as pathological or distinctly termed "inflammatory cells". This misleading terminology was also applied by authors who found them consistently in their normal tissues.[12,18] However, lymphocytes per se and especially those forming a diffuse lymphoid tissue along mucosal

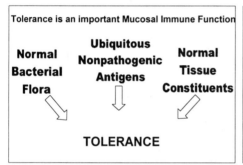

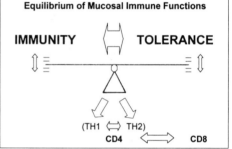

Figure 6. Generation of tolerance (unresponsiveness) is one of the main mucosal immune functions.

Figure 7. Immunity and tolerance must be kept in an equilibrium in order to maintain ocular surface integrity.

surfaces cannot generally be considered as inflammatory. Recent evidence shows that inflammatory reactions are only induced after appropriate activation of lymphocytes in a specific context as determined, for example, by the prevailing cytokines in the tissue and by the presence of adequate co-stimulation.[19–21] In contrast to the misleading term "inflammatory", a basic function of mucosal lymphocytes is the induction of tolerance[3], i.e., un-responsiveness (Fig. 6). This is directed against ubiquitous nonpathogenic antigens to prevent inflammatory reactions that could be destructive for the mucosal tissue and is, besides by a balance of TH1 and TH2 T-helper cells, also achieved by CD8-positive T-cells.[3,8,19,20] The necessity of tissue integrity especially applies to the delicate ocular surface and the mucosal lymphocytes in CALT are in fact predominantly of the CD8 type,[9,22] similar to the situation in LDALT.[7] However, immunity and tolerance must be kept in equilibrium to preserve ocular surface integrity (Fig. 7). The secretory IgA produced by mucosal plasma cells that excludes antigens from access to the tissue, also prevents inflammation and is hence "anti inflammatory".[3] These more recent advances in the understanding of the mucosal immune system should lead to a more differentiated view on lymphoid cells also at the ocular surface.

5. LYMPHOID TISSUE MAY BE AN IMPORTANT REGULATOR OF OCULAR SURFACE DISEASE

If impairment of the epithelial barrier occurs due to tissue damage as in dry eye disease and/or antigens achieve uncontrolled access to the tissue, the mucosal immune tolerance fails. Resident T-helper cells in the subepithelial connective tissue of the conjunctiva may then be activated and immunological reactions shifted towards inflammation, resulting in the secretion of proinflammatory cytokines. In dry eye syndromes, a respective elevation of inflammatory cytokines (IL-1α, IL-6, IL-8, TNF-α) is reported in the tear film and inside the tissue,[23] and a hyperproliferation of the conjunctival epithelium is observed combined with impaired differentiation.[24] Similar events occur in the intestinal mucosa during inflammatory bowel disease (IBD)[20] about which a large body of information is already acquired. At the ocular surface, the production of these cytokines is as yet mainly attributed to the epithelial cells. However, in the intestine it is verified that TNFα and IL1β are secreted by activated lamina propria lymphocytes, resulting in the production of epithelial growth factors by stromal cells.[20] The production of growth factors locally at the ocular surface under inflammatory conditions would also represent an explanation for the epithelial hyperproliferation that occurs in dry eye disease despite of the decline in EGF supply from the lacrimal gland.[24] Subsequent alterations include a release of proteases from epithelial[25] or stromal[26] cells that is upregulated by inflammatory cytokines and results in tissue destruction perpetuating the disease (as indicated in Fig. 9).

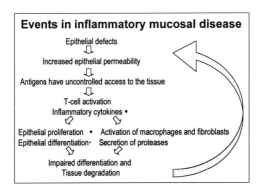

Figure 8. Inflammatory disease of the ocular surface and the bowel shows common events and tends to perpetuate.

A shift of the cytokine profile towards a TH-1 response is reported in several inflammatory ocular surface diseases[21], as similarly found in IBD,[20] and both disorders respond to immunosuppressive treatment. The widespread dry eye syndrome that is increasingly recognized to include an inflammatory component thus resembles disorders in other mucosal organs which are governed by lymphocytes. Hence, the resident lymphatic population localized in the mucosa-associated lymphoid tissue of the ocular surface, which represents a potent source of professional cytokine producing cells, may also be able to act as an important regulator of inflammatory ocular surface disease.

6. CONCLUSION

We found the regular presence of a mucosa-associated lymphoid tissue (MALT) in the conjunctiva (CALT) and the lacrimal drainage system (LDALT). This tissue has all components for a complete immune response. It consists of a diffuse type of lymphocytes and plasma cells that contributes to the secretory immune system and of follicles responsible for antigen uptake and lymphocyte activation. Lymphoid cells are traditionally termed "inflammatory cells" although they occur in every normal tissue. More recent advances in mucosal immunology indicate that mucosal lymphocytes per se do not represent inflammatory cells but rather contribute to mucosal tolerance which preserves tissue integrity. Similarly, the ocular mucosal plasma cells were observed to produce preferably IgA that contributes to secretory immunity by the anti-inflammatory exclusion of antigens from the ocular surface.

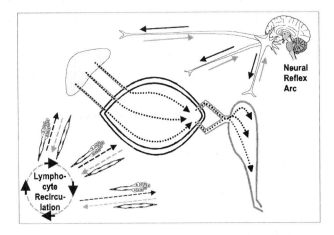

Figure 9. Eye-associated lymphoid tissue (EALT), consisting of the lacrimal gland, conjunctiva and lacrimal drainage system is physically continuous and connected by the flow of tears (dotted arrows), the recirculation of lymphocytes via vessels (interrupted arrows) and probably the neural reflex arc (solid arrows). It forms a functional unit for ocular surface immune defense.

This lymphoid tissue was found to be present from the lacrimal gland along its excretory ducts into the mucosa of the conjunctiva and from there into the lacrimal drainage system (Fig. 9). The ocular mucosal lymphoid tissues are hence not only connected by the flow of tears but also anatomically continuous. They are, furthermore, connected by the regulated traffic of lymphocytes via specialized vessels. Since the mucosa is innervated and neural stimuli influence the preservation of ocular surface integrity, dry eye development[27] and also immunological regulation, these tissues are conceivably also connected by a neural reflex arc. Altogether this leads us to propose the concept of an "Eye-Associated Lymphoid Tissue" (EALT), consisting of the lacrimal gland, conjunctiva, and lacrimal drainage system, and forming a functional unit for the immune defense of the ocular surface.

If ocular surface defense fails, as occurs in several forms of ocular surface disease, inflammatory reactions can occur with an alteration of the cytokine profile. The lymphoid tissue of the ocular surface, which represents a potent source of these factors, may then also be able to act as an important modulator of ocular surface disease similar to events observed in other mucosal organs.

REFERENCES

1. J. Mestecky, J.R. McGhee, S.M. Michalek, R.R. Arnold, S.S. Crago and J.L. Babb. Concept of the local and common mucosal immune response. *Adv.Exp.Med.Biol.* 107:185–192 (1978).
2. T.B. Tomasi, E.M. Tan, A. Solomon and R.A. Prendergast. Characteristics of an immune system common to certain external secretions. *J. Exp. Med. 121*, 101–124 (1965).
3. P. Brandtzaeg. History of oral tolerance and mucosal immunity. *Ann.N.Y.Acad.Sci.* 778:1–27 (1996).
4. D.A. Sullivan. Ocular mucosal immunity. In: Ogra PL, Mestecky J, Lamm ME, Strober W, McGhee J, Bienenstock J, eds. *Handbook of Mucosal Immunology.* 2 ed. Academic Press pp. 1241–1281 (1999).
5. N. Knop and E. Knop. Conjunctiva–associated lymphoid tissue in the human eye. *Invest.Ophthalmol.Vis.Sci.* 41:1270–1279 (2000).
6. N. Knop and E. Knop. The crypt system of the human conjunctiva. *Adv.Exp.Med.Biol* (2001) *in press*
7. E. Knop and N. Knop. Lacrimal drainage associated lymphoid tissue (LDALT): A part of the human mucosal immune system. *Invest.Ophthalmol.Vis.Sci.* 42:566–574 (2001).
8. J.P. Kraehenbuhl and M.R. Neutra. Molecular and cellular basis of immune protection of mucosal surfaces. *Physiol.Rev.* 72:853–879 (1992).
9. H.S.Dua,J.A.Gomes,V.K.Jindal,S.N.Appa,R.Schwarting,R.C.Eagle,L.A.Donoso, P.R. Laibson. Mucosa specific lymphocytes in the human conjunctiva, corneoscleral limbus and lacrimal gland.*Curr.Eye.Res.*13:87–93 (1994).
10. H. Virchow. Mikroskopische Anatomie der äusseren Augenhaut und des Lidapparates. In: Saemisch T, ed. *Graefe-Saemisch Handbuch der gesamten Augenheilkunde, Band 1,* 2 ed. Leibzig: Verlag W. Engelmann (1910).
11. M.R. Allansmith, G. Kajiyama, M.B. Abelson and M.A. Simon. Plasma cell content of main and accessory lacrimal glands and conjunctiva. *Am.J.Ophthalmol.* 82:819–826 (1976).
12. G. Osterlind. An investigation into the presence of lymphatic tissue in the human conjunctiva, and its biological and clinical importance. *Acta Ophthalmol.Copenh.* Suppl. 23:1–79 (1944).
13. A. Gebert, H.J. Rothkotter and R. Pabst. M cells in Peyer's patches of the intestine. *Int.Rev.Cytol.* 1996;167:91–159.
14. E. Knop and N. Knop. High endothelial venules are a normal component of lymphoid tissue in the human conjunctiva and lacrimal sac. *Invest.Ophthalmol.Vis.Sci.* 39:2 (1998).
15. E.Knop and N.Knop. Fine structure of high endothelial venules in the human conjunctiva.*Ophthalmic.Res.*30:169(1998)
16. R.J. Haynes, P.J. Tighe, R.A. Scott and H.S. Dua. Human conjunctiva contains high endothelial venules that express lymphocyte homing receptors. *Exp.Eye Res.* 69:397–403 (1999).
17. E.C. Butcher and L.J. Picker. Lymphocyte homing and homeostasis. *Science* 272:60–66 (1996).
18. M.R. Allansmith, J.V. Greiner and R.S. Baird. Number of inflammatory cells in the normal conjunctiva. *Am.J.Ophthalmol.* 86:250–259 (1978).

19. J.L. Vancott, M. Kweon, K. Fujihashi, M. Yamamoto, M. Marinaro, H. Kiyono and J.R. McGhee. Helper T subsets and cytokines for mucosal immunity and tolerance. *Behring.Inst.Mitt.* 98:44–52 (1997). *Immunol Today* 20:505–510 (1999).

21. M.R. Dana, Y. Quian and Hamrah P. Twenty-five-year panorama of corneal immunology. Cornea 19:625–43 (2000)

22. M. Hingorani, D. Metz and S.L. Lightman. Characterisation of the normal conjunctival leukocyte population. *Exp.Eye Res.* 64:905–912 (1997).

23. S.C. Pflugfelder, D. Jones, Z. Ji, A. Afonso and D. Monroy. Altered cytokine balance in the tear fluid and conjunctiva of patients with Sjogren's syndrome keratoconjunctivitis sicca. *Curr.Eye Res.* 19:201–211 (1999).

24. S.C. Pflugfelder. Tear fluid influence on the ocular surface. *Adv.Exp.Med.Biol.* 438:611–617 (1998).

25. A.A. Afonso, L. Sobrin, D.C. Monroy, M. Selzer, B. Lokeshwar and S.C. Pflugfelder. Tear fluid gelatinase B activity correlates with IL-1alpha concentration and fluorescein clearance in ocular rosacea. *Invest.Ophthalmol.Vis.Sci.* 40:2506–2512 (1999).

26. D. Meller, D.Q. Li and S.C. Tseng. Regulation of collagenase, stromelysin, and gelatinase B in human conjunctival and conjunctivochalasis fibroblasts by IL-1beta and TNF-alpha. *Invest.Ophthalmol.Vis.Sci.*41:2922–29 (2000).

27. M.E. Stern, R.W. Beuerman, R.I. Fox, J. Gao, A.K. Mircheff and S.C. Pflugfelder. The pathology of dry eye: the interaction between the ocular surface and lacrimal glands. *Cornea* 17:584–589 (1998).

ROLE OF TEAR ANTI-*ACANTHAMOEBA* IgA IN *ACANTHAMOEBA* KERATITIS

J. Y. Niederkorn, H. Alizadeh, H. Leher, S. Apte, S. El Agha,
L. Ling, M. Hurt, K. Howard, H. D. Cavanagh, and J.P. McCulley

U.T. Southwestern Medical Center
Dallas, Texas, USA

1. INTRODUCTION

Acanthamoeba keratitis is a sight-threatening corneal disease caused by pathogenic free-living amoebae.[1] The organisms have been isolated from a wide variety of environments and from nasopharyngeal washes of asymptomatic individuals.[1] Contact lens wear, practiced by over 25 million individuals in the United States, is the leading risk factor. Over 85% of the cases of *Acanthamoeba* keratitis occurred in contact lens wearers.[2] Antibodies against *Acanthamoeba* spp. were detected in 52–100% of normal subjects tested in two serological surveys.[3,4] In spite of the ubiquity of *Acanthamoeba* spp., the large number of contact lens wearers, and the apparent frequency of exposure to *Acanthamoeba* antigens, *Acanthamoeba* keratitis is rare. We hypothesized that frequent environmental exposure to *Acanthamoeba* antigens induces an immunity that protects against corneal infection in most contact lens wearers.

The domestic pig and Chinese hamster models of *Acanthamoeba* keratitis have been used to characterize the immune response to ocular infections with *Acanthamoeba* spp. Intramuscular immunization with trophozoite antigens induces strong delayed-type hypersensitivity (DTH) and serum IgG antibody responses in experimental animals, yet fails to protect the hosts against *Acanthamoeba* keratitis.[5,6] Exposure to *Acanthamoeba* antigens can also occur at the mucosal surface and might therefore activate the common mucosal immune system. The lacrimal gland is a component of the common mucosal immune system, and the presence of anti-*Acanthamoeba* IgA antibody in the tears might be an effective immune mechanism for preventing *Acanthamoeba* keratitis. In vitro and in vivo studies in animal models of *Acanthamoeba* keratitis have clearly established that corneal infection requires trophozoite binding to the corneal epithelium. With this in mind, we entertained the hypothesis that activation of the common mucosal immune system and generation of anti-*Acanthamoeba* IgA antibodies in the tears would protect against *Acanthamoeba* keratitis.

2. ROLE OF ANTI-*ACANTHAMOEBA* IgA ANTIBODY IN EXPERIMENTAL *ACANTHAMOEBA* KERATITIS

Experimental *Acanthamoeba* keratitis in domestic pigs and Chinese hamsters fails to induce detectable IgG antibody or DTH responses to *Acanthamoeba* antigens.[7] The apparent failure of corneal infections to elicit either humoral or cell-mediated immunity explains the susceptibility of pigs and Chinese hamsters to repeated corneal infections with *Acanthamoeba* trophozoites.[5,6] Although the normal human population possesses circulating antibodies that react with *Acanthamoeba* antigens, little evidence suggests that serum IgG antibodies affect the course of *Acanthamoeba* keratitis. In both the pig and Chinese hamster, high titers of IgG antibodies and DTH responses against *Acanthamoeba* antigens can be induced by intramuscular (IM) immunization, yet the presence of circulating anti-*Acanthamoeba* IgG antibodies and DTH fails to affect the incidence, severity, chronicity, or clinical features of *Acanthamoeba* keratitis.[1,5–7]

The ocular mucosal epithelium is regularly exposed to the external environment, but is protected against pathogens by a variety of humoral components present in the tears, including lactoferrin, lysozyme, and lactoperoxidase. The tears are also fortified with IgA, the most abundant mammalian immunoglobulin.[8,9] Immunization through mucosal surfaces, such as the gastrointestinal tract, is an effective method for inducing the production of secretory IgA antibody in multiple mucosal secretions, including the tears.[6] Accordingly, we tested the hypothesis that oral immunization with *Acanthamoeba* antigens conjugated with a potent mucosal adjuvant (neutralized cholera toxin; CT) would elicit anti-*Acanthamoeba* IgA antibodies in the tears and protection against experimental corneal infection with *Acanthamoeba* trophozoites. The results demonstrate that oral immunization produces strong resistance to subsequent corneal challenge with *Acanthamoeba* trophozoites in both the pig and Chinese hamster (Fig. 1).

By contrast, IM immunization failed to provide any demonstrable resistance to corneal infection. Tears from orally immunized Chinese hamsters were found to contain high titers of anti-*Acanthamoeba* IgA antibody as detected by ELISA (data not shown). The anti-*Acanthamoeba* IgA antibody in orally immunized hamsters strongly inhibited the binding of trophozoites to corneal epithelial cells in vitro (Fig. 2). To further establish the importance of anti-*Acanthamoeba* IgA antibody in the protection against corneal infection, we produced a hamster anti-*Acanthamoeba* IgA monoclonal antibody. Passive transfer of monoclonal anti-*Acanthamoeba* IgA produced remarkable protection against corneal challenge with trophozoite-laden contact lenses applied to Chinese hamsters (Fig. 3).

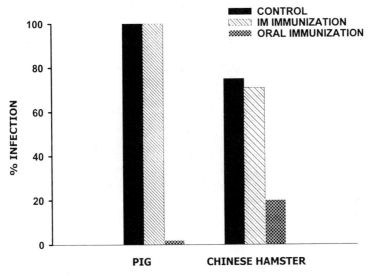

Figure 1. Oral immunization protects pigs and Chinese hamsters against *Acanthamoeba* keratitis. IM = intramuscular immunization with *Acanthamoeba* antigens.

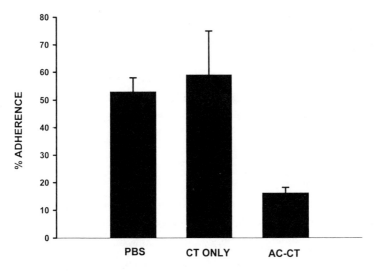

Figure 2. Mucosal extracts from orally immunized Chinese hamsters inhibit the binding of *A. castellanii* trophozoites to Chinese hamster corneal epithelium in vitro. Trophozoites were treated with either phosphate buffered saline (PBS) or intestinal mucosal extracts from Chinese hamsters immunized orally with cholera toxin (CT) or *A. castellanii* antigens conjugated to CT (AC–CT).

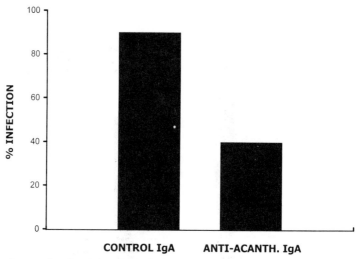

Figure 3. Passive transfer of anti-*Acanthamoeba* IgA antibody protects against *Acanthamoeba* keratitis in Chinese hamsters.

However, the monoclonal anti-*Acanthamoeba* IgA antibody was not directly cytotoxic to trophozoites in vitro, either in the presence or absence of complement (data not shown).

Collectively, the results indicate that the presence of mucosal IgA antibody directed against *Acanthamoeba* antigens produces strong resistance to *Acanthamoeba* keratitis in animal models. Protection correlates with the antibody's capacity to inhibit binding of trophozoites to the corneal epithelium and is not due to direct cytotoxicity.

3. ANTI-*ACANTHAMOEBA* IGG AND IGA ANTIBODIES IN *ACANTHAMOEBA* KERATITIS PATIENTS AND ASYMPTOMATIC INDIVIDUALS

The animal studies summarized above indicated that anti-*Acanthamoeba* IgG antibody in the serum was ineffectual in protecting against *Acanthamoeba* keratitis. However, anti-*Acanthamoeba* IgA in the tears provided strong resistance to corneal infection with trophozoites. Accordingly, we considered the hypothesis that the normal human population is regularly exposed to *Acanthamoeba* antigens and as a result, produces anti-*Acanthamoeba* IgG antibodies in the serum and anti-*Acanthamoeba* IgA antibodies in the tears. The presence of anti-*Acanthamoeba* IgA antibodies in the tears might explain the conundrum of *Acanthamoeba* keratitis. That is, *Acanthamoeba* spp. are found in virtually every environment, including contact lens cases, and the most common risk factor for this disease, contact lens wear, is practiced by over 25 million individuals in the United States. The frequent exposure to *Acanthamoeba* antigens via mucosal surfaces (e.g., nasopharyngeal surfaces) might activate the common mucosal immune system resulting in the appearance of anti-*Acanthamoeba* IgA antibody in the tears and thus, protection against corneal infection. The present study considered this hypothesis.

Sera from normal subjects (N = 25) and *Acanthamoeba* keratitis patients (N = 23) were tested by ELISA for the presence and titer of anti-*Acanthamoeba* IgG antibodies. Tears from 15 normal donors and 15 *Acanthamoeba* keratitis patients were also tested by ELISA for anti-*Acanthamoeba* IgA antibodies. *Acanthamoeba* spp. have been categorized into three distinct subsets based on morphology, isoenzyme analysis, and antigenic phenotype.[10] Accordingly, ELISA's were performed using antigens extracted from *Acanthamoeba* spp. representing each of the three aforementioned subsets. The serum samples from all of the patients and asymptomatic control subjects demonstrated anti-*Acanthamoeba* IgG antibodies against all three antigenic phenotypes (Table 1). However, the IgG antibody titers of the *Acanthamoeba* keratitis patients were significantly higher than the normal control group. By contrast, the anti-*Acanthamoeba* IgA tear antibody titers of the patients were significantly lower than the normal control group.

Table 1. Anti-*Acanthamoeba* IgG and IgA antibodies in *Acanthamoeba* keratitis patients and normal subjects

Group	Anti-*Acanthamoeba* IgG[a]	Anti-*Acanthamoeba* IgA
Normal	2.08 ± 0.65	0.74 ± 0.28
Patients[b]	3.08 ± 0.57	0.25 ± 0.13
Hybridoma Control	0.36 ± 0.3	0.0 ± 0.0

[a]Mean optical density value ± SEM for responses to *A. castellanii*. Similar results were obtained with *A. culbertsoni* and *A. astronyxis* antigens. [b]$P < 0.05$ compared to Normal and Hybridoma control groups.

4. CONCLUSIONS

The results of both the animal and human studies demonstrate that the presence of anti-*Acanthamoeba* IgG antibody does not correlate with protection against corneal *Acanthamoeba* infections. The prospective studies from the pig and Chinese hamster models unequivocally established that anti-*Acanthamoeba* IgA antibody in the tears confers protection against *Acanthamoeba* keratitis and that protection is mediated by inhibiting the binding of the trophozoites to the corneal epithelial surface. The significant reduction in anti-*Acanthamoeba* IgA in the tears of patients is consistent with the results from the animal models and suggests that a deficit in anti-*Acanthamoeba* IgA in the tears, like contact lens wear, may be an important risk factor for the development of *Acanthamoeba* keratitis.

ACKNOWLEDGMENTS

This work was supported in part by NIH grant EY09756 and T32-AIO7520 and an unrestricted grant from Research to Prevent Blindness, Inc., New York, NY.

REFERENCES

1.   H. Alizadeh, J.Y. Niederkorn, and J.P. McCulley, Acanthamoeba keratitis, in: Ocular Infection and Immunity, J.S. Pepose, G.N. Holland, and K.R. Wilhelmus, eds., Mosby, St. Louis, (1996).

2.   M.B. Moore, Acanthamoeba keratitis. Arch Ophthalmol.106:1181 (1988).

3.   L. Cerva, Acanthamoeba culbertsoni and Naegleria fowleri. Occurrence of antibodies in man. J Hyg Epidemiol Microbiol Immunol. 33: 99 (1989).

4.   R.T.M. Cursons, T.J. Brown, E.A. Keys, K.M. Moriarty, and D. Toll, Immunity to pathogenic free-living amoeba: role of humoral antibody. Infect Immun. 29:401 (1980).

5..   H. Alizadeh., Y-G. He, J.P. McCulley, D. Ma, G.L. Stewart, M. Via, E. Haehling, and J.Y. Niederkorn, Successful immunization against Acanthamoeba keratitis in a pig model. Cornea. 14:180 (1995).

6.   H.F. Leher, H. Alizadeh, W.M. Taylor, A.S. Shea., R.S. Silvany, F. van Klink, M.J. Jager, and J.Y. Niederkorn, Role of mucosal IgA in the resistance to Acanthamoeba keratitis. Invest Ophthalmol Vis Sci. 39:2666 (1998).

7.   F. van Klink, H. Leher, M.J. Jager, H. Alizadeh, W. Taylor, and J.Y. Niederkorn, Systemic immune response to Acanthamoeba keratitis in the Chinese hamster. Ocular Immunol. and Inflam. 5:235 (1997).

8.   R.J. Fullard and C. Snyder, Protein levels in nonstimulated and stimulated tears of normal human subjects. Invest Ophthalmol Vis Sci. 31:1119 (1990).

9.   M.B. Mazanec, J.G. Nedrud, C.S. Kaetzel, and M.E. Lamm, A three-tiered view of the role of IgA in mucosal defense. Immunol Today. 14:430 (1993).

10.   G.S. Visvesvara, Classification of Acanthamoeba. Rev Infect Dis. 13 (Suppl. 5):S369 (1991).

EXPRESSION OF VASCULAR ENDOTHELIAL GROWTH FACTOR RECEPTOR-3 (VEGFR-3) IN THE CONJUNCTIVA–A POTENTIAL LINK BETWEEN LYMPHANGIOGENESIS AND LEUKOCYTE TRAFFICKING ON THE OCULAR SURFACE

Pedram Hamrah,[1] Qiang Zhang,[1] and M. Reza Dana[1,2]

[1]Laboratory of Immunology
Schepens Eye Research Institute and the
Department of Ophthalmology
Harvard Medical School
Boston, Massachusetts, USA
[2]Cornea Service
Massachusetts Eye & Ear Infirmary and
Brigham and Women's Hospital
Boston, Massachusetts, USA

1. INTRODUCTION

Although the lymphatic system penetrates most tissues as a dense network, it has received relatively little attention outside the fields of immunology and vascular biology. The normal cornea is devoid of blood and lymphatic vessels, but a number of corneal stimuli such as contact lens wear, infection, surgical invasion or burns can induce penetration of new vessels from the limbus and conjunctiva into the cornea. The process of growth and development of new blood vessels from preexisting ones is termed angiogenesis, and in the case of lymphatic vessels, lymphangiogenesis.

Lymphatic vessels are present in vascularized corneas.   [1-4] Unlike blood vessels, lymphatics do not form a continuous circuit, but rather provide a one-way channel from the tissue (e.g. ocular surface) to the draining lymph nodes. The induction of lymphatic vessels into the cornea facilitates high-volume delivery of antigens and antigen presenting cells (APCs) to T cell reservoirs such as lymph nodes, a process which is hindered under normal

Lacrimal Gland, Tear Film, and Dry Eye Syndromes 3
Edited by D. Sullivan *et al.*, Kluwer Academic/Plenum Publishers, 2002

851

circumstances, but once it is induced, it can contribute to induction of immunogenic inflammation, such as rejection of corneal grafts.[1, 5, 6] The critical relevance of the eye-lymphatic axis in induction of corneal immunity has been demonstrated recently in a study by Yamagami and Dana[7] where hosts of corneal allografts without draining lymph nodes enjoy universal and indefinite survival.

Vascular endothelial growth factor (VEGF) is a secreted polypeptide that was initially identified in tumor cell supernatants by its ability to increase permeability of the vasculature.[8] In addition, it also stimulates endothelial cell migration and promotes survival of the newly formed vessels.[9] VEGF belongs to the platelet-derived growth factor (PDGF)/VEGF family, which is a potent inducer of angiogenesis.[9, 10] The VEGF receptor family consists of 3 members: VEGFR-1 (Flt-1), VEGFR-2 (Flk-1/KDR) and VEGFR-3 (Flt-4). At least 5 ligands (VEGF-A, VEGF-B, VEGF-C, VEGF-D, and VEGF-E) and PDGF bind to 1 or 2 of these receptors (Table 1). VEGFR-1 (FLT-1) and VEGFR-2 (FLK-1/KDR) are receptor tyrosine kinases for VEGF-A.[9] These receptors are largely expressed on vascular endothelial cells. VEGF-C and VEGF-D do not bind to VEGFR-1, although both are ligands for VEGFR-2 and VEGFR-3. Both VEGF-C and VEGF-D primarily affect development of lymphatic vasculature through activation of VEGFR-3, but may also participate in angiogenesis through VEGFR-2 (Table 1).[11–15] VEGF-C elicits a lymphangiogenic response in the chicken embryo chorioallantoic membrane, and transgenic mice overexpressing VEGF-C in the skin are characterized by specific hyperplasia of the lymphatic network, revealing VEGF-C as the first known growth factor for lymphatic endothelium.[16, 17] It can also bind to VEGFR-2 expressed in blood vessel endothelia, and induce capillary endothelial cell migration in culture.[11, 18]

VEGFR-3 is a receptor tyrosine kinase that is similar to the two other VEGF receptors in structure, but does not bind VEGF-A[19] or VEGF-B.[20] VEGFR-3 is initially expressed in all embryonic endothelia, but its expression in the blood vessel endothelium decreases during development, and it becomes largely restricted to the lymphatic endothelium in adult tissues.[21] In the eye, VEGFR-3 has been detected in the human retina.[22] Moreover, VEGFR-3 mRNA was recently found in murine conjunctiva,[23] but the source is unknown. The purpose of this study was to elucidate the expression of VEGFR-3 in the conjunctiva and its possible role in lymphangiogenesis, using highly sensitive confocal microscopy techniques.

2. MATERIALS AND METHODS

2.1. Experimental Animals

Seven to 14-week-old male BALB/c mice (Taconic Farms, Germantown, NY) were used in these experiments. All animals were treated according to the ARVO Statement for the Use of Animals in Ophthalmic and Vision Research.

2.2. Induction of Corneal Neovascularization (NV)

Two weeks before tissue procurement, animals were deeply anesthetized with an intraperitoneal injection of 3 to 4 mg ketamine and 0.1 mg xylazine, and were placed under the operating microscope. Intrastromal sutures were placed in the central cornea of one eye of each mouse to induce inflammatory corneal NV, as described previously.[24] Briefly three pairs of 11-0 nylon sutures (Vanguard, Houston, TX) were placed equidistant from the limbus through the central stroma without perforation of the cornea. Immediately after surgery erythromycin ophthalmic ointment was applied to the eyes.

2.3. Antibodies

The FLT-4 and CD31 antibody (Santa Cruz Biotechnology, Santa Cruz, CA), and the other primary antibodies (all from PharMingen, San Diego, CA), and their specificities are summarized in Table 2. Secondary antibodies were Cy5-conjugated goat anti-hamster IgG (PharMingen) and rhodamine-conjugated goat anti-rat IgG (Santa Cruz Biotechnology). Isotype control antibodies were used as negative controls.

Table 1. The currently known VEGFs and their receptors. While all ligands are capable of inducing angiogenesis, only VEGF-C and VEGF-D induce lymphangiogenesis

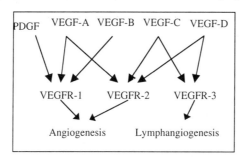

Table 2. Antibodies used for labeling

Primary antibody	Specificity
Flt-4	VEGFR-3
IA^d-FITC[1]	BALB/c mouse class II MHC antigen
CD11c	Integrin DC[2]/LC[3] marker
CD11B-FITC	MAC-1,Integrin monocyte/macrophage marker
CD45	Pan-leukocyte marker and
CD80-FITC	B7.1, costimulatory molecule
CD31	Pan-endothelial marker
GR-1-FITC	Neutrophil marker
CD3e-FITC	T lymphocyte marker

[1]FITC = Fluorescein isothiocyanate
[2]DC = Dendritic cell
[3]LC = Langerhans cell

2.4. Immunohistochemical Studies

Conjunctivae of normal and inflamed eyes were excised from BALB/c mice. The procured tissues were fixed in acetone for 15 min at room temperature (RT). To block nonspecific staining, whole-mounts were incubated in 2% bovine serum albumin (BSA) diluted in PBS (PBS-BSA) for 15 minutes prior to the addition of primary and secondary antibodies. Purified anti-VEGFR-3 antibody, other primary antibodies, or isotype-matched control antibodies were applied to the samples for 2 h, followed by 60 min incubation of the secondary antibodies that were diluted for optimal concentration in PBS-BSA. All staining procedures were performed at room temperature, and each step was followed by three thorough washings in PBS for 5 minutes each. Finally, whole-mounts were covered with mounting medium (Vector, Burlingame, CA) and examined by a confocal microscope (Leica TCS 4D, Lasertechnik, Heidelberg, Germany). Samples were examined at 160 X and 400 X magnification using a X10 eyepiece and X16 or X40 objective lens. Corneas of BALB/c mice were excised and stained for VEGFR-3 alone as described above. In addition, conjunctiva of human eyes obtained from the Florida Eyebank were obtained and stained for VEGFR-3. At least 3 different samples were examined per each double staining experiment. Five to eight different fields were analyzed for each specimen using a grid and the numbers were averaged.

The Student's t-test was used to compare the number of cells with specific surface markers in different parts of the cornea. P-values < 0.05 was considered significant.

3. RESULTS

To characterize the lymphatic vessels of the normal murine conjunctiva, samples were double-stained with anti-VEGFR-3 and anti-CD31. CD31, a marker for all endothelial cells, stained all blood and lymphatic vessels in the conjunctiva. The cornea was avascular as expected. VEGFR-3, a lymphatic marker, stained some CD31$^+$ cells. According to previously published data,[25] VEGFR-3$^+$, CD31$^+$ vascular structures are defined as lymphatic vessels (Fig. 1). Newly sprouting vascular structures were detected in the periphery of inflamed corneas by VEGFR-3 staining (Fig. 1). These VEGFR-3$^+$ vessels are, by definition, lymphatics. Surprisingly, however, VEGFR-3$^+$ non-endothelial cells were detected in addition to the lymphatics, both in the peripheral cornea and conjunctiva of inflamed eyes (Figs. 2A,B), and in the conjunctiva of normal eyes. In the inflamed conjunctiva, where the number of VEGFR-3$^+$ non-endothelial cells was significantly higher, these cells seemed to change their shape depending on their microanatomical localization (Fig. 3). In the conjunctiva the shape was round, but the corneal cells appeared highly dendritic.

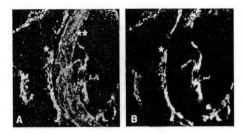

Figure 1. Immunostaining of the normal conjunctiva for CD31 and VEGFR-3. (**A**) All endothelial cells express CD31 by immunstaining. (**B**) Some CD31+ vessels express VEGFR-3. Vessels that express both are defined as lymphatics. * = Lymphatic vessel; ** =Blood Vessel. 1000 X magnification.

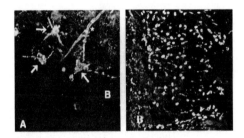

Figure 2. Immunostaining of inflamed, neovascularized cornea. (**A**) 400 X magnification. Sprouting VEGFR-3+ lymphatic vessel (*) and VEGFR-3+ non-endothelial cells (white arrows) in the corneal periphery. (**B**) 1000 X magnification. Large numbers of VEGFR-3+ non-endothelial cells in the inflamed cornea.

To elucidate the phenotype and density of the VEGFR-3⁺ cells, double staining of samples was performed with anti-VEGFR-3 in combination with other antibodies. When corneal tissue was double-stained for VEGFR-3 and CD45, a pan-leukocyte marker, 100% of the VEGFR-3⁺ cells were also CD45⁺, revealing their bone marrow origin. The VEGFR-3⁺ cells comprised 62% of all bone marrow-derived cells in the normal conjunctiva (Fig. 4A and B). In order to delineate the lineage of the VEGRF-3+, bone marrow-derived cells, tissues were double-stained for VEGFR-3 and either CD11b, a monocyte/macrophage marker); CD11c, a dendritic cell (DC)/Langerhans cell (LC) marker; CD3, a T-cell marker; and GR-1, a neutrophil marker. One hundred percent of the VEGFR-3⁺ cells also stained for CD11b, indicating that they are from a monocytic lineage. However, none of these cells were positive for CD11c, GR-1 or CD3, confirming that they are not DC/LC, neutrophils, or T-cells.

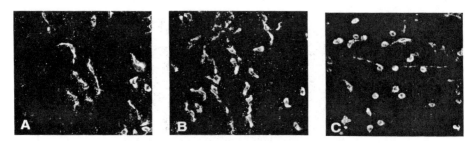

Figure 3. Differential shapes of VEGFR-3⁺ cells in the inflamed eye. (A) Cells in the corneal periphery are more elongated or dendritiform. (B) In the limbal region, cells are oval shaped. (C) In the conjunctiva, cells are round. Magnification 400X.

Almost all CD11b⁺ cells were also positive for VEGFR-3, indicating that all monocytic-derived cells in the conjunctiva are VEGFR-3⁺. Finally, samples were double stained for VEGFR-3 and MHC class II or CD80 (B7.1) to characterize the state of maturation of these monocytic cells. Fifty-four percent of the VEGFR-3⁺ cells were also positive for MHC class II (Fig. 4C, D), and no staining was observed for CD80. Moreover, all samples stained with isotype controls were negative. The results are summarized in Table 3.

Table 3. **Phenotype of VEGFR-3⁺ cells in the conjunctiva**

Antibody	Expression	Percentage
CD11c	--	0%
CD11b	+++	100 %
CD45	+++	100 %
MHC class II	++	54 %
CD80 (B7.1)	--	0%
CD3	--	0%
GR-1	--	0%

4. DISCUSSION

The present study indicates that VEGFR-3 is not only expressed by the lymphatic vasculature as described previously,[21, 25–27] but is also expressed on almost all monocytic-derived cells in the conjunctiva. Using an immunofluorescence double staining technique applied to confocal microscopy, we show that a subpopulation of the vessels positive for the pan-endothelial marker CD31, is identified by anti-VEGFR-3 labeling. VEGF-C and VEGF-D induce endothelial cell migration [11, 14, 15, 18, 28, 29] through VEGFR-3. VEGF-C is upregulated by the proinflammatory cytokines interleukin-1β and tumor necrosis factor-α, and contains an NF-κB binding site in its gene.[30] Proinflammatory cytokines may regulate the lymphatic vessels indirectly via VEGF-C expression. This would explain the

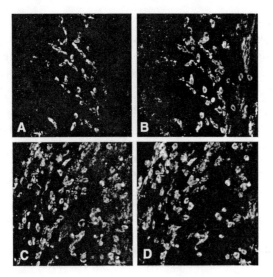

Figure 4. Phenotype of VEGFR-3⁺ cells in the conjunctiva. (**A**) VEGFR-3⁺ controls for B. (**B**) All VEGFR-3⁺ cells also express CD45 and CD11b, indicating that these cells are bone marrow-derived with a monocytic lineage. (**C**) VEGFR-3⁺ controls for D. (**D**) Double labeling of cells for VEGFR-3⁺ and MHC class II demonstrates class II expression by 54% of these cells. These cells did not stain for CD11c, CD3 and GR-1. Magnification 400 X.

sprouting VEGFR-3⁺ vessels observed after the induction of inflammatory neovascularization.

An unexpected finding in this study was the detection of VEGFR-3 on non-endothelial cells of the ocular surface. Wilting et al.[27] have recently described the expression of VEGFR-3 on non-endothelial cells in quail embryos. They also detected VEGFR-3 expression in the developing cornea and podocytes of kidney glomeruli in 4-day-old embryos. Given that these cells collectively exhibit identical expression of multiple cell surface markers, including CD45⁺, CD11b⁺, CD11c ⁻, CD3 ⁻, CD80⁻, CD3 ⁻, with the notable exception of MHC class II antigens, we believe that they are derived from monocytic-derived cells. Negative expression for CD3 and CD11c and GR-1 confirms the absence of T-cells, dendritic cells and neutrophils. Cells `from a monocytic lineage can, however, differentiate into both macrophages and dendritic or Langerhans cells.[31, 32] In the latter case, CD11c is upregulated. Since both ligands of VEGFR-3 induce migration of endothelial cells [11, 14, 15, 18, 28, 29] and Kaposi sarcoma cells, which also express VEGFR-3,[33] it is therefore intriguing to speculate that these ligands may also be involved in the migration of bone marrow-derived cells in the ocular surface. Supporting this hypothesis is the fact that VEGF-A is chemotactic for monocytes, which also express VEGFR-1.[34-36] The function of VEGFR-3 on these monocytic-derived cells is unclear.

Mimura et al.[23] have recently reported upregulation of both VEGF-C and VEGFR-3 gene expression in inflamed corneas. The sensitive confocal microscopy technique we used here revealed that most of the VEGFR-3 expression does not relate to lymphatic endothelial expression, as was suggested by Mimura, but rather, it relates to the expression on monocytic-derived cells. In the normal uniflamed eye, VEGFR-3 could play a role in the trafficking of lymphatic endothelial cells on the ocular surface when these cells secrete VEGF-C. In the inflamed eye, when VEGF-C is additionally secreted by inflammatory cells, VEGF-C can induce the migration of the monocytic-derived cells into the cornea. Finally, it is possible that VEGFR-3 might function as a "trap" or functional "decoy" receptor by binding VEGF-C and VEGF-D growth factors in order to prevent lymphangiogenesis and angiogenesis into the clear cornea. This is an interesting hypothesis that needs to be tested.

REFERENCES

1. H.B. Collin. Corneal lymphatics in alloxan vascularized rabbit eyes. *Invest Ophthalmol Vis Sci*, 5:1 (1966).
2. H.B. Collin. Endothelial cell lined lymphatics in vascularized rabbit cornea. *Invest Ophthalmol Vis Sci*, 5:337 (1966).
3. H.B. Collin. The fine structure of growing corneal lymphatic vessels. *J Pathol*, 104:99 (1971).
4. P.C. Burger, and G.K. Klintworth. Autoradiographic study of corneal neovascularization induced by chemical cautery. *Lab Invest*, 45:328 (1981).
5. H.B. Collin. Lymphatic drainage of ^{131}I-albumin from the vascularized cornea. *Invest Ophthalmol Vis Sci*, 9:146 (1970).
6. M. Fine, and M. Stein. The role of corneal vascularization in human graft rejection. *Corneal graft failure ciba foundation symposium*. Amsterdam, Holland: Assoc. Scientific Publishers; 1973:193.
7. S. Yamagami, and M.R. Dana. The critical role of draining lymph nodes in corneal allosensitization. *Invest Ophthalmol Vis Sci*, 42:1293 (2001).
8. D.R. Senger, S.S. Galli, A.M. Dvorak, C.A. Perruzzi, V.S. Harvey, and H.F. Dvorak. Tumor cells secrete a vascular permeability factor that promotes accumulation of ascites fluid. *Science*, 219:983 (1983).
9. N. Ferrera, and T. Davis-Smyth. The biology of vascular endothelial growth factor. *Endocrine Rev*, 18:4 (1997).
10. H.F. Dvorak, L.F. Brown, M. Detmar, and A.M. Dvorak. Vascular permeability factor/vascular endothelial growth factor, microvascular permeability, and angiogenesis. *Am J Pathol*, 146:1029 (1995).
11. V. Joukov, K. Pajusola, A. Kaipainen, D. Chilov, I. Lahtinen, E. Kukk, O. Saksela, N. Kalkkinen, and K. Alitalo. A novel vascular endothelial growth factor, VEGF-C, is a ligand for the FLT4 (VEGFR-3) and KDR (VEGFR-2) receptor tyrosine kinases. *EMBO J*, 15:290 (1996).
12. E. Kukk, A. Lymboussaki, S. Taira, A. Kaipainen, M. Jeltsch, V. Joukov, and K. Alitalo. VEGF-C receptor binding and pattern of expression with VEGFR-3 suggests a role in lymphatic vascular development. *Development*, 122:3829 (1996).
13. M.G. Achen, M. Jeltsch, E. Kukk, T. Mäkinen, A. Vitali, A.F. Wilks, K. Alitalo, and S.A. Stacker. Vascular endothelial growth factor D (VEGF-D) is a ligand for the tyrosine kinases VEGF receptor 2 (FLK1) and VEGF receptor 3 (FLT4). *Proc Natl Acad Sci USA*, 95:548 (1998).
14. J. Lee, A. Gray, J. Yuan, S.-M. Luoh, H. Avraham, and W.I. Wood. Vascular endothelial growth factor-related protein: A ligand and specific activator of the tyrosine kinase receptor Flt4. *Proc Natl Acad Sci USA*, 93:1988 (1996).
15. Y. Cao, P. Linden, J. Farnebo, R. Cao, A. Eriksson, V. Kumar, J.-H. Qi, L. Claesson-Welsh, and K. Alitalo. Vascular endothelial growth factor C induces angiogenesis *in vivo*. *Proc Natl Acad Sci USA*, 95:14389 (1998).

16. M. Jeltsch, A. Kaipainen, V. Joukov, X. Meng, M. Lakso, H. Rauvala, M. Swartz, D. Fukumura, R.K. Jain, and K. Alitalo. Hyperplasia of lymphatic vessels in VEGF-C transgenic mice. *Science*, 276:1423 (1997).

17. S.J. Oh, M.M. Jeltsch, R. Birkenhäger, J.E.G. McCarthy, H.A. Weich, B. Christ, K. Alitalo, and J. Wilting. VEGF and VEGF-C: Specific induction of angiogenesis and lymphangiogenesis in the differentiated avian chorioallantoic membrane. *Dev Biol*, 188:96 (1997).

18. V. Joukov, T. Sorsa, V. Kumar, M. Jeltsch, L. Claesson-Welsh, Y. Cao, O. Saksela, N. Kalkkinen, and K. Alitalo. Proteolytic processing regulates receptor specificity and activity of VEGF-C. *EMBO J*, 16:3898 (1997).

19. K. Pajusola, O. Aprelikova, G. Pelicci, H. Weich, L. Claesson-Welsh, and K. Alitalo. Signalling properties of Flt4, a proteolytically processed receptor tyrosine kinase related to two VEGF receptors. *Oncogene*, 9:3545 (1994).

20. B. Olofsson, M. Jeltsch, U. Eriksson, and K. Alitalo. Current biology of VEGF-B and VEGF-C. *Curr Opin Biotechnol*, 10:528 (1999).

21. A. Kaipainen, J. Korhonen, T. Mustonen, V.W.M. van Hinsbergh, G.H. Fang, D.J. Dumont, M. Breitman, and K. Alitalo. Expression of the fms-like tyrosine kinase 4 gene becomes restricted to lymphatic endothelium during development. *Proc Natl Acad Sci USA*, 92:3566 (1995).

22. G. Smith, D. McLeod, D. Foreman, and M. Boulton. Immunolocalisation of the VEGF receptors Flt-1, KDR, and Flt-4 in diabetic retinopathy. *Br J Ophthalmol*, 83:486 (1999).

23. T. Mimura, S. Amano, T. Usui, Y. Kaji, T. Oshika, and Y. Ishii. Expression of vascular endothelial growth factor C and vascular endothelial growth factor receptor 3 in corneal lymphangiogenesis. *Exp Eye Res*, 72:71 (2001).

24. M.R. Dana, and J.W. Streilein. Loss and restoration of immune privilege in eyes with corneal neovascularization. *Invest Ophthalmol Vis Sci*, 37:2485 (1996).

25. A. Lymboussaki, T.A. Partanen, B. Olofsson, J. Thomas-Crusells, C.D.M. Fletcher, R.M.W. de Waal, A. Kaipainen, and K. Alitalo. Expression of the vascular endothelial growth factor C receptor VEGFR-3 in lymphatic endothelium of the skin and in vascular tumors. *Am J Pathol*, 153:395 (1998).

26. L. Jussila, R. Valtola, T.A. Partanen, P. Salven, P. Heikkilä, M.T. Matikainen, R. Renkonen, A. Kaipainen, M. Detmar, E. Tschachler, R. Alitalo, and K. Alitalo. Lymphatic endothelium and kaposi's sarcoma spindle cells detected by antibodies against the vascular endothelial growth factor receptor-3. *Cancer Res*, 58:1599 (1998).

27. J. Wilting, A. Eichmann, and B. Christ. Expression of the avian VEGF receptor homologues *quek1* and *quek2* in blood-vascular and lymphatic endothelial and non-endothelial cells during quail embryonic development. *Cell Tiss Res*, 288:207 (1997).

28. L. Marconcini, S. Marchio, L. Morbidelli, E. Cartocci, A. Albini, M. Ziche, F. Bussolino, and S. Oliviero. C-*fos*-induced growth factor / vascular endothelial growth factor D induces angiogenesis *in vivo* and *in vitro*. *Proc Natl Acad Sci USA*, 96:9671 (1999).

29. B. Witzenbichler, T. Asahara, T. Murohara, M. Silver, I. Spyridopoulos, M. Magner, N. Principe, M. Kearney, J.S. Hu, and J.M. Isner. Vascular endothelial growth factor-c (vegf-c/vegf-2) promotes angiogenesis in the setting of tissue ischemia. *Am J Pathol*, 153:381 (1998).

30. D. Chilov, E. Kukk, S. Taira, M. Jeltsch, J. Kaukonen, A. Palotie, V. Joukov, and K. Alitalo. Genomic organization of human and mouse genes for vascular endothelial growth factor C. *J Biol Chem*, 272:25176 (1997).

31. C. Caux, B. Vanbervliet, C. Massacrier, C. Dezutter-Dambuyant, B. de Saint-Vis, C. Jacques, K. Yoneda, S. Imamura, D. Schmitz, and J. Banchereau. CD34+ hematopoietic progenitors from human cord blood differentiate along two independant dendritic cell pathways in response to GM-CSF + TNFα. *J Exp Med*, 184:695 (1996).

32. C. Caux, C. Massacrier, B. Vanbervliet, B. Dubois, I. Durand, M. Cella, A. Lanzavecchia, and J. Banchereau. CD34+ hematopoietic progenitors from human cord blood differentiate along two independant dendritic cell pathways in response to granulocyte-macrophage colony-stimulating factor plus tumor necrosis factor α: II. Functional analysis. *Blood*, 90:1458 (1997).

33. S. Marchiò, L. Prima, M. Pagano, G. Palestro, A. Albini, T. Veikkola, I. Cascone, K. Alitalo, and F. Bussolino. Vascular endothelial growth factor-C stimulates the migration and proliferation of Kaposi's sarcoma cells. *J Biol Chem*, 274:27617 (1999).

34. M. Clauss, M. Gerlach, H. Gerlach, J. Brett, F. Wang, P.C. Familetti, Y.C.E. Pan, J.V. Olander, D.T. Connolly, and D. Stern. Vascular permeability factor: A tumor-derived polypeptide that induces endothelial cell and monocyte procoagulant activity, and promotes monocyte migration. *J Exp Med*, 172:1535 (1990).

35. H. Shen, M. Clauss, J. Ryan, A.M. Schmidt, P. Tijburg, L. Borden, D. Connolly, D. Stern, and J. Kao. Characterization of vascular permeability factor / vascular endothelial growth factor receptors on mononuclear phagocytes. *Blood*, 81:2767 (1993).

36. M. Clauss, H. Weich, G. Breier, U. Knies, W. Röckl, J. Waltenberger, and W. Risau. The vascular endothelial growth factor receptor Flt-1 mediates biological activities. *J Biol Chem*, 271:17629 (1996).

HUMAN LACRIMAL DRAINAGE-ASSOCIATED LYMPHOID TISSUE (LDALT) BELONGS TO THE COMMON MUCOSAL IMMUNE SYSTEM

Erich Knop and Nadja Knop

Department for Cell Biology in Anatomy
Hannover Medical School
Hannover, Germany

1. INTRODUCTION

The lacrimal drainage system constitutes the part of the ocular adnexa where the tear flow drains after production in the lacrimal gland and subsequent to bathing the ocular surface. It is lined by a mucosa that continues that of the conjunctiva. The lacrimal drainage system begins at the two lacrimal puncta of the lid margin that lead via the lacrimal canaliculi into the cavernous lacrimal sac and via the nasolacrimal duct into the nose (Fig. 1).

The tear flow enables the tissue of the lacrimal drainage system to share soluble factors with the upstream ocular surface and also with the lacrimal gland. These factors can be of a protective nature (e.g. IgA, antimicrobial peptides and proteins, growth factors etc.) but can also be potentially harmful as inflammatory cytokines or as the antigenic materials and microbes that reach the ocular surface via the open palpebral fissure.

Therefore, the lacrimal drainage system also shows evidence of a specific

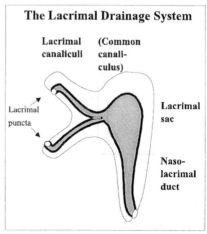

Figure 1. Anatomy of the lacrimal drainage system showing the different regions.

Lacrimal Gland, Tear Film, and Dry Eye Syndromes 3
Edited by D. Sullivan *et al.*, Kluwer Academic/Plenum Publishers, 2002

861

immune protection by a mucosa-associated lymphoid tissue (MALT)[1-4] as reported by us previously.[5-8] It was identified in studies on the upstream conjunctiva-associated lymphoid tissue (CALT)[8] that it continues and was accordingly later termed lacrimal drainage-associated lymphoid tissue (LDALT).[9]

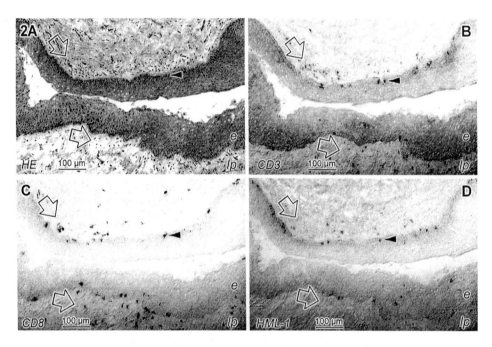

Figure 2. Lymphoid cells (**A**) in the lamina propria (open arrows, lp) and basal epithelial layers (arrowheads, e) consist mainly of CD3 positive T-cells (**B**), CD8-positive cells are frequent (**C**), HML-1-positive cells (**D**) always occur. Plasma cells are rare in this initial part of the lacrimal canaliculi but increase towards the sac.

Although the presence of lymphoid cells in the lacrimal drainage system had been reported for a long time,[9] no detailed description and no conclusive analysis of their functional significance had been given. The lymphoid cells and, in particular, their follicular accumulations which had been occasionally described,[10-12] were usually considered as pathological. Meanwhile our results on the presence of LDALT are supported by other studies. Some, focusing on follicles, consider it in a pathological context,[13] others as a *local* immune system.[14] However, due to the technical difficulty to obtain complete preparations of the total lacrimal drainage system, most investigations are restricted to parts of the system such as sac or nasolacrimal duct and hence also lose the continuity with CALT. In order to clarify these points, we aimed here to investigate the presence, distribution, histology and immunohistochemical characterization of lymphoid tissue only in complete lacrimal drainage systems

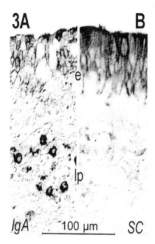

Figure 3. IgA-positive plasma cells are seen in the lamina propria (lp) and IgA (**A**) and SC (**B**) are in the epithelium (e) of the lacrimal drainage system.

2. Material and Methods

Complete normal human lacrimal drainage systems (n = 18) were obtained from cadavers of body donors (average age 81.7 ±16.1 years) at the Department of Anatomy, Hannover Medical School. For histological and immunohistological analysis, the tissue was either fixed in 4% formaldehyde solution and embedded in paraffin or fresh frozen in OCT. Sections of all regions (Fig. 1) were stained with hematoxylin and eosin (HE) or immunohistochemistry. For the latter, sections were rehydrated, incubated with primary antibodies for components of the secretory immune system (IgA, IgM, SC), for lymphocyte subtypes (CD3, CD8, HML-1) or for macrophages (CD68). Primary antibodies were marked with secondary biotinylated ones and visualized with streptavidin-peroxidase and DAB.

3. RESULTS

A narrow layer of diffusely arranged lymphoid cells was observed along the lacrimal canaliculi (Fig. 2), consisting mainly of CD3-positive T-lymphocytes in the lamina propria and in the basal epithelial layers. CD8 positive lymphocytes were frequent and lymphocytes carrying the human mucosa lymphocyte antigen (HML-1) were always seen. Small vessels were located in the lamina propria including occasional specialized high endothelial venules (HEV). Plasma cells were rare in the initial part of the lacrimal canaliculi but increased in their terminal part towards the lacrimal sac. In the common canaliculus, in the lacrimal sac (Fig. 3) and in the nasolacrimal duct a dense zone of diffuse lymphoid tissue was observed, including frequent plasma cells among the lymphocytes. Most plasma cells were positive for IgA (Fig. 3), few for IgM. IgA was also seen in the epithelium, diffusely or as deposits. The IgA transporter molecule SC was strongly positive in the pseudostratified columnar epithelium of all parts of the drainage system; however, it was restricted to patches in the outermost layers of the stratified squamous epithelium in the lacrimal canaliculi. Staining for both IgA and SC was also observed in associated glands and in the drainage lumen.

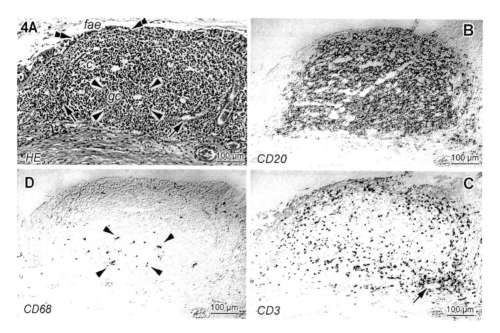

Figure 4. A secondary follicle in the human lacrimal sac with a germinal center (gc, arrowheads) and a denser lymphocyte corona (c) has a flat overlying follicle-associated epithelium (fae, double arrowheads) and is surrounded by numerous parafollicular vessels (arrows) including high endothelial venules (**A**). It is composed of B-cells (**B**) and has parafollicular T-cells, clustering here around one of the vessels (**C, arrow**). In the germinal center, not very obvious in HE-staining, CD68-positive macrophages (arrowheads) are detected (**D**).

Lymphoid follicles (Fig. 4) were observed in 56% of these complete drainage systems and occurred in all regions, rarely also in the terminal canaliculi. They had a roundish shape, an average diameter around 0.5 mm, and were usually composed of accumulations of B-cells (Fig. 4B) and accompanied by parafollicular T-cells. In the parafollicular T-cell areas, HEV were regularly observed, sometimes with lymphocytes clustering around them (Fig. 4C). Secondary follicles with a germinal center were seen. The follicle-associated epithelium had a flat cell shape and frequent intraepithelial lymphocytes. In the germinal center, CD68 positive macrophages were characterized (Fig. 4D) which may indicate phagocytosis of newly formed lymphocytes with unsuitable antigen specificity. Most follicles were primary; sometimes lymphoid accumulations were seen without a flat overlying epithelium.

4. DISCUSSION AND CONCLUSION

LDALT o f the diffuse and follicular type regularly occurs in the normal human lacrimal drainage system similar to the conjunctiva[8] and other organs,[1-4] and contains mucosa specific lymphocytes, further supporting its integration into the common mucosal immune system.

Follicles were found in the majority (56%) of specimens in this study which contained only complete lacrimal drainage systems that were analyzed in all regions. This supports the previous assumption[9] that their amount may easily be underestimated if incomplete tissue is included or a section approach used, and further underlines that follicles represent a normal tissue component here. IgA-positive plasma cells in the lamina propria and secretory component in the epithelium characterize LDALT as a part of the secretory immune system.[2] This represents the efferent limb of immunity and is able to exclude antigens from the mucosa mainly by the production of secretory IgA. Follicles with a germinal center indicate that antigens had been presented here to lymphoid cells resulting in their activation,[3,4] and give evidence that LDALT can also perform an afferent immune function. This may be one of its main functions because antigens that are washed away from the ocular surface have a higher probability of contact with the mucosal immune system due to the slower flow of tears in the drainage system and can hence prime immune responses here that are also relevant for the ocular surface. Antigen transporting M-cells[15] were observed in ultrastructure.[9]

LDALT is physically continuous with CALT and is connected with it by tear flow and by the regulated recirculation of lymphocytes[16] as suggested by the presence of high endothelial venules and discussed in detail in another contribution to this book[17]. We propose that the lacrimal drainage system, conjunctiva, and lacrimal gland form a functional unit for ocular surface immune defense, to be termed "Eye-Associated Lymphoid Tissue" (EALT).

REFERENCES

1. J. Mestecky, J.R. McGhee, S.M. Michalek, R.R. Arnold, S.S. Crago and J.L. Babb. Concept of the local and common mucosal immune response. *Adv.Exp.Med.Biol.* 107:185–192 (1978).
2. J.P. Vaerman. The secretory immune system. *Antibiot. Chemother.* 39:41–50 (1987).
3. J.P. Kraehenbuhl and M.R. Neutra. Molecular and cellular basis of immune protection of mucosal surfaces. *Physiol.Rev.* 72:853–879 (1992).
4. P. Brandtzaeg, A.E. Berstad, I.N. Farstad, G. Haraldsen, L. Helgeland, F.L. Jahnsen, F.E. Johansen, I.B. Natvig, et al. Mucosal immunity-a major adaptive defence mechanism. *Behring.Inst.Mitt.* 98:1–23 (1997).
5. E. Knop and N. Knop. MALT tissue of the conjunctiva and nasolacrimal system in the rabbit and human. *Vision Research* 36:60 (1996).
6. N. Knop and E. Knop. The MALT tissue of the ocular surface is continued inside the lacrimal sac in the rabbit and human. *Invest.Ophthalmol.Vis.Sci.* 38:126 (1997).

7. N. Knop and E. Knop. Mukosa-assoziiertes lymphatisches Gewebe im Tränensack von Mensch und Kaninchen. *Verh.Anat.Ges.(Anat.Anz.Suppl.)* 180:132 (1998).

8. N. Knop and E. Knop. Conjunctiva-associated lymphoid tissue in the human eye. *Invest.Ophthalmol.Vis.Sci.* 41:1270–1279 (2000).

9. E. Knop and N. Knop. Lacrimal drainage associated lymphoid tissue (LDALT): A part of the human mucosal immune system. *Invest.Ophthalmol.Vis.Sci.* 42:566–574 (2001).

10. Fr. Merkel and E. Kallius. Makroskopische Anatomie des Auges. In: Saemisch T, ed. *Graefe-Saemisch Handbuch der gesamten Augenheilkunde,* 2 ed. Leibzig: Verlag W. Engelmann (1910).

11. R. Warwick. *Eugene Wolff's anatomy of the eye and orbit.* London: Lewis (1976).

12. S. Duke-Elder and K.C. Wybar. *The Anatomy of the visual system.* London: Henry Kimpton (1961).

13. F.P. Paulsen, J.I. Paulsen, A.B. Thale and B.N. Tillmann. Mucosa-associated lymphoid tissue in human efferent tear ducts. *Virchows.Arch.* 437:185–189 (2000).

14. P.Sirigu, C.Maxia, R.Puxeddu, I.Zucca, F.Piras and M.T.Perra. The presence of a local immune system in the upper blind and lower part of the human nasolacrimal duct.*Arch.Histol.Cytol* 63:431–439 (2000).

15. A. Gebert, H.J. Rothkotter and R. Pabst. M cells in Peyer's patches of the intestine. *Int.Rev.Cytol.* 167:91–159 (1996).

16. E.C. Butcher and L.J. Picker. Lymphocyte homing and homeostasis. *Science* 272:60–66 (1996).

17. E. Knop and N. Knop. A Functional Unit for Ocular Surface Immune Defense formed by the Lacrimal Gland, Conjunctiva and Lacrimal Drainage System . *Adv. Exp. Med. Biol* (2001) in press.

THE CRYPT SYSTEM OF THE HUMAN CONJUNCTIVA

Nadja Knop and Erich Knop

Department for Cell Biology in Anatomy
Hannover Medical School
Hannover, Germany

1. INTRODUCTION

The normal human conjunctiva usually shows a smooth surface in clinical examination; however, it contains a three-dimensional morphology with cryptal formations. In a study on the conjunctiva-associated lymphoid tissue (CALT)[1] we also investigated these crypts.

In the upper and lower lid, the conjunctiva contains a system of invaginations as mentioned in the old literature. Henle[2] described "tubular glands" with mouth-like openings whereas Stieda[3] reported larger longish and interwoven "crossing furrows". Both were later termed crypts and those of Stieda also clefts. The nomenclature is inconsistent as the openings of goblet cells are also termed crypts by some authors, and the morphology and location of the crypts of Henle and Stieda are frequently confused. Their regular presence, normality and probable function have been under controversial discussion since then.[4] In recent literature crypts were rarely mentioned and their functional significance is largely unclear. A comprehensive investigation was carried out by Kessing[4] who focused on their goblet cell content like Greiner;[5] their role in fluid transport processes was considered by Dark.[6]

The crypts also demonstrated a relation to the mucosa-associated lymphoid tissue of the conjunctiva. In the present study, we analysed conjunctival wholemounts by light and electron microscopy and also applied immunohistochemistry, which is still missing in the characterization of the conjunctival crypts, to focus on their supposed immunological function for human ocular surface immune defense.

Lacrimal Gland, Tear Film, and Dry Eye Syndromes 3
Edited by D. Sullivan *et al.*, Kluwer Academic/Plenum Publishers, 2002

867

2. MATERIAL AND METHODS

Macroscopically normal conjunctival sacs (n = 53) were obtained from human cadavers (n = 27) at the Department of Anatomy, Hannover Medical School. The average age was 76.1 years (±12.3 years). The complete conjunctival sac was excised, placed on plastic board, and fixed as described before.[1] Flat wholemounts (n = 36) were stained en bloc in undiluted Mayer´s hematoxylin, cleared by embedding in anise oil, in methacrylate resin or in epoxy resin. Two of these and four others were used for transmission electron microscopy (TEM) and two additional specimens were prepared for scanning electron microscopy (SEM). Seven conjunctival sacs were embedded in paraffin and four fresh frozen in OCT, sectioned at 5–10 μm thickness and stained with hematoxylin-eosin (HE). Primary antibodies for CD3, CD20, IgA and the IgA transporter molecule secretory component (SC) were incubated with rehydrated sections. Biotinylated secondary antibodies from the goat detected binding of the primary ones. This complex was marked by streptavidin-conjugated peroxidase and visualised by diaminobenzidine (DAB).

3. RESULTS

In the stained and cleared conjunctival wholemounts cryptal invaginations were visible (Fig. 1) that varied in number, shape and size. They were in general more frequent in the conjunctiva of the upper lid compared to the lower, restricted to the tarsal and beginning orbital area and were also prominent around the lacrimal puncta (Figs. 2, 3).

In the midtarsal region of the upper lid (Fig. 1), there were isolated tubular epithelial infoldings into the lamina propria. These had mouth-like openings of about 10–50 μm in diameter and could reach 50–100 μm in depth, representing Henle´s crypts (Fig. 4A). Towards the tarsal rim they increased in frequency, assumed an oblique orientation and tended to form subepithelial tunnels (Fig. 4B). Single crypts became aligned and finally formed the large, interwoven, net-shaped furrows, representing Stieda´s clefts; these were deeper and sometimes interconnected by tunnels. Towards the orbital zone the crypts terminated by forming large shallow grooves. These net-shaped crypt furrows were more pronounced and broader at the temporal and nasal sides and reduced in the medial zone of the conjunctiva (Fig. 3). Similar crypt furrows occurred at the lid margin and crypts of both locations were nasally and temporally continuous. In the lower lid crypt furrows were not located at the tarsal rim, which is distinctly different in shape from the upper lid but were more restricted to the marginal area and also occurred in a broadened zone nasally and temporally.Crypts in the lower lid were frequently plumper in shape than in the upper lid.

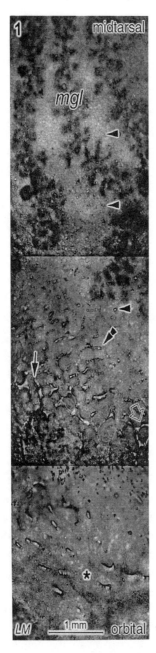

Figure 1. Light microscopy (LM) of a stained and cleared upper palpebral conjunctiva. Besides meibomian glands (mgl), isolated Henle crypts (arrowheads) are seen; they are confluent (double arrowhead) at the tarsal rim and continuous with the furrows (arrow) of Stieda´s net-shaped crypts that encircle roundish plateaus (open arrow) and vanish orbitally (asterisk).

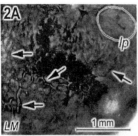

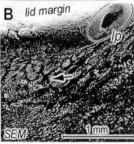

Figure 2. Around the lacrimal punctum (lp, encircled), here the lower, crypt furrows (arrows) occur, as triple lines (bright in the middle) in LM (**A**) or dark lines between bright plateaus in SEM (**B**). The dark central structure in A is an unrelated underlying large vessel; the tilted lid margin is seen as a slope in B; the location of Figs. 1 and 2 is indicated in Fig. 3.

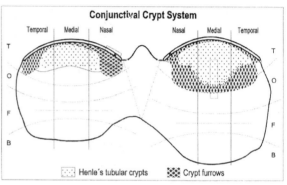

Figure 3. Crypts occur in the tarsal area (T) of both lids, more profuse in the nasal and temporal zone. In the midtarsal area the roundish tubular crypts of Henle occur. Along the tarsal rim of the upper lid, Stieda´s net-shaped crypt furrows form a broad zone nasally and temporally and vanish orbitally (O). Similar furrows occur in the marginal area and both are continuous; fornical (F), bulbar (B).

A variety of secretory cells were found in all types of crypts (Fig. 5). Besides frequent known goblet cells, there were slender cells with dense apical secretory granules and others with a broad apical cytoplasm or luminal protrusions which may suggest an apocrine mechanism of secretion. The secretory cells were not homogeneously distributed as some crypts were found free of goblet cells which might reflect the existence of different types of crypts regarding their secretory function.

Conjunctival crypts were associated with the lymphoid cells of the CALT. Lymphocytes were found in the lamina propria around the crypts in a pericryptal fashion (Fig. 6A) or encircled between furrows of the net-shaped crypts in an intercryptal arrangement (Fig. 6B). Here they occasionally formed dense roundish clusters of lymphocytes (Figs. 1, 6B) with follicular characteristics as a composition of B- and T-cells. Intraepithelial lymphocytes (IEL) were present in the cryptal epithelium and high endothelial venules (HEV) were observed. The lymphocytes were mostly CD3-positive T-cells (Fig. 7A); CD20-positive B-cells were less frequent (Fig. 7B) but more numerous

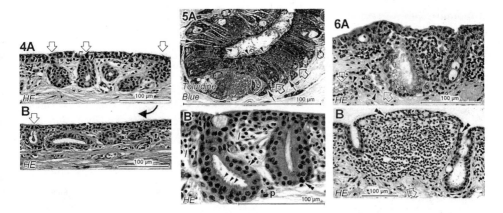

Figure 4. Histology shows straight tarsal Henle crypts (**A,B** open arrows) that become oblique towards the tarsal rim (bent arrow in **B**). **Figure 5.** In the crypt epithelium, besides goblet cells (open arrows in **A**), slender cells (dark arrows) occur with apical secretory granules (arrowheads in **A**), others have apical protrusions (arrowheads in **B**); note plasma cells (p), lymphocytes (l) and IEL (double arrowheads). **Figure 6.** Lymphocytes form a pericryptal arrangement (**A**) or follicular accumulations (**B**) between crypt furrows containing IEL (double arrowheads); HEV (open arrows) are seen with associated lymphocytes (arrowheads) in and around their wall.

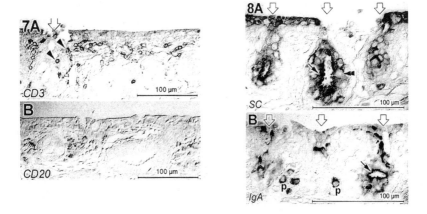

Figure 7. T-cells are frequent around the crypts (open arrow) and inside their epithelium (**A**, arrowheads); B-cells are not as frequent (**B**). **Figure 8.** Crypt epithelium (open arrows) is strongly positive for SC (**A**) and IgA (**B**) which accumulate apically in the cryptal cells (arrows in A and B). Goblet cells do not express SC (double arrowhead in A); IgA-positive plasma cells (p) occur around crypts; tissue defect in A.

than in the surrounding conjunctiva. The crypt epithelium was strongly positive for SC and IgA (Fig. 8), and IgA-positive plasma cells occurred around crypts. IgA and SC accumulated apically in the crypt epithelium and were seen luminally.

4. DISCUSSION

The normal human conjunctiva shows a distinct three-dimensional morphology represented by a multitude of crypts. In addition to previous descriptions, it was observed that the net-shaped crypt system broadens nasally and temporally along the tarsal rim of the upper conjunctiva and finally fuses with marginal crypts. Crypt furrows are also prominent around the lacrimal puncta of both lids. Thereby critsypts largely increase the surface area of the conjunctiva which conceivably increases functional capacity to maintain ocular surface integrity. An assumed secretory function of crypts could be confirmed. However, in addition to the known goblet cells, other types of secretory cells occur in the crypt epithelium. They are not homogeneously distributed and might reflect the existence of different secretory types of crypts.

Crypts are intimately associated with CALT. IgA-positive plasma cells and SC indicate a role in the production and distribution of protective IgA. Follicles enclosed between crypts suggest an afferent immune function for antigen detection and lymphocyte activation including the priming of plasma cells that may populate the conjunctiva and lacrimal gland.[7] We hypothesize that crypts represent sources of enriched antigens that accumulate here via the wiping action of the lids which allows the detection of antigens but also requires defense measures as provided by secretory IgA.

The topographical distribution of the crypt system in the upper and lower lid appears almost completely congruent with the position of the cornea during eye closure (Fig. 3). Crypt furrows then lie on the limbal area while Henle crypts cover the corneal surface. This would represent a suitable situation for the conjunctiva to assist corneal integrity, e.g. by supplying IgA or by bringing the CALT in contact to antigens that adhere to the cornea.

REFERENCES

1.   N. Knop and E. Knop. Conjunctiva-associated lymphoid tissue in the human eye.*Invest.Ophthalmol.Vis.Sci.* 41:1270–1279 (2000).
2.   J. Henle. *Handbuch der systematischen Anatomie des Menschen.* Bd.2, Braunschweig: F.Vieweg, (1866).
3.   L. Stieda. Ueber den Bau der Augenlidbindehaut des Menschen. *Archive mikroskopischer Anatomie* 3:357–365 (1867).
4.   S.V. Kessing. Mucous gland system of the conjunctiva. A quantitative normal anatomical study. *Acta Ophthalmol.Copenh.* Suppl 95:1–133 (1968).
5.   J.V. Greiner, H.I. Covington and M.R. Allansmith. Surface morphology of the human upper tarsal conjunctiva. *Am.J.Ophthalmol.* 83:892–905 (1977).
6.   A.J. Dark, T.E. Durrant, F. McGinty and J.R. Shortland. Tarsal conjunctiva of the upper eyelid. *Am.J.Ophthalmo.* 77:555–564 (1974).
7.   E. Knop and N. Knop. A Functional Unit for Ocular Surface Immune Defense formed by the Lacrimal Gland, Conjunctiva and Lacrimal Drainage System. *Adv.Exp.Med.Biol* (2001) in press.

ORGANIZED MUCOSA-ASSOCIATED LYMPHOID TISSUE IN HUMAN NASOLACRIMAL DUCTS

Friedrich P. Paulsen,[1] Jens I. Paulsen,[2] Andreas B. Thale,[3]
Ulrich Schaudig,[4] and Bernhard N. Tillmann[1]

[1]Department of Anatomy
[2]Department of Otorhinolaryngology
Head and Neck Surgery
[3]Department of Ophthalmology
Christian Albrecht University of Kiel
Kiel, Germany
[4]Department of Ophthalmology
University Hospital Eppendorf
Hamburg, Germany

1. INTRODUCTION

The evidence of primary extranodal marginal zone B-cell lymphoma of the efferent tear ducts[1-3] indicates the presence of pre-existent organized lymphoid tissue at that site. This is a low-grade B-cell lymphoma of the mucosa-associated lymphoid tissue (MALT) type. The mucosa of the lacrimal sac and the nasolacrimal duct consists of a double-layered epithelium with scattered lymphocytes or groups of lymphocytes. Although the mucosa of the efferent tear ducts plays an important role in defense against foreign antigens and various types of microbial pathogens, organized MALT may or may not be a normal component of the human efferent lacrimal system.[4] In the present study, human lacrimal sac and nasolacrimal ducts were examined for lymphoid tissue with the structure and immunophenotype of organized MALT.

Lacrimal Gland, Tear Film, and Dry Eye Syndromes 3
Edited by D. Sullivan *et al.*, Kluwer Academic/Plenum Publishers, 2002

2. METHODS

Both nasolacrimal systems, each consisting of the lacrimal sac and the nasolacrimal duct, were excised in their entirety from the heads of 41 randomly selected adult body donors (18 male, 23 female, age range 25–93 years) from the Department of Anatomy, Christian Albrecht University of Kiel, Germany. Limited information on the specimens was available; however, individuals were free of recent trauma, eye or nasal infections and diseases potentially involving or affecting lacrimal function. After formalin fixation the specimens were embedded in paraffin. Sections (7-μm thick) were stained with toluidine blue and hemalaun. Immunohistochemical staining was performed using the following antibodies: CD20 (L26), CD3, IgM, IgA, IgG, follicular dendritic cells (KiM4), CD45RA (KiB3) and CD68 (KiM1). They were applied using a standard peroxidase-labeled streptavidin-biotin technique.

3. RESULTS

Organized lymphoid tissue was identified with the cytomorphologic and immunophenotypic features of MALT in 17 of the 41 adult specimens. No significant differences existed between females and males. Table 1 displays distribution by percentage and Fig. 1 by age group.

Table 1. Distribution by age and percentage of cases with organized MALT[1]

Age	>20	>30	>40	>50	>60	>70	>80	>90
Cases	1	1	2	3	4	3	2	1
Percentage	2.4	2.4	4.9	7.3	9.8	7.3	4.9	2.4

[1]Table 1 is taken from Paulsen FP, Paulsen JI, Thale AB, Tillmann BN. Mucosa-associated lymphoid tissue in human efferent tear ducts. Virch Arch 2000;437:185–189. Permission for publication is granted by Springer-Verlag, Heidelberg, Germany.

Similar to other mucosal sites, MALT in the efferent tear ducts was characterized by reactive germinal centers containing tingible-body macrophages, a network of KiM4-positive follicular dendritic cells and CD3-positive T-cells. Mantle zones and marginal zone cells (MZC) surrounded the germinal centers. The mantle zones consisted of small CD20-positive lymphocytes expressing CD45RA and IgM. These lymphocytes merged in a population of small-to-medium-sized B-cells with moderately abundant cytoplasm and irregular nucleus outlines, features typical of MZC. The MZC exhibited the following immunophenotype: CD20+, CD45RA-, CD3-, IgM+ and IgA+. Some cells also expressed IgG, but these were rare. The MZC extended into the overlying epithelium to form a characteristic lymphoepithelium. In the parafollicular area, T lymphocytes and B lymphocytes were present as well as high endothelial venules. In no case were features suggestive of lymphoma observed. In the 24 cases in which no organized lymphoid tissue was found, a diffuse infiltrate of variable intensity existed within the lamina propria of the lacrimal sac and the nasolacrimal duct.

4. DISCUSSION

Specific secretory immunity depends on sophisticated cooperation between the mucosal B-cell system and an epithelial glycoprotein called the secretory component.[5] Initial stimulation of Ig-producing B-cells may occur mainly in organized MALT.[6] Considerable regionalization or compartimentalization exists in MALT, perhaps determined by different cellular expression profiles of adhesion molecules and/or the local antigenic repertoire. Antigenic stimulation of the B-cells results in generation of predominantly IgA-synthesizing blasts that leave the mucosae via efferent lymphatics, pass through the associated lymph nodes into the thoracic duct and enter the circulation. Then the cells return selectively to the lamina propria as plasma cells or memory B-cells[7] by means of homing mechanisms.

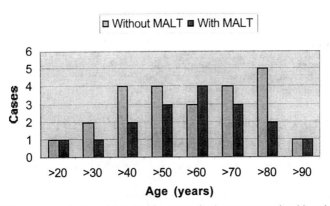

Figure 1. Distribution by age of cases with and without organized mucosa-associated lymphoid tissue. Fig. 1 is taken from Paulsen FP, Paulsen JI, Thale AB, Tillmann BN. Mucosa-associated lymphoid tissue in human efferent tear ducts. Virch Arch 2000;437:185–189. Permission for publication is granted by Springer-Verlag, Heidelberg, Germany.

The lamina propria within the efferent lacrimal pathways normally lacks organized lymphoid structures, suggesting its immunological effector functions depend on generation of immune responses elsewhere. In this context, experimental work has focused on the paired lymphocytic cell aggregates present at the entrance to the nasopharyngeal duct in rodents.[8] This organized lymphatic tissue resembles gut-type MALT, and its inductive function may be crucial to the immune response of mucosal surfaces in the head and perhaps beyond this region. On the other hand, caution is advised when extrapolating animal experimental data to humans, since the distribution and structures of MALT vary among different species.[9]

In this study, we found organized MALT in 41% of the efferent tear ducts from randomly selected body donors with unknown previous history of disease regarding the eye, efferent lacrimal pathway or nose. This suggests MALT is not ubiquitous in the efferent lacrimal system but is acquired in response to antigenic stimulation. It may develop in response to bacterial or viral infections or allergies. In a terminologic analogy to other regions of the human body, organized MALT of the efferent tear ducts is designated tear duct-associated lymphoid tissue (TALT). The development or acquisition of TALT in individual human efferent tear ducts remains unclarified, but once present it can provide the basis from which primary low-grade B-cell lymphoma of the MALT type may arise.

ACKNOWLEGMENTS

The authors thank Karin Stengel for technical assistance. We gratefully acknowledge gifts of monoclonal KiM4, KiB3 and KiM1 antibodies from Professor Dr. med. Dr. h. c. R. Parwaresch (Department of Pathology, Christian Albrecht University, Kiel, Germany). The study was supported by DFG grant Pa 738/1-1--a program of the German Research Foundation.

REFERENCES

1. Kheterpal S, Chan SY, Batch A, Kirkby GR. Previously undiagnosed lymphoma presenting as recurrent dacryocystits. *Arch Ophthalmol.* 1994;112:519–520.
2. White WL, Ferry JA, Harris NL, Grove AS Jr. Ocular adnexal lymphoma. *Ophthalmology.* 1995;102:1994–2006.
3. Tucker N, Chow D, Stockl F, Codere F, Burnier M. Clinically suspected primary acquired nasolacrimal duct obstruction. *Ophthalmology.* 1997;104:1882–1886.
4. Paulsen F, Thale A, Kohla G, Schauer R, Rochels R, Parwaresch R, Tillmann B. Functional anatomy of human lacrimal duct epithelium. *Anat Embryol.* 1998;198:1–12.
5. Brandtzaeg P. Humoral immune response patterns of human mucosae: induction and relation to bacterial respiratory tract infections. *J Infect Dis.* 1992;165:167–176.
6. Butcher EC, Picker LJ. Lymphocyte homing and homeostasis. *Science.* 1996;272:60–66.
7. Isaacson PG. Extranodular lymphomas: the MALT concept. *Verh Dtsch Ges Pathol.* 1992;76:14–23.
8. Kuper CF, Loornstra PJ, Hameleers DMH, Biewenga J, Spit BJ, Duijvestijn AM, van Breda Vriesman PJ, Sminia T. The role of nasopharyngeal lymphoid tissue. *Immunol Today.* 1992;13:219–224.
9. Wotherspoon AC, Ortiz-Hidalgo C, Falzon MR, Isaacson PG. *Helicobacter pylori*-associated gastritis and primary B-cell gastric lymphoma. *Lancet.* 1991;338:1175–1176.

Contact Lenses: Impact on the Tear Film and Ocular Surface

CONTACT LENSES AND TEAR FILM INTERACTIONS

Mark Willcox, Damon Pearce, Maxine Tan, Gulhan Demirci, and Fiona Carney

Cooperative Research Centre for Eye Research and Technology
University of New South Wales
Sydney, New South Wales, Australia

1. INTRODUCTION

Contact lenses are a successful form of vision correction. However, under certain circumstances adverse inflammatory responses can occur during lens wear. These inflammatory responses can be divided into those that involve the cornea and conjunctiva, and those that are primarily conjunctival. The former category includes microbial keratitis, contact lens-induced acute red eye, contact lens-induced peripheral ulcers, infiltrative keratitis, asymptomatic infiltrative keratitis and asymptomatic infiltrates. These conditions have recently been described in detail, and many are produced by bacterial contamination of contact lenses.[1-4] The major contact lens-related inflammatory conditions primarily associated with the conjunctiva are contact lens-induced papillary conjunctivitis (CLPC), and the more severe form of this disease, giant papillary conjunctivitis (GPC). However, the etiology of these conditions is poorly understood. The aim of the current study is to examine the effect of contact lens wear on tear proteins and determine which and how much specific tear proteins are adsorbed to the contact lens during wear.

2. METHODS

Specific enzyme-linked immunosorbent assays (ELISAs) were employed to determine whether contact lens wear alters the concentration of specific proteins (Table 1) in the tear film.[5-8] Contact lens protein deposits were analyzed qualitatively to determine which proteins were adsorbed, and quantitatively to determine the amount of specific proteins that adsorb onto a contact lens surface during wear. For these analyses, proteins were either

Lacrimal Gland, Tear Film, and Dry Eye Syndromes 3
Edited by D. Sullivan *et al.*, Kluwer Academic/Plenum Publishers, 2002

879

removed from the lenses and immunoblotted with specific antibodies,[9,10] or analyzed directly on the lenses by lens surface ELISAs. For immunoblotting, proteins were removed from the lenses with urea, guanidine and SDS in the presence of heat, separated on the basis of molecular weight by SDS-PAGE gel electrophoresis, transferred to nitrocellulose and immunoblotted for specific proteins. For surface ELISA assays, known concentrations of complement C3 were added to nitrocellulose discs and used as standards. Non-specific binding was calculated from unworn lenses or control discs. Standards and lenses were blocked with either 0.2% Tween-20/PBS (PBST) or 3% BSA/PBST. All discs and lenses were incubated with anti-human complement C3 peroxidase-conjugated antibodies at 1/1000 in PBST. ABTS peroxidase substrate was added for 10 min with gentle agitation and the absorbance of a 200-μl aliquot was read at 405 nm. Standard curves were produced and the lens absorbance values converted into absolute amounts of complement C3.

Table 1. Changes in protein concentrations in tears during lens wear [a]

Protein type	No lens wear	Lens wear
Total protein (mg/ml; open-eye tears)	11 ± 3	13 ± 5
Lysozyme (mg/ml; open-eye tears)[10]	1 ± 1	1 ± 0
Lactoferrin (mg/ml; open-eye tears)[10]	1 ± 1	1 ± 1
sIgA (mg/ml; closed-eye tears)[11]	3 ± 2	1 ± 1[b]
Specific sIgA to *Pseudomonas aeruginosa* (%; open-eye tears)[11]	100	20[b]
Albumin (μg/ml; open-eye tears)[10]	13 ± 5	24 ± 11
Fibronectin (ng/ml; open-eye tears)[8]	21 ± 25	120 ± 160[b]
Complement C3 (μg/ml; open-eye tears)[12]	4 ± 2	5 ± 2

[a]Contact lenses worn by subjects were Etafilcon A on a 6-night extended wear (6N EW) schedule with weekly replacement.
[b]Statistical difference, $P < 0.05$. Only sIgA, specific secretory sIgA and fibronectin concentrations were significantly altered during lens wear, with the former two decreasing and the latter increasing.

Lysozyme and lactoferrin are termed regulated proteins, i.e., their concentration does not alter during reflex tearing. sIgA is a constitutive protein; its concentration drops during reflex tearing. Albumin is probably derived from plasma leakage from conjunctival blood vessels. Fibronectin and complement C3 are probably derived from plasma and locally synthesised by epithelial cells. Complement C3 is a potent inflammatory mediator that on activation can cause conjunctival redness and activation of white blood cells.

3. RESULTS AND DISCUSSION

No change occurred in the concentration of the tear proteins lactoferrin, lysozyme or albumin (Table 1)[10] during wear of hydrogel lenses (Etafilcon A lenses) on a 6N EW schedule over 6 months, even though this lens and others were able to adsorb large amounts of proteins during wear (Table 2).

Table 2. Types of proteins that adsorb to lenses during wear

Protein	Amount deposited on lens
Total protein	
Group IV	
Etafilcon A	23 ± 5 ng/lens
Vifilcon A	3 ± 1 ng/lens
Group I	
Tefilcon	1 ± 1 ng/lens
Complement C3 (Etafilcon A)[12]	35 ± 37 ng/lens
sIgA (Etafilcon A)[12]	1 ± 0 µg/lens
Fibronectin (Etafilcon A)[8]	60 ± 53 pg/lens

Group IV lenses adsorbed more protein than group I lenses (P < 0.05). These results suggest the decrease in sIgA in tears (Table 1) was not due to adsorption to lenses. The increase in fibronectin in tears was probably due to plasma leakage into tears as the result of conjunctival hyperaemia during lens wear of Etafilcon A lenses, and release from epithelial cell surfaces.

This implies the turnover of tears rapidly replenishes those proteins adsorbed onto the lens surface during wear. However, other proteins found in tears are affected by wearing hydrogel lenses on an extended wear basis (Table 1). In particular, the concentration of sIgA, specific for bacteria, is reduced during lens wear.[7] In contrast, the concentration of fibronectin, a mammalian protein that has many functions including adhesion of mammalian cells during wound healing and the ability to bind bacterial cells, increased during wear.[8] Although contact lenses adhere sIgA, the amount that adheres is not equal to the amount lost from the tear film (Table 2).

Different types of hydrogel lenses adsorb different types of proteins (Fig. 1). Etafilcon A lenses, a group IV hydrogel lens, adsorbs a large amount of protein (Table 2), though most of this protein is lysozyme (Fig. 1). On the other hand, even though Polymacon lenses (group I hydrogels) do not adsorb as much total protein (Table 2), the protein they do adsorb is largely albumin (Fig. 1). The differences in the types of proteins that adsorb to lenses may correlate with differences in the types of inflammatory conditions resulting from extended lens wear. For example, the higher frequency of CLPC (GPC) that occurs after extended wear of group I hydrogel lenses compared to group IV lenses in prospective clinical trials[13] correlates with the amount of albumin bound to the surface of the lenses.

Using the technique of surface ELISA[14,15] in the current study, we demonstrate complement C3 protein is also adsorbed onto the lens surface (Fig. 2). C3 adsorbed to group IV hydrogel lenses in greater amounts than to silicone hydrogel lenses, and a direct correlation existed between complement adsorption and increasing length of hydrogel lens wear. Complement C3 is a highly inflammatory protein, and its activation to C3c may lead to the release of the anaphylatoxin C3a that can cause increased hyperaemia and activation of white blood cells.

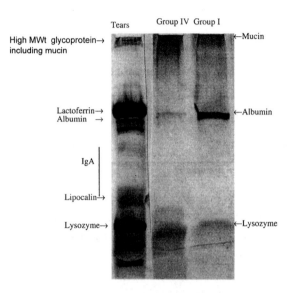

Figure 1. SDS-PAGE analysis of protein extracts from worn contact lenses. Equal amounts of protein (mg/ml) were added to each lane, although initially higher amounts of protein were extracted from the Group IV lens. Albumin, lysozyme and mucin were the major proteins that bound to contact lenses during wear, with group I lenses binding predominantly albumin and group IV lenses binding predominately lysozyme. The darker the band appears the more protein is present.

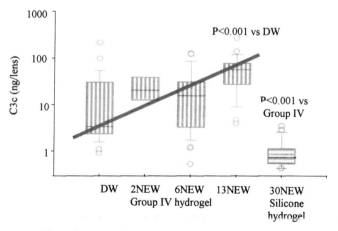

Figure 2. Amount of C3 bound to a lens surface during wear.

In summary, contact lens wear reduces the concentration of sIgA in the tear film but increases the concentration of fibronectin. The mechanism for the reduction of sIgA in tears in unknown, but it is possible a feedback mechanism exists between the corneal/bulbar conjunctival surface and the lacrimal gland that alters the constitutive production of sIgA in tears. The likely mechanism for increased fibronectin in tears is through capillary leakage. These changes in sIgA and fibronectin may predispose the eye to colonization by bacteria. The deposition of complement on the surface of contact lenses may facilitate inflammation in the tissues. . Interestingly, the new silicone hydrogel lenses worn on a 30 night extended wear schedule adsorbed less C3 than HEMA-based hydrogel group IV lenses which may indicate that the former lenses would produce a less inflammatory state during wear. Further studies correlating deposits with clinical characteristics of the lenses need to be performed to correlate the nature of deposit formation with clinical performance

REFERENCES

1. B.A. Holden, P. Sankaridurg, and I. Jalbert. Adverse responses. Which ones and how many, in: *Silicone Hydrogels. The Rebirth of Continuous Wear Contact Lenses*, D.F. Sweeney, ed., Butterworth Heinmann, Oxford (2000).
2. I. Jalbert, M.D.P. Willcox, and D.F. Sweeney. Isolation of *Staphylococcus aureus* from a contact lens at the time of a contact lens-induced peripheral ulcer: case report, *Cornea.* 19: 116 (2000).
3. B.A. Holden, D. La Hood, T. Grant, J. Newton-Howes, C. Baleriola-Lucas, M.D.P. Willcox, and D.F. Sweeney. Gram-negative bacteria can induce contact lens related acute red eye (CLARE) responses, *CLAO Journal.* 22: 47 (1996).
4. P.R. Sankaridurg, M.D.P. Willcox, S. Sharma, U. Gopinathan, D. Janakiraman, S. Hickson, N. Vuppala, D.F. Sweeney, G.N. Rao, and B.A. Holden. *Haemophilus influenzae* adherent to contact lenses associated with production of acute ocular inflammation, *J Clin Microbiol.* 34: 2426 (1996).
5. J. Lan, M.D.P. Willcox, and G.D. Jackson. Detection and specificity of anti-*Staphylococcus intermedius* secretory IgA in human tears, *Aust NZ J Ophthalmol.* 25(Suppl 1): s17 (1997).
6. J. Lan, M.D.P. Willcox, and G.D. Jackson. Effect of tear-specific immunoglobulin A on the adhesion of *Pseudomonas aeruginosa* I to contact lenses, *Aust N Z J Ophthalmol.* 27: 218 (1999).
7. D.J. Pearce, G. Demirci, and M.D.P. Willcox. Secretory IgA epitopes in basal tears of extended-wear soft contact lens wearers and in non-lens wearers, *Aust NZ J Ophthalmol.* 27: 221 (1999).
8. C. Baleriola-Lucas, M. Fukuda, M.D.P. Willcox, D.F. Sweeney, and B.A. Holden. Fibronectin concentration in tears of contact lens wearers, *Exp Eye Res.* 64: 37 (1997).
9. R.A. Sack, S. Sathe, L.A. Hackworth, M.D. Willcox, B.A. Holden, and C.A. Morris. The effect of eye closure on protein and complement deposition on Group IV hydrogel contact lenses: relationship to tear flow dynamics, *Cur Eye Res.* 15: 1092 (1996).
10. F.P. Carney, C.A. Morris, and M.D.P. Willcox. Effect of hydrogel lens wear on the major tear proteins during extended wear, *Aust N Z J Ophthalmol.* 25(Suppl 1): s36 (1997).
11. M.D.P. Willcox, and J. Lan. Secretory immunoglobulin A in tears: functions and changes during contact lens wear. *Clin Exp Optom.* 82: 1 (1999).
12. F. Stapleton, M.D.P. Willcox, C.A. Morris, and D.F. Sweeney. Tear changes in contact lens wearers following overnight eye closure, *Cur Eye Res.* 17: 183 (1998).

13. G.N. Rao, T.J. Naduvilath, P.R. Sankaridurg, L. Ramachandran, R. Gora, V. Kalikivay, N. Vuppula, D.F. Sweeney, B.A. Holden. Contact lens related papillary conjunctivitis in a prospective randomised clinical trail using disposable hydrogels. *Invest Ophthalmol Vis Sci.* 37: s1129 (1996).

14. D.J. Pearce, G. Demirci, and M.D.P. Willcox. A novel method used for the determination of adsorbed proteins on *ex vivo* extended-wear soft contact lenses. Australian Society for Biomaterials. 9th Annual meeting, Canberra. Proceedings. p 32 (1999).

15. D. Pearce, G. Demirci, and M.D.P. Willcox. Estimation of complement component C3c on daily and extended wear hydrogel contact lenses. *Optom Vis Sci.* 76: 176 (1999).

EFFECTS OF O$_2$ TRANSMISSIBILITY ON CORNEAL EPITHELIUM AFTER DAILY AND EXTENDED CONTACT LENS WEAR IN RABBIT AND MAN

Patrick M. Ladage, Kazuaki Yamamoto, Ling Li, David H. Ren,
W. Matthew Petroll, James V. Jester, and H. Dwight Cavanagh

Department of Ophthalmology
The University of Texas Southwestern Medical Center at Dallas
Dallas, Texas, USA

1. INTRODUCTION

The role of the tear film and its adherence to the underlying ocular surface is dependent on a healthy corneal epithelium. A vital feature of the corneal epithelium is a continuous cycle of renewal and exfoliation as a protective mechanism against injury and invasion of infectious organisms. The corneal epithelial renewal rate is relatively rapid[1] and dependent on the homeostatic balance of epithelial proliferation rate, centripetal and vertical cell migration and organized surface cell deletion/exfoliation.[2] During daily and overnight contact lens wear, however, normal tear film stability and corneal epithelial surface integrity are compromised, which may lead to minor and occasionally more serious complications such as infectious corneal ulceration. This article discusses the preliminary results of our recent studies on the effects of contact lens oxygen transmissibility, lens type, wearing schedules and non-lens related hypoxia on human and rabbit corneal epithelium. The following outcome measures were assessed: bacterial binding to exfoliated corneal epithelial surface cells, corneal epithelial thickness and surface cell exfoliation in human subjects. Concomitantly, we examined the proliferation rate of the central corneal epithelium, corneal epithelial basal cell turnover rate and corneal epithelial surface cell apoptosis in the standard rabbit model of lens wear.

2. METHODS

2.1 Human Studies[3,4]

The 177 patients completed a daily and extended lens wear phase of a prospective, randomized, double-masked, single-center, parallel treatment groups clinical trial. Test lenses included high O_2 soft (n = 27), hyper O_2 soft (n = 99) and hyper O_2 rigid gas permeable (n = 51) (Table 1). Patients used one of the three assigned test lenses for 1-month daily wear and 12-month overnight wear on either a 6 or 30 continuous nights wearing regime. The following outcome measures were examined at pre-lens baseline and multiple time points during lens wear as published by previous methods[5]: *Pseudomonas aeruginosa* (PA) binding to exfoliated corneal epithelial surface cells, central corneal epithelial thickness and corneal epithelial surface exfoliation.

Table 1. Description of test lenses

Lens type Study group	Material	Dk/L[1]	EOP[2]	Base Curve (mm)	Diameter (mm)
High O_2 soft (*human/rabbit*)	Etalfilcon A[3]	32.5	12.42	8.4/8.8	14.00
Hyper O_2 soft (*human/rabbit*)	Balafilcon A[4]	110	19.9	8.6	14.00
Hyper O_2 RGP (*human/rabbit*)	Tisilfocon A[5]	97	19.13	7.00–8.00	9.00–9.80 14.00 (*Rabbit*)
Low O_2 RGP (*rabbit*)	NA	10	5.76	7.60–8.00	14.00

[1] Dk/L = oxygen transmissibility measured in saline at 35°C by the polarographic method/edge effect correction (International Standards Organization #9913). 10^{-9}(cm/sec)(mL O_2/mL mmHg)
[2] EOP - equivalent oxygen percentage (21% = normal at sea level)
[3] Vistakon Inc., Jacksonville, FL
[4] Bausch & Lomb, Inc., Rochester, NY
[5] Menicon Ltd., Nagoya, Japan

2.2. Rabbit Studies

2.2.1. Corneal Epithelial Proliferation Rate.[6,7] To assess the effect of hypoxia and contact lens type on the normal epithelial proliferation rate of the central cornea, 24 rabbits, weighing 2.5–3.5 kg, were divided into four groups: (1) low O_2 RGP lens (n = 4), (2) hyper O_2 RGP lens (n = 4), (3) hyper O_2 silicon-hydrogel lens (n = 8) or (4) eyelid suturing (n = 8). To fit the RGP lenses, partial nictitating membranectomy was performed 1 week before the experiment; control experiments with nictitating membrane on/off indicated no effect of the surgery on the outcome measures. In each individual, one eye was randomly chosen as the experimental eye while the other served as control. After 24 h of lens wear or eyelid suturing, 5-bromo-2-deoxyuridine (BrdU) was injected intravenously (200 mg/kg in phosphate buffered saline) at 9:00 a.m. to label proliferating corneal epithelial cells. Animals were sacrificed 24 h after injection. Corneas were fixed in situ and prepared for immunocytochemistry according to previously published methods.[8] Digital images (588 x 984 μm) of the whole-mount epithelium were taken from superior limbus to central cornea and BrdU-labeled corneal epithelial cell pairs were manually counted with specialized software. Data was expressed as percentage change compared to the contralateral control eye.

2.2.2. Corneal Epithelial Basal Cell Turnover Rate.[9] To exclusively label corneal epithelial basal cells in both eyes and monitor the number of labeled cells leaving the basement membrane (becoming post-mitotic wing and squamous cells), a low O$_2$ contact lens was placed on one eye while the other served as control. Because only corneal epithelial basal cells have the capacity to divide[10], the proliferation marker BrdU was used as described above. One low O$_2$ RGP lens (Dk 10) was randomly fitted to one eye 24 h later. The rabbits were sacrificed 4 days after BrdU labeling (3 days after the contact lens fit). Corneas were fixed in situ and double-stained with monoclonal anti-BrdU antibody/FITC-conjugated secondary antibody to identify the BrdU-labeled cells and propidium iodide to label all epithelial nuclei. BrdU-labeled cells in the basal cell layer and upper wing and squamous cell layers were quantified three dimensionally with a laser-scanning confocal microscope (Leica, Deerfield, MI). Data were expressed as the ratio of BrdU-labeled cells still attached to the basement membrane versus labeled cells that had migrated to the upper corneal epithelial layers.

2.2.3. Corneal Epithelial Surface Cell Apoptosis.[11,12] Apoptotic cell death was quantified by TUNEL and Annexin V labeling on the corneal epithelial surface in normal eyes and during hypoxia and overnight lens wear. Rabbits were assigned to either the control group (n = 5, 10 eyes), low O$_2$ RGP (n = 3, 6 eyes) or hyper O$_2$ RGP lens wear (n = 3, 6 eyes) and sacrificed following 3 days of lens wear. Corneas were then fixed in situ, placed in frozen embedding media, sectioned into 8-μm-thick slices and mounted on gelatin-coated slides. TUNEL assay to detect distinctive apoptotic DNA defragmentation was used according to manufacturer protocol. The number of TUNEL positive cells on the epithelial surface of each cornea was counted under the epifluorescence microscope and recorded as positive cells/mm.

For Annexin-V labeling, one eye of each rabbit was randomly assigned to one of the following experimental groups while the other eye served as control: (1) eyelid suturing (n = 6), (2) low O$_2$ RGP lens (n = 6), (3) hyper O$_2$ RGP lens (n = 6) and (4) high O$_2$ soft lens (n = 3). Rabbits were sacrificed 24 h after treatment and non-fixed whole-mount corneas were processed for Annexin-V staining. Annexin-V selectively binds to cells with a compromised cell membrane and routinely serves as a marker for early and late apoptosis; cell nuclei were identified separately by double-labeling with propidium iodide. The number of Annexin-V positive cells at the central corneal surface was counted under the epifluorescence microscope.

3. RESULTS

3.1. Clinical Studies

3.1.1. Bacterial binding. PA binding to exfoliated corneal epithelial cells significantly increased in daily[3] (data not shown) and overnight lens wear[4] (first 3 months) for both soft test lenses (P < 0.01, one-way repeated measure ANOVA), while RGP lens wear did not induce any significant binding increases (Fig. 1). The lower O_2 transmissible soft lens showed significantly higher binding than the hyper O_2 soft lens (P < 0.01, two-way repeated measure ANOVA). PA-binding activity with all test lenses gradually decreased over time to pre-lens baseline levels at the conclusion of the study.

3.1.2. Central Corneal Epithelial Thickness. Daily lens wear with both soft test lenses for 1 month did not reveal any striking changes in the epithelial thickness of the central cornea; however, RGP lens wear showed a rapid significant decrease compared to its baseline (P < 0.001, one-way repeated measure ANOVA) (Fig. 2). Extended lens wear demonstrated a significantly larger decrease in central epithelial thickness with the high O_2 6-night soft lens than the hyper O_2 lens 6-night test group (P = 0.0189, two-way repeated measure ANOVA) and 30-night group (P = 0.02, two-way repeated measure ANOVA). No significant difference occurred between the 6-night and 30-night hyper O_2 lens groups (P = 0.60, two-way repeated measure ANOVA). Epithelial thickness remained significantly decreased with hyper O_2 RGP lens wear compared to baseline, although the corneal epithelial thickness demonstrated adaptive recovery over time.

3.1.3. Corneal epithelial cell exfoliation rate. All lens wear in all test groups demonstrated significant suppression of surface cell exfoliation (cells/min) in both daily and extended lens wear (P < 0.001, one-way repeated measure ANOVA) (Fig. 3).

3.2. Rabbit Studies

3.2.1. Proliferation Rate. BrdU-labeled cells declined significantly (P < 0.05, one-way ANOVA) in the central corneal epithelium compared to their contralateral controls after 24 h of eyelid suturing, low and hyper O_2 RGP lens wear and hyper O_2 silicone hydrogel lens wear (Fig 4).

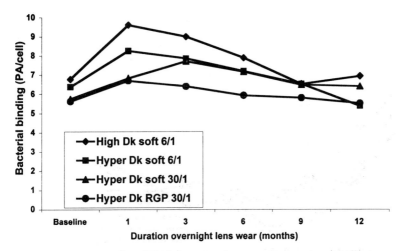

Figure 1. (Human studies)Bacterial binding *(Pseudomonas aeruginosa)* to exfoliated corneal epithelial cells during daily and overnight lens wear for the different test lenses and wearing schedules. Notice the largest increase in PA binding at 1 and 3 months overnight lens wear with a full recovery for all lenses to baseline values 12 months into the study.

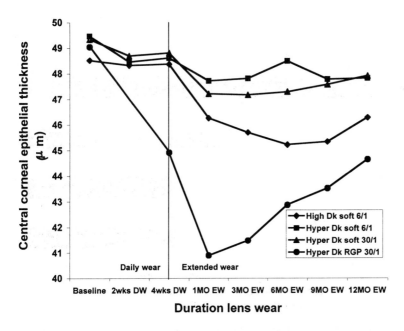

Figure 2. (Human studies) Thickness of the central corneal epithelium during daily and overnight lens wear for the different test lenses and wearing schedules. The rapid decrease in thickness for the hyper Dk RGP lens seems to have a mechanical origin. The epithelial thinning associated with soft lens wear during overnight wear may have a physiological cause.

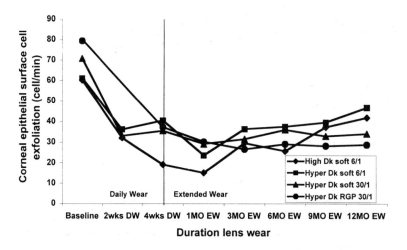

Figure 3. (Human studies) Surface corneal epithelial cell exfoliation rate during daily and overnight lens wear.

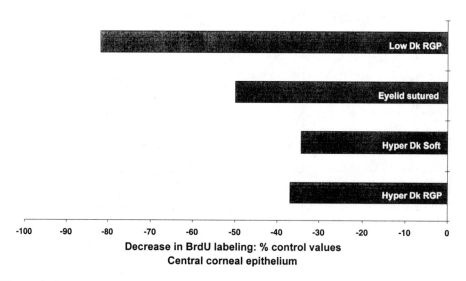

Figure 4. (Rabbit studies) Decreased BrdU labeling (basal cell proliferation) of the central corneal epithelium.

3.2.2. Basal Cell Turnover Rate. The central corneal epithelium of the low Dk lens group retained 37.7% more labeled cells in the basal cell layer 4 days after labeling compared to control ($P = 0.0088$ paired t-test).

3.3.3. Corneal Epithelial Surface Cell Apoptosis. Both low and hyper O_2 RGP contact lens wear significantly decreased the total number of TUNEL-labeled cells on the central corneal surface epithelium after short-term overnight lens wear (Table 2).[11] Short-term eyelid suturing and all lens wear, with the exception of the low O_2 RGP lens, reduced the total number of Annexin-V positive cells on the central corneal surface epithelium[12] (Fig.5).

Table 2. TUNEL-positive corneal epithelial surface cells

	TUNEL-positive cells /mm[a]
Control (n = 9)	3.5 ± 1.7
Low O_2 RGP (n = 6)	0.5 ± 0.3[b]
Hyper O_2 RGP (n = 6)	0.6 ± 0.5[b]

[a]Mean ± SD; [b]($P < 0.05$, one way ANOVA on ranked data)

4. DISCUSSION

A striking finding of the human clinical study was the adaptive response of the corneal epithelium to overnight contact lens wear; all measures properties showed some degree of partial and even full recovery (PA-binding) over time as the study progressed.[4] This new and unexpected observation reveals the importance of adding a time factor (duration overnight lens wear) to future epidemiological studies establishing incidence or relative risk estimates for ulcerative-infectious keratitis in man.

As reported earlier, overnight contact lens wear causes the corneal epithelium to thin.[5,13] In this prospective, long-term, double-masked clinical study we observed that this thinning is related to oxygen transmissibility and lens type. After 1 month of daily lens wear, no significant changes in central corneal epithelial thickness were noted between the two soft test lenses[3]; however, as soon as the subjects initiated overnight lens wear, a significantly larger decrease in corneal epithelial thickness occurred with the lower O_2 soft lens compared to the hyper O_2 silicone-hydrogel soft lens.[3]

Either mechanical compression, cell redistribution or overall cell loss could explain thinning of the central corneal epithelium during overnight lens wear. In the case of RGP lens wear, mechanical reformation of the corneal epithelium may be the principal cause for the observed central epithelial thinning. Of greater concern, however, is overall cell loss, which may occur when the total number of epithelial cells leaving the corneal surface is higher than the combined number of epithelial cells formed by basal cell proliferation and/or migrating centripetally from the peripheral epithelium. In the normal cornea, epithelial cells leave the surface in an orderly manner (exfoliation), which appears to be an

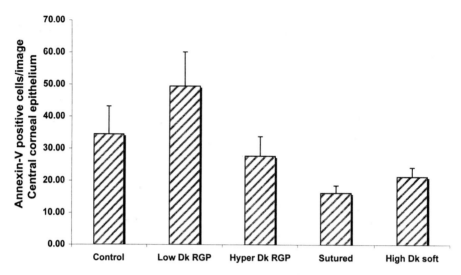

Figure 5. (Rabbit studies) Total number of Annexin-V (apoptosis) positive surface cells in the central corneal epithelium for the four experimental groups. The low Dk RGP data may also contain a necrotic cell component associated with sloughing of some surface cells.

apoptotic event.[14,15] During lens wear, however, epithelial surface exfoliation as measured with the collection irrigation chamber is significantly downregulated, regardless of material oxygen transmissibility and wearing schedule.[3-5]Our current rabbit studies with TUNEL[11] and Annexin-V labeling[12] confirmed and extended the observation that RGP lens wear significantly decreased corneal epithelial surface cell apoptosis. Thus, RGP overnight lens wear decreased apoptotic surface cell death independent of lens material oxygen transmissibility. Paradoxically, this produced a concomitant thinning of central corneal epithelial thickness that appeared, at least during soft lens wear in the human studies, oxygen dependent.

In agreement with these findings, the BrdU-labeling studies on the rabbit cornea established a significant relationship between lens oxygen levels and the epithelial proliferation rate. Hypoxia alone, as induced by the eyelid suturing experiment, decreased the total number of proliferating cells in the central corneal epithelium by 50%.[7] Overnight soft and rigid contact lens wear with hyper O_2 transmissibility materials decreased proliferation by approximately 35%[6,7] while a low O_2 rigid lens did so by 80%.[6,8] Taken together, these results predict overnight lens wear with lower O_2 materials will reduce central corneal epithelial thickness the most. In addition, low O_2 contact lens wear significantly decreased the vertical migration of basal epithelial cells toward the corneal surface as BrdU-labeled basal cells remained longer in the basal layer compared to the non-lens wearing control.[9]In summary, the decreased basal cell proliferation, apoptotic exfoliation and vertical migration demonstrate extended contact lens wear initially slows down the epithelial renewal rate in the central cornea. However adaptive recovery by unknown mechanisms occurs in the human cornea.

ACKNOWLEDGMENTS

This study was supported in part by EY10738 (HDC), The Pearle Vision and Chilton Foundation, Dallas, TX, Senior Scientist Awards (JVJ, HDC), Olga Keith Weiss Scholar Award and an unrestricted research grant from Research to Prevent Blindness, Inc., New York, NY. Part of the work was performed while Patrick Ladage was a recipient of a William C. Ezell Fellowship from the American Optometric Foundation, Rockville, MD.

REFERENCES

1. C. Hanna and J.E. O'Brien, Cell production and migration in the epithelial layer of the cornea, *Arch Ophthalmol.* 64:536 (1960).
2. R.A. Thoft and J.Friend, The x,y,z hypothesis of corneal epithelial maintenance, *Invest Ophthalmol Vis Sci.* 24:1442 (1983).
3. P.M. Ladage, K. Yamamoto, D.H. Ren, et al., Effects of rigid and soft contact lens daily wear on corneal epithelium, tear lactate dehydrogenase and bacterial binding to exfoliated epithelial cells, *Ophthalmology* 108:in press (2001).
4. D.H. Ren, K. Yamamoto, P.M. Ladage, et al., Adaptive effects of 30-day extended wear of new hyper O₂ transmissible RGP and soft contact lenses on bacterial binding and corneal epithelium, *Ophthalmology 108*:in press (2001).
5. D.H. Ren, W.M. Petroll, J.V. Jester, J. Ho-Fan, H.D. Cavanagh, The relationship between contact lens oxygen permeability and binding of Pseudomonas aeruginosa to human corneal epithelial cells after overnight and extended wear. *CLAO J.* 25:80 (1999).
6. P.M. Ladage, K. Yamamoto, D.H. Ren, et al., Low dk versus ultra-high dk RGP extended contact lens wear: effects on proliferation rates of the corneal epithelium of rabbit cornea. *Optom Vis Sci.* 76:S239 (1999).
7. P.M. Ladage, K. Yamamoto, D.H. Ren, et al., Effects of overnight hyper dk soft contact lens wear and eyelid closure on epithelial proliferation rates of the cornea, *Invest Ophthalmol Vis Sci.* 42:S936 (2001).
8. D.H. Ren, W.M. Petroll, J.V. Jester, H.D. Cavanagh, The effect of rigid gas permeable contact lens wear on proliferation of rabbit corneal and conjunctival epithelial cells, *CLAO J.* 25:136 (1999).
9. P.M. Ladage, K. Yamamoto, H.D. Ren, et al., Basal epithelial turnover following RGP extended contact lens wear in the rabbit cornea, *Invest Ophthalmol Vis Sci.* 41:S75 (2000).
10. N.G. Joyce, B. Meklir, S.J. Joyce, J.D. Zieske, Cell cycle protein expression and proliferative status in human corneal cells, *Invest Ophthalmol Vis Sci.* 37:645 (1996).
11. K. Yamamoto, P.M. Ladage, D.H. Ren, W.M. Petroll, J.V. Jester, H.D. Cavanagh, Bcl-2 expression and apoptosis in the rabbit corneal epithelium during low and hyper dk rigid gas permeable overnight lens wear, *CLAO J.* 27:in press (2001).
12. L. Li, D.H. Ren, P.M. Ladage, et al., The effect of overnight contact lens wear and eyelid closure on expression of Annexin-V at the rabbit corneal epithelial surface, *Invest Ophthalmol Vis Sci.* 42:S592 (2001).
13. B.A. Holden, D.F. Sweeney, A. Vannas, K.T. Nilson, N. Efron, Effect of long-term extended contact lens wear on the human cornea, *Invest Ophthalmol Vis Sci.* 26:1489 (1985).
14. H. Ren, G. Wilson, Apoptosis in the corneal epithelium, Invest Ophthalmol Vis Sci. 37:1017 (1996).
15. K. Yamamoto, P.M. Ladage, D.H. Ren, et al., Bcl-2 Expression in the human cornea, *Exp Eye Res.* 71:in press (2001).

INFLUENCE OF THE TEAR FILM COMPOSITION ON TEAR FILM STRUCTURE AND SYMPTOMATOLOGY OF SOFT CONTACT LENS WEARERS

Michel Guillon,[1,2] Cécile Maissa,[1] Karine Girard-Claudon,[1] and Philip Cooper[1]

[1]Contact Lens Research Consultants
London, United Kingdom
[2]Centre Optométrique
Université de Paris-Sud
Orsay, France

1. INTRODUCTION

The lipid layer has a major influence on tear film stability during the interblink period. The prelens tear film lipid layer pattern, formed by the mixing of the different classes of lipids present, affects stability. However, the clinical techniques to investigate the lipid pattern are specialized techniques not accessible to all clinicians in routine practice. The aim of the current investigation was to determine if composition of the lipid layer influences the lipid layer pattern and symptomatology with contact lenses.

2. MATERIALS AND METHODS

The prehydrogel contact lens tear film was analyzed using clinical and analytical techniques. The analysis was carried out on 124 prehydrogel contact lens tear films. The tear film lipid structure and its level of contamination visible during the clinical evaluation were classified using wide diffuse lighting produced by the Tearscope (CLRC, London UK) with a biomicroscope observation system. The lipid layer mixing patterns observed correspond to lipid structures with different uniformity of mixing and different thickness.[1] Further using this lighting and observation system, the tear breakup time was measured in a non-invasive manner and non-invasive breakup time (NIBUT) was recorded. NIBUT was

Lacrimal Gland, Tear Film, and Dry Eye Syndromes 3
Edited by D. Sullivan *et al.*, Kluwer Academic/Plenum Publishers, 2002

895

elapsed time in seconds between a full eye opening after a blink and the first visible break
or initiation of a reflex blink. Three readings were taken for each measurement, and the
minimum and median NIBUT values were calculated for each eye and analyzed
statistically.

The tear film composition was analyzed using a sensitive high-performance liquid
chromatography technique on individual unstimulated tear film samples (2 μl) developed
by our group.[2] This technique previously identified 22 individual types of lipids forming
five general classes. The tear samples were taken from the temporal part of the lower tear
prism while the contact lens was worn. The results obtained were analyzed statistically by
ANOVA multifactorial analysis and Chi Square Automated Interaction Detector
(CHAID)[3,4] for the effect of the lipid composition on the tear lipid layer structure, tear film
stability and patient symptomatology.

3. RESULTS

3.1. Compositional Characteristics

The results for the prehydrogel tear film (PHTF) lipid composition revealed important
intersubject differences. Overall, cholesterol and cholesterol esters dominated the PHTF
lipid composition, and only low levels of polar lipids such as phospholipids were detected.
These compositional characteristics agree with the tear lipid model of a thin polar lipid
inner layer at the aqueous interface and a thicker outer layer of nonpolar lipids acting as a
protective barrier.

3.2. Lipid Composition and PHTF Characteristics

The results revealed a significant influence of the tear lipid composition on the lipid-
mixing pattern within the PHTF and the tear film stability as measured by the NIBUT. The
lipid layer pattern (Fig. 1) was characteristic of soft contact lens wearers. In most cases it
was a meshwork type (open and tight) pattern (56.5%), and the second most common was
a wave pattern (24.2%). These two patterns are characteristics of a relatively thin lipid
layer with a thickness ranging from 10 to 70 nm. The observed lipid-mixing pattern was
correlated with the level of cholesterol esters present in tears. A significantly higher (P =
0.001) level of cholesterol esters was associated with the thinnest lipid layer observable
(none to transient pattern, Table 1).

For NIBUT, the level of phospholipids in the tear film was the most influential factor
detected by CHAID (P = 0.015). The population was divided into two groups based on the
level of phospholipids measured. A low level of phospholipids (0.00 mg/ml to 0.06 mg/ml)
and a low NIBUT (a mean of 4.5 sec compared to the overall population mean of 5.7 sec)
characterized the first group. The second group, with phospholipid levels higher than 0.06
mg/ml, had a significantly higher NIBUT with a mean of 6.5 sec (Fig. 2).

ady,

t is Friday morning & I am between
ams, with time to kill. I could and
robably should be trying to cram some
st bit of knowledge into my skull, but
simply can't bear it. Instead choosing
 let my mind wander.

enjoyed seeing you the other day!
ame that we didn't have a chance to
tch up. It occurs to me that we
ve never had a conversation that
usnt separated by plexiglass. which,
m sure is a great safety feature
t terrible for dialog. so I thought
d write.

m curious how your last few weeks
ve been. Last I heard you were in
me sort of kidney failure - I'm happy
see you are feeling better - but that was
eks ago, so I'd really like to hear
f any exciting exploits since then,
opefully you will write back & let me
ow.

y end of things has been a nonstop.
in-train of higher education.

They have us on a 21 hour schedule
so that every day is cram-packed. cum
I'm in the middle of a (6 exam) test week, and
am fairly certain it's killing me.

I once had a professor tell me that a test is
simply a celebration of learning and we shoul
relish the opportunity... That lady was batshi
crazy but it is a lovely sentiment.

Speaking of... The bell tolls and soon it will be ti
for another joyous exam, I have to go now, b
I do hope you write back. If you do, h
the note in the same book you found this
one in. I'll check for a reply on monday.

Later on —
            ethan

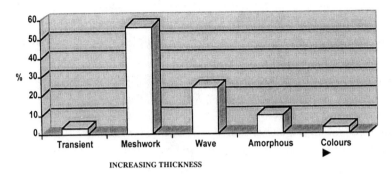

Figure 1. PHTF lipid layer mixing pattern - distribution.

Table 1. Effect of cholesterol esters on the PHTF lipid layer mixing pattern. Comparative analysis by ANOVA (P = 0.038) and (Student Newman Keuls, Multiple Comparisons) (*values joined by a continuous line are not statistically significantly different)

	Amorphous	Meshwork	Wave	Color	Transient
Cholesterol esters	0.39	0.45	0.54	0.77	1.39
SNK (5%)*					

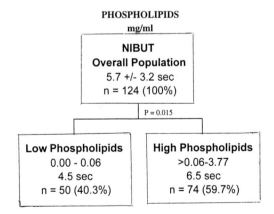

Figure 2. Effect of lipid composition on tear film stability (NIBUT) analyzed by CHAID.

3.3. Lipid Composition and Symptomatology

The McMonnies questionnaire modified for contact lens wearers was used to assess subjects' symptomatology and subdivide the subjects between asymptomatic and symptomatic. The questionnaire was based on the original questionnaire designed for non-contact lens wearers by McMonnies and Ho[5] and modified for contact lens wearers by Guillon et al.[6] No direct association existed between the PHTF tear lipid composition and the calculated overall McMonnies score. However, associations were detected between individual symptoms and specific environmental conditions in the questionnaire.

The subjects were asked to rate the frequency of their dryness symptoms using a 4-point scale (1 = never, 2 = sometimes, 3 = often, 4 = constantly). The most common answer was "sometimes" (66% of cases) followed by "often" (26% of cases). Only 6% of the population reported "never" suffering from dryness symptoms, whereas 2% complained of constant dryness. The level of cholesterol esters most affected the dryness symptomatology: a higher level of cholesterol esters was associated with greater symptoms. CHAID divided the population into two groups according to the level of cholesterol esters detected. Dryness symptomatology increased significantly with a higher level of cholesterol esters (P = 0.048). In the higher cholesterol ester group, 40% reported suffering from dryness "often" to "constantly," whereas only 25% reported such severity of symptoms in the low cholesterol esters group. Further, the percentage of cases never experiencing dryness symptoms decreased from 17% in the low cholesterol ester group to 0% in the high cholesterol ester group (Fig. 3).

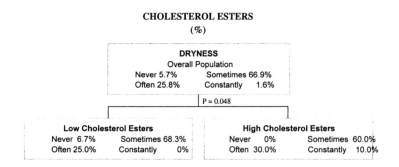

Figure 3. Symptomatology and PHTF lipid composition analyzed by CHAID.

Symptoms experienced in the company of smokers were associated with a lower level of fatty acids (mean level: symptomatic = 0.09 mg/ml; asymptomatic = 0.22 mg/ml, P = 0.044) and monoglycerides (mean level: symptomatic = 0.03 mg/ml; asymptomatic = 0.40 mg/ml, P = 0.019), higher proportion of cholesterol (symptomatic = 24.0%; asymptomatic = 17.5%, P = 0.046) and lower overall level of tear lipid (symptomatic = 1.27 mg/ml; asymptomatic = 1.73 mg/ml, P = 0.047). Finally, an increase in cholesterol esters was associated with problems when driving or as a passenger in a car (yes = 0.42, no = 0.32, P = 0.044), and when in buildings with air conditioning (yes = 0.42, no = 0.36, P = 0.042).

4. DISCUSSION

The role of the lipid layer in preventing evaporation is well known, and the bipolar model of a thin inner polar lipid layer and a thicker outer nonpolar lipid layer well accepted. In contrast, relationships between the lipid layer composition, its structure as observed clinically and efficacy in preventing evaporation as illustrated by longer NIBUT have not been reported. Of interest also were the associations identified between the lipid layer compositional characteristics and symptomatology with contact lenses. Because clinical evaluation of the tear film is difficult and evaluation of symptoms is a reactive approach to contact lens fitting, an assay to identify the compositional characteristics of the tear film lipid layer could be developed to identify potential problem contact lens wearers prior to fitting.

In conclusion, the results show (1) the level of cholesterol esters influences the prelens tear film pattern, e.g., an excessive level of cholesterol esters is detrimental to a stable, homogeneous lipid layer; (2) the level of polar lipids influences the tear film stability, and lower levels of phospholipids/triglycerides are associated with a lower tear film stability and (3) a higher concentration of cholesterol esters is associated with greater symptomatology. Analysis of the composition of the tear film lipid layer could be the basis for a useful in vitro test to identify potential problematic contact lens wearers prior to fitting.

REFERENCES

1.   J.P.Guillon, M.Guillon. The role of tears in contact lens performance and its measurement. In M. Ruben,M. Guillon eds. *Contact Lens Practice*. London. Chapman & Hall:452–83 (1994).

2.   C.Maissa. Biochemical markers and contact lens wear. PhD Thesis. University of Aston, UK (1999).

3.   D.Biggs, B.de Ville, E.Suen. A method of choosing multiway partitions for classification and decision trees. *J Applied Statistics* 18:49–62 (1991).

4.   J.M. Zar. *Biostatistical Analysis* 2nd Ed. Englewood Cliffs, NJ, Prentice-Hall (1984).

5.   C.W.McMonnies, A.Ho. Marginal dry eye diagnosis: history versus biomicroscopy, in Holly FJ (Ed): *The Pre-Ocular Tear Film in Health, Disease and Contact Lens Wear*. Lubbock, TX, Dry Eye Institute:32–40 (1986).

6.   M.Guillon, J.C.Allary, J.P.Guillon, G.Orsborn. Clinical management of regular replacement: Part 1. Selection of replacement frequency. *ICLC* 19:104–120 (1992).

EVALUATION OF THE PRE-LENS TEAR FILM FORMING ON THREE DISPOSABLE CONTACT LENSES

Jean-Pierre Guillon,[1] Judith Morris,[2] and Brenda Hall[3]

[1]Lions Eye Institute
Perth, Australia
[2]Institute of Optometry
London, England
[3]Biocompatibles LTD

1. INTRODUCTION

The most common cause of complaint from contact lens wearers is that of dryness. This induces discomfort, reduced wearing time and is one of the reasons of cessation of contact lens wear. The reduced comfort is caused by evaporation of the pre-lens tear film, leading to surface dehydration and loss of tear fluid from the lens matrix. This fluid flow towards the surface produces pervaporation staining which is commonly seen in patients presenting with symptoms of dryness. The lens surface with reduced tear coverage is prone to increased protein deposition thus reducing the effective life of the contact lens. The most common solution to those problems has been to fit thicker lenses to limit pervaporation staining or to accelerate the rate of replacement of disposable lenses to limit the effect of surface deposits. This is of limited use for patients with poor tear films who experience discomfort at the outset and can only wear lenses for 2 to 3 hours. The aim of the study was to study the pre-lens tear film structure of three monthly disposable contact lenses and to find out if it supported an adequate superficial lipid layer as it would forms a barrier to evaporation.

2. METHODS

2.1. Imaging the Tear Film

The pre-ocular tear film (POTF) and pre-lens tear film (PLTF) are made visible with the keeler tearscope plus. It consists of a cold cathode illuminated conical tube that provides 360 degree specular illumination. This produces a white background against

Figure 1. Imaging the tear film.

which the tear films can be observed non-invasively. It is the only instrument that permits the observation of both the pre-ocular and pre-lens tear film with the same technique. It is the only clinical way to observe the lipid layer over the whole corneal surface. It is the only clinical technique that makes visible successively the aqueous phase, the mucous coverage and the contact lens surface during the process of pre-lens tear film breakup. The tearscope plus allows the non-invasive examination of each layer of the pre-lens tear film. The lipid layer at the surface, the intermediate aqueous phase and the underlying mucous when a breakup occurs.

The tearscope plus produces interference fringes within the aqueous phase. The thinner the aqueous phase, the brighter the fringes. The visibility of these fringes is dependant on the presence of a lipid layer above and the nature of the mucous coating underneath. A thick lipid layer will reduce the fringes visibility. A thin lipid layer will allow the fringes to be visible but dim. An absent lipid layer will allow the fringes to be fully visible and bright. The worse the lens surface mucous coating the brighter the aqueous fringes

A keeler microlens camera mounted on the ocular of the slit lamp is linked to a S-VHS VCR and a display television screen. The camera is also linked to an image capture device connected to the parallel port of a portable computer. This system allows the capture of high definition individual images as well as short movies in the MPEG format. All the images were saved in BMP format and coded to enable grading. Each tear film was photographed at three magnifications. Each photograph was graded for lipid presence, lipid visibility, lipid class and lipid contamination for the POTF. It was also graded for aqueous phase visibility and thickness, mucous coverage and lens surface visibility for the PLTF. When all images had been graded, the median of the three values was chosen and the code revealed for analysis.

2.2. Subjects

Nine subjects took part in the study, They were separated in three groups according to their POTF lipid classification

3 groups	3 subjects per group
Group 1	Best POTF lipid layer
Group 2	Normal POTF lipid layer
Group 3	Reduced POTF lipid layer

2.3. Lenses

Three lens types were used.

Lens 1:	PROCLEAR compatibles
Lens 2:	Soflens 66
Lens 3:	Acuvue

The study lens was PROCLEAR Biocompatible (Omafilcon A) and the control lenses were Acuvue (Etafilcon A) and Baush & Lomb Soflens 66 Medalist (Alphafilcon A).

PROCLEAR compatibles is a monthly replacement lens made with Omafilcon A. PROCLEAR is the only line of contact lenses with clearance from the US Food and Drug Administration (FDA) to carry the claim "may provide improved comfort for contact lens wearers who experience mild discomfort from symptoms relating to dryness during lens wear" (due to aqueous or evaporative tear deficiency, non- Sjögren's only). The claim is based on the use of phosphorylcholine (PC) technology. PC technology is incorporated into biocompatibles eyecare's soft contact lenses sold under the PROCLEAR brand name. The lens material has the ability to retain water within its matrix and thus limits the effect of pervaporation. Previous studies have shown favorable subjective results for the use of PROCLEAR compatibles when compared to other lenses.

2.4. Routine

POTF examination	POTF	vis 1
Lens insertion	PLTF	vis 2
30 min wear	PLTF	vis 3
6 hrs wear	PLTF	vis 4
Lens removal	POTF	vis 5

2.5. Lens Wear

Each subject wore one PROCLEAR Compatible Lens in one eye and one of the control lenses in the other. Each subject attended twice. All the pictures were coded and the grading was taken as the mode of the three images captured for each test time. For analysis purposes, the mean and standard deviation were calculated for all the combinations tested.

2.6. Lipid Layer Classification

2.6.1. Lipid Grading. The lipid layer appearance has been graded in 16 categories.

It has been shown in the past that the lipid layer appearance of the POTF was linked to its stability as measured by the NIBUT (non-invasive breakup time) (1). The latest research findings are now linking their appearance to their biochemical composition.

2.7. Lipid Layer Classification

The lipid layer has been classified according to its presence, integrity, visibility and appearance. Fifteen individual subclasses were determined for grading purposes to take into account differences of appearance in the inferior and superior part of the cornea ,variations induced by blinking and the presence of contact lenses. All the images were coded and graded as follows:

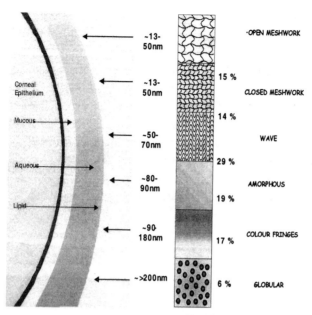

Figure 2. Preocular tear film lipid coverage.

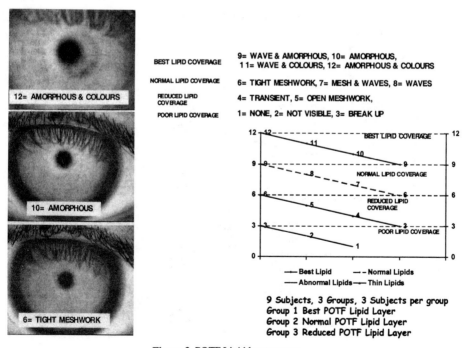

Figure 3. POTF Lipid layer groups.

Pre ocular tear film grading:

- Lipid Presence: Complete coverage is necessary to limit evaporation. An incomplete coverage has been shown to increase evaporation four fold. (1=absent (no visible lipid layer), 2=isolated coverage, 3=partial coverage, 4= complete coverage)
- Lipid Layer Visibility: The lower the visibility the thinner the lipid layer. (1=not visible, 2=low visibility , 3=average visibility, 4=high visibility)
- Lipid Type: The Tearscope plus appearance of the lipid layer is graded from the photographs. (1=none, 2=not visible, 3=breakup, 4=transient, 5=open, 6=tight, 7=mesh & waves, 8=waves, 9=wave & amorphous, 10=amorphous, 11=wave and colours, 12=amorphous and colours, 13=colours, 14=globular, 15=more than two, 16=other)

3. RESULTS

3.1. Lipid Coverage by Patient Groups

This displays the lipid coverage present at the surface of the pre-ocular tear film (POTF) and pre-lens tear film (PLTF) for the three patients groups. The data pooled all lens groups within the patient groups.

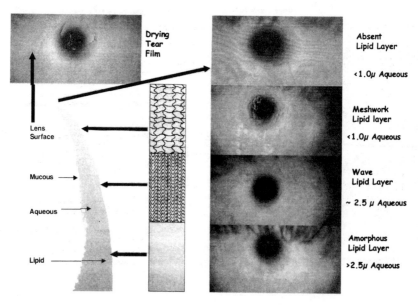

Figure 4. Pre lens tear film photgraphs.

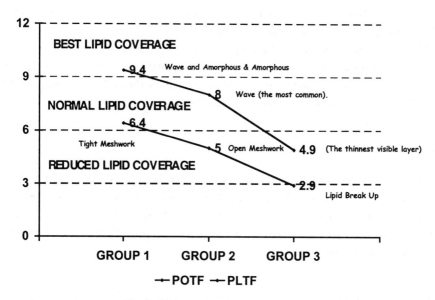

Figure 5. Lipid coverage by patient group.

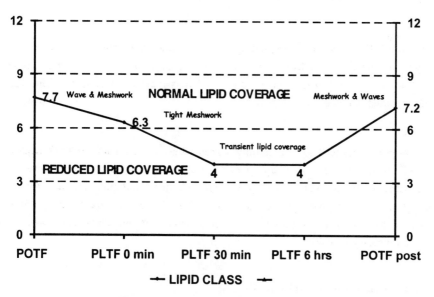

Figure 6. Lipid coverage: Effect of lens wear.

3.1. POTF Results

The data demonstrate the validity of the technique and grading for the determination of POTF quality for potential contact lens wearers.

Group 1 demonstrated the best lipid coverage (average grade: 9.4), wave and amorphous & amorphous. Group 2 demonstrated a normal lipid coverage (average grade: 8.0). Wave (the most common). Group 3 demonstrated a reduced lipid coverage (average grade: 4.9). Open meshwork (the thinnest visible layer).

3.2. PLTF Results

This displays an average decrease of three categories in lipid coverage for all three patient groups and all visits. A decrease is present for all lenses but its extent is determined by the lens material. This demonstrates one of the changes in the tear film structure induced by the use of contact lenses. It also explain why dryness symptoms are present in a greater percentage of patients during contact lens wear.

3.2.1. Group 1 Patients. The lipid grading decreased in average from 9.4 (wave and amorphous & amorphous = best coverage) to 6.4 (tight meshwork & wave = average coverage). Successful contact lens wear should be achieved in this patient group with most lens type and long wearing schedule.

3.2.2. Group 2 Patients. The lipid grading decreased in average from 8.0 (wave = average coverage) to 5.0 (open meshwork = reduced coverage). Contact lens wear should be achieved in this patient group but the choice of lens material, wearing conditions and wearing schedule becomes very important.

3.2.3. Group 3 Patients. The lipid grading decreased in average from 4.9 (open meshwork = reduced coverage) to 2.9 (breakup =poor coverage). Successful contact lens wear may only be achieved in this patient group with the best lens material that can provide an adequate lipid coverage and reduced fluid losses from the lens matrix. Contact lens wear will also be limited in environmental conditions that promote evaporation which will reduce their wearing schedule.

3.3. Lipid Coverage: Effect of Lens Wear

This displays the average lipid coverage for each visit using the pooled data from all patient groups and all lenses.

3.4. Visits

The lenses are worn for 6 hours, and the visit times are as follows:

Visit 1: Pre insertion: examination of the POTF.

Visit 2: Post insertion: examination of the PLTF.

Visit 3: 30 min post insertion: examination of the PLTF.

Visit 4: 6 hours post insertion: examination of the PLTF.

Visit 5: Post removal: examination of the POTF.

The average lipid grade of the POTF was 7.7 (waves = normal coverage). It decreased to 6.3 (tight meshwork = normal coverage) on the PLTF immediately after insertion. At the 30 minute visit it had further decreased to 4.0 (transient pattern = reduced coverage). This stayed at the same level throughout the day wear until the 6 hours visit. Following removal the lipid coverage returned towards its baseline value: 7.2 (meshwork & waves = normal coverage).

3.5. Lipid Coverage Effect of Lens Material

Three lens type were used: Acuvue, PROCLEAR compatibles and Baush & Lomb Soflens 66. The data were pooled for all visits and all patient groups and compared with the POTF average lipid cover. POTF average lipid layer: 7.7 (waves = normal coverage).

PLTF:

- PROCLEAR compatibles 5.9 (tight meshwork = normal coverage)
- Baush & Lomb Soflens 66 average lipid layer: 4.3 (transient layer = reduced coverage)
- ACUVUE average lipid layer: 3.0 (breakup = poor lipid coverage)

This demonstrate the advantages of PROCLEAR compatibles as it supports near normal lipid layer at the surface of the PLTF and provides protection from evaporation at the PLTF level. A transient lipid layer (as seen in the B&L 66) will provide only limited protection. The occurrence of a breakup of the lipid layer (as seen on the Acuvue) has been linked to a 4 fold increase in tear evaporation. (Craig and Tomlinson, 1997).

3.6. Lipid Coverage: Effect of Lens Wear

This graph follows the individual effect of each lens group throughout the study visit schedule. The PROCLEAR compatibles lens maintains a better lipid coverage at all visits.

- Visit 2: Post insertion. Average lipid layer: 6.9 (meshwork & wave = normal coverage).

- Visit 3: 30 min post insertion.average lipid layer: 5.1 (open meshwork = reduced coverage).

- Visit 4: 6 hours post insertion.average lipid layer: 5.6 (open & tight meshwork=normal coverage).

The Baush & Lomb Soflens 66 lens presents a poorer lipid coverage at all visits.

- Visit 2: post insertion. Average lipid layer: 6.4 (tight meshwork = normal coverage).

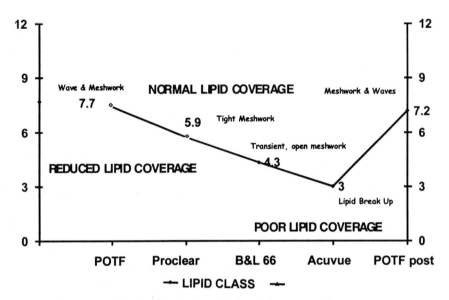

Figure 7. Lipid coverage: Effect of lens material.

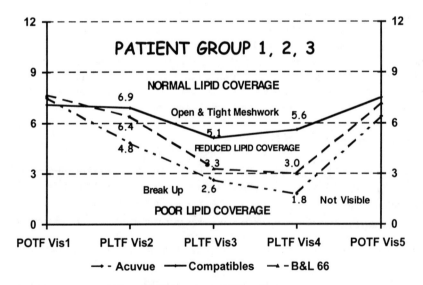

Figure 8. Lipid coverage: Effect of lens wear.

- Visit 3: 30 min post insertion. Average lipid layer: 3.3 (breakup = poor coverage).
- Visit 4: 6 hours post insertion. Average lipid layer: 3.0 (breakup = poor coverage).

The ACUVUE lens presents the poorest lipid coverage at all visits.
- Visit 2: post insertion. Average lipid layer: 4.8 (open meshwork = reduced coverage).
- Visit 3: 30 min post insertion. Average lipid layer: 2.6 (breakup = poor coverage).
- Visit 4: 6 hours post insertion. Average lipid layer: 1.8 (not visible = poor coverage).

3.7. Lipid Coverage for Group 1

The PROCLEAR compatibles lens supports throughout the 6 hours a very good lipid coverage.
- Visit 2: post insertion. Average lipid layer: 9.2 (wave & amorphous = best coverage).
- Visit 3: 30 min post insertion. Average lipid layer: 6.3 (tight meshwork = normal coverage).
- Visit 4: 6 hours post insertion. Average lipid layer: 8.2 (waves = normal coverage).

The Baush & Lomb Soflens 66 lens after the 30 min visit supports a reduced to poor lipid coverage.
- Visit 2: post insertion. Average lipid layer: 11..3 (wave & colours = best coverage).

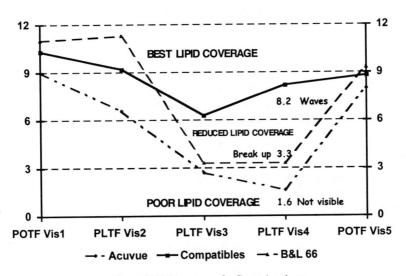

Figure 9. Lipid coverage for Group 1 patients.

- Visit 3: 30 min post insertion. Average lipid layer: 3.3 (breakup = poor coverage).
- Visit 4: 6 hours post insertion. Average lipid layer: 3.3 ((breakup = poor coverage).

The ACUVUE lens after the 30 min visit supports a poor lipid coverage that worsens with wear.
- Visit 2: post insertion. Average lipid layer: 6.6 (meshwork & wave = normal coverage).
- Visit 3: 30 min post insertion. Average lipid layer: 2.7 (breakup = poor coverage).
- Visit 4: 6 hours post insertion. Average lipid layer: 1.6 (not visible = poor coverage).

3.8. Lipid Coverage for Group 2 Patients

The PROCLEAR compatibles supports a near normal lipid coverage throughout the 6 hours.
- Visit 2: post insertion. Average lipid layer: 6.2 (tight meshwork = normal coverage).
- Visit 3: 30 min post insertion. Average lipid layer: 5.8 (tight meshwork = normal coverage).
- Visit 4: 6 hours post insertion. Average lipid layer: 6.0 (tight meshwork = normal coverage).

The Baush & Lomb Soflens 66 lens after the 30 min visit supports a reduced lipid coverage.
- Visit 2: post insertion. Average lipid layer: 5.0 (wave & colours = best coverage).
- Visit 3: 30 min post insertion. Average lipid layer: 3.3 (breakup = poor coverage).
Visit 4: 6 hours post insertion. Average lipid layer: 3.7 (transient = poor coverage).

The ACUVUE lens after the 30 min visit supports a poor lipid coverage that worsens with wear.
- Visit 2: post insertion. Average lipid layer: 5.0 (meshwork & wave = normal coverage).
- Visit 3: 30 min post insertion. Average lipid layer: 4.0 (breakup = poor coverage).
- Visit 4: 6 hours post insertion. Average lipid layer: 2.7 (not visible = poor coverage).

3.9. Lipid Coverage for Group 3 Patients

The PROCLEAR compatibles lens supports a reduced or partial lipid presence throughout the 6 hours.
- Visit 2: post insertion. Average lipid layer: 3.3 (open meshwork = reduced coverage).

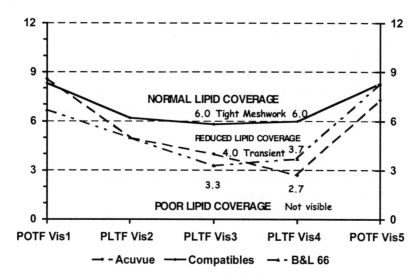

Figure 10. Lipid coverage for Group 2 patients.

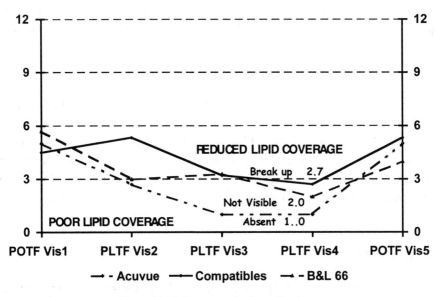

Figure 11. Lipid coverage for Group 3 patients.

914 J-P. Guillon *et al.*

- Visit 3: 30 min post insertion. Average lipid layer: 3.2 (breakup = poor coverage).
- Visit 4: 6 hours post insertion. Average lipid layer: 2.7 (breakup = poor coverage).

The Baush & Lomb Soflens 66 lens after the 30 min visit supports a partial lipid coverage that worsen with wear.
- Visit 2: post insertion. Average lipid layer: 3.0 (breakup = poor coverage).
- Visit 3: 30 min post insertion. Average lipid layer: 3.2 (breakup = poor coverage).
- Visit 4: 6 hours post insertion. Average lipid layer: 2.0 (not visible = poor coverage).

The ACUVUE lens does not support a lipid layer after the 30 min visit. Without this protection, evaporation of the aqueous phase will be accelerated, dryness symptoms will increase and failure to wear is the likely outcome.
- Visit 2: post insertion. Average lipid layer: 2.7 (breakup = poor coverage).
- Visit 3: 30 min post insertion. Average lipid layer: 1..0 (layer absent = no protection).
- Visit 4: 6 hours post insertion. Average lipid layer: 1.0 (layer absent = no protection).

3.10. Reduction in Lipid Coverage Grading

The reduction in lipid layer coverage induced by lens wear is greater for both the Baush & Lomb 66 lens and the ACUVUE lens than for the PROCLEAR compatibles lens.

3.11. Aqueous Layer Visibility

The PROCLEAR compatibles pre-lens tear film aqueous phase is not visible in most cases. This situation closely resembles that found in the normal pre-ocular tear film where a superficial lipid layer limits its visibility and a thick mucous layer inferiorly limits the formation of fringes. The Baush & Lomb Soflens 66 lens and ACUVUE lenses demonstrate aqueous fringes of higher visibility which are usually produced when the layer is thin and the inferior interface with the lens presents a poorer mucous coverage.

4. CONCLUSIONS

PLTF structure is both material and patient dependant. The grading of the POTF lipid layer is necessary when testing the wettability of contact lens materials and their resistance to evaporation. Patient with poor lipid layer should be fitted with materials resistant to evaporation that also provide a good support for a full pre-lens tear film that include a superficial lipid layer.

REFERENCES

1. Craig J.P. and Tomlinson A. Importance of the lipid layer in human tear film stability and evaporation. *Optometry and visual science*. 1997, 74(1): 8–13.

AN EVALUATION OF MUCIN BALLS ASSOCIATED WITH HIGH-DK SILICONE-HYDROGEL CONTACT LENS WEAR

Jennifer P. Craig, Trevor Sherwin, Christina N. Grupcheva,
and Charles N. J. McGhee

Discipline of Ophthalmology
University of Auckland
Auckland, New Zealand

1. INTRODUCTION

Conventional hydrogel contact lenses worn on an extended wear basis can cause many adverse responses,[1-3] which can be described as non-significant, significant and sight-threatening. Non-significant findings, such as asymptomatic infiltrative keratitis (AIK), occur almost as frequently in spectacle wearers.[4] Significant, specific symptomatic responses include contact lens acute red eye (CLARE), contact lens peripheral ulcer (CLPU), infiltrative keratitis (IK), superior epithelial arcuate lesion (SEAL) and contact lens papillary conjunctivitis (CLPC). The most serious and sight-threatening response is microbial keratitis (MK).[3] A risk of developing MK exists for any contact lens wearer, but with conventional extended wear soft lenses, this risk increases 10–20 times when compared to rigid gas permeable lenses.[2,5] The increased level of corneal hypoxia in the closed eye situation, in the presence of a soft contact lens, may play an important role in the development of MK. The stressed superficial cells may desquamate prematurely, leaving microscopic gaps on the corneal surface and providing a route of entry for infection.[6,7] This is of particular concern in the case of continuous lens wear, in which the corneal epithelium may never fully recover from repeated hypoxia.

The material properties of the contact lens dictate its oxygen permeability. Until recently, the aim was to design soft lens materials with increasingly high water content and make the lenses progressively thinner to enhance oxygen transmissibility. However, this is limited by increased lens pervaporation and difficulty in lens handling. Evolution of the new generation of extended wear lenses has witnessed development of a new material that has successfully combined silicone and hydrogel monomers to produce a lens which is both wettable and highly oxygen permeable. The addition of a vinyl carbamate group has facilitated copolymerization of the naturally hydrophobic silicone component with the hydrogel monomer, to allow increased material hydrophilicity. The outcome of this careful balance of silicone and hydrogel monomers is a transparent material with optimal transmissibility and wettability. In the final stages, an oxidizing plasma surface process enhances the hydrophilicity of the surface and maximizes its biocompatibility.[8]

The resultant material has recovery characteristics similar to conventional soft lenses, allowing good lens movement on the eye, but with a relatively low water content, typically 24% (Lotrafilcon A) or 36% (Balafilcon A) in current commercially available materials. This reduces lens dehydration during wear and provides good deposit resistance. To date, these lenses have been very successful,[9] with fewer adverse responses and good patient tolerance. Only one reported, but dubious, case of MK has occurred. Significant responses such as CLPU, CLPC, IK, CLARE and SEAL have occurred, but at relatively low levels with no long-term, irreversible complications or associated loss of vision.

However, one feature that appears to be peculiar to, or certainly much more common in, the new generation materials is mucin balls, or micro-spheroidal post-lens tear film-derived debris.[10] These mucin balls were previously noticed following overnight wear and appear innocuous from a clinical viewpoint,[11,12] but relatively little is known about them. In a controlled, prospective manner, this study investigates the course of development of mucin balls during periods of continuous wear and considers the structure and composition of this unique debris.

2. METHODS

Seven subjects (three females, four males) with a mean age of 33.9 ± 7.3 years were recruited from university staff to take part in the study. No subjects had worn contact lenses within the previous 3 months, except for one who was a (daily) rigid gas permeable lens wearer. Subjects had no history of ocular or systemic disease, and slit-lamp biomicroscopy prior to the study highlighted no corneal or tear-film abnormalities. Subjects were randomly assigned contact lenses of either Balafilcon A (PureVision™, Bausch and Lomb, USA) (n = 4), or Lotrafilcon A (Focus Night & Day™, Novartis, USA) (n = 3) material for wear on a continuous basis. Following lens fitting, visual acuity was recorded on a LogMAR acuity chart, and a biomicroscopic assessment was performed at 4 h, 1 day, and then at regular 4-day intervals over the period of wear to evaluate the clinical response of the eye to lens wear and determine variation in mucin ball number, size and distribution. Slit-lamp photographs were taken at each visit to document variations over a maximum of 30 days as recommended by the manufacturers. A careful history was taken at each time point, with attention to any discomfort, visual disturbance, lens removal or displacement, or use of rewetting drops.

Where significant numbers of mucin balls were detected, in vivo confocal microscopy was performed using the Confoscan II (Fortune Technologies, USA) and an x40 contact objective lens. No anaesthetic was used to avoid flushing out the mucin balls, and the examination was performed with the contact lens in situ, using a drop of transparent gel (Viscotears, Novartis) as a contact medium. Subjects observed a fixation target while the objective lens was brought toward the eye. When the gel was in contact with the contact lens surface, and good centration of the image was established, a series of up to 300 images was collected and archived to hard disk. On completing the period of wear, visual acuity was checked once more, the lenses were removed for laboratory examination and the corneas were examined and documented photographically, with and without instilled fluorescein sodium (Fig. 1).

To examine the contact lenses for mucin balls in vitro, the removed whole lenses were immediately fixed with gluteraldehyde 2.5% and viewed by epifluorescence microscopy at 514, 580 and 670 nm, and with transmitted light by phase contrast microscopy. In one case, the lenses were mounted in an embedding medium (OCT compound, Tissue-Tek, IN, USA), and 50-μm transverse cryosections were taken for analysis by phase contrast microscopy. To attempt to collect any mucin balls adherent to the ocular surface following lens removal, the surface was flushed with saline (NaCl 0.9%) and the fluid collected on filter paper strips. These strips were then air-dried and examined by epifluorescence and phase contrast microscopy.

3. RESULTS

Two subjects discontinued wear after one week. In one case this was due to a slightly flat-fitting lens falling out (Lotrafilcon A), and in the other, the lenses were removed for emotional reasons (Balafilcon A). One subject had the lenses removed at 26 days due to anticipated difficulty with a final follow-up appointment. All other subjects wore their lenses on a continuous basis for 30 days. No significant or non-significant adverse responses were noted in any subjects, and, other than two reports of lens displacement following overnight wear, subjects were asymptomatic throughout and did not use rewetting drops. All subjects at some point exhibited translucent, greyish mucin balls, which displayed reversed illumination on slit-lamp biomicroscopy, lying between the posterior surface of the contact lens and the cornea in one or both eyes (Fig.1a). However, the time of first appearance varied between 4 h and 8 days.

Variation in mucin ball number occurred across the period of wear, with up to 230 mucin balls beneath the lens of one subject. A tendency also existed for the number to increase with time in the absence of lens displacement. The reported lens displacement by subject 2 on day 18 on the morning of examination, and by subject 7 on day 7 prior to examination on day 8, coincided with a dramatic decrease in the number of mucin balls observed at these visits. The mucin balls varied in size in all eyes at each time period (estimated on slit-lamp biomicroscopy to range from 10 to 200 μm), but a tendency existed

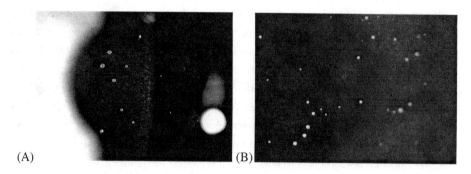

(A)   (B)

Figure 1. (A) Appearance of mucin balls beneath a silicone hydrogel contact lens after 15 days of continuous lens wear. **(B)** Typical fluorescein pooling in the corneal indentations created by mucin balls immediately following silicone hydrogel lens removal at 30 days.

for the size to increase with time, again excepting cases in which the lens was displaced. In two individuals, more mucin balls were found beneath the upper portion of the lens, and preponderance existed for the medial aspect in three subjects. Neither mucin ball count nor distribution was completely symmetrical, but the tendency was for mucin ball count to increase in both eyes with increasing length of wear (Fig. 2). Visual acuity, LogMAR range 0 to -0.2 (6/6 to 6/3.8), did not alter over the wear period. In the small group studied, no differences could be established between the two materials, either subjectively or in terms of development or distribution of mucin balls.

In vivo confocal microscopy revealed the mucin balls were essentially spherical in shape and were as small as 5 μm in diameter. They displayed highly reflective cores with a more poorly reflective, apparently translucent, outer layer (Fig. 3a). The diameter of the central core relative to the outer coating varied among mucin balls (Fig. 3b). From some of the transverse optical sections, the mucin balls extended in depth to around the level of the superficial nerve plexus, immediately adjacent to the basal epithelium. However, in no cases did epithelial cells overlie mucin balls.

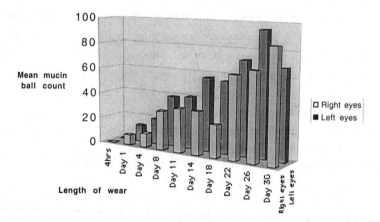

Figure 2. Mean mucin ball count over 30-day continuous wear period for the right and left eyes (n = 5).

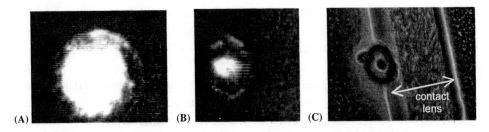

Figure 3. (A) and (B) Mucin ball appearance on in vivo confocal microscopy. (C) Mucin ball appearance on the back surface of a fixed, mounted and cryosectioned lens (in cross-section) by phase contrast microscopy.

On removal of the lenses, the majority of mucin balls remained on the ocular surface. These were easily dislodged by subsequent blinking, but left indentations in the corneal surface which showed pooling of fluorescein (Fig. 1b) and took between 2 and 24 h to fully resolve. In one individual several mucin balls adhered to the palpebral conjunctiva on eversion of the upper eyelid.

The scarcity of mucin balls on the whole lens surface made in vitro analysis difficult. En face, mucin balls could not be identified by epifluorescence microscopy at 514-, 580- or 670-nm wavelengths or by phase contrast microscopy. From the 50-μm cross-sectional cryosections of the lenses mounted in the embedding medium, apparently spherical structures were identified (between 5 and 50 μm in diameter), which appeared flattened on the aspect closest to the lens surface and slightly embedded in the contact lens surface. Several of these structures showed a dense central core (Fig. 3c). When examining the air-dried filter paper strips, very few structures resembling mucin balls were observed on transmission microscopy, and those that were visible exhibited no autofluorescence.

4. DISCUSSION

From clinical observations it appears development of spheroidal mucin balls beneath highly oxygen permeable silicone hydrogel contact lenses is a common but innocuous phenomenon.[11,12] Differential diagnoses include epithelial microcysts, although these microcysts tend to be smaller, are less commonly associated with the new generation highly oxygen permeable lenses than those with poorer permeability and do not exhibit fluorescein pooling on lens removal.[13] Conversely, epithelial vacuoles and bullae show unreversed illumination on slit-lamp examination due to their relative optical clarity; in addition, they also do not induce surface depressions.[13] Similar to mucin balls, trapped air bubbles beneath a contact lens can induce fluorescein pooling, described clinically as dimple veiling, but air bubbles are transparent and disappear on lens removal.[14] Very rarely, in the absence of an intact epithelium, topically applied ointment can form inclusions within the cornea.[15]

The current study showed development of mucin balls between the cornea and contact lens surfaces in all subjects, but highlighted no clinically significant adverse responses when the contact lenses were worn continuously for up to 30 days without the use of rewetting drops. The rate of mucin ball development was highly subject-dependent and exhibited only moderate symmetry of distribution between eyes. Such differences may be related to inter-subject variations in tear film composition and different cornea-lens interfacial characteristics.

Mucin balls have been found within the body of the cornea as epithelial inclusions.[11] No such inclusions were identified in the current study. The mucin balls did create indentations in the epithelium such that fluorescein pooling occurred on slit-lamp biomicroscopy following lens removal, and the base of some mucin balls could be detected at the level of the superficial nerve plexus on examination by in vivo confocal microscopy. However, in no cases did the epithelium cover the most superficial aspect of the mucin balls. This implies that while mucin balls may have produced a degree of corneal invagination in the subjects studied, they did not become entirely encapsulated by epithelial cells. It does, however, suggest superior rigidity of the mucin ball structure relative to the corneal epithelium, thus allowing preferential redistribution or compression of the superficial epithelial cells.

The study showed a trend for increased mucin ball numbers and size over the period of lens wear, suggesting that, in the absence of any lens displacement, the debris continues to accumulate. In addition to lens displacement, the use of rewetting drops during an extended period of lens wear may decrease the number of mucin balls (J. P. Craig, unpublished data).

To date, the composition of this post-lens debris has not been established, hindered not only by the primary difficulties in sample collection but probably also by the material's remarkable deposit resistance. Indeed, mucin balls may develop in the first instance as a result of this deposit resistance.[11] In conventional hydrogel materials, the lower resistance to deposition allows absorption of some tear film-derived deposits into the lens matrix. In the new generation lenses, this may not occur, and the cornea-lens surface sheer forces during eye movements may encourage debris to become enveloped by a translucent mucin coating, particularly overnight when the mucin content is proportionally increased relative to the aqueous phase.

Very few mucin balls existed on the contact lens surface post-wear following removal, fixing, embedding and cryosectioning. Those that were visible appeared less spherical than those observed by in vivo confocal microscopy, but this may be due to the sampling technique. It may be that only those mucin balls that become slightly embedded in the lens surface adhere to the lens on removal. Interestingly, however, the darkly shaded, dense cores of the spheroidal debris observed on the lens surfaces in the in vitro investigations using transmitted light correlated directly with the dense, highly reflective cores of the mucin balls observed with reflected light by in vivo confocal microscopy. The filter paper strips, which were used to collect debris flushed from the ocular surface after lens removal in an attempt to trap debris in the fiber meshwork, unfortunately also yielded few mucin balls. Further collection techniques and tests are being devised to ascertain the composition of this unique debris.

This study shows that new generation, highly oxygen permeable contact lenses intended for up to 30-day continuous wear are commonly associated with space-occupying, spheroidal post-lens debris. This debris appears with variable frequency, often asymmetrically, and has, at the very least, a short-term effect on the corneal epithelium, post-removal. The assumption to date is that the debris is composed predominantly of mucin; however, the exact aetiology, nature and consequence of these mucin balls remain to be fully explained.

REFERENCES

1. S.D. Zantos, and B.A. Holden, Ocular changes associated with continuous wear of contact lenses, *Aust J Optom.* 61:418–426 (1978).
2. O.D. Schein, R.J. Glynn, E.C. Poggio, J.M. Seddon, and K.R. Kenyon, The relative risk of ulcerative keratitis among users of daily-wear and extended-wear soft contact lenses. A case-control study. Microbial Keratitis Study Group. *New Eng J Med.* 321(12):773–778 (1989).
3. J.K. Dart, F. Stapleton, and D. Minassian, Contact lenses and other risk factors in microbial keratitis, *Lancet* 338(8775):1146–1147 (1991).
4. P.R. Sankaridurg, D.F. Sweeney, R. Gora, M. Naduvilath, M.K. Aasuri, B.A. Holden, and G.N. Rao, Adverse events with daily disposable contact lens and spectacle wear from a prospective clinical trial, *Invest Ophthalmol Vis Sci.* 41(4):S74 (2000).
5. K.H. Cheung, S.L. Leung, H.W. Hoekman, W.H. Beekhuis, P.G. Mulder, A.J. Geerards, and A. Kijlstra, Incidence of contact-lens-associated microbial keratitis and its related morbidity, *Lancet* 354(9174):181–185 (1999).
6. B.A. Holden, D.F. Sweeney, A. Vannas, K.T. Nilsson, and N. Efron, Effects of long-term extended contact lens wear on the human cornea, *Invest Ophthalmol Vis Sci.* 26(11):1489–1501 (1985).
7. G.A. Stern, Pseudomonas keratitis and contact lens wear: the lens/eye is at fault, *Cornea* 9(Suppl. 1):S36–38 (1990).
8. L. Alvord, J. Court, T. Davis, C.F. Morgan, K. Schindhelm, J. Vogt, and L. Winterton, Oxygen permeability of a new type of high Dk soft contact lens material, *Optom Vis Sci.* 75(1):30–36 (1998).
9. B. Long, S. Robirds, and T. Grant, Six months of in-practice experience with a high Dk Lotrafilcon A soft contact lens, *Contact Lens and Anterior Eye* 23(4):112–118 (2000).
10. S. Bourassa, and W.J. Benjamin, Transient corneal surface "microdeposits" and associated epithelial surface pits occurring with gel contact lens extended wear, *Int Contact Lens Clin.* 15:338–340 (1988).
11. N. Pritchard, L. Jones, K. Dumbleton, and D. Fonn, Epithelial inclusions in association with mucin ball development in high-oxygen permeablility hydrogel lenses, *Optom Vis Sci.* 77(2): 68–72 (2000).
12. J. Tan, L. Keay, I. Jalbert, D.F. Sweeney, and B.A. Holden, Tear microspheres (TMSS) with high Dk lenses, *Optom Vis Sci.* 76(12s):226 (1999).
13. J.E. Josephson, S. Zantos, B. Caffery, and J.P. Herman, Differentiation of corneal complications observed in contact lens wearers, *J Am Optom Assoc.* 59(9):679–685 (1988).
14. K. Zadnik, A case of dimple veiling/staining, *Contact Lens Forum* 13(2):68–69 (1988).
15. J.H. Krachmer, and D.A. Palay, *Cornea: Colour Atlas*, Mosby , St. Louis (1995).

CELLS COLLECTED FROM THE CORNEAL SURFACE IN SJÖGREN'S SYNDROME, DRY EYE, AND NORMALS

Salih Al-Oliky, Carolyn Begley, Gerald Lowther, and
Graeme Wilson

School of Optometry
Indiana University
Bloomington, Indiana, USA

1. INTRODUCTION

Cells collected by contact lens cytology (CLC) have been described as nucleated or ghost, based on the presence of a nucleus in fluorescence microscopy.[1] Cell ghosts resemble cells in the appearance of their cytoplasm, but are without nuclei. It is hypothesized that the instability of the tear film in the dry eye states increases the accumulation of both categories of cells on the corneal surface. Hence, more cells will be collected from the dry eye than from the normal eye. The purpose of this study was to determine if cells on the corneal epithelial surface are harvested more easily in the dry eye states.

2. MATERIALS AND METHODS

Subjects consisted of three groups of four women matched according to age. The four subjects in the first group suffered from Sjögren's syndrome (SS), confirmed either by biopsy of the salivary glands or blood tests. The second group consisted of subjects who reported dry eye (DE) symptoms and were positive for rose bengal and lissamine green staining. The third group consisted of control subjects with no history of DE, and no recent contact lens wear, eye surgery, or ocular medication use. The age of the subjects in each group had a range from 44 to 76 years.

Lacrimal Gland, Tear Film, and Dry Eye Syndromes 3
Edited by D. Sullivan *et al.*, Kluwer Academic/Plenum Publishers, 2002

For data collection each subject had two visits, one in the morning between 8:00 and 9:00 AM and the other in the afternoon between 3:00 and 4:00 PM. Each subject was fitted with a new Acuvue® disposable soft lens (58% H_2O) that was worn for a period of 2 min. Using sterile gloves, the lens was removed directly from the cornea without decentration onto the conjunctiva. The contact lens was draped over the end of a glass test tube with the front surface of the contact lens against the convex surface of the glass. The exposed back surface of the contact lens was then rinsed vigorously with 0.9% NaCl directed onto the lens from a syringe. These procedures (contact lens insertion, removal, and irrigation) were repeated three times in each subject using the same eye, making a total of four CLC collections on each visit.

The cell suspension irrigated from the contact lens was stained with acridine orange (AO) and Hoechst. Stained cells in suspension were filtered through a polycarbonate filter and examined using fluorescence microscopy. The combined AO-Hoechst stain permitted the differentiation of nucleated cells and cell ghosts.

Statistical comparisons between groups were made by one-tailed Mann-Whitney U-Test.

3. RESULTS

When viewed with AO staining, both nucleated cells and ghost cells showed a green fluorescence of the cytoplasm. Sometimes the nucleus was visible with AO. However, nucleated cells could be clearly distinguished from cell ghosts using a Hoechst filter. This filter enabled the nuclei to be visualized. The absence of a nucleus caused a structure to be classified as a cell ghost.

The great majority of shed cells were cell ghosts that fluoresced uniformly green in all groups of subjects. There were no differences regarding the appearance of these cells among all groups. However, there were differences in nucleated cells between the normal subjects and the other two groups. In the normal subjects, almost all Hoechst positive cells had intact nuclei and cytoplasm and they were present as single cells. Some nucleated cells in all groups of subjects had a nucleus with an irregular outline. These nuclei stained densely with Hoechst, but occasionally had dark circular spots, which could have been pores or areas that had not taken up Hoechst. Some nuclei had lost their staining and were scarcely visible. Other nuclei had disintegrated, leaving scattered fragments of chromatin throughout the cells. In both SS subjects and DE subjects groups of cells attached to each other were observed, and, occasionally, a nucleated cell attached to a ghost cell.

There was no statistically significant difference in the number of cells collected in the morning compared with the afternoon and, therefore, the morning and evening data have been averaged. For the normal versus the DE group and the SS group, there were statistically significant differences ($P < 0.05$) in the number of cells collected (Fig. 1). There were no such differences when the DE group was compared directly with the SS group ($P > 0.05$).

4. DISCUSSION

More cells were collected from the DE group and the SS group than from the normal group (Fig. 1). The epithelium of a patient with a compromised tear film, such as occurs in DE states, could experience a greater blink-induced shear force.[2] When the lubricating ability of the tear film is reduced, this blink-induced shear force may be enough to cause changes at the surface of the corneal epithelium. In DE states, the lack of proper lubrication no doubt contributes to discomfort and leads to damage to the ocular surface.

Cells collected from the back of a contact lens using CLC revealed large numbers of cell ghosts. These cell-like structures have the shape of squamous epithelial cells but do not have nuclei that stain with Hoechst. A similar finding has been detected previously in corneal epithelial cells using CLC.[1] They described ghost cells as dead cells that still had the overall shape of cells, although they lacked nuclei. This result is strikingly similar to the proposal of Teng[3] who stated that surface cells lose their nuclei prior to being removed from the cornea by tears and eyelid movement.

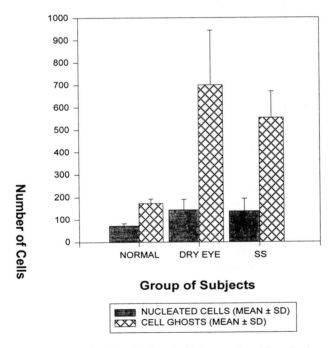

Figure 1. Shows the number of nucleated cells and cell ghosts collected from the three groups of subjects.

In an experiment on normal subjects in which samples were collected twenty times by repeated insertion and removal of a contact lens (CLC), Estil et al.[1] suggested that the increase in cell ghosts was due to the repeated frictional forces of the lens on the epithelial surface. In the experiment reported here, the lenses were inserted and removed only four times. The same procedure was applied to all groups of subjects, so the large number of cells in the DE subjects cannot be attributed only to the frictional forces from the lens.

The lack of any significant difference between the number of ghost cells in SS and DE subjects suggests that the compromise to the epithelium is equal, even if the etiology of the two conditions may be different. For example, SS could be due to damage to the lacrimal gland, perhaps because of lymphocytic infiltration; on the other hand there is a strong belief that a large number of DE states are due to deficiency of meibomian secretion. It might be that both of these very different pathological states manifest changes on the ocular surface that are similar. Although cell counts show potential for the diagnosis of DE states in general, they may not be helpful in the differential diagnosis of DE and SS.

REFERENCES

1. S. Estil, E.J. Primo, G. Wilson. Apoptosis in shed human corneal cells. Invest Ophthalmol Vis Sci. 41:3360–3364 (2000).
2. K. Nakamori, M. Odawara, T. Nakajima, T. Mizutani, and K. Tsubota. Blinking is controlled primarily by ocular surface conditions. Amer J Ophthalmol 124:24–30 (1997).
3. G.C. Teng. The fine structure of the corneal epithelium and basement membrane of the rabbit. Amer J Ophthalmol 51:278–97 (1961).

THE THICKNESS OF THE POST-LENS TEAR FILM MEASURED BY INTERFEROMETRY

Jason J. Nichols and P. Ewen King-Smith

College of Optometry
The Ohio State University
Columbus, Ohio, USA

1. INTRODUCTION

Lin et al.[1] measured the thickness of the tear film behind a soft contact lens by combining pachometry and mechanical measurement of contact lens thickness. They found an average thickness of 11.5 μm. Fogt et al.[2] measured the thickness of the tear film in front of a soft contact lens by analyzing oscillations in the reflectance spectrum from the front of the eye. These "spectral oscillations" are caused by interference of light waves reflected from different surfaces. This method has also been used to measure the thickness of the precorneal tear film, the corneal epithelium, Bowman's layer and the complete cornea.[3] The purpose of this study is to determine whether this method can also be used to measure the post-lens tear thickness, and, if so, to compare our results with those of Lin et al.[1]

2. METHODS

The left side of Fig. 1 represents, schematically, the relative strengths of reflections from 4 surfaces of the front of an eye wearing a contact lens. The right side of Fig. 1 represents 6 single or multiple layers that could yield interference effects. At some wavelengths, the reflections from any two surfaces will be in phase, yielding a maximum of reflectance, whereas at other wavelengths, the two reflections will be out of phase, yielding a minimum of reflectance. Thus, the reflectance spectrum from the front of the eye may show a series of maxima and minima called "spectral oscillations" (see Fig. 2). The frequency of these oscillations is proportional to the thickness of a single (or multiple) layer.[2-4] Note that layers B and E are predicted to give relatively weak interference effects, with contrasts below 0.5%.

Details of the experimental method were reported previously.[3] Twenty reflection spectra were recorded from each of 9 subjects wearing contact lenses. Six subjects wore Acuvue II, -0.50 D lenses with a back surface radius of 8.3 mm; these lenses have nearly parallel surfaces which should make it easier to record interference fringes. A further three subjects wore Acuvue I, -3.00 D lenses with a back surface radius of 8.8 mm, similar to those used by Lin et al.[1] The average age of the test subjects was 34 ± 9 y (mean $\pm$ sd). The exposure duration was 0.5 sec and the measurement area was nominally 33 x 35 μm. The average temperature and humidity were 25° C and 60%, respectively. Informed consent was obtained from each subject.

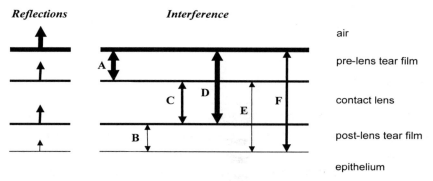

Figure 1. Reflections and interference effects from the pre- and post-lens tear films, and a contact lens.

3. RESULTS

Fig. 2 shows a reflection spectrum from the front of an eye wearing a contact lens. The spectrum has been analyzed by a least squares method[3] into interference contributions from the lipid layer[3,5] and from four of the layers in Figure 1. Because the contrast of these components are all greater than 0.5%, they are too large to come from layers B and E and thus must come from A, C, D and F. Fig. 3 is the Fourier transformation of this spectrum, showing corresponding peaks. It should be noted that because thickness is proportional to frequency, the horizontal scale has been given in units of thickness. This interpretation is confirmed by the finding that the direct estimate of pre-lens tear film thickness, A = 2.24 μm, agrees well with an indirect estimate D–C = 2.26 μm. An indirect estimate of post-lens tear thickness is given by F–D = 2.55 μm. These values are based on the least squares fit of Fig. 2.

Fig. 4a is a plot of indirect (D–C) versus direct (A) estimates of pre-lens tear thickness for the subject in Figs. 2 and 3. There is good agreement between the two estimates. Altogether, 119 spectra in the 9 subjects showed contributions from 4 layers, as in Figs. 2 and 3. Fig. 4b is a similar plot to Fig. 4a, showing average results for all 9 subjects. Again, there is good agreement between the two estimates. These findings support the validity of determining the thickness of a tear layer by indirect estimation (D–C or F–D).

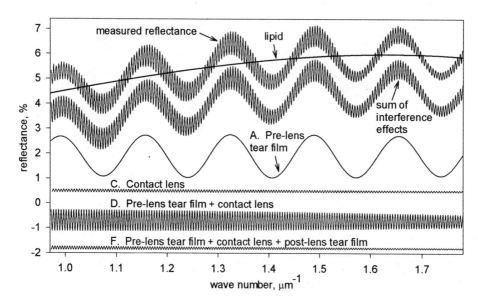

Figure 2. Reflectance spectrum from the front of an eye wearing a contact lens, showing contributions from interference effects.

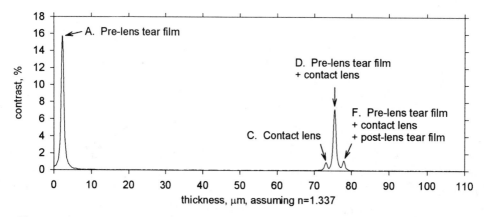

Figure 3. Fourier transform of the reflection spectrum in Fig. 2, showing 4 peaks due to interference effects.

To support the interpretation that F–D is the post-lens tear thickness, observation of interference from layers B and E (Fig. 1) would be helpful. However, these interference effects are difficult to observe because they are weak, and if the pre- and post-lens tear thicknesses are similar, as in Figs. 2 and 3, B and E are often masked by the strong interference effects A and D. In one subject, the post-lens tear film was often considerably thicker than the pre-lens tear film, making it possible to observe the direct interference from layer B in 8 spectra. A plot of indirect F–D versus direct B estimates of the post-lens tear thickness demonstrated reasonable agreement between the two estimates (Spearman R = 0.929, P < 0.001), supporting the conclusion that F–D is an estimate of post-lens tear thickness. For one spectrum in this subject, all 6 interference effects were observed; good agreement was found between the three estimates of post-lens tear thickness, B = 4.05 μm, E–C = 4.03 μm, and F–D = 4.25 μm, again supporting the validity of our analysis.

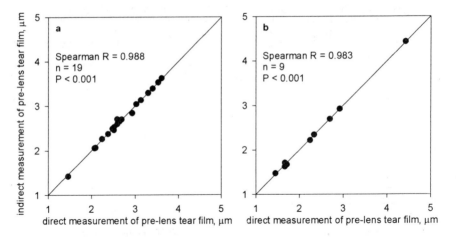

Figure 4. Comparison of direct and indirect measurement of pre-lens thickness. **(a)** For the subject of Figs. 2 and 3. **(b)** Averages for all 9 subjects.

The above evidence demonstrates that F–D can be used as an estimate of the post-lens tear film. This method yields a mean post-lens tear thickness of 2.44 ± 0.48 μm between subjects which is similar to the mean pre-lens tear thickness of 2.35 ± 0.93 μm derived from the direct estimate (layer A). There was no significant difference between the -0.50 D and -3.00 D lenses for either pre- or post-lens tear thickness.

4. CONCLUSIONS

Reflection spectra from the front of an eye wearing a contact lens, typically show interference effects from 4 (single or multiple) layers corresponding to: A, the pre-lens tear film; C, the contact lens; D, the pre-lens tear film + contact lens; F, the pre-lens tear film + contact lens + post-lens tear film. In one subject, a fifth interference effect, from B, the post-lens tear film was observed in 8 spectra, and in one spectrum, a sixth interference effect was observed, corresponding to E, the contact lens + post-lens tear film.

Direct A and indirect D–C estimates of pre-lens tear thickness showed good agreement with each other. Mean pre-lens tear thickness was 2.35 ± 0.93 μm between subjects (n = 9). This is similar to our previous estimate[3] of precorneal tear thickness, 2.7 ± 0.4 μm. Similarly, direct B and indirect F–D and E–C estimates of post-lens tear thickness showed good agreement with each other. Mean post-lens tear thickness was 2.44 ± 0.48 μm. This value of post-lens tear thickness is considerably smaller than the value of 11 to 12 μm found by Lin et al.,[1] even when the contact lenses were the similar. This discrepancy deserves further investigation.

ACKNOWLEDGMENTS

We thank the following for advice and encouragement: Drs. J.T. Barr, B.A. Fink, N. Fogt, R.M. Hill, M.C. Lin, K.K. Nichols, K.A. Polse, J.P. Schoessler, and G.S. Wilson. This work was supported by the Ohio Lions Eye Research Foundation. We thank Dean J.P. Schoessler for equipment funds from the College of Optometry.

REFERENCES

1. M.C. Lin, A.D. Graham, K.A. Polse, R.B. Mandell, and N.A. McNamara, Measurement of post-lens tear thickness, *Invest. Ophthal. Vis. Sci.* 40:2833 (1999).
2. N. Fogt, P.E. King-Smith, and G. Tuell, Interferometric measurement of tear film thickness by use of spectral oscillations, *J. Opt. Soc. Am. A* 15:268 (1998).
3. P.E. King-Smith, B.A. Fink, N. Fogt, K.K. Nichols, R.M. Hill, and G.S. Wilson, The thickness of the human precorneal tear film: evidence from reflection spectra. *Invest. Ophthal. Vis. Sci.* 41:3348 (2000).
4. Y. Danjo, M. Nakamura, and T. Hamano, Measurement of the precorneal tear film thickness with a non-contact optical interferometry film thickness measurement system, *Jpn. J. Ophthalmol.* 38:260 (1994).
5. T. Olsen, Reflectometry of the precorneal tear film, *Acta Ophthalmol.* 63:432 (1985).

TEAR LIPID COMPOSITION OF HYDROGEL CONTACT LENS WEARERS

Cécile Maïssa,[1] Michel Guillon,[1,2] Karine Girard-Claudon,[1]
and Philip Cooper[1]

[1]Contact Lens Research Consultants
London, United Kingdom
[2]Centre Optométrique
Université de Paris-Sud
Orsay, France

1. INTRODUCTION

The lipid layer plays a major role in retarding the evaporation of the tear film during the inter-blink period and in influencing the stability of the tear film. Previous investigations have shown that the stability of the tear film is influenced by the differential mixing of lipids. This is observed by clinicians as different visible patterns within the lipid layer.[1,2] However, the visualization of the patterns, while possible with specialized biomicroscopic techniques such as the Tearscope[TM], is a difficult clinical technique. Furthermore both the tear lipid pattern and tear film stability are different for preocular and the prehydrogel contact lens tear film.[1-3] It is, therefore, postulated that the composition of the lipid layer is influential on tear film stability and different for the preocular and prehydrogel contact lens tear film.

2. METHODOLOGY

Tear samples were taken from the temporal part of the lower tear prism and prepared for analysis immediately. For contact lens wearers, tear samples were taken while the lens was worn. The tear film composition was analyzed using a sensitive high performance liquid chromatography (HPLC) technique on individual unstimulated tear film samples (2 μl). The technique had previously identified 22 types of lipids forming five general lipid classes. For each sample, both the percentage of the different lipids present and their concentrations were calculated using internal lipid standards.

The results obtained were analyzed statistically by the Mann-Whitney exact test to identify significant differences between the preocular and the prehydrogel lens tear films, as well as differences between gender and age group. Multifactorial analysis of the combined effects of gender, age and contact lens wear on tear lipid composition was also carried out using a Chi-Square Automated Interaction Detector (CHAID)[TM 4,5].

3. RESULTS

3.1. General Compositional Characteristics

Important intersubject differences in the percentage of different classes of lipids and the amount of lipid present in PHTF and POTF (Table 1) were present. However, in all cases non-polar lipids (cholesterol esters: 30.2% and 0.47 mg/ml and cholesterol: 25.7% and 0.32 mg/ml) dominated the polar lipids (phospholipids/triglycerides: 14.0% and 0.33 mg/ml; fatty acid: 11.5% and 0.24 mg/ml).

Table 1. Descriptive statistics of tear lipid composition [a]

	Composition (%)		Concentration (mg/ml)	
	Mean	Median	Mean	Median
Cholesterol Esters	30.2	25.0	0.47	0.36
Cholesterol	25.7	21.0	0.32	0.27
Monoglycerides	18.6	10.5	0.33	0.15
Fatty Acids	11.5	8.1	0.28	0.12
Phospholipids/Triglycerides	14.0	10.4	0.55	0.14

[a]For PHTF, n = 124; for POTF, n = 34

3.2. PHTF and POTF Lipid Composition

PHTF and POTF lipid compositions revealed significantly different lipid levels and profiles. The PHTF had a lower percentage of phospholipids and triglycerides (8.0 % vs. 18.5%, $P < 0.001$) and monoglycerides (4.8% vs. 22.4%, $P = 0.012$) than the POTF. Conversely, PHTF tear samples were characterized by a higher percentage ($P = 0.024$) of cholesterol (22.8%) than POTF (17.6%).

A tendency towards lower tear lipid levels was also detected in the PHTF compared to the POTF. The decrease in lipid levels was especially significant for phospholipids and triglycerides (POTF = 0.26 mg/ml; PHTF = 0.09 mg/ml, $P < 0.001$) and monoglycerides (POTF = 0.27 mg/ml; PHTF = 0.07 mg/ml, $P = 0.010$) lipid types.

3.3. Effect of Gender

The effect of gender was more evident for the POTF than for the PHTF. It was characterized by significantly higher concentrations of cholesterol esters (Male = 0.27mg/ml; female = 0.49mg/ml, P = 0.002) and phospholipid/triglyceride esters (Male = 0.20mg/ml; female = 0.34mg/ml, P = 0.038) for the females than males. However, POTF samples from male subjects revealed a statistically significant, higher percentage of cholesterol than female subjects (Male = 20.5%; female = 11.8%, P = 0.022).

For the PHTF, the differences between genders were limited to a statistically significant, higher concentration of cholesterol for males than females (male = 0.37mg/ml; female = 0.28mg/ml, P = 0.041). The diminished effect of gender on the PHTF lipid composition may have been due to the imbalance in sample size between the two groups (male n = 32; female n = 92) or simply due to contact lens wear, which could be the more dominant factor.

3.4. Effect of Age

No significant effect of age was found for the POTF lipid composition. However, significant compositional differences were observed between the two subgroups of subjects aged 40 or less and those over 40. The differences were limited to two lipid families: cholesterol esters and fatty acids. The percentage of fatty acids in the tear film was significantly lowered for subjects over 40 years of age (≤40 years = 9.5%; >40 years = 1%, P = 0.049), whereas the concentration of cholesterol esters was significantly higher (≤40 years = 0.32 mg/ml; >40 years = 0.39 mg/ml, P = 0.019).

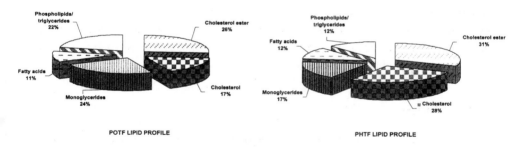

Figure 1. PHTF and POTF Lipid Profiles.

3.5. Multifactorial analysis

Multifactorial analysis with CHAID™ provided a unique way to examine data and to identify important groupings of cases. It allowed the detection of key factors that identified group membership and summarized statistically significant patterns or relationships. The most influential factors among contact lens wear, gender and age, were identified for each lipid class: (1) Contact Lens wear was the determinant factor for both the percentage and concentration of phospholipids/triglycerides; (2) Gender was the most influential factor for monoglycerides; (3) Both contact lens wear and gender were found to be discriminant factors for cholesterol, creating three significant grouping of cases (Nonwearer female< Nonwearer male < soft contact lens wearer).

4. DISCUSSION

The current technique using HPLC of individual non-reflex tear samples was sensitive to the main lipid classes. The results obtained confirmed that the tear lipid composition is highly variable between subjects, but that, on average, cholesterol esters and cholesterol dominate the composition. This tends to confirm the model of a dual lipid layer with a thin polar lipid layer at the interface with the aqueous layer and a thicker non-polar outer layer.

The prehydrogel tear film has different lipid compositional characteristics than the preocular tear film. In particular, it has lower levels of polar lipids that enhance tear film spreading and stability. It is postulated that the differential lipid composition of the prelens tear film and preocular tear film is a contributing factor of the difference in stability of that layer and of the overall tear film.

In conclusion, the results obtained in this investigation led us to conclude that tear lipid composition is affected by contact lens wear. For lens wearers, the PHTF has a lower concentration of polar lipids and greater level of nonpolar cholesterol based lipids. Gender and age are secondary factors that influenced the tear lipid composition, both in contact lens wearers and nonwearers. The study results are consistent with the lipid layer influence on the stability and spreading characteristics of the tear film observed clinically.. The polar lipids that facilitate the spreading of the lipid layer over the aqueous, and nonpolar lipids, forming the outer hydrophobic lipid layer, were the two key factors to assess. The preocular tear film has different characteristics than the prehydrogel tear film. It is postulated that the differential lipid composition of the prelens tear film and tear film is a contributing factor of the difference in stability.

REFERENCES

1.   M.Guillon, C.Maissa, E.Styles. Relationship between POTF structure and stability. In *Lacrymal Gland, Tear Film and Dry Eye Syndromes 2*. Ed. D A Sullivan, Plenum Press, Boston:401–405 (1998).
2.   M.Guillon, C.Maissa. Relationship between hydrogel pre lens tear film structure and stability. *Optom. Vis. Sci,* 73 (12S): 161 (1996).
3.   M.Guillon, E.Styles, J.P.Guillon, C.Maissa. Pre ocular tear film characteristics of non wearers and soft contact lens wearers. *Optom. Vis. Sci,* 74:273–279 (1997).
4.   G.Kass. An exploratory technique for investigating for large quantities of categorical data. *Applied Statistics*, 29(2), 119–127 (1980).
5.   D.Biggs, B.De Ville, E.Suen. A method of choosing multiway partitions for classification and decision trees. *J. Applied Statistics*, 18:49–62 (1991).

CONJUNCTIVAL CHARACTERISTICS OF CONTACT LENS WEARERS AND NONWEARERS AND THEIR ASSOCIATION WITH SYMPTOMATOLOGY

Michel Guillon,[1,2] Karine Girard-Claudon,[1] Cécile Maïssa,[1] and Philip Cooper[1]

[1] Contact Lens Research Consultants
London, UK
[2] Centre Optométrique
Université de Paris-Sud
Orsay, France

1. INTRODUCTION

Unlike corneal staining, conjunctival staining has not been extensively investigated. However, interest in the association between conjunctival anomalies and dry eye symptomatology recently has increased. Histopathological studies have pointed to a modified goblet cell secretion in dry eye pathology[1] and a reduction in goblet cell count in long-term contact lens wearers.[2]

Rose bengal has been the main vital stain used to clinically evaluate conjunctival tissue damage. The main drawback is significant ocular discomfort associated with instillation, which has limited its broad use in clinical practice. Several groups have used lissamine green (LG) and proposed it as a substitute. LG produces a similar staining profile but does not cause ocular discomfort.[3-5]

The aims of our study were to characterize conjunctival staining by using sodium fluorescein (SFL) and LG and assess their association with dry eye symptomatology. The hypotheses tested were as follows: (1) dry eye symptomatology is associated with a bulbar conjunctival anomaly, and (2) LG is a better vital stain than SFL to evaluate conjunctival staining and its association with dry eye symptomatology.

Lacrimal Gland, Tear Film, and Dry Eye Syndromes 3
Edited by D. Sullivan *et al.*, Kluwer Academic/Plenum Publishers, 2002

939

2. MATERIALS AND METHODS

All patients attending the clinic at Contact Lens Research Consultants (CLRC) for the first time are examined using a battery of tests. These include quantification of their symptomatology by McMonnies questionnaires[6,7] and evaluation of conjunctival integrity with SFL and LG. The subjects in this investigation were 79 nonwearers (46 males, 33 females, age 37 ± 15 years, mean±SD) and 102 soft contact lens wearers (36 males, 66 females, age 33 ± 10 years) who were free of ocular anomalies or active pathology when first seen at CLRC. Nonwearers and soft contact lens wearers attended with their own spectacle correction; soft contact lens wearers did not wear their contact lenses on the day of the visit.

LG was first instilled with a sterile strip hydrated with saline. LG staining was observed with the slit-lamp biomicroscope using white light and graded in four positions (nasal, temporal, inferior, superior). The same procedure was repeated with SFL using a blue filter and a yellow enhancing filter to grade the staining. The stainings observed were graded with a forced-choice 5-point scale (0 = No staining, 1 = Slight staining, 2 = Mild staining, 3 = Moderate staining, 4 = Severe staining).

Comparative statistics between symptomatic and asymptomatic subjects for nonwearing and contact lens-wearing groups were carried out using the Mann-Whitney test. This test was also used to establish differences between contact lens wearers and non-contact lens wearers. Chi Square Automated Interaction Detector (CHAID) analysis[8,9] was used to identify discriminant factors of dryness symptomatology among clinical findings.

3. RESULTS

The results comparing nonwearers and soft contact lens wearers showed a low incidence of clinically significant staining (Grade 2 - mild or higher) for both groups. However, soft contact lens wearers presented more staining than nonwearers (Table 1).

Table 1. Conjunctival staining - maximum - descriptive statistics and distribution for soft contact lens wearers and non-contact lens wearers

		Median (Range)	Mode	None (0.0) (%)	Slight (0.5→1.0) (%)	Mild (1.5→2.5) (%)	Moderate (3.0) (%)	Severe (4.0) (%)
SFL	Nonwearers	0.5 (0.0→2.5)	0.0	46	51	3	0	0
	SCLW	0.5 (0.0→3.0)	0.0	38	47	15	1	0
LG	Nonwearers	0.0 (0.0→3.0)	0.0	60	29	10	1	0
	SCLW	0.0 (0.0→3.0)	0.0	51	35	11	3	0

LG and SFL stainings were significantly greater for the symptomatic than asymptomatic group for nonwearers and soft contact lens wearers (nonwearers: LG P = 0.001, SFL P < 0.001; soft contact lens wearers: LG P = 0.003, SFL P = 0.018). Moreover, the incidence of Grade 2 (mild) or higher staining was greater for the symptomatic than asymptomatic group with both vital stains (nonwearers: SLF 13% vs. 2%, LG 25% vs. 5%; soft contact lens wearers: SFL:15% vs. 8%, LG 18% vs. 5%).

The difference between symptomatic and asymptomatic groups was greater with LG than SFL. However, LG and SFL stainings were significantly correlated, but the degree of correlation ($r^2 = 0.48$ for nonwearers, $r^2 = 0.57$ for soft contact lens wearers) was too low to have a good predictive value.

CHAID analysis identifies the factors (discriminants) that best predict the characteristics of a population (response), in the present, case symptomatology. For the nonwearing group, higher levels of both LG and SFL staining were associated with symptomatology. The discriminant parameter was the maximum staining observed for any quadrant for both stains (Figs. 1–3), and the discriminant value was Grade 1.5 staining. For the soft contact lens-wearing group only LG was associated with symptomatology; the discriminant value was Grade 1 staining (Fig. 3).

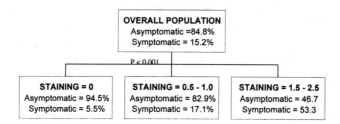

Figure 1. Symptomatology and maximum SFL staining (scale 0–4) for nonwearers' analysis by CHAID.

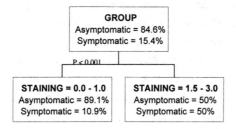

Figure 2. Symptomatology and maximum LG staining for nonwearers' analysis by CHAID.

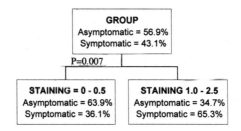

Figure 3. Symptomatology and maximum LG staining for soft contact lens wearers' analysis by CHAID.

4. DISCUSSION

The results showed conjunctival staining was greater for soft contact lens wearers than nonwearers, confirming previous findings that the conjunctiva, even away from the contact lens edge, is affected by contact lens wear.[1] However, both groups produced a low incidence of clinically significant staining.

The greater incidence of clinically significant conjunctival staining for the symptomatic than asymptomatic group confirmed conjuctival involvement in dry eye symptomatology for nonwearers and soft contact lens wearers. The inclusion of the assessment of the conjunctiva in any investigation of patients complaining of dry eye symptoms is essential. Failure to do so ignores a very important component of the histopathological process. It is our contention that dry eye symptoms are associated with increased friction due to poor lubrication, and the major site of the abnormal friction, particularly for contact lens wearers, is in the bulbar conjunctival area.

Although LG and SFL were statistically correlated, a greater difference in the incidence of Grade 2 (mild) or higher staining was recorded with LG than SFL between symptomatic and asymptomatic groups. In the nonwearing population, a grade equal to or above 1.5 (maximum value) is a discriminant value between symptomatic and asymptomatic with both vital stains. In the soft contact lens wearers, only LG staining was associated with symptomatology, with a discriminant value of Grade 1 or above. Therefore, both LG and SFL staining should be evaluated to enhance the quality of dry eye diagnosis as previous authors advised.[5]

In conclusion, our study shows dryness symptoms are associated with increased signs of conjunctival damages. The evaluation of conjunctival staining must be included as a test routinely performed during evaluation and diagnosis of patients with dry eye symptoms. Both LG and SFL stains should be used.

REFERENCES

1.   A.J.Bron. Non-Sjögren's dry eye: pathogenesis diagnosis and animal models. In *Lacrimal Gland, Tear Film and Dry Eye Syndromes.* Ed DA Sullivan, Plenum Press, New York: 471–488 (1994).

2.   E.Knop, H.Brewitt. Morphology of the conjunctival epithelium in spectacle and contact lens wearers. A light and electron microscopic study. *Contactologia* 14E: 108–120 (1992).

3.   F.J.Manning, S.R.Wehrly, G.N.Foulks. Patient tolerance and ocular surface staining characteristics of lissamine green vs. rose bengal. *Ophthalmology* 102(12): 1953–7 (1995).

4.   A.K.Khurana. et al. Tear film profile in dry eye. *Acta Ophthalmologica* 69: 79–86 (1991).

5.   K.A.Kinney. Detecting dry eye in contact lens wearers. *Contact Lens Spectrum* 5 : 21–28 (1998).

6.   C.W.McMonnies, A.Ho. Marginal dry eye diagnosis: history versus biomicroscopy, in Holly FJ (Ed): *The Pre-Ocular Tear Film in Health, Disease and Contact Lens Wear.* Lubbock, TX, Dry Eye Institute:32–40 (1986).

7.   M.Guillon, J.C.Allary, J.P.Guillon, G.Orsborn. Clinical management of regular replacement: Part 1. Selection of replacement frequency. *ICLC* 19: 104–120 (1992).

8.   G. Kass. An exploratory technique for investigating large quantities of categorical data. *Applied Statistics* 29: 119–127 (1980).

9.   D.Biggs, B.de Ville, E.Suen. A method of choosing multiway partitions for classification and decision trees. *J Applied Statistics* 18: 49–62 (1999).

DRY EYE SYMPTOMATOLOGY OF CONTACT LENS WEARERS AND NONWEARERS

Michel Guillon,[1,2] Philip Cooper,[1] Cécile Maïssa,[1]
and Karine Girard-Claudon[1]

[1]Contact Lens Research Consultants
London, United Kingdom
[2]Centre Optométrique
Université de Paris-Sud
Orsay, France

1. INTRODUCTION

Dry eye is a common symptom in optometric practice. It is defined as a disorder of the tear film, due to tear deficiency and/or tear evaporation that causes damage to the ocular surface and is associated with symptoms of ocular discomfort.[1] McMonnies et al.[2] reported a significantly higher frequency of dry eye symptoms in contact lens wearers than in nonwearers, indicating contact lens wear is a provocative condition for tear dysfunction. Caffery et al.,[3] in a study of the prevalence of dry eye symptoms in a large population, reported dry eye symptoms in 21.7% of noncontact lens wearers and 50.1% of contact lens wearers. The aims of this investigation were (1) to determine the frequency and nature of dry eye symptoms in nonwearers and soft contact lens wearers presenting in an optometric practice, and (2) to identify the questions that best predict dry eye symptomatology for the development of a short but efficient screening routine for use in practice

2. METHODS

The relevant McMonnies questionnaires were administered to 502 soft contact lens wearers and 309 nonwearers to determine symptomatology (Table 1). The original McMonnies[2] questionnaire to detect dry eye pathologywas later modified by Guillon[4] to deal more specifically with contact lens wearers. Both versions are fairly long but were

Lacrimal Gland, Tear Film, and Dry Eye Syndromes 3
Edited by D. Sullivan *et al.*, Kluwer Academic/Plenum Publishers, 2002

945

administered as applicable in the current investigation. The data obtained for each
population were compared by Kruskal-Wallis, and the relevant diagnostic value of the
individual questions was determined by Chi square Automated Interaction Detector
(CHAID) analysis.[5,6]

Table 1. Population demographics descriptive statistics

	n	Gender		Age
		Male	Female	years
Noncontact lens wearers	309	135	174	36 ± 13^a
Soft contact lens wearers	502	166	336	34 ± 10

[a]mean ± sd

3. RESULTS

The prevalence of dry eye symptomatic patients, as detected by the McMonnies
questionnaire, was significantly greater ($P < 0.001$) for soft contact lens wearers (43%)
than nonwearers (15%). Higher scores were recorded for soft contact lens wearers than
nonwearers (Figs. 1 – 2).

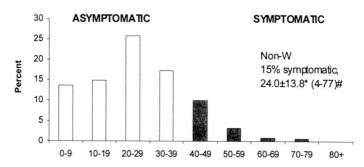

Figure 1. McMonnies score distribution for nonwearers. Cutoff point for dry eye symptomatology: score =
40.*mean ± sd, #range.

For the nonwearers, dry skin (68% of symptomatic subjects vs. 27% of asymptomatic
subjects) and dryness of the mouth or nose (85% of symptomatic subjects vs. 47% of
asymptomatic subjects) were significantly more frequent ($P < 0.001$) for the symptomatic
subjects pointing toward an association with systemic conditions as possible etiology. For
the soft contact lens wearers problems with air conditioning (71% of symptomatic subjects
vs. 30% of asymptomatic subjects), central heating (60% of symptomatic subjects vs. 19%

of asymptomatic subjects), and smoky environments (79% of symptomatic subjects vs. 36% of asymptomatic subjects) were significantly more frequent (P < 0.001) for the symptomatic subjects, supporting the hypothesis of the evaporative etiology of contact lens-related dry eyes.

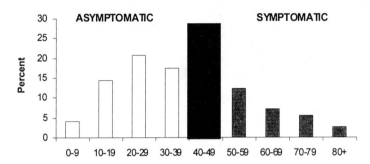

Figure 2. McMonnies score distribution for soft contact lens wearers. Cutoff point for dry eye symptomatology: score = 40.*mean ± sd, #range.

The aim of CHAID analysis was to identify the factors that best predict the characteristic of a population, in the present case, whether or not the patients were symptomatic according to the overall McMonnies score. For both groups the symptom of ocular dryness was the best detector of the symptomatology as per the McMonnies score (Fig. 3). For the nonwearers a patient that reported "never" experiencing dryness symptoms was in 95% of cases asymptomatic. Further, if burning symptoms were "never" experienced, in 99% of cases the patient was asymptomatic. Conversely, if the patients experienced dryness symptoms "often" or "sometimes," the percentage of symptomatic subjects more than doubled, compared to the reference population (34% vs. 15%).

For soft contact lens wearers a patient who reported never experiencing "dryness" and "scratchiness" was in 94% of cases symptomatic, according to the McMonnies questionnaire. Conversely, patients that reported "often" or "constantly" experiencing dryness symptoms and problems in the presence of smokers were in 94% of cases symptomatic. For the patients that reported "sometimes" experiencing symptoms of dryness, the question was not discriminant, but if they also had no problems in the presence of smokers, they were in 80% of cases asymptomatic.

The best predictive associated conditions were use of makeup and medications for nonwearers and use of makeup and dry eye treatment for soft contact lens wearers (Fig. 4). Nonwearers that did not use makeup or medications were in 95% of cases asymptomatic according to McMonnies. Soft contact lens wearers that use makeup were symptomatic in 77% of cases.

NON WEARERS

EFFECT OF SYMPTOMS

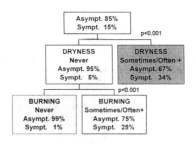

SOFT CONTACT LENS WEARERS

EFFECT OF SYMPTOMS

Figure 3. Identification of best predicting symptoms of overall symptomatology for nonwearers and soft contact lens wearers analysis by CHAID.

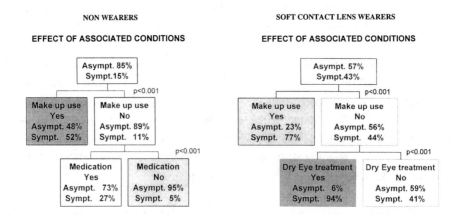

Figure 4. Effect of associated conditions for nonwearers and soft contact lens wearers analysis by CHAID.

4. DISCUSSION

A re-classification of McMonnies scores into three classes seems more clinically appropriate and involves two cutoff points, 40 and 50: (1) Asymptomatic score < 40, (2) Marginal "mild" symptomatology score 40–50, and (3) Confirmed moderate to severe symptomatology score > 50. This classification produces an incidence of mild symptoms of 10% for nonwearers and 16% for contact lens wearers. For moderately symptomatic subjects, the incidence is 5% for nonwearers and 28% for contact lens wearers.

The relative diagnostic value of the individual questions determined by CHAID showed that the key question related to the symptom of dryness. For the nonwearers a further discriminant factor was burning. For the soft contact lens wearers if dryness was absent, the discriminant factor was scratchiness; if dryness was present "more than often," the discriminant factor was problems in smoky environments. For soft contact lens wearers and nonwearers makeup use was a discriminant factor. For the nonwearers, if makeup was not used, the discriminant factor was the use of medication. For the soft contact lens wearers, if makeup was not used, the discriminant factor was a history of dry eye treatment. The use of makeup may have a role in tear film contamination and/or an association with meibomian gland dysfunction.

In conclusion, dry eye is a significant problem with current soft contact lenses. This may be due to excessive evaporation or disruption of the tear film through contamination. The question that best predicts dry eye symptomatology in routine practice without administering a full dry eye questionnaire is that relating to the incidence of ocular dryness. The other symptomatology questions that should be included are scratchiness and burning, and the associated factors are problems in the presence of smokers and use of makeup.

REFERENCES

1. M.Rolando, A.Macri, T.Carlandrea, G.Calabria. Use of a questionnaire for the diagnosis of tear film related ocular surface disease, in Sullivan et al (Ed): *Lacrimal Gland, Tear Film, and Dry Eye Syndromes 2*: Plenum Press, New York:821–825, (1998).

2. C.W.McMonnies, A.Ho. Marginal dry eye diagnosis: history versus biomicroscopy, in Holly FJ (Ed): *The Pre-Ocular Tear Film in Health, Disease and Contact Lens Wear*. Lubbock, TX, Dry Eye Institute:32–40 (1986).

3. B.E.Caffery, D.Richter, T.Simpson, D.Fonn, M.Doughty, K.Gordon. The Canadian Dry Eye Epidemiology Study, in Sullivan et al (Ed): *Lacrimal Gland, Tear Film, and Dry Eye Syndromes 2*: Plenum Press, New York: 805–806 (1998).

4. M. Guillon, J.C.Allary, J.P.Guillon, G.Orsborn. Clinical management of regular replacement: Part 1. Selection of replacement frequency. *ICLC*, 19:104– 120 (1992).

5. G.Kass. An exploratory technique for investigating large quantities of categorical data. *Applied Statistics*, 29:119–127 (1980).

6. D.Biggs, B.De Ville, E.A.Suen. A method of choosing multiway partitions for classification and decision trees. *J Applied Statistics*, 18:49–62 (1991).

CONTACT LENS-INDUCED PAPILLARY CONJUNCTIVITIS IS ASSOCIATED WITH INCREASED ALBUMIN DEPOSITS ON EXTENDED WEAR HYDROGEL LENSES

Maxine E. Tan, Gulhan Demirci, Damon Pearce, Isabelle Jalbert, Padmaja Sankaridurg, and Mark D. P. Willcox

Cooperative Research Centre for Eye Research and Technology
University of New South Wales
Sydney, Australia

1. INTRODUCTION

Contact lens papillary conjunctivitis (CLPC) may be a combination of immunological and mechanical responses to contact lenses and their deposits, resulting in a hypersensitivity reaction.[1] Enlarged papillae on the tarsal conjunctiva, redness, itching, increased mucus, and decreased contact lens tolerance are typical symptoms.[2] CLPC is a major reason for discontinuing contact lens wear;[3] however, the link between lens deposits, mechanical trauma, the immunological response and ultimate discontinuation of lens wear is still unclear. In view of the known immunological aspects of CLPC, our aim was to determine the levels of human serum albumin[4] (HSA), secretory IgA (sIgA, a constitutive protein) and lactoferrin (a regulated protein[5]) on extended wear contact lenses (EWCL).

2. METHODS

2.1. Lenses

EWCL from our current studies were collected at the end of 1 month of wear or during CLPC. The lenses were stored in 2 ml sterile PBS at 4°C until analysis. Previous studies have demonstrated no loss of immunoreactivity with this procedure. These lenses were then randomly assigned to each of the enzyme-linked immunosorbant assays (ELISAs) detailed below.

Lacrimal Gland, Tear Film, and Dry Eye Syndromes 3
Edited by D. Sullivan *et al.*, Kluwer Academic/Plenum Publishers, 2002

951

2.2. ELISAs

ELISAs for HSA, sIgA and lactoferrin were performed directly on the lenses. Known concentrations of each protein were added to nitrocellulose discs and used as standards. Non-specific binding was calculated from unworn lenses or control discs. Standards and lenses were blocked with either 0.2% Tween-20/PBS (PBST) or 3% BSA/PBST. All discs and lenses were incubated with either α–human HSA, α–human sIgA or α–human lactoferrin peroxidase conjugated antibodies at 1/1000 in PBST. 2,2'-AZINO-bis(3-ETHYLBENZ-Thiazoline-6-SULFONIC ACID) peroxidase substrate was added for 10 min with gentle agitation and the absorbance read at 405 nm, from a 200-ul aliquot. Standard curves were produced and the lens absorbance values converted into absolute amounts of each protein.

2.3. Data Analysis

The non-parametric Mann-Whitney test was used to examine differences between lenses from subjects with and without CLPC.

3. RESULTS

HSA surface deposits were significantly higher on EWCL from subjects with CLPC ($P < 0.0001$, Fig. 1). In addition, lenses from six of the CLPC subjects were analysed prior to the event. No significant difference existed in the amount of HSA on these lenses (median: 0.075 µg/lens, range: 0–0.82 µg/lens, n = 11) compared to the CLPC lenses (median: 0.11 µg/lens, range: 0–1.0 µg/lens, n = 32). The amounts of sIgA and lactoferrin were not significantly different between the CLPC and non-CLPC groups (Figs. 2 and 3).

4. DISCUSSION

Significant HSA deposits on EWCL may be the end product of an inflammatory response cascade. The increase in conjunctival permeability during CLPC may be due to inflammatory mediators, such as leukotrienes as seen in guinea pigs.[6] LTC_4 is elevated in the tears of CLPC subjects[7] and is probably produced by the mast cells, which are abundant during CLPC. The effect of increased LTC_4 may result in redness, conjunctival edema and increased mucus secretion. Increased immunoglobulins in tears and on lenses of CLPC subjects may also reflect leakage of conjunctival vasculature.[5,8] Tears from CLPC subjects have reduced lactoferrin levels,[9] possibly leading to activation of the complement pathway, in turn causing activation of mast cells and basophils, contraction of smooth muscle and increased capillary permeability. In addition, an increase in the complement proteins C3 and factor B in the tears of CLPC subjects has been reported.[10]

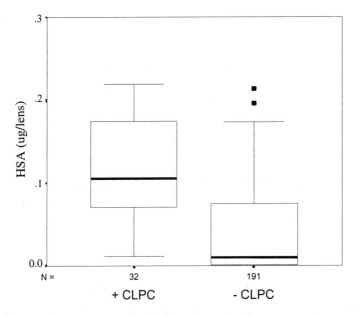

Figure 1. Boxplot of surface-bound HSA on EWCL from subjects with and without CLPC. The dark bars represent the median values and the black dots represent outliers.

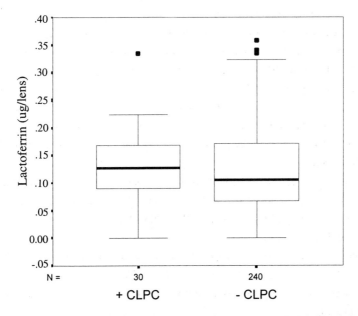

Figure 2. Boxplot of surface-bound lactoferrin on EWCL with and without CLPC. The dark bars represent the median values and the black dots represent outliers.

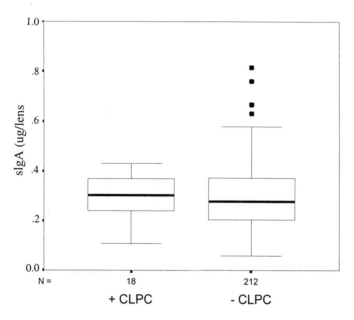

Figure 3. Boxplot of surface-bound sIgA on EWCL with and without CLPC. The dark bars represent the median values and the black dots represent outliers.

For the 6 subjects who developed CLPC during our study, no significant difference in HSA deposits occurred before and during CLPC, although the values follow the same trend as the data in Fig. 1. This may be due to the smaller number of lenses obtained prior to a CLPC event. There was also no difference in the lactoferrin deposits in subjects with and without CLPC, an observation that is consistent with a previous study..[11] This supports our finding of no difference in the amounts of lactoferrin on the surface of CLPC lenses. Lactoferrin in the tears of CLPC subjects is decreased, and may lead to increased coating of lenses with bacteria and their products.[9] Why we do not see decreased lactoferrin deposits on the CLPC lenses remains unclear; perhaps it is preferentially bound to lenses during a CLPC event.

No significant difference in sIgA levels existed between the groups. This agrees with a report showing no difference in tear IgA levels between people with and without CLPC.[4] However, IgE, IgG and IgM increased in the tears of CLPC subjects[5,8] and deposits of IgM[11] increased during CLPC.

Our future work will involve prospective studies to determine if increased HSA on the surface of contact lenses precedes clinical signs of CLPC. We will determine if other proteins, in particular, other immunoglobulins, are altered on the lenses or in the tears of subjects with CLPC.

REFERENCES

1. Z.K. Larzarou, Contact lens associated giant papillary conjunctivitis, *Clin Eye Vis Care*, 11:33–35 (1999).

2. M.R. Allansmith, D.R Korb, J.V. Greiner, A.S. Henriquez, M.A. Simon, and V.M. Finnemore, Giant papillary conjunctivitis in contact lens wearers, *Am J Ophthalmol.*, 83:697–708 (1977).

3. P.R. Sankaridurg, D.F. Sweeney, S. Sharma, R. Gora, T. Naduvilath, L. Ramachandran, B.A. Holden, and G.N. Rao, Adverse events with extended wear of disposable hydrogels: the first 13 months of lens wear, *Ophthalmol.*, 106:1671–1680 (1999).

4. P.C. Donshik and M. Ballow, Tear immunoglobulins in giant papillary conjunctivitis induced by contact lenses, *Am J Ophthalmol.*, 96:460–466 (1983).

5. R.J. Fullard and C. Snyder, Protein levels in nonstimulated and stimulated tears of normal human subjects, *Invest Ophthalmol Vis Sci.*, 31:1119–1126 (1990).

6. R.K. Gary, D.F. Woodward, A.L. Nieves, L.S. Williams, J.G. Gleason, and M.A. Wasserman, Characterization of the conjunctival vasopermeability response to leukotrienes and their involvement in immediate hypersensitivity, *Invest Ophthalmol Vis Sci.*, 29:119–125 (1988).

7. M.T. Irkec, M. Orhan, and U. Erdener, Role of tear inflammatory mediators in contact lens-associated giant papillary conjunctivitis in soft contact lens wearers, *Ocular Immunol Inflamm.*, 7:35–38 (1999).

8. Y. Barishak, A. Zavaro, Z. Samara, D. Sompolinsky, An immunological study of papillary conjunctivitis due to contact lenses, *Curr Eye Res.*, 3:1161-1168 (1984).

9. M. Ballow, P.C. Donshik, P. Rapacz, and L. Samartino, Tear lactoferrin levels in patients with external inflammatory ocular disease, *Invest Ophthalmol Vis Sci.*, 28:543–545 (1987).

10. M. Ballow, P.C. Donshik, and L.Mendelson, Complement proteins and C3 anaphylatoxin in the tears of patients with conjunctivitis, *J Allergy Clin Immunol.*, 76:473–476 (1985).

11. N.R. Richard, J.A. Anderson, Z.G. Tasevska, and Binder P.S., Evaluation of tear protein deposits on contact lenses from patients with and without giant papillary conjunctivitis, *CLAO*, 18:143–147 (1992).

SURFACE PROTEIN PROFILE OF EXTENDED-WEAR SILICON HYDROGEL LENSES

Damon Pearce, Maxine E. Tan, Gulhan Demirci, and
Mark D. P. Willcox

Cooperative Research Centre for Eye Research and Technology
University of New South Wales
Sydney, Australia

1. INTRODUCTION

Protein deposition onto the surfaces of contact lenses during wear is problematic since it promotes bacterial adhesion[1] and growth.[2] For instance, when Etafilcon A or Polymacon contact lenses are surface-coated with albumin, adhesion of *P. aeruginosa* increases.[1] When Etafilcon A lenses are coated with an artificial tear fluid, the adhesion of some strains of *Serratia marcescens* is enhanced,[2] and when lysozyme is adsorbed to these lenses, they exhibit greater adhesion of *Staphylococcus aureus*.[3] Lens deposits may also be involved in the production of giant papillary conjunctivitis.[4] Furthermore, over a long time period, protein deposition can result in reduced visual acuity.

Recently, silicon hydrogel contact lenses that can be worn for extended periods of up to 30 nights have become available. In general, protein adsorbs more readily to less hydrophilic surfaces compared to surfaces with anionic charges such as the Etafilcon A lens surface.[5] One exception to this generality is lysozyme, which adsorbs more readily to anionic surfaces.[6,7] The silicon hydrogel lenses are generally hydrophobic and require surface modification to increase biocompatability and hydrophilicity. Therefore, it is of interest to determine the types of protein that adsorb onto the surface of these lenses during wear. In this study we evaluate the adsorption of lactoferrin, lysozyme, secretory IgA (sIgA), and albumin. Lactoferrin and lysozyme are two of the major regulated tear proteins of the lacrimal gland, sIgA is the main immunoglobulin in tears, and albumin is derived from plasma leakage into tears.

Lacrimal Gland, Tear Film, and Dry Eye Syndromes 3
Edited by D. Sullivan *et al.*, Kluwer Academic/Plenum Publishers, 2002

2. METHODS

2.1. Lenses

Silicon hydrogel extended wear contact lenses (Focus Night and Day, CIBA Vision, Georgia, USA) were collected from study participants (n = 45) at the end of one month of wear, and stored in 2 ml of sterile PBS at 4°C until analysis. Previous studies[8,9] have demonstrated no loss of immuno-reactivity with this procedure. All lens wearers were asymptomatic at the time of lens collection. All subjects signed informed consent forms, and human ethics committee approval was granted.

2.2. Enzyme-Linked Immuno-sorbant Assay (ELISA) for Surface-bound Protein

ELISAs for albumin, sIgA, lysozyme and lactoferrin were performed directly on the lenses. Known concentrations of each protein were added to nitro-cellulose discs and used as standards. Non-specific binding (NSB) was calculated from unworn lenses or control discs. Standards and lenses were blocked with either 0.2% Tween-20/PBS (PBST) or 3% BSA/PBST. All discs and lenses were incubated with either α–human HSA, α–human sIgA α-lysozyme or α-human lactoferrin peroxidase conjugated antibodies at 1/1000 in PBST. 2,2'-azino-di-[3-acyl-benzthiazolinsulphonate] peroxidase substrate was added for 10 min and gently agitated, and the absorbance of a 200 μl aliquot was read at 405 nM. Standard curves were produced, and the lens absorbance values were converted into absolute amounts of each protein.

3. RESULTS

The silicon hydrogel lenses bound all four proteins studied (Fig.1). The number of samples, median amount of protein, and range of individual protein amounts adsorbed onto lenses is shown in Table 1. A strong contra-lateral correlation was observed for all proteins (R^2 = 0.26–0.70), suggesting that the large variation was due to inter-individual variance, rather than intra-individual, or assay variation.

Secretory IgA appears to have a particular affinity to the surface of silicon hydrogel contact lenses. Although large amounts of lysozyme adsorb onto ionic, high water, hydrogel lenses,[4] the same was not true for silicon hydrogel lenses. Of all the proteins evaluated, albumin bound least to the surface of these lenses.

4. DISCUSSION

The technique used in the present study was chosen for its ability to measure only surface bound proteins, since the large antibodies are unable to penetrate the lens matrix. However, the use of antibodies can be problematic, yielding varying results depending on the degree of protein degradation after adsorption. In an attempt to minimise this effect, polyclonal antibodies were used.

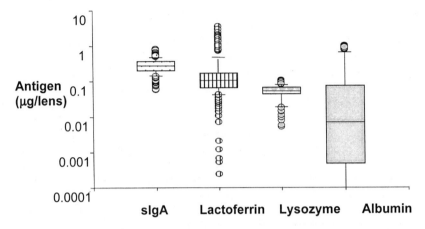

Figure 1. Total exposed surface protein on silicon hydrogel lenses.

Table 1. Amounts of individual proteins adsorbed onto silicon hydrogel lenses after 30 days of extended wear

Protein	n	Median (ng)	Range (ng)
sIgA	182	282	58–814
Lactoferrin	209	113	0–1970
Lysozyme	68	54	5–106
Albumin	170	4	0–1006

The variations in the levels of lactoferrin and albumin were considerably higher than the variations in the levels of lysozyme and sIgA. This may indicate that the lenses themselves are variable, especially given the need to surface coat these lens types. However, it is more likely that these variations demonstrate inter-subject variability. Indeed, in another study from our laboratory, also presented within these procedings, we demonstrated that increased albumin adsorption onto the lenses is associated with contact lens-induced peripheral conjunctivitis (CLPC; Tan, Demirci, Pearce, Jalbert, Sankaridurg, Willcox). Furthermore, variations in the levels of lactoferrin in tears are associated with giant papillary conjunctivitis, which is a more severe form of CLPC.[10]

This is the first report evaluating the types and levels of proteins that deposit onto the surface of silicon hydrogel lenses. The signficance of these data may be clarified as more data is generated concerning the types of adverse responses associated with the wear of this type of contact lens.

REFERENCES

1.   R.L. Taylor, M.D.P. Willcox, T. Williams and J. Verran Modulation of bacterial adhesion to hydrogel contact lenses by albumin. Optom Vis Sci. 75: 23–29 (1998).

2.   E.B.H. Hume and M.D.P. Willcox. Adhesion and growth of *Serratia marcescens* on artificial closed eye tears (ATF) soaked hydrogel contact lenses. Aust NZ J Ophthalmol. 25: s39–s41 (1997).

3.   A. Thakur, A. Chauhan and M.D.P. Willcox. Effect of lysozyme on adhesion of *Staphylococcus aureus* strains, isolated from contact lens induced peripheral ulcers, to contact lenses and toxin release. Aust NZ J Ophthalmol. 27: 224–227 (1999).

4.   M. Refojo and F. Holly Tear protein adsorption on hydrogels: a possible cause of contact lens allergy. Contact Lens J. 3: 23–35 (1977).

5.   J.L. Bohnert, T.A. Horbett, B.D. Ratner and F.H. Royce Adsorption of proteins from artificial tear solutions to contact lens materials. Invest Ophthalmol Vis Sci 29: 362–373 (1988).

6.   X.M. Deng, E.J. Castillo and J.M. Anderson. Surface modification of soft contact lenses: silanization, wettability and lysozyme adsorption studies. Biomaterials 7: 247–251 (1986).

7.   R.A. Sack, S. Sathe, L.A. Hackworth, M.D.P. Willcox, B.A. Holden and C.A. Morris. The effect of eye closure on protein and complement deposition on Group IV hydrogel contact lenses: relationship to tear flow dynamics. Curr Eye Res 15: 1092–1100 (1996).

8.   D. Pearce, G. Demirci and M.D.P. Willcox. Estimation of complement component C3c on daily and extended wear hydrogel contact lenses. Optom Vis Sci. 76 (12s). 223 (1999).

9.   D.J. Pearce, G. Demirci and M.D.P. Willcox. A novel method used for the determination of adsorbed proteins on ex-vivo extended-wear soft contact lenses. Aust Soc Biomaterials. 9th Annual meeting, Canberra. Proceedings. p 32 (1999).

10.   M.J. Velasco Cabrera, J. Garcia Sanchez and F.J. Bermudez Rodriquez. Lactoferrin in tears in contact lens wearers. CLAO J. 23: 127–129 (1997).

THE DETECTION OF KININ ACTIVITY IN CONTACT LENS WEAR

Aisling M. Mann and Brian J. Tighe

Biomaterials Research Unit
Aston University
Birmingham, United Kingdom

1. INTRODUCTION

Many factors can be, and have been, attributed to the appearance of complications in lens wear, but the greatest is associated with deposition. Reduced acuity, irritation and inflammatory responses are often referred to as adverse reactions arising as a result of deposition. In this study, particular attention was paid to the potential role of adsorbed proteins in activating, mediating and/or stimulating a host immune response, i.e., the hypothesis that the adsorption of certain proteins from the tears and ocular surfaces may actively affect successful lens wear. In particular, the purpose of this study was to investigate the presence of a group of proteins previously undiscovered in the ocular environment. The intention was to target a family of proteins/glycoproteins that have become prominent recently in a variety of inflammatory responses and disorders at many other mucosal associated sites around the body, e.g. in nasal rhinitis and in joint inflammation. The protein cascade in question is the kinin family of inflammatory mediators. The aim was to investigate their presence in the ocular environment, specifically in relation to contact lens wear, and consequently assess the implications of their discovery. High molecular weight kininogen (HMWK), with its central role in kinin responses, was investigated initially as the marker of kinin activity, with subsequent members examined thereafter.

1.1. The Kinin System

The kinins are a family of potent bioactive peptides that directly mediate inflammation and produce the end product bradykinin, a nonapeptide with many pro-

Lacrimal Gland, Tear Film, and Dry Eye Syndromes 3
Edited by D. Sullivan *et al.*, Kluwer Academic/Plenum Publishers, 2002

961

inflammatory functions. The consequences of kinin activation and bradykinin generation include an increase in vascular permeability, vasodilation, pain stimulation, smooth muscle contraction and an ability to stimulate arachidonic acid metabolism.[1] Initiation is achieved on contact with a variety of negatively charged species, both natural and synthetic, which result in Hageman factor (coagulation factor XII) activation and cleavage.[2] This cleavage induces a kinin response through the activation of kallikrein,[3] induces the activation of the intrinsic coagulation system and initiates fibrinolysis. Hageman factor is the core protein in kinin activation; other important proteins are HMWK and prekallikrein. A summary of the kinin pathway and its interactions with other key host reaction cascades are illustrated in Fig. 1.

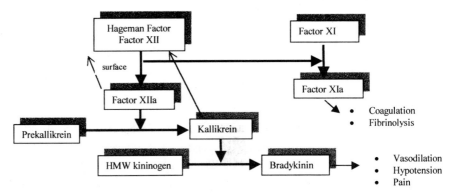

Figure 1. A simple guide to the kinin system.

1.2. The Kinins in Mucosal Disorders

The kinins have been reported to be involved in edema formation of hereditary angioedema, evidenced by an increase in plasma kallikrein levels found in blisters induced on the skin[4] and during allergic rhinitis responses. HMWKs have also been detected in nasal secretions.[5] Recent studies have also demonstrated kinin responses in bronchoalveolar fluids in asthmatics,[6] and bradykinin causes an increase in pulmonary inflation pressure, mucus secretion and breathlessness.[7] The physiological role of the kinin system in these responses remains unclear; it has yet to be seen whether kinin activation, in some or all instances, is a direct cause of the resultant, sometimes problematic symptoms. Alternatively, its activation and involvement could be a consequence of an excessive or natural host response. Whatever the cause, the end result of these aforementioned kinin implicated disorder studies, among others, highlight the increasing significance of the kinins, particularly at mucosal surfaces and secretions. It further dictates the importance of assessing the presence of kinins at the ocular surfaces and in relation to contact lens wear.

1.3. The Search for Kinin Activity

The inspiration and motivation to look for HMWK and thus kinin involvement in this way was three-fold. First, with the ongoing detection of a host of new plasma derived proteins found in tears, the aim was to gain a greater understanding regarding the extent of plasma protein leakage into the tears and consider their potential consequences. Second, HMWK, a key protein in the mediation of inflammation, has been implicated in a variety of disorders and allergic responses at other mucosal body sites. Finally, particularly in relation to our interest in contact lens wear, is the interaction between the lens and tear film in vivo. One point of interest is the fact that the kinin system can be initiated on contact with a variety of negatively changed surfaces one of which may be the anionic contact lens.

2. MATERIALS AND METHODS

No methods were previously available for the study of the deposition of individual proteins (with the exception of lysozyme) on single lenses worn over a variety of wear modalities. However, a recently developed procedure was employed in the investigation of HMWK in lens wear.[8,9] This technique combined the use of an efficient protein extraction technique and the application of an immunoassay used to investigate the resultant eluate. Ultraviolet spectrophotometry (UV) was used to measure the levels of total protein deposition and calculate the quantities of protein extracted. A highly sensitive double electro-immunodiffusion assay counter immunoelectrophoresis (CIE) was used to analyse the resultant eluate and exact a profile of the individual proteins involved and, specifically, to assess the kinin proteins in the lens resultant eluate extracts. CIE was also used to investigate the presence of kinin activity in normal, non-stimulated tears.

2.1. Laboratory Protocol

Total pre-extraction protein deposition in the bulk and matrix of the lens was measured by UV spectrophotometry. Proteins were extracted from the lenses utilising a urea/SDS/DTT/Tris solution at 90°C for 3 h. Post-extraction proteins levels were determined by UV to ascertain extraction efficiency. CIE-based protein assay was used to analyze the extracted protein deposits for high molecular weight kininogen and kallikrein.

2.2. Clinical Protocol

Etafilcon A is peculiar, because it is the only material type capable of being used in all three wear regimes and, thus, it was used to compare the three different modalities: daily disposable wear, conventional daily wear and extended wear. This allowed us to estimate the effect of wear time and modality on kinin activity, avoiding possible material variance.

The effect of material differences was provided in the daily disposability modality where three material types were employed. (1) Daily Wear Lenses: Patients (n = 10) wore Group IV (Etafilcon A) Acuvue lenses for 28 days on a daily wear regime, using ReNu as the multipurpose cleaner. (2) Extended Wear Lenses: A different set of patients (n = 22) wore Group IV (Etafilcon A) Acuvue (Vistakon) for a 7 day extended wear time period. (3) Daily Disposable Lenses: In a preliminary study patients (n = 4) wore the 3 disposable types: Group IV (Etafilcon A) Acuvue (Vistakon), Group II (Nelfilcon A) Focus Dailies (CibaVision), and Group II (Hilafilcon A) Soflens (Bausch & Lomb). All of these patients were entered into a controlled, randomised, cross over clinical study, which analysed all the lenses for deposited kinin components comparing wear time and modality.

3. RESULTS AND DISCUSSION

The kinin activity marker protein HMWK was not found in open eye basal tears, which suggests that it is not present in normal tears of non-lens wearers. It may alternately indicate its presence at levels below the sensitivity of the assay utilized. It is anticipated that further studies with greater sensitivities will fail to detect HMWK at levels of any clinical significance in the tears of the non-lens wear eye. In marked contrast, the anticipated lens related HMWK influx was clearly observed. HMWK, a glycoprotein previously undetected in tears and in lens wear, was discovered in the lens deposits of daily disposable, conventional daily and extended wear lenses. Whether HMWK is absent or present at low levels in basal tears, this study specifically shows that the levels of HMWK are enhanced to significant levels under certain lens wear conditions. The results are summarised in Fig. 2 and Table 1.

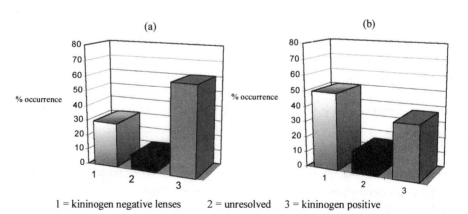

Figure 2. The detection rates for high molecular weight kininogen on Etafilcon A (**A**) daily wear lenses and (**B**) extended wear lenses.

It is important to understand that the distinction between the presence and the absence of HMWK accumulation in the patient population was clear (except in the cases of the small number of unresolved samples) and that it demonstrated definite HMWK positive and HMWK negative population subsets. The difference between positive and negative detection of HMWK in the same material type may be due to a number of factors, but the most important must be patient variation, and their response to the lens and applied stress. The difference in patient response is a possible warning of existing or potential distress at a certain threshold.

Kinin activity was observed in all three wear modalities, demonstrating that kinin activity is independent of duration of wear and regime on the Etafilcon A material. The results suggest patient variability is of key importance. The outcome of the daily disposable study utilising different materials demonstrated that the kinin response was primarily material dependent and secondarily patient dependant.

Table 1. A comparison of the deposition of HMWK and kallikrein onto three daily disposable contact lenses

		1 Day Acuvue (Vistakon) Etafilcon A	Focus Dailies (Ciba Vision) Nelfilcon A	Soflens (B&L) Hilafilcon A
Subject 1	Kininogen	—	+	+
	Kallikrein	—	+	—
Subject 2	Kininogen	—	+	—
	Kallikrein	—	—	—
Subject 3	Kininogen	—	+	+
	Kallikrein	—	—	—
Subject 4	Kininogen	+	+	No Data
	Kallikrein	—	+	No Data

The significance of this work is the detection of HMWK and, thus, kinin activity in lens wear. The implications of the detection of kinin activity in the ocular environment and, particularly, in lens wear are varied. The detection of new entities, in this case HMWK, provides a greater scope of targets for the development of therapeutics, drugs and pain management in a variety of disorders. The discovery of the presence of kinin proteins in relation to contact lens wear further emphasises the potential inflammatory responses that can occur during lens wear. In particular, in allergic responses, the kinins may be at the fore in creating adverse complication. The origin and purpose of its presence has yet to be resolved, yet its occurrence during lens wear in particular may be related to the potential of lens wear to cause irritation and other irregularities. Some, presently unknown episodes are transpiring in lens wear which give rise to an upregulation in the incidence of HMWK. A further understanding of the mode of action of kinins in general is required in order to clarify their participation in adverse complications in host responses.

4. CONCLUSIONS

This study demonstrates a new analytical technique to detect HMWK in tear and lens extracts. Within the limits of the sensitivity of the assay, no HMWK was discovered in the tears of normal open eye, non-lens wear subjects. However, when the lens deposits of daily disposable, conventional daily and extended wear lenses were analysed, they demonstrated the presence of HMWK in a proportion of extracted lens deposits. This overwhelmingly points to the accumulation of HMWK in the ocular environment as a result of, or in association with, lens wear.

REFERENCES

1.   M.M. Frank, Complement & kinin, in: *Basic and Clinical Immunology*, D.P. Stites, and A.L. Terr, 7th ed., Appleton & Lange, Connecticut (1991).
2.   J. Margolis, Activation of plasma by contact with glass: Evidence for a common reaction which releases plasma kinin and initiates coagulation. *J Physiol.* 144:1 (1958).
3.   R.W. Colman, Surface-mediated defense reactions: The plasma contact activation system. *J. Clin. Invest.* 73:1249 (1984).
4.   D. Proud, and A.P. Kaplan, Kinin formation: Mechanisms and role in inflammatory disorders. *Ann. Rev. Immunol.* 6:49 (1988).
5.   D. Proud, A. Togias, R.M. Naclerio, S.A. Crush, P.S. Norman, and L.M. Lichtenstein, Kinins are generated in vivo following nasal airway challenge of allergic individuals with allergen. *J. Clin. Invest.* 72:1678 (1983).
6.   H. Lukjan, J. Hofman, B. Kiersnoroska, M. Bielawiec, and S. Chyrek-Borowska, The kinin system in allergic states. *Allerg. Immunol. Band.* 18:25 (1972).
7.   S.G. Farmer, Role of kinins in airway diseases. *Immunopharmacology* 22:1 (1991).
8.   A.M. Mann, and B.J. Tighe, The development of a CIE based assay to investigate the impact of wear regime on the deposition of immunoregulatory proteins. *J. Am. Acad. Optom.* 76: suppl. 224 (1999).
9.   A.M. Mann, L. Jones, and B.J. Tighe, The effect of wear time on the stimulation of immunoregulation protein deposition on anionic hydrogel contact lenses. *Awaiting publication*

IMMUNOBLOTTING AND TEAR SAMPLING TECHNIQUES FOR THE STUDY OF CONTACT LENS-INDUCED VARIATIONS IN TEAR PROTEIN PROFILES

Helena C. Peach, Aisling M. Mann, and Brian J. Tighe

Biomaterials Research Unit
Aston University
Birmingham, United Kingdom

1. INTRODUCTION

More information on the biochemical interactions taking place between the tear film and the contact lens is required to further our understanding of the causative mechanisms behind the symptoms of dryness and grittiness often experienced by contact lens wearers. These symptoms can often lead to an intolerance to contact lens wear.

Tear proteins are critical to immunological reactions. However, deposition of protein debris onto the surface of contact lenses remains a serious, yet unresolved, problem that hinders contact lens wear. In this study, we examine changes to the tear protein profile of contact lens wearers following a period of contact lens wear using one-dimensional sodium dodecyl sulphate polyacrylamide gel electrophoresis (SDS-PAGE) and a modified version of Western blotting. The ultimate aim of these tear profiling experiments is to provide some insight into the causes of contact lens induced discomfort, particularly end-of-day dryness

2. MATERIALS AND METHODS

Three distinct sampling methods were used to study tears and tear-lens interactions. The first was direct tear sampling with a 5 μl microcapillary tube (Drummond) to collect unstimulated basal tears. The second was the use of a contact lens as a probe to collect and remove the reversibly bound proteins that had accumulated on the lens in the wear period until the time of removal. The proteins deposited on the lens were extracted using a

Lacrimal Gland, Tear Film, and Dry Eye Syndromes 3
Edited by D. Sullivan *et al.*, Kluwer Academic/Plenum Publishers, 2002

967

urea/SDS/DTT/Tris solution for 3 hours at 90°C. Extraction efficiency was measured by ultraviolet (UV) spectrophotometry and fluorescence spectroscopy before and after extraction. The third was the novel use of the lens to instantaneously sample the tear film by collecting the tear "envelope" associated with the lens at the time of its removal. Briefly, the freshly removed lens was submerged in 40 μl of treatment buffer containing SDS and 2-mercaptoethanol in a small microcentrifuge tube. A small incision was made in the base of the microcentrifuge tube, which was then placed into a larger microcentrifuge tube. The tear film was then eluted from the contact lens by ultracentrifugation at 5000 r.p.m. for 10 minutes, and the resulting eluate was examined by SDS-PAGE. In a series of preliminary experiments the "tear envelope" technique was compared with conventional tear samples collected by glass microcapilliary tubes.

In a randomised, controlled preliminary study, four subjects were asked to wear three different types of daily disposable lenses, one per day for 3 days, to compare the effects of lens material on the tear film and the deposition of tear components on the lens. The lens types were as follows: Acuvue (Vistakon, etafilcon-A, FDA Group IV), Focus Dailies (CibaVision, nelfilcon A, FDA Group II), and Soflens (Bausch & Lomb, hilafilcon A, FDA Group II). The chemical properties of the three lens materials are very distinct. Etafilcon-A is an anionic copolymer of 2-hydroxyethyl methacrylate (HEMA) and methacrylic acid. Nelfilcon A is a neutral copolymer of HEMA and N-vinyl pyrrolidone (NVP). Hilafilcon A is a neutral polymer based primarily on polyvinyl alcohol. Subjects were given a rest period of two days to allow the tear film to recover before the next lens type was tested. Tear samples were collected from the subjects before lens wear, and then again at the completion of the test period, using a 5 μl microcapillary tube. Lenses were collected at the completion of the test period, and proteins were extracted using the techniques described above. The following variables were evaluated in relation to lens material: (a) the amount of protein deposition on the lenses, (b) individual subject response variability, and (c) changes in the protein components of the tear film.

To evaluate the effects of diurnal variations on contact lens wear, subjects wore a lens for one hour in the morning and then a new lens for one hour at night. Proteins were then eluted from the lenses as above, and the eluates were analysed by SDS-PAGE and Western blotting.

3. RESULTS AND DISCUSSION

In the first set of experiments, tear samples were obtained from the subjects before lens wear began, and then again at the end of the three-day wear period. Changes in the tear protein profile following contact lens wear clearly occurred (Fig. 1). The gel scan represents tear samples collected from two non-accustomed contact lens wearers. Changes in the protein profile of the sample collected from subject 1 following 3 days of Soflens daily disposable lenses wear were clearly visible. Many more bands were visible following the test period, indicating the presence of a greater number of proteins in the tear sample. In addition, the bands that were originally present prior to contact lens wear stained more intensely with Coomassie Brilliant Blue after the test period, indicating that higher concentrations of these proteins were secreted into the tear film. The second subject (lanes 4 and 5) wore Acuvue daily disposables for 3 consecutive days. After the test period (lane 5), some bands had increased in intensity, and hence concentration, whereas, bands present before contact lens wear were absent. The changes observed in this subject may ultimately lead to an intolerance to contact lens wear since the integrity of the tear film, of the utmost importance for comfortable contact lens wear,[1] appears to have been compromised. Any departure from the normal tear film composition, whether it be excessive amounts of proteins or the absence of certain proteins, results in a disrupted tear film which is likely to be detrimental to the subject.

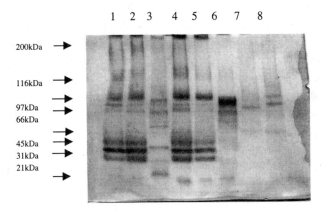

Figure 1. SDS-PAGE of tears from pre- and post-lens wear. Lane1: Subject 1 before lens wear, Lane 2: Subject 1 after 3 days of contact lens wear - Soflens), Lane 3: Broad band molecular weight marker, Lane 4: Subject 2 before contact lens wear, Lane 5: Subject 2 after 3 days of contact lens wear - Acuvue), Lane 6: Albumin (2 mg/ml), Lane 7: IgG (1 mg/ml), Lane 8: IgA (1 mg/ml).

Marked differences in protein profiles were observed in samples collected after 3 different daily disposable lenses were worn by the subjects. Of the 3 types of daily disposable lenses extracted, the most protein was extracted after the use of Acuvue daily disposable and the Focus Dailies. The Soflens yielded an extraction efficiency of 31.5% as compared to 86.5% with the Focus Dailies. The non-ionic, high water content of the Soflens prevented these lenses from sustaining a tear film as thick as that of the Focus Dailies, and hence the tear film broke up more quickly.

As part of the initial lens analysis, fluorescence was also performed on the lenses to determine the amount of lipids that had been deposited on them. Lipid deposition can affect the amount of protein deposited onto the surface of the lens, and an excess of lipids can reduce the efficiency of protein extraction. The Soflens material retained a greater amount of lipid than either the Acuvue or Focus Dailies materials, and consequently, the efficiency of protein extraction from these lenses was drastically reduced. Acuvue daily disposables are ionic, high water content, Group IV lenses, and adsorb a great deal of lysozyme, both onto the surface and into the lens matrix, which supports protein deposition. Therefore, clearer protein profiles were obtained from the Focus Dailies and the Acuvue lenses than from the Soflens.

The protein profiles obtained from lenses worn late at night in the diurnal variation study stained more intensely with Coomasie Blue than those worn in the morning (data not shown), indicating greater protein deposition. The increased protein deposition observed in this study could be a contributing factor to the end of day discomfort experienced by contact lens wearers.

Using our new technique for collecting the tear envelope from a contact lens, clear profiles were obtained for both the Acuvue lens and the Focus Dailies lens, though differences in the profiles from these two lenses were apparent (Fig. 2). It was not possible, however, to obtain a clear profile from the Soflens daily disposable lenses. Lysozyme does not appear to be present in the tear envelopes obtained with the Acuvue daily disposable lenses. The Acuvue lenses adsorb a great deal of lysozyme and, as a result, the tear film appears to be depleted of this protein. In contrast, lysozyme is visible in the protein profiles of tear envelopes collected with the Focus Dailies and the Soflens lenses, since these two lens types do not adsorb as much lysozyme onto the lens surface. Proteins with molecular weights of approximately 29 kDa are absent from tear envelopes collected with the Softlens lenses. Again, this may be due to the chemical properties of the lens material since this band is clearly visible in profiles of tear envelopes obtained with the Acuvue lenses and the Focus Dailies. Further modification of this technique may be required to remedy this inconsistency.

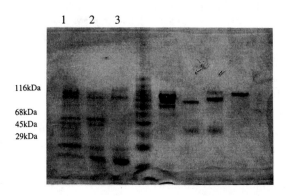

Figure 2. SDS-PAGE of tear envelopes from lenses. Lane 1: Acuvue lens, Lane 2: Focus Dailies lens, Lane 3: Soflens lens.

A modified Western blot was performed to identify the nature of the individual proteins within the profiles. Following SDS-PAGE and Western blotting, the polyvinylidene-difluoride (PVDF) membranes were immunostained for IgE. IgE, chosen for study because of its documented role in hypersensitivity reactions, is present in tears[2,3] in concentrations that range from 0.02 mg/100ml to 0.003 mg/100ml. This protein was detected in extracts of the Acuvue and Focus Dailies lenses, indicating the sensitivity of the technique. The high degree of sensitivity of this technique should enable the detection and identification of a wide range of tear proteins.

REFERENCES

1. N. Effron, Contact lens-associated tear film dysfunction. *Optician* 5676: 216: 16 – 26 (1998)
2. M. Allansmith, Immunology of the tears. In Holly, F.J. and Lemp, M.A., (Eds) *The Preocular Tear film and dry eye syndromes.* Little, Brown and Co., Boston: *Int. Ophthalmol. Clinic.*, 13: 1: 47–72 (1973)
3. M. Tragakis, L. Economidis, S. Pollalis, and S. Gartaganis,. Tear immunoglobulin levels in normal persons and in patients with staphylococcal, trachomatous and allergic disease. *VI th Congress Eur. Soc. Ophthalmol.*, 281 –285 (1980)

INTERACTION BETWEEN THE CONTACT LENS AND THE OCULAR SURFACE IN THE ETIOLOGY OF SUPERIOR EPITHELIAL ARCUATE LESIONS

Nicole O'Hare, Fiona Stapleton, Thomas Naduvilath, Isabelle Jalbert, Deborah F. Sweeney, and Brien A. Holden

The Cornea and Contact Lens Research Unit
School of Optometry and the Cooperative Research
Centre for Eye Research and Technology
The University of New South Wales
Sydney, Australia

1. INTRODUCTION

Superior epithelial arcuate lesions (SEALs) develop in the cornea in areas covered by the eyelid, and are associated with soft contact lens wear.[1] Generally, SEALs are arc-shaped, and are located between 10 and 2 o'clock, within 2 to 3 mm of the limbus.[2-4] The lesions are white or opalescent with a clear zone between the SEAL and the limbus, and can appear either as an arcuate split or a band lesion after staining with sodium fluorescein. The arcuate split extends the full thickness of the epithelium and has jaggedly stained borders that are surrounded by an area of superficial and punctate staining. The band lesion is generally more superficial and is of constant width and depth.[2]

Although SEALs were first described in the 1970s,[5-7] most of the evidence for their etiology is anecdotal.[4-8] It has been suggested that characteristics of both the contact lens wearer[2,3,7,8] as well as the contact lens design[2-4,8] are risk factors for SEALs. The current hypothesis is that mechanical chaffing of the limbal area by the back surface of some lens in combination with inward pressure from the upper lid is the major cause of SEALs.[3,7,8] By extension, subjects with the additional risk factor of abnormal tear films may be even more susceptible to SEALs.

High Dk silicone hydrogel lenses, made of higher modulus materials than conventional soft contact lenses,[9] have recently been released to market. These lenses are

Lacrimal Gland, Tear Film, and Dry Eye Syndromes 3
Edited by D. Sullivan *et al.*, Kluwer Academic/Plenum Publishers, 2002

973

designed for extended wear for up to 30 nights. The combination of a full-thickness epithelial break SEAL while wearing high Dk lenses for extended lenths of time could potentially result in corneal infiltration and subsequent scarring. Therefore, it is important to fully understand how SEALs develop and to identify which patient or lens design characteristics are risk factors for these lesions.

The aim of this study is to clarify the etiology of SEALs by identifying subject characteristics, lens surface characteristics, and fitting factors associated with SEALs. This study is a retrospective analysis of a group of subjects that had worn high Dk silicone hydrogel lenses for extended periods.

2. MATERIALS AND METHODS

2.1. Study Design

All experimental protocols were reviewed and approved by the Committee for Experimental Research Involving Human Subjects at the University of New South Wales, Australia. Subjects were over 18 years old and had given informed consent.

This study is a retrospective case-control analysis of lens and subject characteristics in individuals that developed SEALs while wearing prototype extended wear high Dk soft contacts. Subjects with one eye that had experienced a SEAL were termed the Case group and matched subjects that had not experienced a SEAL in the corresponding eye were termed the Control group.

The subject characteristics evaluated in this study include central corneal curvature at baseline and ethnicity. The lens surface characteristics that were evaluated include wettability and back surface film deposits, and the lens fitting characteristics that were evaluated include lens tightness and primary gaze movement.

The lens surface and fitting characteristics of each subject were measured at regular intervals both before and after the development of a lesion. The means of each lens surface and fitting characteristic from (i) the Case and Control groups before SEALs developed were compared to those from (ii) the Case group before and after the development of lesions.

At a single follow up visit, contact lens cytology was performed for a subset of the SEALs cases (n = 6) and matched controls (n = 15).

2.2. Subjects and Lenses

Twenty subjects that developed SEALs were matched with subjects that had not developed SEALs for lens type, wear schedule, duration of extended wear with high Dk lenses, experience of prior lens wear, spectacle refraction, gender and ages. Subject demographics are presented in Table 1. Subjects were enrolled in ongoing clinical trials at the Cornea and Contact Lens Research Unit, School of Optometry, The University of New South Wales and wore high Dk lenses on 6 or 30 night (N) lens removal and disinfection schedules with monthly replacement.

Subjects wore prototype silicone hydrogel lenses made from balafilcon A (water content = 35%, Dk = 110 barrers, modulus = 1.1 MPa) or lotrafilcon B (water content = 24%, Dk = 140 barrers, modulus = 1.2 MPa) materials.

2.3. Procedures

SEALs were detected after the instillation of sodium fluorescein using a slit-lamp biomicroscope (Zeiss SL-30, Oberkochen, Germany) with a cobalt blue filter in the illumination system and a yellow fluorescein enhancement filter (Eastman Kodak, Rochester, NY, USA) over the objective lens.

Central corneal curvature was measured with an autokeratometer (Canon RK-3, Japan). Lens wettability was graded on a decimalised scale (0.1) from 0 to 5 where 0 corresponds to a surface that is non-wetting, 2 corresponds to the wetting equivalent of a hydroxyethylmethacrylate (HEMA) lens and 5 corresponds to wetting that is equivalent to a healthy cornea. The film-like deposits on the back surface of the lens (back surface film) was graded on a decimalised scale (0.1) of 0 to 4 where 0 corresponds to the absence of deposites, 1 corresponds to very slight deposites, 2 corresponds to slight deposites, 3 corresponds to moderate deposites and 4 corresponds to severe deposites. Any grade of film-like deposits greater than 0 was considered clinically significant. Lens tightness was assessed using the Push up Test[10] where 100% corresponds to total tightness (no movement), 50% corresponds to optimum tightness and 0% corresponds to a lens that slides from the eye when the lid is retracted. Primary gaze movement of the lens (mm) was assessed during normal blinking with diffuse white light and biomicroscopy at 16x magnification with graticule. Lens movement greater or equal to 0.3 mm was considered clinically ideal.

Table 1. Biometric details of Case (SEAL) and Control group subjects

	Case group	Control group	P-Value
Number of subjects	20	20	
Mean age (years)	29 ± 5	30 ± 4	0.576
Gender (M:F)	10:10	10:10	1.000
Mean Rx[a], sphere (DS)	-2.71 ± 1.07	-2.66 ± 1.16	0.888
Mean Rx, cylinder (DC)	-0.26 ± 0.26	-0.34 ± 0.27	0.381
Mean EW[b] experience (months)	19 ± 7	19 ± 6	0.713
Wear schedule (6N:30N)	6:14	6:14	1.000

[a]Rx: Refractive error, [b]EW: Extended wear

2.4. Contact Lens Cytology

At a single follow up visit, the habitual lens was removed directly from the cornea and irrigated. The harvested epithelial cells and lens were stained with Hoescht 33342 and propidium iodine. Irrigated cells were then filtered, and both the lens and the filtered irrigating fluid were visualised under fluorescent microscopy.

2.5. Statistics

This study was a preliminary analysis of potential indicators of SEAL etiology. Results are presented as mean values ± the standard deviation (SD). Due to the small sample, and to improve the statistical power, all differences were considered to be statistically significant at $P \leq 0.10$.[11] Differences between the Case and Control groups were measured using an independent two-tailed t-test. Differences within the Case group before and after development of lesions were measured using the paired two-tailed t-test. Categorical data were compared using the Chi squared test. Contact lens cytology data were compared using the Mann Whitney test. The statistical package for social sciences (SPSS) version 10 was used to analyse differences between groups.

3. RESULTS

3.1. Subject Characteristics

No significant differences in central corneal curvature were found between the Case and Control groups after lesion development. Mean K-Flat keratometry readings were 43.0 ± 1.2 D and 43.0 ± 1.3 D (P = 0.965) from the Case and Control groups, respectively, and K-Steep readings were 43.7 ± 1.2 D and 44.0 ± 1.2 D (P = 0.428) from the Case and Control groups, respectively. The proportion of subjects of Asian origin in each group was determined to evaluate whether tight lids were involved the etiology of SEALs. Of the 20 subjects in each group, 2 and 3 subjects were Asian in the Case and Control groups, respectively. Differences in the number of Asian subjects were not significant (P > 0.1).

3.2. Comparison of lens surface and fitting characteristics in Case and Control groups before lesion development.

The lens surface and fitting characteristics were compared in the Case and Control groups before lesion development to identify any factors that may have predisposed subjects to the development of SEALs. Lenses from the Case group were less wettable and developed more back surface film prior to lesion development, although the overall film deposits on the prototype lenses was less than very slight (Table 2). No differences were found in lens tightness or movement in the Case and Control groups before lesion development (Table 2).

Table 2. Comparison of the mean lens surface and fitting characteristics between Case and Control groups before development of SEALs

	Case group		Control group		P-value
Surface wettability	1.7 ± 0.5^a	63^b	1.9 ± 0.4	68	0.018
Back surface film	0.13 ± 0.40	50	0.02 ± 0.10	56	0.048
Lens tightness (%)	46 ± 9	63	45 ± 6	68	0.672
Lens movement (mm)	0.32 ± 0.15	63	0.36 ± 0.17	68	0.159

^{a}mean $\pm$ sd; ^{b}n

3.3. Comparison of Lens Surface and Fitting Characteristics in the Case Group Before and After Lesion Development

The lens surface and fitting characteristics were compared before and after lesion development within the Case group to identify any factors that occur concurrently with lesion development that may be involved in the etiology of SEALs. During lesion development, lenses fitted more tightly to the cornea, with lower primary gaze movement in comparison to lenses worn before lesion development (Table 3). There were no differences in lens wettability or the degree of back surface film before and during lesion development.

Table 3. Comparison of the mean lens surface and fitting characteristics before and during SEAL development within the Case group

	Before event		At event		P-value
Surface wettability	1.7 ± 0.5^a	63^b	1.8 ± 0.5	17	0.557
Back surface film	0.13 ± 0.36	50	0.11 ± 0.4	14	0.839
Lens tightness (%)	46 ± 9	63	50 ± 6	18	0.076
Lens movement (mm)	0.32 ± 0.15	63	0.25 ± 0.11	18	0.063

^{a}mean $\pm$ sd; ^{b}n

3.4. Contact Lens Cytology

Cell counts and viability were compared at a single follow up visit to the SEAL event to identify any difference in interaction of the back surface of the lens with the epithelium between the two groups. Total cell counts, loosely adherent cells (collected during irrigation) and tightly bound cells (those remaining attached to the lens) were compared. Significantly higher numbers of total cells and tightly adherent cells were recovered from the lenses of the Case group compared to the Controls, indicating greater interaction between the back surface of the lens and the epithelium in these subjects (Table 4).

Table 4. Comparison of cell counts and viability between Case and Control groups at a single follow up visit to the SEAL event

Cell count	Case group[a]	Control group[b]	P-Value
Total[c]	22[d]	7	0.006
	6–42[e]	3–21	
Tightly adherent	19	5	0.045
	2–42	0–21	
Loosely adherent	3	1	0.205
	0–8	0–5	

[a]n = 6, [b]n = 15, [c]Total = tightly + loosely adherent cells, [d]median, [e]range

4. DISCUSSION

Poor lens wettability before development of SEALs, and tighter fitting lenses during SEAL development are factors that increase the likelihood of develop in SEALs when wearing high Dk soft contact lenses during extended wear. These results suggest that poor wettability and tight fitting lenses lead to greater shear forces that disrupt the ocular surface in the superior cornea during extended wear.

Previous reports have indicated that subjects with steep corneas[3,7] or tight upper lids[8] may be predisposed to develop SEALs. Our results indicate that there is no difference in central curvature of the cornea between subjects who develop SEALs and those who do not. Although these results are preliminary, they also indicate that the proportion of Asian subjects that developed SEALs was not greater than the proportion of Asian subjects that did not develop SEALs. This is unexpected because subjects of Asian origin are more likely to have tight upper lids. Increased pressure from the upper lid is believed to exacerbate the shear forces between a misaligned lens and the superior cornea.[3,7,8] Since the sample size of Asian subjects in this study was small, future studies that incorporate larger sample sizes are required to better determine the roles of tight upper lids and Asian ethnicity in SEAL development.

Lower lens wettability and greater build up of film-like deposits on the back surface of the lens predisposed subjects to SEAL development. Although the level of back surface film was less than very slight, it was more than 6-fold greater in the Case subjects. High Dk silicone hydrogel lenses are made from more hydrophobic bulk materials than conventional low Dk materials and are therefore treated to render the lens surface more hydrophilic.[9] Inadequate lens treatment may cause the surface to become less wettable. However, we found no significant differences in lens wettability and build up of film deposits in Case subjects at the time of lesion development, whereas significant differences in these factors were found between the Case and Control groups before lesion development. If SEALs occurred as a result of inadequate surface treatment, we would not expect to find differences in lens surface characteristics in the Case and Control groups before lesion development. Therefore, it is more likely that subjects with poor tear film characteristics developed SEALs in this study. A large prospective trial that looks more closely at subject tear film characteristics and other subject characteristics at the beginning of lens wear, and before and after SEAL development is now required.

Subjects with SEALs had tighter fitting lenses with less primary gaze movement at the time of lesion development, but their lens fitting characteristics were no different from the control subjects prior to lesion development. These results indicate different lenses of the same type may have inconsistent fitting characteristics for individual subjects.

The greater cell recovery from lenses worn by individuals in the Case group compared to those worn by individuals in the Control group indicates that there is more interaction between the back surface of the lens and the epithelium in these subjects.

Subjects with SEALs that are refitted with a lens of the same type as previously worn do not always have recurrent events.[2,4] Yet some subjects that are refitted with a lens of the same type or with improved fit or design have recurrent SEALs.[2,4,8,12] This evidence supports our hypothesis that subjects with poor tear film characteristics and ill-fitting lenses are more likely to develop SEALs.

Overall our findings support the hypothesis that mechanical interaction or shear forces created by a thin post-lens tear film in the limbal zone can contribute to the formation of SEALs.[3] Silicone hydrogel contact lenses are made from materials with higher modulus than conventional hydrogel lenses.[9] The increased modulus of silicone hydrogels may exacerbate the mechanical chaffing of the lens in the superior region of the cornea.

ACKNOWLEDGEMENT

This work was supported by the Australian Federal Government through the Cooperative Research Centres programme, CIBA Vision and Bausch and Lomb. The authors thank Serina Stretton for assistance in preparation of this manuscript.

REFERENCES

1. B.A. Holden, A. Stephenson, S. Stretton, P. Sankaridurg, N. O'Hare, I. Jalbert, et al. Superior epithelial arcuate lesions with soft contact lens wear. *Optom Vis Sci.* In press (2001).

2. N. Hine, A. Back and B. Holden. Aetiology of arcuate epithelial lesions induced by hydrogels. *Trans Brit Cont Lens Assoc Conf.* 48–50 (1987).

3. G. Young and D. Mirejovsky. A hypothesis for the aetiology of soft contact lens-induced superior arcuate keratopathy. *ICLC.* 20:177–180 (1993).

4. V. Malinovsky, J. Pole, N. Pence and D. Howard. Epithelial splits of the superior cornea in hydrogel contact lens patients. *ICLC.* 16(9 & 10):252–255 (1989).

5. L. Kline and T. DeLuca. An analysis of arcuate staining with the B&L SOFLENS® - Part I. *J Am Optom Assoc.* 46(11):1126–1129 (1975).

6. L. Kline and T. DeLuca. An analysis of arcuate staining with the B&L SOFLENS® - Part II. *J Am Optom Assoc.* 46(11):1129–1132 (1975).

7.   L. Kline and T. DeLuca. Pitting stain with soft contact lenses - Hydrocurve® thin series. *J Am Optom Assoc.* 48(3):372–376 (1977).

8.   G. Horowitz, J. Lin and H. Chew. An unusual corneal complication of soft contact lenses. *Am J Ophthalmol.* 100:794–797 (1985).

9.   B. Tighe. Silicone hydrogel materials - how do they work? in: *Silicone Hydrogels: The Rebirth of Extended Wear Contact Lenses.* D. Sweeney, ed., Butterworth-Heinemann, Oxford (2000).

10. G. Young, B. Holden and G. Cooke. Influence of soft contact lens design on clinical performance. *Optom Vis Sci.* 70:394–403 (1993).

11. B. Winer. *Statistical Principles in Experimental Design*, McGraw-Hill, New York (1962).

12. J. Josephson. A corneal irritation uniquely produced by hydrogel lathed lenses and its resolution. *J Am Optom Assoc.* 49(8):869–870 (1978).

CLINICAL BENEFITS AND PHYSICAL PROPERTIES OF ADDITION OF HYDROXYPROPYL METHYLCELLULOSE TO A MULTI-PURPOSE CONTACT LENS CARE SOLUTION

Peter A. Simmons, William Kelly, William Prather.
and Joseph Vehige

Allergan, Inc.
Irvine, California, USA

1. INTRODUCTION

Contact lens-induced dry eye is a common side effect of contact lens use[1] and primarily due to the well-established disruption of the tear film by the contact lens.[2-4] Eye care practitioners may address the problem by changing the lens material, modifying the fit or recommending rewetting drops.[5] Another possibility for treatment may be selection of an alternative lens care solution.

Multi-purpose contact lens care solutions are used increasingly worldwide because of ease of use, convenience and cost-effectiveness. Clinical performance and subjective acceptance of these products are satisfactory, and most clinicians consider the brands generally similar. While all multi-purpose solutions (MPS) contain chemical agents for buffering, cleaning and disinfection, additional ingredients may be added to enhance performance in many ways. This study investigates the potential benefits of adding the cellulosic ocular lubricant hydroxypropyl methylcellulose (HPMC) to an MPS.

HPMC has been widely used in artificial tears and comfort drops such as Opti-Tears (Alcon), Re-wetting Drops (Bausch & Lomb) and Tears Naturale II (Alcon).[6,7] HPMC is a high molecular weight (80–100 kD), viscoelastic soluble polymer that exhibits a low surface tension[8,9] and is classified as an ophthalmic demulcent for the treatment of dry eye.[10-12] It is an additional ingredient in one MPS (Complete, Allergan) in which it is included specifically for its lubrication and wetting properties. In this report the subjective responses of patients using an MPS containing HPMC in comparison with one without

Lacrimal Gland, Tear Film, and Dry Eye Syndromes 3
Edited by D. Sullivan *et al.*, Kluwer Academic/Plenum Publishers, 2002

981

were recorded along with laboratory studies of the physical properties of MPS with and without HPMC.

2. METHODS

2.1. Clinical Study

Subjects (n = 147) who were regular MPS users wearing conventional and 2-week replacement hydrogel lenses were given two multi-purpose lens care regimens: a standard one containing buffer, surfactant and disinfectant, and the other an identical formula with HPMC. The study was a masked, randomized, crossover design in which all subjects used both solutions and were evaluated for their subjective responses to each solution alone and in comparison with each other.

2.2. Laboratory Study

Wetting characteristics of MPS with and without HPMC were studied. The extent of the fluid layer formed by the solutions on a standard hydrophobic plastic (polystyrene) was measured over time in terms of weight loss due to evaporation and surface area loss due to drainage and evaporation. Dynamic contact angle measurements were made using an automated instrument. The contact angle for each piece of plastic was measured in distilled water, dipped into the test solution and then retested in successive containers of fresh distilled water. Improved wetting was expressed as a decrease in dynamic contact angle and reported as a percentage change from the initial value. Wetting of hydrogel lenses was also measured as weight of adsorbed fluid upon removal from each solution during air drying. Release of HPMC into successive changes of saline following overnight adsorption onto lenses was measured using a chemical assay.[13]

3. RESULTS

3.1. Clinical Study

The subjects responded to eight questions related to lens-wearing comfort by providing preference scores in which the two solutions were compared. In each case, the solution with HPMC scored significantly better than the one without it, including the duration-related questions regarding keeping lenses moist and lubricated during wear (Table 1).

Table 1. Patient preferences for MPS with and without HPMC

Preference Question	% Prefer MPS with HPMC	% Prefer MPS without HPMC
Overall Preference	53	33
How solution feels in the hand	48	28
How solution feels in the eye	49	30
Overall comfort for the eyes	48	31
Amount of time for lens to settle on eye	39	21
Keeps lenses moist in eye	45	23
Keeps lenses lubricated in eye	50	26
Solution is soothing to the eyes	47	27

3.2. Laboratory Study

Addition of HPMC to the MPS solution produced a thicker (heavier), longer-lasting coating on the polystyrene plastic surface. This is seen in surface area covered by the fluid (Fig. 1A) and weight of fluid on the plastic surface (Fig. 1B). Improvements in contact angle wettability after a single exposure to the HPMC solution were maintained through multiple rinses in distilled water (Fig. 1C).

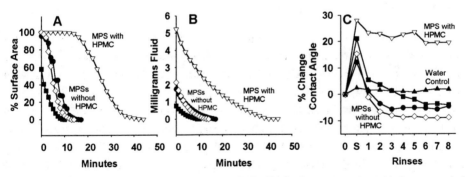

Figure 1. Wetting properties of MPS on polystyrene plastic. **(A)** Surface area coating with time on vertical plastic surface. **(B)** Weight of fluid with time on vertical plastic surface. **(C)** Change in wetting as percent change in dynamic contact angle following single exposure to solutions (at S) and successive water rinses. MPS with HPMC from Allergan. MPS without HPMC from Allergan, Alcon and Bausch & Lomb.

Fig. 2 shows the interactions of the solutions with hydrogel lenses. The HPMC solution produces a thicker and longer-lasting fluid layer on the lens (Fig. 2A), and HPMC is released continuously for at least 12 to 24 h from the lens surface following an overnight soak in the HPMC solution (Fig. 2B).

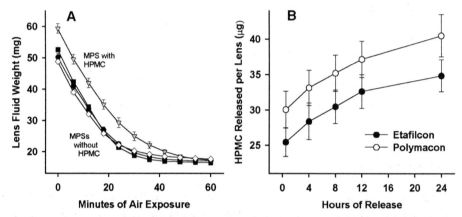

Figure 2. Properties of MPS on hydrogel lenses. **(A)** Weight of fluid over time on lenses exposed to air on one surface. **(B)** Release of HPMC from hydrogel lenses of two polymer types into successive changes of saline following overnight soaking in an MPS containing 0.15% HPMC. MPS with HPMC from Allergan. MPS without HPMC from Allergan, Alcon and Bausch & Lomb.

3. DISCUSSION

These studies demonstrate the addition of the lubricant HPMC to a multi-purpose contact lens care solution produces a more comfortable wearing experience. The enhanced wetting properties of the MPS with HPMC demonstrated in the laboratory, as well as the binding of HPMC to hydrogel lenses and its subsequent release following the overnight soak, are the probable mechanisms for the strong patient preference for the HPMC formula. These results are consistent with numerous reports documenting increased on-eye retention of fluid with the use of HPMC and similar substances.[14,15] Use of an MPS containing HPMC therefore reduces patient symptoms of dryness and discomfort resulting from lens wear, allowing increased wearing time and patient satisfaction.

REFERENCES

1. C.G. Begley, B. Caffery, K.K. Nichols, and R. Chalmers, Responses of contact lens wearers to a dry eye survey, *Optom Vis Sci.* 77:40 (2000).
2. F. Holly, Tear film physiology and contact lens wear: II. Contact lens tear film interaction, *Am J Optom Physiol Opt.* 58:331 (1981).
3. A. Tomlinson and T. Cedarstaff, Tear evaporation from the human eye: the effects of C.L. Wear, *J Br Contact Lens Assoc.* 5:141 (1982).

4.   A. Sharma and E. Ruckenstein, Mechanism of tear film rupture and its implication for contact lenses tolerance, *Am J Optom Physiol Opt.* 62:246 (1985).

5.   A. Tomlinson, *Complications of Contact Lens Wear* Mosby–Year Book, St. Louis (1992).

6.   M. Doughty, Re-wetting, comfort, lubricant and moisturizing solutions for the contact lenses wearers, *Contact Lens and Ant Eye.* 22:116 (1999).

7.   P. White and C. Scott, Contact lenses and solution summary, *Cont Lens Spect.* (1999).

8.   D.A. Benedetto, D.O. Shah, and H.E. Kaufman, The instilled fluid dynamics and surface chemistry of polymers in the preocular tear film, *Invest Ophthalmol.* 14:887 (1975).

9.   A. Ludwig, N.J. van Haeringen, V.M. Bodelier, and M. Van Ooteghem, Relationship between precorneal retention of viscous eye drops and tear fluid composition, *Int Ophthalmol.* 16:23 (1992).

10.   Ophthalmic demulcents, 21 CFR 349.12, 4-1-00 ed., U.S. Government Printing Office, Washington, D.C. (2000).

11.   I. Toda, N. Shinozaki, and K. Tsubota, Hydroxypropyl methylcellulose for the treatment of severe dry eye associated with Sjøgren's syndrome, *Cornea.* 15:120 (1996).

12.   P. Versura, M.C. Maltarello, F. Stecher, R. Caramazza, and R. Laschi, Dry eye before and after therapy with hydroxypropyl methylcellulose. Ultrastructural and cytochemical study in 20 patients, *Ophthalmologica.* 198:152 (1989).

13.   U.S. Pharmacopeial Convention, *United States Pharmacopeia 24, National Formulary 19, 2000* United States Pharmacopeial Convention, Rockville, MD (1999).

14.   F.C. Bach, J.B. Adam, H.C. McWhirter, and J.E. Johnson, Ocular retention of artificial tear solutions. Comparison of hydroxypropyl methylcellulose and polyvinyl alcohol vehicles using an argyrol marker, *Ann Ophthalmol.* 4:116 (1972).

15.   G.R. Snibson, J.L. Greaves, N.D.W. Soper, J.M. Tiffany, C.G. Wilson, and A.J. Bron, Ocular surface residence times of artificial tear solutions, *Cornea.* 11:288 (1992).

Epidemiology and Pathogenesis of Dry Eye Syndromes
(e.g. Sjögren's Syndrome)

EPIDEMIOLOGY OF DRY EYE SYNDROME

Debra A. Schaumberg,[1] David A. Sullivan,[2] and M. Reza Dana[2]

[1]Division of Preventive Medicine
Brigham and Women's Hospital
Harvard Medical School
Boston, Massachusetts, USA
[2]Department of Ophthalmology
Schepens Eye Research Institute
Harvard Medical School
Boston, Massachusetts, USA

1. INTRODUCTION

Dry eye syndrome (DES) represents a heterogeneous group of conditions that share inadequate lubrication of the ocular surface as their common denominator. DES is characterized by symptoms of ocular dryness and discomfort due to insufficient tear quantity or quality caused by low tear production and/or excessive tear evaporation. Symptoms can be debilitating[1] and, when severe, may affect psychological health and ability to work. No cure exists for DES, which is one of the leading causes of patient visits to ophthalmologists and optometrists in the United States. Because of the presumed high prevalence of DES and the attendant health care burden, the National Eye Institute (NEI) has identified tear film and dry eye research as important areas in need of further study.

As recently as 1993–1994 at a workshop co-sponsored by NEI and industry, it was recognized that very little information was available on the prevalence of DES in the overall population and the accompanying demographic characteristics such as age, sex and race.[2] Since that time, investigators have collected information about the epidemiology of DES in many relatively large epidemiological studies, including the Salisbury Eye Evaluation,[3] the Beaver Dam Eye Study,[4] the Melbourne Visual Impairment Project[5] and the Physicians' and Women's Health Studies (PHS and WHS).[6]

Four types of global tests for DES for use in research were proposed at the NEI/Industry workshop: (1) validated questionnaire of symptoms, (2) objective

Lacrimal Gland, Tear Film, and Dry Eye Syndromes 3
Edited by D. Sullivan *et al.*, Kluwer Academic/Plenum Publishers, 2002

989

demonstration of ocular surface damage by vital dye staining, (3) demonstration of tear instability by decreased tear breakup time and (4) demonstration of tear hyperosmolarity.[2] Nonetheless, largely due to inadequacies of currently available tests and disease variability, standardizing the diagnosis of DES for clinical and epidemiological studies has proven difficult and no reliable standard exists. A report from the Salisbury study showed minimal overlap within individuals in the results of some of these tests,[7] but this must be interpreted in light of poor repeatability, particularly for the clinical measures, which would result in predictably poor correlations. Without a consensus definition of DES, studies have assessed dry eye in several ways, though all epidemiological studies have assessed dry eye symptoms, which have very good reproducibility.[8] Studies centering on dry eye symptoms may provide useful information about the public health importance of DES because (1) a major goal of therapy for DES is the relief of symptoms, and (2) clinically important degrees of ocular surface damage rarely occur without symptoms. In two recent studies, assessment of dry eye symptoms was the single most important test for DES identified by clinicians in practice.[9, 10] This article summarizes the information on DES currently available from epidemiological studies, concentrating on the data regarding dry eye symptoms.

2. SUMMARY OF STUDIES AND METHODS

The dry eye component of the Salisbury Eye Evaluation included 2,420 men and women 65 years of age or older from a community on the eastern shore of Maryland.[3] This represented 96% of all subjects examined for the overall study. Fifty-eight percent of the dry eye study cohort was female. The majority of subjects were white and 26% were black. Dry eye was assessed by questionnaire and clinical tests. The questionnaire queried subjects about six symptoms: dryness, grittiness/sandiness, burning, redness, crusting on the eyelashes and eyes being stuck shut in the morning. Each subject was asked to describe the frequency with which he or she experienced each symptom as being "all of the time", "often", "sometimes" or "rarely". Based on symptoms alone, a subject was considered to have DES if he or she indicated experiencing at least one of the six symptoms at least "often". In addition to symptom assessment, rose bengal staining and Schirmer testing were also performed. In this manuscript, however, we will confine our comments to comparisons of dry eye symptoms, which were assessed in each study.

The Melbourne Visual Impairment Project dry eye substudy included a population-based sample of 926 men and women aged 40 years and older (mean 59.2 years).[5] The study population represented 82.3% of eligible subjects; 46.8% of subjects were male. Subjects were asked in an interview to rate the severity of six dry eye symptoms including discomfort, foreign body sensation, itching, tearing, dryness and photophobia. Ratings were "absent/none", "mild", "moderate" or "severe", and the subjects were provided with the investigators' definitions of these severity levels. Severe symptoms generally had to present constantly. A subject was determined to have dry eye if he or she considered at least one of the six symptoms severe and not attributable to hay fever.

Dry eye was assessed at the second Beaver Dam Eye Study examination in 3,722 men and women who were at least 48 years of age (mean ± SD, 65 ± 10 years).[4] Ninety-nine percent of subjects were white, and 43% were male. Dry eye was assessed by a single question: "For the past 3 months or longer have you had dry eyes?" If needed, the subject was given a description of "foreign body sensation with itching, burning, sandy feeling, not related to allergy." Subjects who answered this question affirmatively were considered to have DES.

WHS, a randomized trial designed to assess the benefits and risks of aspirin and vitamin E in the primary prevention of cardiovascular disease and cancer in healthy women, comprised 39,876 health professionals aged 45 to 84 in 1992,[11] and aged 49 and older at the time of dry eye assessment.[6] PHS was a randomized trial of aspirin and beta-carotene in the prevention of cardiovascular disease and cancer conducted among 22,071 U.S. male physicians.[12] PHS participants were aged 55 and older at the time of dry eye ascertainment.[6] DES was ascertained on the 4-year WHS and the 14-year PHS follow-up questionnaires by three questions pertaining to diagnosis or symptoms of DES: (1) "Have you ever been diagnosed by a clinician as having dry eye syndrome?" (2) "How often do your eyes feel dry (not wet enough)?" (3) "How often do your eyes feel irritated?"

The two questions pertaining to symptoms were chosen from a questionnaire developed by Oden et al.[13] Possible answers to these questions included "constantly", "often", "sometimes" or "never". These two questions were found to have a sensitivity of approximately 60% with a specificity of 94% compared to clinical diagnosis of dry eye. Although those with DES may certainly report other symptoms (especially when asked), Oden et al. found the questions pertaining to dry and irritated eyes taken together provided nearly the same predictability in correctly identifying subjects with DES as a longer 14-item questionnaire.[13] Three primary outcome measures for DES were defined for WHS and PHS. Clinically diagnosed DES was defined as a self-reported diagnosis of DES by a clinician. Severe DES symptoms were defined as a report of both dryness and irritation either constantly or often. Because symptomatic women would not have been diagnosed, and some diagnosed women would be using therapies that might affect their symptoms, a composite end point of either a previous clinical diagnosis or severe symptoms of DES (cDES/Sx) was formed. In this review, preliminary data are presented from the first 35,370 women in WHS and 12,720 men in PHS who filled out and returned the dry eye questionnaire. Final analyses of data from these studies are forthcoming.

3. OVERALL PREVALENCE OF DES

The first large-scale epidemiological study to report the prevalence of DES was the Salisbury study. In this·study, 41% reported no symptoms, 45% reported one or more symptoms at least rarely or sometimes and 14% reported one or more symptoms at least often.[3] In the Melbourne study, dry eye had been previously diagnosed in less than 1% of subjects, and Sjögren's syndrome in only 2 people (0.2%). The most commonly reported severe symptom of the six was photophobia, which was reported by 4.9% of the population. Severe discomfort and itching were each reported by 2.3% of the population, but the majority (90.5%) of the severe itching symptoms were attributable to hay fever. Severe tearing was reported by 1.9% of subjects, severe foreign body sensation by 1.5% and severe dryness by 0.4% (none of which was attributable to hay fever). Overall, 5.5% of subjects reported at least one severe symptom not attributable to hay fever.[5] Among the 3,703 subjects in the Beaver Dam Eye Study with data on dry eye, symptoms were reported by 14.4% (95% confidence interval = 13.3% to 15.6%).[4]

In preliminary findings from WHS standardized to the age-specific prevalences to the age distribution of U.S. women aged 55 and older, the prevalence of diagnosed DES was 6.7%. With regard to symptoms of both dryness and irritation constantly or often the age-standardized prevalence was 3.2%. For cDES/Sx, the prevalence among women aged 55 and older was 8.1% after adjusting the estimate to the age distribution of U.S. women in 1999. This translated into an estimated 2.2 million women aged 55 and older in the United States with dry eye. In preliminary findings from PHS, the age-standardized prevalences were 2.3% for clinically diagnosed DES, 1.9% for severe symptoms and 3.5% for cDES/Sx among men aged 55 and older. Based on the estimate for cDES/Sx, 900,000 men in this age group may have dry eye.

4. PREVALENCE OF DES BY AGE

Clinical dogma has long suggested DES becomes more common with age, and some evidence suggests an age-related decrease in tear production.[14, 15] However, until recently, epidemiological data to support an age-related increase in DES have been lacking. We have summarized data on the relationship of dry eye prevalence with age in the four large epidemiological studies conducted to date (Fig. 1). The association between age and the prevalence of DES was not reported to be statistically significant in the Salisbury study.[3] The results of all studies taken together are most consistent with a trend toward higher prevalence of DES in the older age groups, though with a possible leveling off in the oldest groups (Fig. 1). Further, the overall relationship of age with dry eye prevalence appears rather linear. A linear relationship would be expected if the incidence of new cases of dry eye were constant with age and no relationship existed between dry eye and survival in the population. In the case of dry eye, currently no information is available on the incidence either generally or with regard to age. In other words, even given these data on prevalence, it remains unknown whether a 65-year-old person who has survived without developing dry eye has any greater risk of developing it over a given time interval compared to a 40-year-old person who is also free of the disease. It is also unknown whether any association exists between dry eye and survival. Other age-related eye diseases such as cataract have been associated with higher mortality in some studies. In the Salisbury study dry eye was associated with a higher probability of having more co-morbid conditions,[3] which appears consistent with the possibility that people who develop dry eye may not survive as long as their peers.

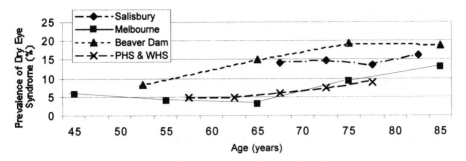

Figure 1. Prevalence of DES by age.

5. PREVALENCE OF DES BY GENDER

A sound biological basis exists for the proposal that DES is more common in females and may be influenced by sex hormone levels in both women and men. Androgens probably account for many gender differences observed in lacrimal glands of numerous species.[16] The putative heightened state of immunological activity in females compared to males with differential actions of sex steroids on the immune system could be one factor that accounts for the higher prevalence and increased severity of autoimmune conditions such as Sjögren's syndrome in females.[17, 18] Sjögren's syndrome and other autoimmune conditions have been associated with altered profiles of circulating sex hormones. Estrogens appear to play a major role in the pathogenesis and progression of Sjögren's syndrome,[19, 20] while androgens may actually lessen various immunopathologies of the disease.[21] Serum concentrations of testosterone may be reduced in Sjögren's syndrome,[22] and an attenuated testosterone/estrogen ratio with increased levels of estrogen metabolites has been observed in lupus.[22, 23] With regard to evaporative DES, the meibomian gland, a large sebaceous gland, is probably controlled by androgens, which control lipid production of sebaceous glands throughout the body.[24] Consistent with this theory, human meibomian glands contain androgen receptors and androgen receptor mRNA as well as mRNAs for both Types 1 and 2 5-alpha-reductase,[25] and sebaceous gland activity declines with age in concert with an age-related decline in serum androgen levels.[26] Evidence has also emerged that androgen deficiency in humans (e.g., in patients taking anti-androgen medication) is associated with meibomian gland dysfunction and DES.[25] Other investigators have shown topical administration of dehydroepiandrosterone to animals stimulates the production and release of meibomian gland lipids and improves tear film stability.[27]

Consistent with the basic research, clinical observations have long suggested DES is more common in women, particularly after menopause. However, clinical observations are subject to possible biases such as surveillance bias that may occur if women seek eye care in greater numbers than men, or diagnostic bias if women are more likely to be given a diagnosis of DES. Until recently, few epidemiologic data were available to address this issue. One study, in which Canadian optometrists were contacted by mail and asked to

administer a survey to 30 consecutive patients, concluded symptoms of severe dry eye
were more common among females by a ratio of 4.6 to 1 (55% of subjects were 21 to 50
years of age, and 61% were female); however, only 15% of optometrists responded to the
survey.[28] Moreover, data from this clinical population also tend to reflect women are more
likely to visit health care providers compared to men. Nevertheless, in the Melbourne
study, which is population-based and therefore does not reflect this limitation, women
were nearly two times more likely to report severe symptoms of dry eye (odds ratio =
1.85).[5] Similarly, in the Beaver Dam study the prevalence of DES was higher in women
(17.0%) compared to men (11.1%, P < 0.001).[4] In preliminary analyses from the two
Harvard-based cohorts, WHS and PHS, the prevalence of DES was higher in women than
in men in each age group (Fig. 2). In contrast, investigators of the Salisbury study observed
no significant gender-related difference in the prevalence of dry eye symptoms among
women (15.6%) and men (13.3%) aged 65 years and older.[3]

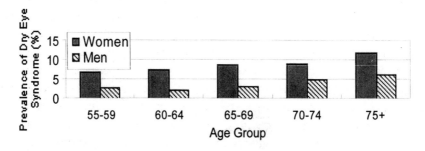

Figure 2. Prevalence of DES in WHS and PHS by gender and age (preliminary findings).

6. PREVALENCE OF DES BY RACE

As pointed out in the NEI/Industry workshop report published in 1995,[2] very little
information was available on the prevalence of dry eye by race. Unfortunately this is still
true, as only two studies have addressed this issue. The Salisbury study included 650
blacks and 1,832 whites from the same community in Maryland and compared the
prevalence of dry eye between these groups. The prevalence of at least one of six
symptoms at least "often" was 13.5% among blacks and 15.0% among whites, with no
statistically significant difference between them.[3] In the WHS, 95% of the women
indicated they were white, 2.2% black, 1.4% Asian/Pacific islander, 1.0% Hispanic, 0.3%
Native American and 0.2% other or unknown race/ethnicity.[11] Given the large size of the
study, sufficient numbers of women existed in most minority race/ethnicity groups to
estimate differences in prevalence. In preliminary analyses, the prevalence of dry eye was
similar among these groups of women, and no significant differences occurred in the
prevalence of clinically diagnosed DES, severe symptoms or cDES/Sx among blacks and

whites. However, both Hispanic and Asian women reported a higher prevalence of severe dry eye symptoms compared to whites (DA Schaumberg, et al., submitted manuscript).

7. OTHER RISK FACTORS FOR DES

Unlike the case with other prevalent age-related eye diseases such as cataract, macular degeneration and glaucoma, relatively little study to date has assessed potential risk factors for DES. In the Melbourne study,[5] investigators examined relationships of several factors with the prevalence of dry eye and found no significant associations with the use of contact lenses, cigarette smoking, history of previous eye injury, definite symptoms of dry mouth, menopause or the use of hormone replacement therapy (HRT) among postmenopausal women. Arthritis was one factor significantly associated with a higher prevalence of two or more clinical signs of dry eye. However, the usefulness of these analyses is limited because some exposures were rare in the population (for example only 2% of the study population, 21 subjects, wore contact lenses) and no estimates of the magnitude or confidence intervals for these relationships were provided. In the Salisbury study,[3] dry eye symptoms were significantly associated with self-reported poor health and the number of medical co-morbidities. Investigators in the Beaver Dam study[4] performed a screening (i.e., non-hypothesis driven) type of analysis of a long list of factors (Table 1) to determine whether any would show a statistical relationship with dry eye. In these analyses, statistically significant independent associations were observed for arthritis history (Odds Ratio [OR] = 1.91, $P < 0.001$), past (OR = 1.22) and current (OR = 1.82) cigarette smoking (P for trend < 0.001), caffeine use (OR = 0.75, $P < 0.005$), thyroid disease (OR = 1.41, $P < 0.01$), history of gout (OR = 1.42, P = 0.03), total/high-density lipoprotein cholesterol ratio (OR = 0.93, P = 0.03), diabetes mellitus (OR = 1.38, P = 0.03) and past (OR = 1.35) and current (OR = 1.41) use of multivitamins (P for trend = 0.03).

Table 1. Factors examined for statistical associations with DES in the Beaver Dam Eye Study

Blood Pressure	Heavy drinking	ACE inhibitors
Body Mass Index	Cigarette smoking	Antihistamines
Diabetes	Caffeine consumption	Antiemetics
Arthritis	Heating season	Reserpine
Fractures	Calcium channel blockers	Antidepressants
Osteoporosis	Parasympathetics	Multivitamins
Gout	Aspirin	Glasses or contacts
Thyroid disorders	Beta blockers	Macular degeneration
Cardiovascular disease	Antianxiety agents	Cataract
Menstrual status	Diuretics	Lens surgery
Oophorectomy	Methyldopa	Glaucoma
HRT		

In WHS and PHS, a hypothesis-driven approach is being taken for the examination of potential risk factors for DES. Given the basic research supporting the hypothesis that sex steroid hormones modulate the production and release of tear components, the potential relationship of HRT with DES is of particular interest. In WHS, preliminary analyses on the relationship of HRT with DES[8, 29] have indicated a 66–70% higher prevalence of severe dry eye symptoms, clinically diagnosed DES or cDES/Sx (each P ≤ 0.0001) among postmenopausal women who use estrogen alone compared to women who never used HRT, controlling for age, education, income and frequency of eye examinations. Use of estrogens with progesterone/progestins was associated with a 25–31% (each P ≤ 0.005) higher prevalence of dry eye, but this risk was significantly lower than the risk with estrogens alone (each P ≤ 0.001).

8. SUMMARY AND FUTURE DIRECTIONS

This review of epidemiological studies shows DES is a relatively common condition among middle-aged and older adults. Although its prevalence clearly varies among the different studies, this is likely due in large part to differences in the definition of DES used in each study, as prevalence varies in a predictable way according to the restrictiveness of the definition. Evidence from epidemiological studies indicates DES increases with age. However, a higher prevalence among older individuals would be expected even if the incidence of DES remained constant with age, unless a tendency existed for earlier mortality among those with DES, which, given the apparent association between DES and more co-morbidities, may indeed be the case. Data on how the incidence rate of DES might vary with age are still lacking; therefore, based on prevalence data alone, it remains unclear whether the incidence of new cases of dry eye is truly higher in older age groups. Future studies on the incidence of DES would be very informative in this regard.

With the exception of the Salisbury study,[3] epidemiological studies of dry eye have consistently shown a higher prevalence of DES among women than men.[4–6] Androgen levels may decline with age in both women and men, and some older men may be subjected to anti-androgen therapy for treatment of benign prostatic hypertrophy or prostate cancer, two highly prevalent conditions in older men. Anti-androgen treatment is associated with dry eye.[25] Furthermore, at least one study suggested use of HRT may be associated with a higher prevalence of DES.[8, 29] Differences in the various study populations with regard to these gender-related factors might explain the inconsistent results of the Salisbury study.

Based on the limited evidence to date, the prevalence of DES does not vary greatly by race/ethnicity, with the notable exception in WHS of more prevalent symptoms among Hispanic women and those of Asian/Pacific Islander descent. The reasons are unclear but may be related to several factors. For example, some groups of women with DES may obtain less relief from dry eye symptoms. In addition, the higher prevalence of symptoms in some minority groups could be related to a greater number of health problems in these women, either by themselves or by virtue of their treatments. It is possible that DES is underdiagnosed among some minority groups. Further studies are needed to clarify these issues.

Current information on other risk factors for DES is limited. Although each study reviewed here attempted to examine potential risk factors, the power to do so was often limited and the data were often not fully analyzed or presented. Nevertheless, some interesting hypotheses have been raised, in particular by the data presented from the Beaver Dam Eye Study[4] and ongoing basic research. Future studies should examine compelling hypotheses in more detail, and the apparent adverse effect of HRT on the prevalence of DES in WHS deserves further study. Since exposure to HRT was not randomly allocated in WHS and investigators could not determine if HRT use preceded the onset of DES, the observed relationship might reflect a higher tendency of women with DES to be prescribed HRT. Although prescription of HRT specifically for the relief of DES symptoms is not likely to be common, to the extent symptoms of DES arose with other menopausal symptoms for which HRT is beneficial,[30] women with symptoms of DES may simply have been prescribed HRT in greater numbers. Arguing against this possibility, however, is the fact that the association differed significantly for estrogen alone versus estrogen plus progesterone. Moreover, as would be expected if the relationship of HRT use with DES were real rather than spurious, the WHS results consistently showed a higher prevalence of DES among women who used HRT, regardless of how DES was defined. This evidence with plausible biology in support of an adverse effect of estrogens or altered estrogen-androgen balance on the tear film[16, 20, 21, 31–33] supports a true association, but further study is clearly needed.

REFERENCES

1. McMonnies, C.W. and H. Ho, Patient history in screening for dry eye conditions. J Am Optom Assoc, 58: p. 296–301 (1987).
2. Lemp, M.A., Report of the National Eye Institute/Industry workshop on Clinical Trials in Dry Eyes. Clao J, 21(4): p. 221–32 (1995).
3. Schein, O.D., et al., Prevalence of dry eye among the elderly. Am J Ophthalmol, 124(6): p. 723–8 (1997).
4. Moss, S.E., R. Klein, and B.E. Klein, Prevalence of and risk factors for dry eye syndrome. Arch Ophthalmol, 118(9): p. 1264–8 (2000).
5. McCarty, C.A., et al., The epidemiology of dry eye in Melbourne, Australia. Ophthalmology, 105(6): p. 1114–9 (1998).
6. Schaumberg, D.A., et al., The epidemiology of dry eye syndrome. Cornea, 19 (Suppl): p. S120 (2000).
7. Schein, O.D., et al., Relation between signs and symptoms of dry eye in the elderly. A population-based perspective. Ophthalmology, 104(9): p. 1395–401 (1997).
8. Schaumberg, D.A., et al., Hormone replacement therapy and the prevalence of dry eye symptoms in women. Invest Ophthalmol Vis Sci, 41: p. S740 (2000).
9. Nichols, K.K., J.J. Nichols, and K. Zadnik, Frequency of dry eye diagnostic test procedures used in various modes of ophthalmic practice. Cornea, 19(4): p. 477–82 (2000).
10. Korb, D.R., Survey of preferred tests for diagnosis of the tear film and dry eye. Cornea, 19(4): p. 483–6 (2000).
11. Rexrode, K.M., et al., Baseline characteristic of participants in the Women's Health Study. J Women's Health & Gender–Based Med, 9(1): p. 19–27 (2000).

12. Steering Committee of the Physicians' Health Study Research Group. Final report on the aspirin component of the ongoing Physicians' Health Study [see comments]. N Engl J Med, 321(3): p. 129–35 (1989).

13. Oden, N.L., *et al.*, Sensitivity and specificity of a screening questionnaire for dry eye. Adv Exp Med Biol, 438: p. 807–20 (1998).

14. Mathers, W.D., J.A. Lane, and M.B. Zimmerman, Tear film changes associated with normal aging. Cornea, 15(3): p. 229–34 (1996).

15. Henderson, J.W. and W.A. Prough, Influence of age and sex on flow of tears. Arch Ophthalmol, 43: p. 224–31 (1950).

16. Sullivan, D.A., *et al.*, Influence of gender, sex steroid hormones, and the hypothalamic- pituitary axis on the structure and function of the lacrimal gland. Adv Exp Med Biol, 438: p. 11–42 (1998).

17. Sullivan, D.A., Sex hormones and Sjogren's syndrome. J Rheumatol, 24 Suppl 50: p. 17–32 (1997).

18. Homo-Delarche, F., *et al.*, Sex steroids, glucocorticoids, stress and autoimmunity. J Steroid Biochem Mol Biol, 40(4–6): p. 619–37 (1991).

19. Ahmed, S.A., *et al.*, Estrogen induces the development of autoantibodies and promotes salivary gland lymphoid infiltrates in normal mice. J Autoimmun, 2(4): p. 543–52 (1989).

20. Sato, E.H. and D.A. Sullivan, Comparative influence of steroid hormones and immunosuppressive agents on autoimmune expression in lacrimal glands of a female mouse model of Sjogren's syndrome. Invest Ophthalmol Vis Sci, 35(5): p. 2632–42 (1994).

21. Sullivan, D.A., *et al.*, Androgens and dry eye in Sjogren's syndrome. Ann N Y Acad Sci, 876: p. 312–24 (1999).

22. Cutolo, M., *et al.*, Estrogens, the immune response and autoimmunity. Clin Exp Rheumatol, 13(2): p. 217–26 (1995).

23. Lahita, R.G., The connective tissue diseases and the overall influence of gender. Int J Fertil Menopausal Stud, 41(2): p. 156–65 (1996).

24. Imperato-McGinley, J., *et al.*, The androgen control of sebum production. Studies of subjects with dihydrotestosterone deficiency and complete androgen insensitivity. J Clin Endocrinol Metab, 76(2): p. 524–8 (1993).

25. Sullivan, D.A., *et al.*, Androgen regulation of the meibomian gland. Adv Exp Med Biol, 438: p. 327–31 (1998).

26. Bologna, J.L., Aging skin. Am J Med Suppl, 1A: p. 99S–102S (1995).

27. Zeligs, M.A. and K. Gordon, Dehydroepiandrosterone therapy for the treatment of dry eye disorders, in Int Patent Application. 1994.

28. Doughty, M.J., *et al.*, A patient questionnaire approach to estimating the prevalence of dry eye symptoms in patients presenting to optometric practices across Canada. Optom Vis Sci, 74(8): p. 624–31 (1997).

29. Schaumberg, D.A., D.A. Sullivan, and M.R. Dana, The relationship of hormone replacement therapy with the prevalence of dry eye syndrome in a large cohort of women. Ophthalmology, 107 (Suppl.): p. 105 (2000).

30. Belchetz, P.E., Hormonal treatment of postmenopausal women. N Engl J Med, 330(15): p. 1062–71 (1994).

31. Thody, A.J. and S. Shuster, Control and function of sebaceous glands. Physiol Rev, 69(2): p. 383–416 (1989).

32. Gurwood, A.S., *et al.*, Idiosyncratic ocular symptoms associated with the estradiol transdermal estrogen replacement patch system. Optom Vis Sci, 72(1): p. 29–33 (1995).

33. Brennan, N.A. and N. Efron, Symptomatology of HEMA contact lens wear. Optom Vis Sci, 66(12): p. 834–8 (1989).

IMMUNOGENETICS OF AUTOIMMUNE EXOCRINOPATHY IN THE NOD MOUSE: MORE THAN MEETS THE EYE

M. G. Humphreys-Beher,[1] J. Brayer,[1] S. Cha,[1] H. Nagashima,[1] S. Diggs,[1] and A. B. Peck[2]

Departments of [1]Oral Biology and [2]Pathology
University of Florida
Gainesville, Florida, USA

1. INTRODUCTION

Sjögren's syndrome is defined as a chronic inflammatory autoimmune disorder primarily affecting the lacrimal and salivary exocrine glands. However, the disorder has also been associated with numerous extraglandular manifestations affecting such diverse tissues as the cardiovascular, central nervous, and renal systems.[1] Diagnostic criteria include oral and/or ocular dryness, focal lymphocytic infiltrates in the exocrine tissues, and the presence of autoantibodies in patient sera, particularly against the nuclear factors SS-A/Ro and SS-B/La. Due to inconsistencies in the various diagnostic scales, evaluation of disease prevalence worldwide is difficult and complicated further by the chronic and inconsistent nature of the symptoms in patients. [1,2]

Typically, Sjögren's syndrome is diagnosed in its later stages when focal lymphocytic infiltration and glandular destruction has already occurred. For this reason, the initial and early events in the disease pathogenesis remain largely unknown. Conversely, the persisting immunological response in the context of the minor salivary glands has been studied to a much greater depth. This is, in part, a reflection of the relative accessibility of minor salivary glands in obtaining tissue biopsies. This has allowed for the assessment of cytokine activity, as well as the categorization of infiltrating leukocyte populations in the glands of patients exhibiting an ongoing disease process. During the past couple of decades, investigators have established several inbred mouse strains as applicable animal models for the further study of Sjögren's syndrome pathogenesis. Among others, the MRL/lpr, NFS/sld and the NOD and NOD.B10.$H2^b$ mouse strains have emerged as the

Lacrimal Gland, Tear Film, and Dry Eye Syndromes 3
Edited by D. Sullivan *et al.*, Kluwer Academic/Plenum Publishers, 2002

999

most popular animal models for Sjögren's syndrome based on immune infiltration of the exocrine tissues, cytokine profiles, and serology.[3-6] Additionally, the NOD and NOD.B10.$H2^b$ mice exhibit an immune-dependent loss of secretory function in both the lacrimal and salivary glands. Assuming the appropriateness of these animal models, our understanding of both initiation and development of Sjögren's syndrome has advanced markedly, from a reasonable appreciation of the later stages of the disease, to the determination of numerous additional autoantigens and, more importantly, insights into the initiating events in the disease.

2. ETIOLOGY OF SJÖGREN'S SYNDROME

Like other autoimmune disorders, Sjögren's syndrome appears to be a multifactorial disease, with its onset dependent on both intrinsic and extrinsic parameters. From a genetic standpoint, there appears to be an association with certain MHC haplotypes in patient populations and genomic analysis in animal models has already uncovered regions on chromosomes 1, 3, 4, 10, and 18 associated with susceptibility.[7,8] Since these chromosomal regions are polygenic, it is impossible at this time to definitively state whether the regulatory elements encoded in these loci are immunological in nature.

More than any other body of evidence, the studies in the NOD-*scid* mouse have promoted the idea that the exocrinopathy may not be entirely dependent on the immune infiltrate.[9,10] This animal model exhibits gross morphological restructuring and elevated levels of apoptosis in the salivary and lacrimal epithelial cells despite the absence of glandular lymphocytic foci. Furthermore, while proapoptotic proteins, such as caspase-3 and Bax, are elevated in acinar and ductal tissues along with bcl-2 (an anti-apoptotic factor) in NOD and NOD-*scid* mice, the infiltrating lymphocytic populations found in the NOD exocrine glands appear to upregulate bcl-2 only, while maintaining normal expression levels of caspase-3 and bax.[11] The presence of constitutively expressed FasL and induced Fas expression on the salivary epithelial cell surfaces at progressive stages of disease, along with the increase in TUNEL staining, indicative of excessive programmed cell death, encourages the belief that an underlying disturbance in glandular homeostasis could potentially facilitate or even promote immunopathological activity. Thus, disease initiation may pass through an asymptomatic phase leading to subsequent immune cell targeting and activation resulting in clinical presentation of disease.

Non-immunological factors influencing the development of Sjögren's syndrome also include hormonal regulation, as suggested by the sexual dimorphism (9:1 in favor of females over males) found in the patient population. This has been supported by numerous lacrimal studies in models such as the MRL and NZB/NZW mouse strains where androgens tend to exert a protective influence for dacryoadenitis. Interestingly, the NOD mouse presents a reversal of androgen influence, where the presence of androgens correlates to increased lacrimal destruction. In the case of the influence of the sex hormones on disease pathogenesis, studies have demonstrated the abilities of estrogens and androgens to regulate the immune response.[12-14]

3. CHARACTERIZATION OF INFILTRATING LYMPHOCYTES

The focal lymphocytic infiltrates found in the salivary and lacrimal glands of Sjögren's patients and of mouse models are comprised predominantly of CD4$^+$ $\alpha\beta$ T-cells. In the human disease, lymphocytes comprise approximately 80% of the focal infiltrates.[15] Plasma B-cells present in the lesions produce mostly IgM and IgG and have a greater than normal tendency to mutate into lymphomas. The salivary infiltrates in the NOD mouse are comprised of approximately 45% T$_H$ cells, 15–20% B-cells, and 10–15% CD8$^+$ T-cells, while in the lacrimal glands, the populations are skewed slightly in favor of the B-cells (25–30%), the T$_H$ cells constituting approximately 25–30%.[16] While focal infiltrates develop at different times depending on the animal model, comparisons of the NOD and MRL/lpr strains indicate that the autoimmunity in the NOD model, but not the MRL-lpr model, is preceded by an influx of CD11c$^+$ dendritic cells.[17]

4. CYTOKINE PROFILES IN SJÖGREN'S SYNDROME

Cytokine expression in the salivary glands of Sjögren's syndrome patients, plus that observed in various mouse models, suggests a pro-inflammatory profile. Cytokine mapping in the salivary and lacrimal glands has relied largely on the measurement of stable mRNA levels via reverse transcriptase polymerase chain reaction. Evaluation of patient populations reveal the expression of IL-1β, IL-2, IL-6, IL-10, IL-12, IL-18, TNF-α, TGF-β, and IFN-γ in the minor labial glands.[18] Similarly, studies in the mouse models report the expression of IL-1β, IL-2, IL-6, IL-7, IL-10, IL-12, TNF-α, TGF-β, iNOS and IFN-γ in the submandibular and lacrimal glands.[16] Rarely, IL-4 can be detected but is generally considered absent in the infiltrated glands. More recent studies in both human and mouse salivary glands have indicated that this comprehensive pro-inflammatory pattern of cytokine expression can be detected in the salivary tissues of healthy donors, as well as non-autoimmune mouse strains at levels similar to disease state values. Therefore, rather than being a defining hallmark of the disease process, these cytokine mRNA expression profiles are more a reflection of tissue specific micro-regulation of the immune responses favoring a pro-inflammatory response in these glands in the context of localized immune activation. Recent reports indicate that IFN-γ can induce upregulation of IL-1β, IL-6 and TNF-α, along with HLA-DR and ICAM-1 in HSG cells.[19]

5. CYTOKINE KNOCKOUTS FOR THE STUDY OF AUTOIMMUNITY

Ongoing studies using cytokine knockout mice are revealing interesting new insights regarding the autoimmune pathogenesis. Although the detection of specific cytokine expressions in target tissues may reveal an underlying role for that cytokine, this approach operates under the assumption that the entire pathological course of events occurs in the target tissues. Studies employing cytokine-deficient congenic animals emphasize an essential consideration that critical aspects of the pathogenic processes in Sjögren's syndrome occur outside the environment of the exocrine glands.

Just as there are limitations to looking at cytokine expression levels in the target tissues, it can be misleading to define the importance of a cytokine based on changes in disease pathogenesis in its absence or overexpression. It has been shown that the activity of a cytokine is not simply controlled by its presence or absence, but also by the level and timing of expression. Additionally, many cytokines, for instance IL-4 and IL-13, have overlapping effects such that the elimination of an individual cytokine from an in vivo system may not exhibit as dramatic an effect, despite the distinct role of the factor in the normal disease process. Regardless, gene-knockout models eliminating either cytokines or cell populations provide unique new perspectives into the questions or relevance with regard to these immune components.

Analysis of the NOD.IL-4$^{-/-}$ mouse exemplifies the highly significant contribution of studies introducing cytokine deficiencies into a disease-susceptible mouse to map cytokine functions. IL-4 is rarely, if ever, detected in salivary or lacrimal gland homogenates in studies of mouse models, or in labial gland biopsies of patients; yet this cytokine is absolutely critical in driving the suppression of secretory function in the autoimmune exocrinopathy in the context of the NOD mouse (Table 1). The NOD.IL-4$^{-/-}$ mouse develops massive focal lymphocytic infiltrates in the exocrine glands, despite maintaining normal secretory capacity throughout its life. Further elucidation of the autoantibody composition reveals an inability to generate an anti-IgG1 subtype antibody to the muscarinic receptor. This finding supports accumulating evidence that the secretory suppression in the lacrimal and salivary glands is a direct result of autoantibodies acting on the fluid secreting acinar cells.[20,23]

Table 1. Biological markers of autoimmune exocrinopathy in the cytokine knockout mice

STRAIN	ANA/Anti-M$_3$	Lac./Sal. Gland Infiltrate	TUNEL	Saliva Volume[a]	Saliva Protein[b]	Amylase Activity
C57BL/6	No/No	No/No	-	229 ± 17	2.2 ± 1.3	2764 ± 137
IL-4$^{-/-}$(8wk)	No/No	Yes/No	-/+	200 ± 18	3.1 ± 2.0	2833 ± 246
IL-4$^{-/-}$(12Wk)	Yes/No	Yes/Yes	+	235 ± 24	4.9 ± 2.1	1522 ± 248
IL-4$^{-/-}$(20wk)	Yes/No	Yes/Yes (>12 foci/gl.)	++	216 ± 27	5.7 ± 2.3	1188 ± 189
IFNγ$^{-/-}$(9 wk)	No/No	No/No	-	222 ± 21	3.4 ± 1.5	2966 ± 208
IFNγ$^{-/-}$ (13 wk)	No/No	No/No	-	212 ± 19	4.5 ± 1.7	2885 ± 174
IFNγ$^{-/-}$ (20 wk)	No/No	Yes/No (>10 foci/gl)	-	225 ± 24	3.1 ± 2.0	2801 ± 237
IL-10$^{-/-}$(20 wk)	Yes/Yes	Yes/Yes (>10 foci/gl)	++	123 ± 20	6.1 ± 2.6	844 ± 201
NOD	Yes/Yes	Yes/Yes (3-4 foci/gl.)	+	115 ± 17	6.8 ± 1.7	971 ± 257

a, volume unit of measure (μl/10 min) ; b, protein concentration unit of measure (μg/μl).

Another cytokine that has received increasing attention is IL-10. IL-10 expression increases notably over time in the course of disease progression. Recent studies in which a transgene construct introducing IL-10 under the control of the salivary amylase promoter to induce exocrinopathy show that the induction of IL-10 expression in the salivary glands can lead to focal infiltration and secretory suppression in the C57BL/6 genetic background.[24] However, in the context of the Sjögren's syndrome-like pathology, introduction of the IL-10-deficiency mutation into the NOD mouse contradicts the findings derived from IL-10 overexpression; the lack of IL-10 allows for normal focal infiltration and disease progression (Table 1). This should not necessarily be interpreted as a contradiction, but rather as a prime example of the importance of timing of expression as opposed to strict cytokine presence in directing an immune response. Together, these data begin to establish parallels to the other autoimmune diseases like diabetes, where more extensive studies into the role of IL-10 in disease onset in the NOD model imply that IL-10 acts in a distinctly time-dependent fashion, such that it promotes and even accelerates diabetes initiation when expressed constitutively in the pancreatic tissues, but serves to ameliorate pancreatitis when administered later in the course of the disease.[25,26] With respect to Sjögren's syndrome, the NOD.IL-10[-/-] mouse appears to progress normally through the development of autoimmune exocrinopathy, indicating that IL-10 is not an absolutely critical component driving the pathogenesis at the time of disease onset. Studies on overexpression and deficiency of IL-10 suggest that this cytokine is capable of contributing to the underlying pathology but is not absolutely necessary for autoimmune disease progression.

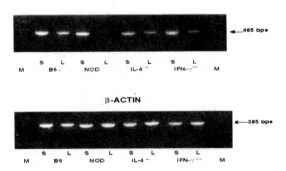

Figure 1. Detection of reduced EGF and β-actin mRNA in the submandibular (SMX) and lacrimal (LAC) glands of C57BL/6 control mice, NOD parental and NOD cytokine knockout mice by RT-PCR amplification.

Most surprising, are the findings of the Th-1 cytokine knockout, the NOD.IFNγ[-/-] strain. As indicated in Table 1, a temporal analyses of the secretory response, as well as a number of other biochemical characteristics of autoimmune sialoadenitis are more closely identified with the levels of healthy C57BL/6 control mice than with autoimmune exocrinopathy of the parental NOD. Interestingly, while the submandibular gland remains free of detectable focal infiltrates, even out to 35 weeks of age, a normal progression of immune cell infiltrates is observed in the lacrimal glands. The composition of these infiltrates is distinctly different from the NOD and NOD.IL4-/- mice in that there is a decline of $CD8^+$ T-lymphocytes in the focal infiltrates (Table 2). The ratio of $CD4^+$ to $CD8^+$ cells goes from 3:1 to 6–7:1 (P < 0.01). As with the NOD-*scid* and NOD.Igμ^{null} mice, the lacrimal glands of NOD.IFN$\gamma^{-/-}$ retain full secretory capacity and protein concentrations, comparable to the C56BL/6 control mice (Table 2). NOD.IFN$\gamma^{-/-}$ mice show some of the underlying pathophysiology of epithelial cells, as do the parental NOD, such as declining synthesis of EGF, elevated MMP activity, and caspase 3 activation (Fig. 1 and Table 2). Densitometric analysis of the submandibular gland shows a 20–25% decline in EGF mRNA, while the decline in the lacrimal gland is approximately 30–50% less than C57BL/6 controls. The decline in lacrimal gland steady state levels is consistent with the reports of lower levels of growth factor in the tears from Sjögren's patients.[27,28] Previous studies by Wu and colleagues[27,29] indicate that IFNγ has not only a toxic effect on salivary cells in vitro, but can also perturb gene expression, increasing levels of MMP-2 and MMP-9 activity.[29–31]

Table 2. Markers of autoimmune pathophysiology of lacrimal glands in NOD cytokine knockout mice

STRAIN	Flow Rate (μl/5 min)	Protein (ug/ml)	MMP Activity (U/min/mg prot) SMX	LAC	CD4$^+$, CD8$^+$, and B-Cells[a] (%)
C57BL/6	3.2 ± 0.5	12.3 ± 1.7	0.410 ± 0.055	0.020±0.005	0.2%, 0.2%, 0.1%
NOD	2.3 ± 0.6	24.7 ± 2.9	0.890 ± 0.032	0.045 ± 0.008	46%, 14.5%, 25%
NOD.IL-4-/-	3.5 ± 0.6	8.5 ± 2.6	0.630 ± 0.017	0.030 ± 0.005	19%, 6.7%, 35%
NOD.IFN$\gamma^{-/-}$	3.3 ± 0.5	10.2 ± 2.3	0.630 ± 0.004	0.073 ± 0.005	9.3%, 1.3%, 19%

a, percent of CD3$^+$ cells. B-cells determined by CD19 staining in Flow Cytometry.

6. CONCLUSION

Our appreciation of the immunological component in Sjögren's syndrome has greatly improved based on the studies of several mouse models of autoimmune exocrinopathy. However, in our attempts to dissect the immunological attack on the exocrine tissues, there is a tendency to focus on a specific gland, typically either the lacrimal or salivary glands, rather than assessing the global effects in all of the affected tissues *in toto*. While the symptoms of keratoconjunctivitis sicca or the oral manifestations of Sjögren's syndrome may share many similar underlying factors, there are also some fundamental differences in such basic components as ratios of cell populations and cytokine expression within the glands. The development of Sjögren's syndrome in patients may affect both the lacrimal or salivary glands, or target one of these exocrine glands independently, suggesting that the events leading to the precipitation of the autoimmune attack in each gland are not entirely identical and are, to a certain degree, independent. As seen with the NOD.IFNγ$^{-/-}$ mice, despite the appearance of autoimmune dacryoadenitis in the lacrimal glands and underlying loss of differentiated function in the epithelial cells of the gland, activation of the immune response, leading to the loss of secretory function, does not occur. In this regard, the pathology of NOD.IFNγ$^{-/-}$ are similar to NOD.Igμ^{null} mice in that, despite a histopathology reminiscent of Sjögren's syndrome, without the maturation of the immune response and ensuing production of autoantibodies, secretory dysfunction does not take place. Thus, the IFNγ knockout begins to provide us with an understanding of a duel role for cytokines in affecting not just the lymphocyte compartment of the autoimmune response but also the target epithelial cells.

To understand the Sjögren's syndrome pathogenicity is to begin to discern not only what role the immune system plays in the inflammatory lesion but also what critical events are occurring outside the microenvironment of the exocrine lesion, e.g. in draining lymphatics leading to the classic systemic symptoms affecting not only the primary target tissues (the lacrimal and salivary exocrine tissues), but also the extraglandular manifestations of Sjögren's syndrome. The intriguing findings developed by the IL-4 knockout emphasize the point that simply looking at the pathogenesis within the exocrine glands fails to take into consideration many essential factors occurring on a more global systemic immunological level. In fact, of the autoimmune pathologies reliant on an effector autoantibody, pemphigus vulgaris is one of the few that is associated with the detection of IL-4 at the site of immune-mediated damage while autoimmune diseases such as Grave's disease and Hashimoto's thyroiditis, dependent at least in part on the presence of effector autoantibodies similar to Sjögren's syndrome, are associated with a pro-inflammatory cytokines such as IFN-γ in the glands.[32] Many of the critical events involved in orchestrating the immunological attack occur in the draining lymphatics, not in the tissue lesion. The antibodies shown to mediate exocrine suppression are produced by plasma B-cells that likely reside in the lumenal regions of the draining lymph nodes or in the bone marrow, despite the evidence supporting locally produced IgM and IgG by infiltrating B cells. This does not rule out the importance of the exocrine lymphocytic infiltrates in the overall disease state, but it also implies that important effector populations reside in extraglandular locations. Therefore, conclusions drawn from the analysis of the disease state strictly within the exocrine tissues, such as cytokine profiles, do not provide enough information to truly understand the mechanisms of the disease.

ACKNOWLEDGEMENT

The authors would like to thank Ms. Joy Nanni and Janet Cornelius for technical assistance. This research was supported by NIH grant DE 13290. SC is supported by funds for graduate education from the Department of Oral Biology.

REFERENCES

1. R.I. Fox, T.J. Tornwall, T. Maruyami, and M. Stern. Sjögren's syndrome: Evolving concepts of diagnosis, pathogenesis and therapy. Curr Op Rheum 10:446(1998).
2. P.C. Fox and P.M. Speight. Current concepts of autoimmune exocrinopathy: immunologic mechanisms in the salivary pathology of Sjögren's syndrome. Crit Rev Oral Biol Med. 7:144(1996).
3. R.W. Hoffman, M.A. Alspaugh, K.S. Waggie, J.B. Durham, and S.E. Walker. Sjögren's syndrome in MRL/l and MRL/n mice. Arthritis Rheum. 27:157(1984).
4. N. Haneji, H. Hamano, K. Yanagi, and Y. Hayashi,. A new animal model for primary Sjögren's syndrome in NFS/sld mutant mice. J Immunol. 153:2769(1994).
5. M.G. Humphreys-Beher, Y. Hu, Y. Nagakawa, P.-L. Wang, and K.R. Purushotham. Utilization of the NOD mouse as a model for the study of secondary Sjögren's syndrome. *Adv Exp Med Biol.* 1994;350:631-637.
6. C.P. Robinson, S. Yamachika, D.I. Bounous, J. Brayer, R. Jonsson, R. Holmdahl, A.B. Peck, and M.G. Humphreys-Beher. A novel NOD-derived murine model of primary Sjögren's syndrome. Arthritis Rheum. 41:150(1998).
7. J. Brayer, J. Lowry, S. Cha, C.P. Robinson, A.B. Peck, and M.G. Humphreys-Beher. Alleles from chromosomes 1 and 3 of NOD mice combine to influence Sjögren's syndrome-like autoimmune exocrinopathy. J Rheum. 27:1896(2000).
8. M. Nishihara, M. Terada, J. Kamogawa, Y. Ohashi, S. Mori, S. Nakatsuru, Y. Nakamura, and M. Nose. Genetic basis of autoimmune sialadenitis in MRL/lpr lupus-prone mice- additive and hierarchical properties of polygenic inheritance. Arthritis Rheum. 42:2616(1999).
9. H.J. Garchon, P. Bedossa, L. Eloy, and J.F. Bach. Identification and localization on chromosome 1 of a gene controlling the occurrence of periinsulitis in NOD diabetic mice. C-R-Acad-Sci-III. 312:377(1991).
10. L. Kong, C.P. Robinson, A.B. Peck, N. Vela-Roch, K.M. Sakata, H. Dang, N. Talal, and M.G. Humphreys-Beher. Inappropriate apoptosis of salivary and lacrimal gland epithelium of immunodeficient NOD-scid mice. Clin Exp Rheumatol. 16:675(1998).
11. R. Masago, S. Aiba-Masago, N. Talal, F. Jiminez Zuluaga, I. Al-Hashimi, M. Moody, C.A. Lau, M.G. Humphreys-Beher, and H. Dang,. Elevated levels of bax and caspase-3 activation in the NOD-*scid* model of Sjögren's syndrome. Arthritis Rheum. 44:693(2001)..
12. D.A. Sullivan, H. Ariga, A.C. Vendramini, F.J. Rocha, M. Ono, and E.H. Sato. Androgen-induced suppression of autoimmune disease in lacrimal glands of mouse models of Sjögren's syndrome. Adv Exp Med Biol. 350:683(1994).
13. R.E. Hunger, C. Carnaud, I. Vogt, and C. Mueller. Male gonadal environment paradoxically promotes dacryoadenitis in nonobese diabetic mice. J Clin Invest. 101:1300(1998).
14. I. Toda, B.D. Sullivan, L.A. Wickham, and D.A. Sullivan. Gender- and androgen-related influence on the expression of proto-oncogene and apoptotic factor mRNAs in lacrimal glands of autoimmune and non-autoimmune mice. J Steroid Biochem Mol Biol. 71:49(1999).
15. R.I. Fox, T.C. Adamson, S. Fong, C. Young, and F.V. Howell. Characterization of the phenotype and function of lymphocytes infiltrating the salivary gland in patients with primary Sjögren's syndrome. Diagn Immunol. 1:233(1983).
16. C.P. Robinson, J. Cornelius, D.E. Bounous, H. Yamamoto, M.G. Humphreys-Beher, and A.B. Peck. Characterization of the changing lymphocyte populations and cytokine expression in the exocrine tissues of autoimmune NOD mice. Autoimmunity. 27:29(1998).
17. S.C.A. van Blokland, C.G. van Helden-Meeuwsen, A.F. Wierenga-Wolf, H.A. Drexhage, H. Hooijkaas, J.P. van de Merwe, and M.A. Versnel. Two different types of sialoadenitis in the NOD- and MRL/lpr mouse models for Sjögren's syndrome: A differential role for dendritic cells in the initiation of sialoadenitis? Lab Invest. 80:575(2000).
18. M. Boumba, F.N. Skopouli, and H.M. Moutsopoulos. Cytokine mRNA expression from patients with primary Sjögren's syndrome. Brit J Rheum. 24:326(1995).

19. H. Hamano, N. Haneji, K. Yanagi, N. Ishimaru, and Y. Hayashi. Expression of HLA-DR and cytokine genes on interferon-gamma-stimulated human salivary gland cell line. Pathobiology. 64:255(1996).

20. O.G. Pankewycz, J.X. Guan, and J.F. Benedict. Cytokines as mediators of autoimmune diabetes and diabetic complications. Endocr Rev. 16:164(1995).

21. C.P. Robinson, J. Brayer, S. Yamachika, T. Esch, A.B. Peck, C. Stewart, E. Peen, R. Jonsson, and M.G. Humphreys-Beher. Transfer of human serum IgG into nonobese diabetic Igμ^{null} mice reveals a role for autoantibodies in the loss of secretory function of exocrine tissues in Sjögren's syndrome. Proc Natl Acad Sci. 95:7538(1998).

22. S. Bacman, L. Sterin-Borda, J. Jose Camusso, R. Arana, O. Hubscher, and E. Borda. Circulating antibodies against rat parotid gland M3 muscarinic receptors in primary Sjögren's syndrome. Clin Exp Immunol. 104:454(1996).

23. S.A. Waterman, T..P Gordon, and M. Rischmueller. Inhibitory effects of muscarinic receptor autoantibodies on parasympathetic neurotransmission in Sjögren's syndrome. Arthritis Rheum 43:1647(2000).

24. I. Saito, H. Kumiko, M. Shimuta, H. Inoue, H. Sakurai, K. Yamada, N. Ishimaru, H. Higashiyama, T. Sumida, H. Ishida, T. Suda, T. Noda, Y. Hayashi, and K. Tsubota. Fas ligand-mediated exocrinopathy resembling Sjögren's syndrome in mice transgenic for IL-10. J Immunol. 162:2488(1999).

25.0 B. Balasa, and N. Sarvetnick. The paradoxical effects of interleukin 10 in the immunoregulation of autoimmune diabetes. J Autoimmun. 9:283(1996).

26. M.E. Pauza, H. Neal, A. Hagenbaugh, H. Cheroutre, and D. Lo. T-cell production of an inducible interleukin-10 transgene provides limited protection from autoimmune diabetes. Diabetes.;48:1948(1999).

27. S.C. Pflugfelder, D. Jones, Z. Ji, A. Afonso, and D. Monroy. Altered cytokine balance in the tear fluid and conjunctiva of patients with Sjögren's syndrome keratoconjunctivitis sicca. Curr. Eye Res.;19:201(1999).

28. D.T. Jones, D. Monroy, Z. Ji, and S.C. Pflugfelder. Alterations in ocular gene expression in Sjögren's syndrome. Adv Exp Bio Med. 438:544(1998).

29. A.J. Wu, R.M. Lafrenie, C. Park, W. Apinhasmit, Z.J. Chen, H. Birkedal-Hansen, K.M. Yamada, W.G. Stetler-Stevenson, and B.J. Baum. Modulation of MMP-2 (gelatinase A) and MMP-9 (gelatinase B) by interferon-γ in a human salivary cell line. J Cell Physiol. 171:117(1997).

30. A.J. Wu, R.H. Kurrasch, J. Katz, P.C. Fox, B.J. Baum, and J.C. Atkinson. Effect of tumor necrosis factor-α and interferon-γ on the growth of a human salivary gland cell line. J Cell Physiol. 161:217(1994).

31. A.J. Wu, Z.J. Chen, B.J. Baum, and I.S. Ambudkar. Interferon-γ induces persistent depletion of internal Ca^{2+} stores in a cultured human salivary gland cell line. Am J Physiol. 270:C514(1996).

32. B. Balasa, and N. Sarvetnick. Is pathogenic humoral autoimmunity a Th1 response? Lessons from (for) myasthenia gravis. Immunol Today. 21:19(2000).

RESULTS OF A DRY EYE QUESTIONNAIRE FROM OPTOMETRIC PRACTICES IN NORTH AMERICA

Carolyn G. Begley,[1] Barbara Caffery,[2] Kelly Nichols,[3] Gladys Lynn Mitchell,[3] and Robin Chalmers[4]

[1]Indiana University
School of Optometry
[2]Toronto, Ontario, Canada
[3]The Ohio State College of Optometry
[4]Atlanta, Georgia, USA
and the DREI study group

1. INTRODUCTION

Symptoms of dry eye are relatively prevalent in the population.[1-6] However, current clinical tests for dry eye are relatively non-specific, poorly repeatable,[7,8] and show only a weak association with symptoms.[2,9,10] The dearth of information about general ocular surface symptoms and knowledge of their daily fluctuations in the population further limits our understanding of dry eye symptoms. This study characterized symptoms of ocular irritation from a large multi-centered survey of patients in optometric practices in the United States and Canada.

Clinical researchers in the area do not always agree on what constitutes "typical" dry eye symptoms,[2,9,11] perhaps because there are many proposed categories of dry eye.[12] Patients with Sjögren's Syndrome or keratoconjunctivitis sicca commonly report symptoms of ocular dryness and burning, grittiness, burning, and foreign body sensation.[9,10,13-15] Among patients diagnosed with mild to moderate dry eye, dryness, soreness, and light sensitivity were the most frequently reported symptoms.[16] Dryness was the most frequent symptom among contact lens wearers, along with scratchiness, irritation, grittiness, and blurry, changeable vision.[17-19]

Despite the lack of agreement on the exact nature of dry eye symptoms, the symptoms that are generally attributed to dry eye are relatively prevalent in the

Lacrimal Gland, Tear Film, and Dry Eye Syndromes 3
Edited by D. Sullivan *et al.*, Kluwer Academic/Plenum Publishers, 2002

1009

population.[1,4,6,20] Estimates among clinical populations range from a 17% to 28.7% prevalence of dry eye symptoms.[1,5] A recent survey of more than 2000 elderly residents of Salisbury, Maryland, showed that symptoms of ocular irritation were relatively common, with 15% reporting one or more of six dry eye symptoms often or all the time.[2] However, these symptoms were only weakly associated with clinical test results, underscoring the need for more specific observation techniques, diagnostic tools and a better understanding of the dry eye syndrome.[20]

The lack of data concerning the frequency and diurnal intensity of many ocular symptoms prompted the present study. Two very similar questionnaires, the Dry Eye Questionnaire (DEQ) and the Contact Lens Dry Eye Questionnaire (CLDEQ), were designed to assess the prevalence, frequency and diurnal severity of ocular surface symptoms of patients entering optometric practices in North America.

2. METHODS

From July 1998 to July 1999, patients at six clinical centers were asked to complete a self-administered questionnaire. Participating centers included a large private practice in Toronto, Ontario and schools and colleges of optometry, including Indiana University, The Ohio State University, University of Missouri at St. Louis, the State University of New York, and the University of Waterloo, Ontario. Informed consent was obtained from subjects at all sites. All subjects were 18 years of age or older.

There were two versions of the dry eye questionnaire, the DEQ and the CLDEQ. The two questionnaires were very similar, except that contact lens wearers were asked to report their symptoms while wearing contact lenses. Both questionnaires included categorical scales to measure the prevalence, frequency, diurnal severity, and intrusiveness of common ocular surface symptoms. including discomfort, dryness, visual changes, soreness and irritation, grittiness and scratchiness, foreign body sensation, burning and stinging, light sensitivity, and itching.

After each subject filled out the questionnaire, the survey was placed in a file in the reception area. A separate doctor diagnosis form, linked by number to the questionnaire, was then placed in the subject's clinic file. If the subject was judged to have dry eye, the severity of the dry eye was graded as mild, moderate or severe. The diagnosis of dry eye was the clinical assessment of the examining doctor. The doctor was masked to questionnaire responses. When the clinic visit was completed, each subject's doctor diagnosis form was paired with the corresponding questionnaire by a receptionist or staff person at each site. The association between doctor diagnosis and the frequency and intensity of patient symptoms was examined by contingency table analysis.

3. RESULTS

One thousand fifty four (1054) subjects in six clinical centers completed the DEQ or CLDEQ questionnaires. The numbers of subjects at each site ranged from 93–274, with

687 (65%) filling out the DEQ and 367 (35%) completing the CLDEQ. Subject age ranged from 18 to 94 years, with an average of 46 ± 14 years for DEQ subjects and 39 ± 13 years from CLDEQ subjects. Approximately two thirds (64%) of the sample was female. Twenty-eight subjects did not report gender and are not included in Table 2.

The distribution of the contact lens wearing population was divided among lens types as follows: 32% daily wear soft, 30% disposable soft lenses, 17% rigid gas permeable lenses (RGP), 16% frequent replacement soft lenses, and 5% percent extended wear lenses.

The three symptoms most commonly reported by non-contact lens wearers included ocular discomfort, visual changes, and itching. Ocular discomfort was reported by 64% of non-contact lens wearers and 56–57% of subjects reported visual changes and itching (Fig. 1a). Other symptoms were reported less commonly (32–50% of subjects) and are not illustrated. Foreign body sensation was the least commonly reported symptom (13% of subjects) among non-contact lens wearers. Fig. 1b illustrates the frequency of three symptoms most commonly reported among contact lens wearers, ocular discomfort, dryness, and visual changes. These symptoms were reported by 70–80% of the contact lens wearing population, although mostly on an infrequent basis. Other symptoms were reported by 19–54% of contact lens wearers and are not illustrated. We also asked contact lens wearers about the symptoms of dryness without their contact lenses. Only 38% reported ocular dryness when not wearing their contact lenses.

Subjects were also asked about the intensity of morning, mid-day, and evening symptoms, on a scale of 1 ("not at all intense") to 5 ("very intense"). If categories 1–2 represent less intense symptoms and categories 3–5 represent moderate to intense symptoms, then many symptoms became more intense in the evening. Fig. 2a shows the moderate to very intense (category 3–5) responses of non-contact lens wearers in the morning and evening. Although not all symptoms are illustrated, many symptoms showed a shift toward more intense symptoms in the evening, with discomfort showing the largest shift. Contact lens wearers reported large shifts toward more intense symptoms in the evening (Fig. 2b). Among contact lens wearers, the symptom of discomfort also showed the greatest intensity shift from morning to evening. For both groups, light sensitivity was actually most intense at mid-day (not shown), with little difference between morning and evening intensities.

Twenty-five percent of non-contact lens wearing subjects reported having previously worn contact lenses. The DEQ listed reasons for discontinuing lens wear, and respondents were asked to grade the importance of these reasons on a 1 ("not at all important") to 5 ("very important") scale. The top five reasons for discontinuing lens wear were dryness and discomfort later in the day (median response of "4"), and eye scratchiness and irritation, lens discomfort all day, and difficulty in getting used to contact lenses (median response of "3"). The least important reason by median response (median of "1") was not being able to sleep in the lenses.

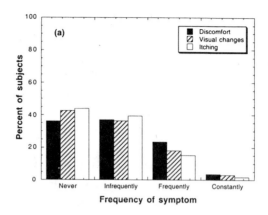

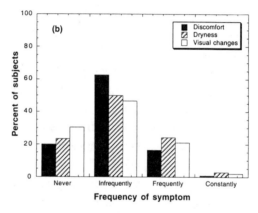

Figure 1. Frequency of the 3 most commonly reported symptoms among (**a**) non-contact lens wearers and (**b**) contact lens wearers.

In this study, 22% of non-contact lens wearers and 15% of contact lens wearers were diagnosed with dry eye, with 98% in the mild to moderate categories. Among non-contact lens wearers, there was a significant difference in symptom frequency between subjects diagnosed with dry eye (DE+) and subjects not diagnosed with dry eye (DE-) for all symptoms but foreign body sensation (contingency table analysis; $P < 0.05$). Figs. 3a and b illustrate the frequency of discomfort and dryness among DE+ and DE- non-contact lens wearing subjects. Among contact lens wearers, only three symptoms, dryness, discomfort, and visual changes, were significantly different between DE+ and DE- subjects (contingency table analysis; $P < 0.05$). The frequency of dryness showed the largest difference(Fig. 3c).

4. DISCUSSION

This study shows that ocular surface symptoms are prevalent among this clinical population. For example, 64% of non-contact lens wearers and 80% of contact lens

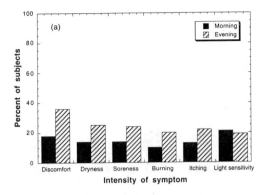

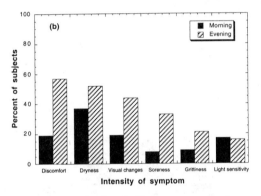

Figure 2. Moderate to intense (categories 3–5) symptoms in the morning and evening among (**a**) non-contact lens wearers and (**b**) contact lens wearers.

wearers reported ocular discomfort, at least on an infrequent basis (Fig. 1a, b). Although the clinical population in this study should not be considered representative of the normal population as a whole, most patients filling out the questionnaires in this study were routine clinical patients and were not diagnosed with an anterior surface pathology.

The most frequent symptom reported by non-contact lens wearers in this study was discomfort, followed by visual changes, and itching (Fig. 1a). These results are similar to McCarty et al.,[4] who found that photophobia, itching, tearing, and discomfort were common ocular symptoms. Others have found that ocular fatigue, blurred vision, itching, and a feeling of sand and gravel in the eye, burning, and light hurting the eyes were prevalent ocular symptoms.[3,9,10,13,20] In this study, dryness was not the most frequent symptom among non-contact lens wearers, but it did serve to differentiate DE+ from DE- subjects.

Among contact lens wearers in this study, eye dryness was a frequent symptom (Fig. 1a). Seventy-eight percent of contact lens wearers reported eye dryness, although

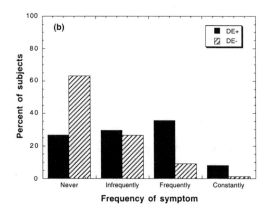

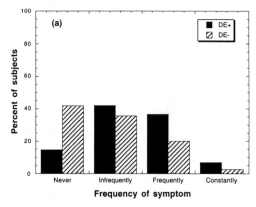

Figure 3. Frequency of symptoms in DE+ and DE- subjects: **(a)** discomfort and **(b)** dryness among non-contact lens wearers.

mostly on an infrequent basis. There are only a few reports in the literature concerning the frequency of common ocular symptoms among contact lens wearers. In two separate clinic-based studies of contact lens wearing patients, Begley et al.[17] and Brennan and Efron[18] reported that eye dryness was the most frequent symptom present in approximately 75% of contact lens wearers. In this study, eye discomfort and visual changes were also very frequently reported symptoms, cited by 70–80% of contact lens wearers.

We found that the intensity of many common ocular symptoms increased over the course of the day, especially among contact lens wearers (Fig. 2a, b). The reason for this increase in intensity is unknown, but may be environmental or task related. As the day progresses, the eyes are open longer, and it is more likely that the ocular surface will become dry or irritated. Some visual tasks, such as reading or computer use may cause a decrease in blinking,[21,22] which may further irritate the eyes.[23] These results suggest that clinicians should specifically ask about ocular symptoms at the end of the day, or examine patients at that time, to fully appreciate the intensity of common ocular symptoms.

Many subjects in this study reported visual changes, described as fluctuating, blurry vision that cleared with a blink. Fluctuating, disturbed, or blurry vision is a common symptom among contact lens wearers[17] and patients with dry eye.[3,6,9,13] While this symptom is not widely considered part of the clinical presentation of dry eye, recent research shows a relationship between tear breakup and subjective reports of blurred vision.[24] It appears that further investigation of the connection between visual disturbances and dry eye is justified.

The prevalence of dry eye in this study was 22% for non-contact lens wearers and 15% for contact lens wearers, with the majority of patients in the mild to moderate categories. This is similar to other clinic-based populations where a general diagnosis of dry eye was used.[3,4] Several studies have shown that dry eye symptoms are relatively prevalent in the general public[2] and among clinical populations.[1,10] Although DE+ subjects reported more ocular symptoms than DE- subjects, there were many subjects with frequent to constant ocular symptoms who were DE- (Fig. 3a, c). Others have found a lack of overlap between clinical tests for dry eye and typical dry eye symptoms.[2,9,10] Unfortunately, there is no "gold standard" clinical test that can definitively diagnose dry eye,[2] in particular in its mild to moderate presentations, which leaves patient symptoms as an important variable in diagnosis. This study provides additional information about ocular surface symptoms in a large population that may be used to design further clinical research.

ACKNOWLEDGMENTS

The authors would like to thank CIBA Vision for their support of this project.

REFERENCES

1. M.J. Doughty, D. Fonn, D. Richter, T. Simpson, B. Caffery, K. Gordon. A patient questionnaire approach to estimating the prevalence of dry eye symptoms in patients presenting to optometric practices across Canada. *Optom Vis Sci*;74(8):624–31 (1997).
2. O.D. Schein, B. Munoz, J.M. Tielsch, B. Bandeen-Roche, S. West. Prevalence of dry eye among the elderly. *Am J Ophthalmol*;124:723–28 (1997).
3. I. Toda, H. Fujishima, K. Tsubota. Ocular fatigue is the major symptom of dry eye. Acta Ophthalmol;71:347–352 (1993).
4. C.A. McCarty, A.K. Bansal, P.M. Livingston, Y.L. Stanislavsky, H.R Taylor. The epidemiology of dry eye in Melbourne, Australia. *Ophthalmology*;105(6):1114–1118 (1998).
5. T. Hikichi, A. Yoshida, M. Ri, K. Araki, K. Horimoto, E. Takamura, K. Kitagawa, M. Oyama, Y. Danjo, S. Kondo, H. Fujishima, I. Toda, K. Tsubota. Prevalence of dry eye in Japanese eye centers. *Graefe's Arch Clin Exp Ophthalmol*;233:555–558 (1995).
6. S. Shimmura, J. Shimazaki, K. Tsubota. Results of a population-based questionnaire on the symptoms and lifestyles associated with dry eye. *Cornea*;18(4):408–411 (1999).
7. P. Cho, M. Yap. Schirmer test II. A clinical study of its repeatability. *Optom Vis Sci*;70:157–159 (1993).
8. P. Cho, B. Brown. Review of the tear breakup time and a closer look at the tear breakup time of Hong Kong Chinese. *Optom Vis Sci*;70:33–38 (1993).

9.   K.B. Bjerrum. Test and symptoms in keratoconjunctivitis sicca and their correlation. Acta Ophthalmol Scand;:436-A.
10. Hajeer, H. Chambers, A.J.Silman . Weak association between subjective symptoms of and objective testing for dry eyes and dry mouth: results from a population based study. *Ann Rheum Dis*;57:20–24 (1998).
11. C.W. McMonnies. Key questions in a dry eye history. *J Am Optom Assoc*;57(7):512–517 (1986).
12. M.A. Lemp. Report of the National Eye Institute/Industry workshop on clinical trials in dry eyes. *CLAO*;21(4):221–232 (1995).
13. C. Vitali, H.M. Moutsopoulos, S. Bombardieri. The European Community Study Group on Diagnostic Criteria for Sjögren's Syndrome. Sensitivity and specificity of tests for ocular and oral involvement in Sjögren's Syndrome. *Ann Rheum Dis*;53:637–647 (1994).
14. J.G. Brun, H. Jacobsen, R. Kloster, M. Cuida, A.C. Johannesen, H.M. Hoyeraal, R. Jonsson. Use of a sicca symptoms questionnaire for the identification of patients with Sjögren's syndrome in a heterogeneous hospital population with various rheumatic diseases. *Clin Exp Rheumatol*;12:649–652 (1994).
15. C.W. McMonnies, A. Ho. Patient history in screening for dry eye conditions. *J Am Optom Assoc*;58(4):296–301 (1987).
16. K.K. Nichols, C.G. Begley, B. Caffery, L.A. Jones. Symptoms of ocular irritation in patients dignosed with dry eye. *Optom Vis Sci*; 76(12): 838–844.
17. C.G. Begley, B. Caffery, K.K. Nichols, R. Chalmers. Responses of contact lens wearers to a dry eye survey. *Optom Vis Sci*;77(1):40–46 (2000).
18. N.A. Brennan, N. Efron. Symptomatology of HEMA contact lens wear. *Optom Vis Sci*; 66:834–838 (1989).
19. S.A. Little, A.S. Bruce. Postlens tear film morphology, lens movement and symptoms in hydrogel lens wearers. *Ophthal Physiol Opt*;14(1):65–69 (1994).
20. K. Bandeen-Roche, B. Muñoz, J.M. Tielsch, S.K. West, O.D. Schein. Self-reported assessment of dry eye in a population-based setting. *Invest Ophthalmol Vis Sci*; 38(12):2469–2475 (1997).
21. S. Patel, R. Henderson, L. Bradley, B. Galloway, H. Hunter. Effect of visual display unit use on blink rate and tear stability. *Optom Vis Sci*; 68(11):888–892 (1991).
22. A.R. Bentivoglio, S.B. Bressman, E. Cassetta, D. Carretta, P. Tonali, A. Albanese. Analysis of blink rate patterns in normal subjects. *Movement Disorders*;12(6):1028–1034 (1999).
23. M.C. Acosta, J. Gallar, C. Belmonte. The influence of eye solutions on blinking and ocular comfort at rest and during work at video display terminals. *Exp Eye Res*; 68(6):663–669 (1999).
24. R. Tutt, A. Bradley, C.G. Begley, L.N. Thibos. Optical and visual impact of tear breakup in human eyes. *Invest Ophthalmol Vis Sci*;: 41:4117–4123 (2000).

THE EFFECT OF HORMONE REPLACEMENT THERAPY ON THE SYMPTOMS AND PHYSIOLOGIC PARAMETERS OF DRY EYE

William D. Mathers,[1] Anne Marie Dolney,[1] and Dale Kraemer[2]

Casey Eye Institute
Oregon Health Sciences University
Portland, Oregon, USA
[2]Division of Medical Informatics and Outcomes Research
Oregon Health Sciences University
Portland, Oregon, USA

1. INTRODUCTION

Symptoms of dry eye frequently occur in women and become more frequent as they age.[1-8] Human and animal data show the lacrimal gland requires sex hormone support for normal function.[9-12] While testosterone appears to be the most important of these hormones, estrogen may play a role, and estrogen receptors have been identified in the lacrimal gland.[13] Many women experience dry eye as they enter menopause, and a previous study reported women on hormone replacement therapy (HRT) had fewer symptoms of dry eye, compared to those who were not.[14] We studied a group of women to determine whether those taking HRT had fewer symptoms or improved physiologic parameters associated with dry eye.

2. METHODS

Women between the ages of 35 and 60 were recruited to participate in this study through public announcements and advertising. Our announcements carefully avoided any indication this was a dry eye study to minimize selection bias in the volunteers. After enrollment, the women were requested to answer a series of questions regarding the presence of dry eye symptoms. They were then given a battery of tests to measure stimulated and unstimulated tear flow, tear volume, decay constant and tear osmolarity.[15] Serum levels of estradiol, prolactin, follicle-stimulating hormone (FSH) and total testosterone were also measured. Blood samples were collected at random times during the day and without regard to the subjects' menstrual cycle.[16] Patients were excluded for wearing contact lenses, using any eye medications other than artificial tears and using medications such as diuretics and antihistamines, which might affect tear flow. Patients were requested to list all their medications and the type of hormone replacement they were using, particularly with regard to the use of progestins with their estrogen.

The analytic approach was conducted in stages. First, associations between dry eye symptoms and menopause and HRT status were evaluated. The second step added the physiologic measurements and the hormones to the best-fitting model from the first step to determine associations between these measurements and dry eye symptoms and between the serum hormone levels. Logistic regression models were fit to these data using SAS Version 7.0 for Windows 95. All possible regression models and stepwise selection of variables were used.

An FSH level greater than 40 mIU/ml was used to denote menopause, unless the patients were taking estrogen. For these women an FSH level of 25 mIU/mlor greater was used to indicate the presence of menopause.[17]

Symptoms of dry eye were recorded from the subject's responses to the questionnaire. Subjects were asked if they experienced symptoms of dry eye including dryness, burning or grittiness. A single positive response indicated dry eye symptoms.

A series of dry eye tests was performed during a single session and prior to any other examination or administration of any eye drops. The tests were performed in a specific order: tear osmolarity, fluorophotometric determination of tear flow, tear volume and decay constant of fluorescein concentration and lastly, Schirmer's test without anesthetic.

3. RESULTS

Table 1 lists summary statistics for all subjects and several subsets. In this study of 110 women, 38 were menopausal and of these 19 (50%) received HRT. Of the 19, 16 were taking progestins with their estrogen, either cycled or continuously. Also included were 72 non-menopausal women not receiving HRT. Non-menopausal women receiving HRT were excluded from these analyses.

In a model containing menopausal status and HRT, we found HRT was not a significant predictor of dry eye symptoms ($P > 0.05$). After eliminating HRT status, the association between menopausal status and dry eye symptoms approached significance ($P = 0.053$). The odds of having dry eye symptoms were approximately 2.2 times greater for woman in menopause. Since the association of dry eye symptoms and menopause status only approached significance, subsequent models did not automatically include the menopause variable. Three women were omitted from this model due to missing data.

Table 1. Summary data

	N	Age	Decay[a]	Tear Vol[b]	Tear Fl[c]	Osmol[c]	Schirm[d]	DE Symp[e]
All Subjects	110	46.9 ± 4.5	0.18 ± 0.7	1.86 ± 1.0	0.34 ± .53	309 ± 7	14 ± 10	39 (35%)
Pre-meno[f] no HRT[g]	72	43.9 ± 5.7	0.16 ± .07	1.90 ± 1.5	0.34 ± .30	309 ± 10	14 ± 11	20 (28%)
Meno[h] on HRT	19	52.7 ± 4.5	0.18 ± .05	1.67 ± 0.9	0.32 ± .23	308 ± 10	14 ± 9	10 (53%)
Meno no HRT	19	51.4 ± 4.6	0.18 ± .07	1.87 ± 1.0	0.34 ± .30	308 ± 7	13 ± 11	9 (47%)

[a]Decay, decay constant; [b]Tear Vol, tear volume (μl); [c]Tear Fl, tear flow (μ/min); [c]Osmol, tear osmolarity (mOsm/ml); [d]Schirm, Schirmer's test without anesthetic (mm/5min); [e]DE Symp, dry eye symptoms; [f]Pre-meno, pre-menopausal; [g]HRT, hormone replacement therapy; [h]Meno, menopause

In the model evaluating the contribution of serum hormone variables or age, we found age was a significant predictor of dry eye symptoms (P = 0.045), and no serum hormone value was significantly associated with dry eye symptoms. Thus, age was a slightly better predictor of dry eye symptoms than menopausal status. The odds of having dry eyes increased by a factor of 1.068 for each additional year (or about 1.94 for each additional decade). Four women were omitted due to missing data in this model.

In the model evaluating the contribution of physiologic variables for dry eye, age and menopausal status to the development of dry eye symptoms, both Schirmer's (P = 0.0152) and menopausal status (P = 0.0123) were significant predictors. The odds of dry eye symptoms decreased per unit of the Schirmer's scale by a factor of 1.058 (and odds ratio equivalent to 1.76 per 10 unit change of the Schirmer's scale). The odds of dry eye symptoms increased by a factor of approximately 3.0 after menopause using this model. No other physiologic variable, osmolarity, decay constant, tear volume or tear flow predicted dry eye symptoms. Seventeen women were omitted from the analysis with this model, which may explain the stronger association with menopausal status, compared to the other models.

Dry eye symptoms were very common in pre-menopausal and menopausal women–35% expressed at least some symptoms. For women in menopause on HRT, the incidence of symptoms was 53%, and for women in menopause not on HRT, the incidence was 47%. These values were not statistically different.

4. DISCUSSION

This study indicates women on HRT did not experience fewer dry eye symptoms or improved aqueous tear function, compared to women who did not use HRT. This was an unexpected finding. We originally hypothesized that women on HRT would have fewer complaints and demonstrate better tear function. Both increased age and menopause status were associated with increased dry eye symptoms.

Our results indicating a lack of improvement with HRT agree with data recently published by Schaumberg et al. They studied a large database of 39,876 subjects taken from the Women's Health Study of Female Health Professionals.[18] They found the use of replacement estrogens without progestins was associated with a 66 – 70% higher prevalence of dry eye symptoms, diagnosed dry eye or a combined endpoint of either or both of these. The combined use of estrogens and progestins was still associated with a 25 – 30% higher prevalence of dry eye, but the risk was significantly lower than the risk of estrogens alone. Our smaller study was unable to differentiate between women on estrogens with progestin and those without, since 85% of our subjects who took HRT used estrogens and progestin.

Schaumberg et al. also found 9.4% of women reported experiencing symptoms of dryness or irritation either constantly or often and 4.7% reported a clinical diagnosis of dry eye. This incidence was much lower than we found. In our study, however, we identified the presence of symptoms using less stringent criteria and did not require subjects to identify the frequency of their symptoms. If the authors of the Schaumberg paper had used the same criteria we used to indicate the presence of symptoms of dry eye, it is likely the results would have been similar (unpublished data from Dr. Schaumberg).

Begley et al. recently reported the incidence of symptoms in a study of 1054 subjects, ages 18–54, 63% of whom were women. The incidence of symptoms related to dry eye was 60% and for contact lens wearers 80%.[19] This is very similar to the incidence we are reporting.

Our study indicates the extreme commonality of dry eye symptoms in this population. It is, therefore, difficult to identify precisely the difference between what is normal or abnormal. In addition, dry eye remains ill defined. No specific symptom indicates the presence of dry eye. We found a poor correlation between dry eye symptoms and each of our physiologic measures of tear function except the Schirmer's test without anesthetic. This poor correlation has been reported previously.[20,21] Schirmer's test of stimulated tear flow, a measure of the reserve capacity of the lacrimal system, emerged as the most useful tear measurement to predict dry eye symptoms. These subjects demonstrated they could tolerate relatively high levels of tear osmolarity, low levels of tear flow or a low decay constant without generating symptoms of dry eye. We hypothesized this poor correlation between physiologic parameters and symptoms supports the conclusion that symptoms of dry eye are generated by ocular surface inflammation, which is not possible to measure directly. Surface inflammation can have many other causes in addition to dry eye, such as allergic response, blepharitis and other ocular immune diseases.

The tear tests used in this study were designed to measure only aqueous tear film components, and we omitted testing or evaluating meibomian gland function, evaporation or clinical signs of blepharitis. Ocular surface lipid derangement may affect the aqueous component of the tear film by altering evaporation or the ocular surface tear flow stimulus.[22-24]

We suggest it is important, when evaluating dry eye patients, to assess both the symptoms and the physiologic parameters creating the dry eye. Since a poor correlation exists between any single dry eye measurement and dry eye symptoms, it is important to perform more than one physiologic test for dry eye to obtain a clear understanding of each patient. No single test represents a gold standard for evaluating dry eye. Tear osmolarity has been studied for many years but is subject to considerable error and has shown only moderate correlation with symptoms.[25,26] Fluorophotometric evaluation of tear flow, tear volume and decay constant has the advantage of assessing the steady state, unstimulated, aqueous component of the tear film.[15,27,28] It is non-invasive, relatively easy to perform and standard commercial instruments are available to perform the test. Damage to the ocular surface, resulting in alteration in goblet cell density, vital dye staining and changes in impression cytology of conjunctival epithelial cells, has also been proposed as a useful marker for the presence of dry eye.[29-31] Each of these assesses a secondary change, which may be affected by blepharitis and other ocular surface inflammation.

Our study demonstrates women in menopause on HRT do not experience a lower incidence of dry eye symptoms. Our study also indicates the age of the subject is the strongest predictor for dry eye symptoms in women between the ages of 35 and 60.

REFERENCES

1. Bjerrum KB: Keratoconjunctivitis sicca and primary Sjøgren's syndrome in a Danish population aged 30–60 years. *Acta Ophthalmologica Scandinavica* 1997; 75: 281–286.

2. McCarty CA, Bansal AK, Livingston PM, Stanislavsky YL, Taylor HR: The epidemiology of dry eye in Melbourne, Australia. *Ophthalmology* 1998; 105: 1114–1119.

3. Caffery BE, Richter D, Simpson T, Fonn D, Doughty M, Gordon K: CANDEES. The Canadian Dry Eye Epidemiology Study. *Advances in Experimental Medicine & Biology* 1998; 438: 805–806.

4. McGill JI, Liakos G, Seal DV, Goulding N, Jacobs D: Tear film changes in health and dry eye conditions. *Trans Ophthalmol Soc.* UK 1983; 103: 313–327.

5. Seal DV: The effect of aging and disease on tear film constituents. *Trans Ophthalmol Soc UK* 1985; 104: 355–361.

6. Jordan A., Baum JL: Basic tear flow: does it exist? *Ophthalmology* 1980; 87: 920–930.

7. Craig JP, Tomlinson A: Age and gender effects on the normal tear film. *Advances in Experimental Medicine & Biology* 1998; 438: 411–415.

8. Schein OD, Munoz B, Tielsch JM, Bandeen-Roche K, West S: Prevalence of dry eye among the elderly. *Am J Ophthalmol.* 1997; 124: 723–728.

9. Sullivan DA, Edwards JA: Androgen stimulation of lacrimal gland function in mouse models of Sjøgren's syndrome. *Journal of Steroid Biochemistry & Molecular Biology* 1997; 60: 237–245.

10. Sullivan DA, Krenzer KL, Sullivan BD, Tolls DB, Toda I, Dana MR: Does androgen insufficiency cause lacrimal gland inflammation and aqueous tear deficiency? *Invest Ophthalmol Vis Sci.* 1999; 40: 1261–1265.

11. Mircheff AK, Warren DW, Wood RL: Hormonal support of lacrimal function, primary lacrimal deficiency, autoimmunity, and peripheral tolerance in the lacrimal gland. *Ocular Immunology Inflammation* 1999; 4: 145–172.

12. Warren DW: Hormonal influences on the lacrimal gland. [Review] [27 refs]. *International Ophthalmology Clinics* 1994; 34: 19–25.

13. Wickham LA, Gao J, Toda I, et al: Identification of androgen, estrogen and progesterone receptor mRNAs in the eye. *Acta Ophthalmologica Scandinavica* 2000; 78: 146–153.

14. Costello LC, Franklin RB: Effect of prolactin on the prostate. [Review] [19 refs]. *Prostate* 1994; 24: 162–166.

15. Mathers WD, Daley TE: Tear flow and evaporation in patients with and without dry eye. *Ophthalmology* 1996; 103: 664–669.

16. Mathers WD, Stovall D, Lane JA: Menopause and tear function—the influence of prolactin and sex hormones on human tear production. *Cornea* 1998; 17: 353–358.

17. Folberg R, Stone EM, Sheffield VC, Mathers WD: The relationship between granular, lattice type 1, and Avellino corneal dystrophies. A histopathologic study. *Archives of Ophthalmology* 1994; 112: 1080–1085.

18. Schaumberg, Sullivan DA, Buring JE, Dana MR: Hormone replacement therapy (HRT) and the prevalence of dry eye symptoms. *Invest Ophthalmol Vis Sci.* 2000; 41: S740–S740.

19. Begley CG, Caffery B, Nichols K, Mitchell GL, Chalmers R: Results of dry eye questionnaire from optometric practices in North America. *Cornea* 2000; 19: S75–S75(Abstract).

20. Hay EM, Thomas E, Pal B, Hajeer A, Chambers H, Silman AJ: Weak association between subjective symptoms and/or objective testing for dry eyes and dry mouth: results from a population based study. *Annals of the Rheumatic Diseases* 1998; 57: 20–24.

21. Pflugfelder SC, Tseng SC, Sanabria O: Evaluation of subjective assessments and objective diagnostic tests for diagnosing tear-film disorders known to cause ocular irritation. *Cornea* 1998; 17: 38–56.

22. Mathers WD, Lane JA: Meibomian gland lipid, evaporation, and tear film stability. In: *Lacrimal Gland, Tear Film, and Dry Eye Syndromes 2*. Sullivan DA, Dartt DA. Meneray M, eds. New York: Plenum Press, 1998; 349–360.

23. Mathers WD: Ocular evaporation in meibomian gland dysfunction and dry eye. *Ophthalmology* 1993; 100: 347–351.

24. Craig JP, Tomlinson A: Importance of the lipid layer in human tear film stability and evaporation. *Optometry & Vision Science* 1997; 74: 8–13.

25. Nelson JD, Wright JC: Tear film osmolarity determination: an evaluation of potential errors in measurement. *Curr Eye Res.* 1986; 5: 677–681.

26. Farris RL, Gilbard JP, Stuchell RN, Mandell ID: Diagnostic tests in keratoconjunctivitis sicca. *CLAO* 1983; 9: 23–28.

27. Gobbels M, Goebels G, Breitbach R, Spitznas M: Tear secretion in dry eyes as assessed by objective fluorophotometry. *German Journal of Ophthalmology* 1992; 1: 350–353.

28. Brubaker RF, Maurice D, McLaren JW: Fluorometry of the anterior segment. In: *Noninvasive Diagnostic Techniques in Ophthalmology*. Masters BR, ed. New York: Springer-Verlag, 1990; 248–280.

29. Pflugfelder SC, Tseng SC, Yoshino K, Monroy D, Felix C, Reis BL: Correlation of goblet cell density and mucosal epithelial membrane mucin expression with rose bengal staining in patients with ocular irritation. *Ophthalmology* 1997; 104: 223–235.

30. Danjo Y, Watanabe H, Tisdale AS, et al: Alteration of mucin in human conjunctival epithelia in dry eye. *Invest Ophthalmol Vis Sci.* 1998; 39: 2602–2609.

31. Kobayashi TK, Ueda M, Nishino T, Ishida Y, Takamura E, Tsubota K: Cytologic evaluation of conjunctival epithelium using Cytobrush-S: value of slide preparation by thin prep technique. *Cytopathology* 1997; 8: 381–387.

IMPAIRED NEUROTRANSMISSION IN LACRIMAL AND SALIVARY GLANDS OF A MURINE MODEL OF SJÖGREN'S SYNDROME

Driss Zoukhri and Claire Larkin Kublin

Schepens Eye Research Institute
Department of Ophthalmology
Harvard Medical School
Boston, Massachusetts

1. INTRODUCTION

A decrease in lacrimal and salivary gland secretion is a primary cause of dry eye and dry mouth. Sjögren's syndrome is the leading cause of the aqueous tear-deficient type of dry eye.[1-3] It is an autoimmune disease occurring almost exclusively in females (> 90%) and involves an extensive lymphocytic infiltration of the lacrimal and salivary glands as well as destruction of epithelial cells.[1-3] No cure exists for this disease. Moreover, its exact etiology is largely unknown, but may involve numerous factors including those of viral, endocrine, neural, genetic and environmental origin.[2-4-5]

The precise mechanism(s) responsible for the decreased tears and saliva secretion in Sjögren's syndrome is unknown.[1-3] The dry eye and dry mouth characteristic of this disease may be due to a progressive lymphocytic infiltration of the lacrimal and salivary glands, an immune-mediated destruction of the epithelial cells and a consequent decline in tears and saliva production.[1-3] However, adjacent to the immunopathological lesions in Sjögren's syndrome, the lacrimal and salivary gland tissues contain apparently normal acinar and ductal epithelia that should be able to secrete enough tears and saliva. A decrease in lacrimal gland innervation and/or an alteration of the signaling pathways of the remaining epithelia could account for the decreased function of the lacrimal and salivary glands associated with Sjögren's syndrome.

Lacrimal Gland, Tear Film, and Dry Eye Syndromes 3
Edited by D. Sullivan *et al.*, Kluwer Academic/Plenum Publishers, 2002

In our studies using MRL/lpr mice, a murine model of Sjögren's syndrome, the lymphocytic infiltration of the lacrimal and salivary glands did not alter the parasympathetic, sympathetic and sensory innervation of the remaining epithelial cells in these tissues.[6] We also found acinar cells isolated from lacrimal and salivary glands of diseased MRL/lpr animals were hyper-responsive to exogenous cholinergic and adrenergic stimulation when compared to age-matched control MRL/+ animals.[7] We hypothesized that the exocrine tissues from diseased animals behaved as denervated ones, i.e., the remaining nerves cannot release their neurotransmitters leading to increased responsiveness of these tissues to exogenous stimulation. Another consequence of the impaired release of neurotransmitters is a loss of tear and saliva secretion from lacrimal and salivary glands, which leads to dry eye and dry mouth. The present study tested this hypothesis.

2. METHODS

2.1. Measurement of Stimulated Salivary Flow

Secretion of saliva was stimulated by intraperitoneal (i.p.) injection (0.1 ml) of 0.02 mg isoproterenol plus 0.05 mg pilocarpine/100 mg body weight. After 5 min, saliva was collected over 10 min from the oral cavity. The volume of saliva was measured using a micropipette.

2.2. KCl-evoked Release of Acetylcholine and Secretion of Peroxidase

Lobules (~2 mm diameter) prepared from lacrimal and submandibular glands were placed in cell strainers. After a 120-min equilibration period in Krebs-Ringer buffer (KRB), the lobules were incubated for 20 min in a total volume of 0.8 ml in normal KRB (spontaneous release) and depolarizing KRB (evoked release) where KCl was increased to 75 mM and NaCl decreased to 55 mM to maintain isotonicity. Both normal and depolarizing KRB solutions contained neostigmine bromide (0.1 mM), an inhibitor of acetylcholine esterase, to allow the accumulation of acetylcholine in the media. In some experiments, lacrimal gland lobules were further incubated for 20 min in 0.8 ml normal KRB containing phenylephrine (an _1-adrenergic agonist, 10^{-4} M). After incubation, the media were collected and centrifuged to remove debris. The lobules were homogenized in 10 mM Tris-HCl, pH 7.5. The amounts of acetylcholine, choline and peroxidase in the media and tissue homogenate were determined using a spectrofluorometric assay.

3. RESULTS

To determine if salivary glands of MRL/lpr mice are hyper-responsive in vivo to exogenous stimulation, secretion of saliva was stimulated by i.p. injection of isoproterenol and pilocarpine. Saliva was then collected over 10 min from 16-week-old male and female MRL/lpr and MRL/+ mice.

Female MRL/lpr mice produced significantly higher amounts of saliva compared to female MRL/+ mice (Table 1). In contrast, the amount of saliva secreted by MRL/lpr male mice was not significantly different from that secreted by MRL/+ male mice, ruling out the possibility this increased salivation was due to the *lpr* mutation rather than inflammation of the salivary glands. Indeed, Toda et al.[8] showed inflammation of the exocrine glands occurs in female but not male MRL/lpr mice. Thus, if the increased salivation observed in female MRL/lpr mice were solely due to the *lpr* mutation, male MRL/lpr mice should have produced more saliva than age-matched male MRL/+ mice. These results suggest exocrine glands from MRL/lpr mice behave as denervated tissues. We hypothesize this is due to the inability of their nerves to release neurotransmitters.

Table 1. Effect of gender and disease on salivary flow in MRL mice

	Male		Female	
	MRL/+	MRL/lpr	MRL/+	MRL/lpr
	108 ± 20^a	138 ± 7	52 ± 6	$167 \pm 31^*$

[a]Values are means ± SEM of flow measured in _l/10 min, n = 4–6.
*Significantly different from female MRL/lpr (P < 0.05)

To directly test our hypothesis that the nerves of exocrine tissues from MRL/lpr are not able to release their neurotransmitters, we measured the spontaneous and evoked release of acetylcholine. High KCl induced a small but significant release of acetylcholine from lacrimal and submandibular glands of 18-week-old MRL/+ mice (Table 2). In contrast, nerves of submandibular and lacrimal glands from 18-week-old MRL/lpr mice were not able to release acetylcholine in response to the depolarizing solution. These results show activation, with high KCl, of nerve endings of lacrimal and salivary glands infiltrated with lymphocytes does not increase the release of acetylcholine.

Table 2. Effect of disease on acetylcholine release in MRL mice

	Lacrimal Gland		Submandibular Gland	
	MRL/+	MRL/lpr	MRL/+	MRL/lpr
Spontaneous	0.11 ± 0.004^a	0.11 ± 0.013	0.33 ± 0.04	0.36 ± 0.04
Evoked	$0.17 \pm 0.005^*$	0.12 ± 0.013	$0.54 \pm 0.05^*$	0.46 ± 0.05

Values are means ± SEM of acetylcholine (μM), n = 3–5, *Significantly different from spontaneous release (P < 0.05)

The main function of the lacrimal gland is to synthesize and secrete proteins as well as water and electrolytes onto the ocular surface. High KCl induced a significant secretion of peroxidase from lacrimal glands of 18-week-old MRL/+ mice (Table 3). In contrast, lacrimal glands from 18-week-old MRL/lpr mice were not able to secrete peroxidase in response to high KCl. This was not due to a defect in the secretory process in MRL/lpr mice lacrimal glands since exogenous addition of phenylephrine stimulated peroxidase secretion from MRL/+ as well as MRL/lpr mice lacrimal glands. These results show activation of nerve endings with high KCl does not stimulate peroxidase secretion from inflamed lacrimal glands.

Table 3. Effect of disease on peroxidase secretion in MRL mice

| | Peroxidase Secretion (% of total) | |
	MRL/+	MRL/lpr
Spontaneous	0.023 ± 0.005	0.30 ± 0.23
Evoked	0.91 ± 0.31*	0.44 ± 0.24
Phenylephrine	2.41 ± 0.28*	1.71 ± 0.21*

Values are means ± SEM, n = 3–5. *Significantly different from spontaneous peroxidase secretion ($P < 0.05$)

4. DISCUSSION

In the present report, we show stimulation of nerves from inflamed, but not non-inflamed, lacrimal and salivary glands with high KCl did not increase the release of acetylcholine. Furthermore, activation of non-inflamed lacrimal gland nerves with high KCl resulted in protein secretion, while activation of those in inflamed glands did not elicit protein secretion. These findings demonstrate that, as suggested by Sullivan,[4] inflammation of exocrine glands in Sjögren's syndrome leads to impaired release of neurotransmitters from nerves and hence impaired fluid secretion.

What are the mechanisms involved in inflammation-induced inhibition of neurotransmitter release? Several studies have shown that suppression of acetylcholine and norepinephrine release from myenteric nerves was mediated by proinflammatory cytokines including interleukin-1β (IL-1β), interleukin-6 and tumor necrosis factor-α (TNFα).[9–12] Moreover, in a rat model of acute colitis, an inflammatory disease of distal colon, IL-1β was implicated in blocking KCl-induced norepinephrine release from the myenteric plexus.[13] IL-1β also decreases the acetylcholine content of rat hippocampus.[12] Another study showed TNFα alters neurotransmitter release in cultured sympathetic neurons.[14] IL-1β and TNFα were also implicated in the pathogenesis of acute graft-versus-host disease.[15] Based on these studies, inflammation-mediated production of cytokines seems to inhibit neurotransmitter release, thus altering the physiology of inflamed tissues.

Given this abundant literature on the role of proinflammatory cytokines, we hypothesize concomitant with inflammation of the lacrimal and salivary glands in Sjögren's syndrome, an increase in production of cytokines occurs that inhibits neurotransmitter release, leading to decreased tear and saliva production from the acinar and ductal cells. In support of this hypothesis, levels of proinflammatory cytokines are elevated in lacrimal and salivary glands of Sjögren's syndrome patients as well as in animal models.[16–19] Moreover, we found the protein level of IL-1β increased in a disease-dependent manner in lacrimal and salivary glands of MRL/lpr, but not MRL/+ or BALB/c, mice and exogenous IL-1β inhibits neurally mediated lacrimal gland secretion (Kublin et al., this series). In summary, our results show nerves of inflamed lacrimal and salivary glands of MRL/lpr mice are not able to release their neurotransmitters, which results in impaired secretion from these glands.

ACKNOWLEDGMENTS

The authors are grateful to Dr. F. Sourie for her invaluable contribution to this work. This research was supported by National Eye Institute grant EY12383.

REFERENCES

1. N. Talal, H.M. Moutsopoulos, and S.S. Kassan. *Sjögren's Syndrome. Clinical and Immunological Aspects,* Springer Verlag, Berlin (1987).
2. R.I. Fox, Pathogenesis of Sjögren's syndrome, *Rheum Dis Clin North Am.* 18:517 (1992).
3. R.I. Fox, J. Törnwall, and P. Michelson, Current issues in the diagnosis and treatment of Sögren's syndrome, *Curr Opin Rheumatol.* 11:364 (1999).
4. D.A. Sullivan. Possible mechanisms involved in the reduced tear secretion in Sjögren's syndrome, in: *Sjögren's syndrome: state of the art,* Homma M., S. Sugai, T. Tojo, N. Miyasaka, and M. Akizuki, eds., Kugler publications, Amsterdam (1994).
5. D.A. Sullivan, L.A. Wickham, K.L. Krenzer, E.M. Rocha, and I. Toda. Aqueous tear deficiency in Sjögren's syndrome: possible causes and potential treatment, in: *Oculodermal Diseases,* Pleyer U., and C. Hartmenn, eds., Aeolus Press, Buren (1997).
6. D. Zoukhri, R.R. Hodges, and D.A. Dartt, Lacrimal gland innervation is not altered with the onset and progression of disease in a murine model of Sjögren's syndrome, *Clin Immunol Immunopathol.* 89:126 (1998).
7. D. Zoukhri, R.R. Hodges, I.M. Rawe, and D.A. Dartt, Ca^{2+} signaling by cholinergic and α_1-adrenergic agonists is up-regulated in lacrimal and submandibular glands in a murine model of Sjögren's syndrome, *Clin immunol Immunopathol.* 89:134 (1998).
8. I. Toda, B.D. Sullivan, E.M. Rocha, L.A. Da Silveira, L.A. Wickham, and D.A. Sullivan, Impact of gender on exocrine gland inflammation in mouse models of Sögren's syndrome, *Exp Eye Res.* 69:355 (1999).
9. S. Hurst, and S.M. Collins, Interleukin-1β modulation of norepinephrine release from rat myenteric nerves, *Am J Physiol.* 264:G30 (1993).
10. C. Main, P.A. Blennerhassett, and S.M. Collins, Human recombinant interleukin 1 beta suppresses acetylcholine release from rat myenteric plexus, *Gastroenterology.* 104:1648 (1993).

11.   S.M. Hurst, and S.M. Collins, Mechanism underlying tumour necrosis factor-α suppression of norepinephrine release from rat myenteric plexus, *Am J Physiol.* 266:G1123 (1994).

12.   P. Rada, G.P. Mark, M.P. Vitek, R.M. Mangano, A.J. Blume, B. Beer, and B.G. Hoebel, Interleukin-1 beta decreases acetylcholine measured by microdialysis in the hippocampus of freely moving rats, *Brain Res.* 550:287 (1991).

13.   K. Jacobson, K. McHugh, and S.M. Collins, The mechanism of altered neural function in a rat model of acute colitis, *Gastroenterology.* 112:156 (1997).

14.   B. Soliven, and J. Albert, Tumour necrosis factor modulates the inactivation of catecholamine secretion in cultured sympathetic neurons, *J Neurochem.* 58:1073 (1992).

15.   J.H. Antin, and J.L.M. Ferrara, Cytokine dysregulation and acute graft-versus-host disease, *Blood.* 80:294 (1992).

16.   I. Saito, K. Terauchi, M. Shimuta, S. Nishiimura, K. Yoshino, T. Takeuchi, K. Tsubota, and N. Miyasaka, Expression of cell adhesion molecules in the salivary and lacrimal glands of Sjögren's syndrome, *J Clin Lab Anal.* 7:180 (1993).

17.   H. Hamano, I. Saito, N. Haneji, Y. Mitsuhashi, N. Miyasaka, and Y. Hayashi, Expression of cytokine genes during development of autoimmune sialadenitis in MRL/lpr mice, *Eur J Immunol.* 23:2387 (1993).

18.   R.I. Fox, H.I. Kang, D. Ando, J. Abrams, and E. Pisa, Cytokine mRNA expression in salivary gland biopsies of Sjögren's syndrome, *J Immunol.* 152:5532 (1994).

19.   K. Koh, S. Sawada, and R.I. Fox. High levels of IL-10 and Th1 cytokine mRNA transcript in salivary gland biopsies from Sjögren's syndrome, in: *Sjögren's syndrome. state of the art,* Homma M., S. Sugai, T. Tojo, N. Miyasaka, and M. Akizuki, eds., Kugler Press, Amsterdam (1994).

MENOPAUSE, HORMONE REPLACEMENT THERAPY AND TEAR FUNCTION

Victoria Evans,[1,2] Thomas J. Millar,[2] John A. Eden,[3] and
Mark D. P. Willcox[1]

[1]Cooperative Research Centre for Eye Research and Technology
The University of New South Wales
Kensington, NSW, Australia
[2]School of Science
University of Western Sydney
Nepean Parramatta, NSW, Australia
[3]Sydney Menopause Centre
Royal Hospital for Women
Randwick, NSW, Australia

1. INTRODUCTION

This pilot study investigates the effect of menopause, hormone replacement therapy (HRT) and gender on tear function. Hormones affect the incidence and course of dry eye conditions,[1] and anecdotal evidence suggests the incidence of dry eye increases in women after menopause. Estrogen HRT alleviates some of the side effects of menopause, but whether systemic HRT benefits dry eye sufferers is not known.[2] HRT may restore hormonal support to the lacrimal gland and tear function to normal levels.

Pre-menopausal women, post-menopausal women, hysterectomized women taking HRT and age-matched men were enrolled. We examined ocular symptomatology, tear film function and tear film composition in the four groups. In particular, we examined tear lipocalin secretion. Tear lipocalin is a marker of lacrimal function[3] and may be important in maintaining tear film stability.[4] A difference in tear lipocalin secretion among groups was expected to reflect differences in signs and symptoms of dry eye.

2. MATERIALS AND METHODS

Men and women were enrolled for a single-visit study that included an ocular surface examination, tear collection and a questionnaire. Informed consent was obtained from all participants. HRT was limited to continuous estrogen therapy in a patch or implant (Table 1). Subjects with ocular disease (other than dry eye), systemic disease with known ocular side effects, recent ocular injury or contact lenses were excluded from the study. Subjects who used androgen therapy, oral contraceptives, diuretics, antihistamines, tranquilizers or sedatives were also excluded. Subjects were asked to refrain from artificial tear use on the day of examination.

Table 1. Subject groups, number enrolled and age range of each group

Group name	Inclusion criteria	n	Mean age, range (years)
Control	Pre-menopausal women	10	47, 42–53
Menopause	Post-menopausal women not using HRT	11	56, 48–66
HRT	Hysterectomized women taking HRT	10	60, 48–71
Men	Men over 40 years	10	52, 43–70

Clinical assessment included three measurements of fluorescein breakup time and maximum forced blink interval. Tear film lipid layer appearance was graded using a tearscope (Guillon Tear Film Grading Scale, Keeler, Berkshire). A symptomatology and medical history questionnaire was modified from the McMonnies[5] dry eye survey. Symptoms and history variables were measured as present or absent.

Tear samples were collected using glass microcapillary tubes. Care was taken not to induce reflex tearing. Samples were immediately placed on ice in Eppendorf vials then stored at -86°C. Tear osmolality was measured with a Wescor vapor pressure osmometer (Amscorp, Sydney, Australia). Total tear protein was determined using a Pierce BCA Protein Assay kit (Pierce Lab Supply, Sydney, Australia). Lipocalin concentration was analyzed using an in-house competitive ELISA. Lipocalin isoform expression was examined using 2-dimensional gel electrophoresis with immobilized pH gradient strips (7 cm, pH 4–7) and 15% acrylamide, Tris-HCL mini-gels (Bio-Rad, Sydney, Australia).

Questionnaire variables were analyzed using Chi-square and Fishers exact testing. Clinical and tear biochemistry results were compared using analysis of variance (ANOVA) with post-hoc Bonferroni testing. The distribution of variables was examined with Pearson Chi-square analysis. Statistical significance for two-tailed p-values was set as $P < 0.05$.

3. RESULTS

The menopausal group reported symptoms of ocular burning and use of artificial tears more frequently than the control and male groups (Table 2, P = 0.01). These symptoms are often associated with dry eye. The HRT group was statistically equivalent to the control group and the male group.

Table 2. Frequency of symptoms and artificial tear use (% of subjects)

Questionnaire variable	Control n = 10	Menopause n = 11	HRT n = 10	Men n = 10	Chi-square P-value
Ocular burning	10%*	82%*,†	50%	30%†	0.01
Artificial tears	30%	64%*	40%	0%*	0.01

*,†indicate statistically significant differences (P <= 0.05) between values with like symbols

No statistical difference occurred between group means for tear lipocalin concentration and lipid layer appearance (Fig. 1, ANOVA, P = 0.132 and P = 0.097, respectively). Notably, the observed power for the ANOVA was 50%. To detect a difference between groups with an observed power of 90%, the sample size needs to be increased to 26 subjects per group. The menopausal group showed high variability and the HRT group little variability in tear lipocalin concentration.

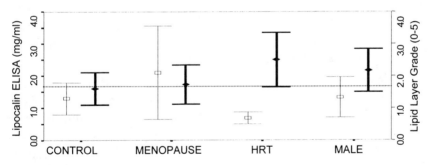

Figure 1. Lipocalin concentration and lipid layer appearance. Legend: ☐Tear lipocalin (mg/ml),◆lipid layer appearance (Grade 0–5). - - - - The reference value for tear lipocalin, 1.68 mg/ml. The mean and 95% confidence intervals are indicated for each group.

To analyse the difference in distribution of the data between groups, the data for each subject group were split into two subsets using reference values for normal lipocalin concentration (1.68 mg/ml)[6] and normal lipid layer appearance (grade 2.0). Table 3 shows Chi-squared comparison of the subsets for each treatment group. The menopausal group had a significantly increased frequency of high lipocalin levels compared to the HRT group. The HRT group had significantly decreased frequency of high lipocalin levels and significantly increased frequency of high lipid levels.

Table 3. Distribution of subjects above reference value for lipocalin concentration and lipid layer appearance (% of subjects)

Variable	Control n = 10	Menopause n = 11	HRT n = 10	Men n = 10	Chi-square P-value
Lipocalin > 1.68 mg/ml	20%	45%*	0%*,†	40%†	0.046
Lipid layer > Grade 2.0	0%*	27%	60%*	22%	0.023

*,† indicate statistically significant differences (P <= 0.05) between values with like symbols

The male group had a significantly higher fluorescein breakup time and higher maximum forced blink interval than the female groups (ANOVA, P < 0.05), indicating a more stable tear film. Total protein and osmolality were not significantly different among groups. The mean total protein for all groups combined was 5.2 ± 2.1 mg/ml and mean osmolality was 322 ± 53 mmol/kg. Although individual variation in lipocalin isoform expression occurred, no significant difference in the number of isoforms between groups was detected.

4. DISCUSSION

Using Chi-square analysis, symptoms indicative of dry eye were significantly higher in the menopausal group than in the male and female control groups. Estrogen HRT improved ocular comfort as frequency of burning and artificial tear use in this group were statistically equivalent to the control groups.

The poor tear film stability seen in dry eye conditions is associated with symptoms of ocular discomfort. Lipocalin increases tear surface tension and has a stabilizing effect on the tear film,[4] suggesting high levels of lipocalin should be associated with good ocular comfort. Our data indicate this is not the case. Chi-square analysis showed the menopausal group had the statistically highest frequency of increased lipocalin levels but also the statistically highest frequency of ocular burning and artificial tear use. This finding correlates with a previous report that high lipocalin levels are associated with contact lens intolerance and symptoms of ocular discomfort.[7]

Tear lipocalin binds lipids of the tear film layer, and increased lipocalin levels may increase lipid layer stability.[8] We found the group with the most frequently high lipid layer grades had the most frequently low tear lipocalin levels (HRT group, Chi-square, P < 0.05). Our data suggest a thicker, more stable lipid layer is associated with low tear lipocalin levels. However, our lipid layer values were based on grading the ocular appearance of the lipid layer. Other tests may be more appropriate in quantifying the relationship.

An unexpected finding is the range of the data measured for tear lipocalin. The variability of tear lipocalin levels in the menopausal group contrasts with the very narrow range in the HRT group. Indeed, the HRT group had a much narrower range than both control groups. Menopause is associated with fluctuating hormone levels, whereas HRT provides a constant level of estrogen. This suggests estrogen may have a controlling effect on tear lipocalin secretion by the lacrimal gland.

The higher fluorescein breakup time and maximum forced blink interval in the male group suggest that males have better tear film stability than females. This agrees with other studies showing males have a more stable tear film. Gender differences in the development and function of the lacrimal gland have been well documented[1] and highlight the complexity of hormonal interactions that influence tear film secretion.

Given the limited number of participants in this pilot study, these findings will need to be confirmed by increasing the sample size. Overall, the data from our study indicate estrogen HRT has a beneficial effect on ocular symptoms, and this effect may be due to its action on tear lipocalin secretion and improvement of the tear film lipid layer.

ACKNOWLEDGMENTS

The authors thank B. Redl for the donation of lipocalin antibodies and K. Willets, J. Sawyer and S. Champion for assistance with patient recruitment.

REFERENCES

1.   D. Sullivan, L. Wickham, E. Rocha, R. Kelleher, L. da Silveira and I. Toda. Influence of gender, sex steroid hormones, and the hypothalamic-pituitary axis on the structure and function of the lacrimal gland, *Adv Exp Med Biol*. 438:11 (1998).

2.   W. Mathers, D. Stovall, J. Lane, M. Zimmerman and S. Johnson. Menopause and tear function: the influence of prolactin and sex hormones on human tear production, *Cornea*. 17:353 (1998).

3.   P. Janssen and O. van Bijsterveld. The relations between tear fluid concentrations of lysozyme, tear specific prealbumin and lactoferrin, *Exp Eye Res*. 36:773 (1983).

4.   R. Schoenwald, S. Vidvauns, D. Wurster and C. Barfknecht. The role of tear proteins in tear film stability in the dry eye patient and in the rabbit, *Adv Exp Med Biol*. 438:391 (1998).

5.   C. McMonnies and A. Ho. Patient history in screening for dry eye conditions, *J Am Optom Assoc*. 58:296 (1987).

6.   R. Fullard and A. Snyder. Protein levels in stimulated and non-stimulated tears of normal human subjects, *Invest Ophthalmol Vis Sci*. 31:12 (1990).

7.   M. Glasson. Understanding the reasons why some patients are intolerant to soft contact lens wear, *Invest Ophthalmol Vis Sci*. 41:s73 (1999).

8.   B. Glasgow, A. Abduragimov, Z. Farahbaksh, K. Faull and W. Hubbell. Tear lipocalins bind a broad array of lipid ligands, *Curr Eye Res*. 14:363 (1995).

IDD3 AND IDD5 ALLELES FROM NOD MICE MEDIATE SJÖGREN'S SYNDROME-LIKE AUTOIMMUNITY

S. Cha,[1] H. Nagashima,[1] A. B. Peck,[2] and M. G. Humphreys-Beher[1]

[1]Department of Oral Biology
[2]Department of Pathology
University of Florida
Gainesville, Florida, USA

1. INTRODUCTION

Sjögren's syndrome (SS) is a chronic autoimmune disorder clinically manifested as dry eyes and dry mouth in human patients. The loss of secretory capacity of the eyes and salivary glands presumably results from the effector functions of immune cells infiltrating the target organs.

The NOD mouse and its congenic partner strains may be an appropriate animal models for the study of SS. NOD mice, a strain that has been used for the study of type I diabetes, demonstrates a temporal lymphocytic infiltration of the exocrine tissues correlated with a loss of secretory function. The NOD-*scid* mouse, which contains no functional B and T cells, provides evidence for a series of physiological and morphological changes that occur even in the absence of adaptive immune response but do not result in the loss of secretory function. These results suggest the involvement of genetic programming in the initiation of the disease.[1]

To identify the genetic susceptibility loci for SS-like autoimmune disease in NOD mice, several NOD and C57BL/6 congenic mice containing single or multiple diabetogenic differential loci have been investigated.[2] Although the autoimmune endocrinopathy and exocrinopathy occurring in NOD mice represent two independent diseases, it is also reasonable to assume that development of SS-like pathology may be influenced by some of the diabetes-susceptibility loci. The data indicate that alleles on chromosomes 1 (*Idd5*) and 3 (*Idd3*) appear to greatly influence susceptibility and resistance to development of autoimmune exocrinopathy. Therefore, we have further investigated the contributions of

Lacrimal Gland, Tear Film, and Dry Eye Syndromes 3
Edited by D. Sullivan *et al.*, Kluwer Academic/Plenum Publishers, 2002

1035

Idd5 and *Idd3* loci of NOD mice in the genetic background of C57BL/6 to determine whether these two loci are sufficient to render the duplication of the disease phenotype in a normal genetic background.

2. MATERIALS AND METHODS

Twenty week-old male and female NOD/Lt and C57BL/6 mice were obtained through the Department of Pathology Animal Colony. C57BL/6.NODc3 (*Idd3,10*) and C57BL/6.NOD c1t (*Idd5*) were kindly provided by Dr. Edward Wakeland, Center for Immunology, University of Texas Southwestern Medical Center (Dallas, TX). NOD.B6*Idd3*B10*Idd5* was provided by Dr. Linda Wicker and Larry Peterson, Merck Research Laboratories (Rahway, NJ). C57BL/6.NOD*Idd3Idd5* was generated using standard genetic backcross protocols and used at 12 weeks of age. Heterozygous C57BL/6.NOD*Idd3Idd5* chromosomal segments were screened by microsatellite markers (IL-2-Tshb for *Idd3*; D1mit8-D1mit15 for *Idd5*).[3] Mice were backcrossed for 3 generations followed by sibling mating to maintain the homozygosity for the markers.

Biochemical analyses for amylase, cysteine protease, and matrix metalloproteinase (MMP) activity were performed by the protocols in our previous study.[2]

3. RESULTS

Our previous study[2] showed that, in contrast to the non-susceptible parental C57BL/6 strain, the C57BL/6.NOD*Idd5* congenic partner strain, containing a genetic region derived from chromosome 1 of the NOD mouse, exhibited pathophysiological characteristics of autoimmune exocrinopathy, except for the loss of secretory function. Replacement of individual diabetes susceptibility intervals *Idd3*, *Idd5*, *Idd13*, *Idd1*, and *Idd9*, as well as a combination of the *Idd3, Idd10* and *Idd17* intervals, with resistance alleles, had little effect on the development of autoimmune exocrinopathy. Conversely, NOD mice, in which the chromosome regions containing both *Idd5* and *Idd3* have been replaced by intervals derived from C57BL mice, exhibit a reduced pathophysiology associated with Sjögren's syndrome-like autoimmune exocrinopathy.

The analysis of disease markers in newly generated C57BL/6.NOD*Idd3Idd5* mouse strain showed a 41% decrease in saliva flow at 12 weeks of age, compared to the C57BL/6 parental inbred partner at 20 weeks of age (Table1). MMP activity in the submandibular glands, which had been shown to be elevated in disease mice[4] and human patients,[5] was similar to the value of the NOD donor strain (Table 2), indicating that these two chromosomal regions from the NOD mouse were able to duplicate the phenotype of the NOD mouse in the submandibular glands of C57BL/6 strain. The focus score in the submandibular glands C57BL/6.NOD*Idd3Idd5* was ≥2.0 (data not shown), whereas no lymphocytic infiltration was detected in the lacrimal glands (Fig. 1).

Table 1. Analysis of C57BL/6 and NOD congenic mice for Sjögren's syndrome-like pathophysiology

Strain	N	Saliva (μl/10 min)	Protein[1] (μg/μl)	Amylase[2] (U/mg protein)	Cysteine protease[3] (μg/mgprotein)
c57BL/6	10	225±33	3.24±0.09	210±6.93	11.35±3.71
c57BL/6.NOD*Idd3Idd5*	10	133.8±6.8	4.79±0.08	101±6.51	N/D
NOD. B6*Idd3*. B10*Idd 5*	11	179±18	3.37±0.09	210±2.14	22.59±4.95
NOD	8	92±19	5.01±0.17	117±14.58	40.62±3.98

All values represent the mean ± standard error for each assay. [1]Protein concentration in saliva by volume; [2]Amylaseactivity in saliva; [3]measurement of cysteine protease activity in submandibular gland lysates.

Table 2. Analysis of MMP activity in double congenic mice

Strains		C57BL/6	C57BL/6.NOD*Idd3Idd5*	NOD.B6*Idd3*. B10*Idd 5*	NOD
MMPActivity	SMX	0.413±0.055	0.9±0.032	0.495±0.004	0.89±0.032
(U/mg protein)	LAC	0.02±0.005	0.03±0.004	0.033±0.004	0.045±0.008

All values represent the mean ± standard error for each assay. Submandibular (SMX) and lacrimal (LAC) glands were collected from 3–4 mice, pooled, and assayed 4 times. 1unit(U) indicates 1_g of collagen degraded/min.

C57BL/6   NOD   NOD.B6*Idd3*B10*Idd5*   C5BL/6.NOD*Idd3Idd5*

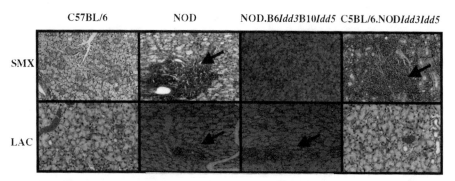

Figure 1. Histology of submandibular (SMX) and lacrimal glands (LAC) of control, NOD, and double congenic mice. 200x . Arrows indicate lymphocytic infiltrates in the glands.

3. DISCUSSION

The major, current approach to determining genetic susceptibility to a certain disease in the mouse consists of screening the genome for genetic markers such as restriction fragment length polymorphisms (RFLPs) and more abundant microsatellites by generating experimental crosses and congenic mouse strains. Screening congenic mice containing diabetes susceptibility loci allowed us to examine the genetic components that influence the initiation and the progress of SS-like autoimmune exocrinopathy.

Several biological disease markers from the congenic mice, including C57BL/6.NOD*Idd3Idd5* and NOD.B6*Idd3*.B10*Idd 5* double congenic mice, suggest that alleles from *Idd5* on chromosome 1 and *Idd3* on chromosome 3 in combination greatly influence the disease pathogenesis. Furthermore, the C57BL/6.NOD*Idd5* exhibiting disease phenotype without losing secretory function, and C57BL/6.NOD*Idd3Idd5* showing dramatic loss of secretory function even at 12 weeks of age, indicate two possibilities. First, the *Idd5* interval on chromosome 1 containing IL-1 receptor, bcl-2, caspase-8, and CTLA-4, greatly influences the ability to duplicate or reverse the disease phenotype of SS in mice. Genetic intervals in the MRL/lpr mouse[6] and the NOD mouse[7] controlling sialadenitis have been mapped to chromosome 1, supporting its contribution as a major locus for the development of SS-like autoimmune exocrinopathy. Second, *Idd3* containing IL-2 from NOD seems to be necessary to manifest clinical hallmark of SS. Allelic variation of IL-2 in the NOD mouse,[8] producing less potent protein, may lead to defective apoptosis of T-cells in response to self-antigens in the thymus, the accumulation of autoreactive T-cells in the periphery, and B-cell activation.

ACKNOWLEDGEMENT

We thank Ms. Joy Nanni and Janet Cornelius for technical assistance and helpful discussions. This research was supported by NIH grant DE 13290. Dr. Seunghee Cha is supported by funds for graduate education from the Department of Oral Biology.

REFERENCES

1. C.P. Robinson, H, Yamamoto, A.B. Peck, , M.G. Humphreys-Beher. Genetically programmed development of salivary gland abnormalities in the NOD (nonobese diabetic)-scid mouse in the absence of detectable lymphocytic infiltration: a potential trigger for sialadenitis of NOD mice. *Clin Immunol Immunopathol.* 79: 50–9 (1996).

2. J. Brayer, J. Lowry, S. Cha, C.P. Robinson, S. Yamachika, A.B. Peck, M.G. Humphreys-Beher. Alleles from chromosomes 1 and 3 of NOD mice combine to influence Sjogren's syndrome-like autoimmune exocrinopathy. *J Rheumatol.* 27: 1896–904 (2000).

3. M.A. Yui , K. Muralidharan, B. Moreno-Altamirano, G. Perrin, K. Chestnut, E.K. Wakeland. Production of congenic mouse strains carrying NOD-derived diabetogenic genetic intervals: an approach for the genetic dissection of complex traits. *Mamm Genome.* 7: 331–4 (1996).

4. S. Yamachika, J.M. Nanni, K.H. Nguyen, L. Garces, J.M. Lowry, C.P. Robinson, J. Brayer, G.E. Oxford, A. da Silveira, M. Kerr, A.B. Peck, M.G. Humphreys-Beher. Excessive synthesis of matrix metalloproteinases in exocrine tissues of NOD mouse models for Sjogren's syndrome. *J Rheumatol.* 25: 2371–80 (1998).

5. Y.T. Konttinen, S. Halinen, R. Hanemaaijer, T. Sorsa, J. Hietanen, A. Ceponis, J.W. Xu, R. Manthorpe, J. Whittington, A. Larsson, T. Salo, L. Kjeldsen, U.H. Stenman, A.Z. Eisen. Matrix metalloproteinase (MMP)-9 type IV collagenase/gelatinase implicated in the pathogenesis of Sjogren's syndrome. *Matrix Biol.* 17: 335–47 (1998).

6. M. Nishihara, M. Terada, J. Kamogawa, Y. Ohashi, S. Mori, S. Nakatsuru, Y. Nakamura, M. Nose. Genetic basis of autoimmune sialadenitis in MRL/lpr lupus-prone mice: additive and hierarchical properties of polygenic inheritance. *Arthritis Rheum.* 42: 2616–23 (1999).

7. H.J. Garchon, P. Bedossa, L. Eloy, J.F. Bach. Identification and mapping to chromosome 1 of a susceptibility locus for periinsulitis in non-obese diabetic mice. *Nature.* 353: 260–2 (1991).

8. P.L. Podolin, M.B. Wilusz, R.M. Cubbon, U. Pajvani, C.J. Lord, J.A. Todd, L.B. Peterson, L.S. Wicker, P.A. Lyons. Differential glycosylation of interleukin 2, the molecular basis for the NOD Idd3 type 1 diabetes gene? *Cytokine.* 12: 477–82 (2000).

DRY EYE ASSOCIATED WITH CHRONIC
GRAFT-VERSUS-HOST DISEASE

Y. Ogawa,[1,2] M. Kuwana,[2] K. Yamazaki,[3] Y. Mashima,[1]
S. Okamoto,[4] K. Tsubota, [1,5] Y. Oguchi,[1] and Y. Kawakami [2]

[1]Department of Ophthalmology
[2]Institute for Advanced Medical Research
[3]Department of Pathology
[4]KEIO BMT Program
Department of Medicine
Keio University School of Medicine
[5]Department of Ophthalmology
Tokyo Dental College
Tokyo, Japan

1. INTRODUCTION

Hematopoietic stem cell transplantation (SCT) is an established treatment for a variety of hematologic malignancies and some cancers. Because the survival rate in SCT recipients has significantly improved, quality of life and the late complications after SCT in long-term survivors have become increasingly important in the clinical setting.[1,2] One of the most common late complications in SCT recipients is chronic graft-versus-host disease (cGVHD) that affects skin, lung, lacrimal and salivary glands.[2,3] Chronic GVHD-associated dry eye is prevalent and sometimes leads to blindness.[4] However, clinical and histological characteristics, as well as the pathogenic process of lacrimal gland involvement in cGVHD are not well understood. The purpose of our ongoing studies is to clarify some of these issues.

Lacrimal Gland, Tear Film, and Dry Eye Syndromes 3
Edited by D. Sullivan *et al.*, Kluwer Academic/Plenum Publishers, 2002

2. CLINICAL FEATURES OF CGVHD ASSICIATED-DRY EYE

We initially conducted a prospective study to evaluate dry eye in SCT recipients.[3] The purpose of this study was to determine the incidence, natural course, and severity of dry eye occurring or worsening after SCT.

Fifty-three patients who received allogeneic or autologous SCT were prospectively enrolled and followed at least 180 days. Clinical variables examined included grading of the symptoms of dry eye, evaluation of the ocular surface, tear breakup time, and Schirmer tests with and without nasal stimulation. Meibomian gland secretion was also examined using a slit-lamp while applying steady digital pressure.

Half of the allograft recipients (n = 22) developed dry eye, or their pre-existing dry eye worsened after SCT, while none of 9 autograft recipients developed dry eye. The mean onset of dry eye was 171 ± 59 days after SCT. Based on clinical characteristics, dry eye was classified into two types. In the severe type, 10 patients had severe dry eye with ocular surface findings resembling Sjögren's syndrome (SS), and reduction of reflex tearing occurred soon after onset. In these patients, severe dry eye progressed rapidly after the onset. In contrast, in the mild type, 12 patients had dry eye without impaired reflex tearing. Meibomian gland dysfunction (MGD) was more frequent and severe in patients with severe type than those with mild type. Therefore, the presence of both impaired reflex tearing and MGD may predict severe dry eye early in the course of the disease.

Dry eye associated with cGVHD frequently developed when the dosage of immunosuppressants was reduced or shortly after immunosuppressants were stopped.[3] This suggests that systemic administration of immunosuppressants is effective in preventing the onset of dry eye. In 2 patients, systemic FK506 with or without corticosteroids was an effective treatment for severe dry eye.[5] One of these patients was a 43-year-old woman with myelodysplastic syndrome who underwent allogeneic SCT. The patient had mild dry eye before SCT, but the dry eye significantly worsened in conjunction with the onset of oral and intestinal cGVHD on day 150 while tapering FK506. PSL (1 mg/kg/day) was then initiated and the dosage of FK506 was increased. By day 240, symptoms of dry eye and the ocular surface findings markedly improved, and cGVHD in other organs was completely resolved. However, dry eye recurred when FK506 was tapered again. In another case, dry eye also flared up when FK506 dosage was decreased. These case reports suggest that long-term administration of FK506 is required to achieve curable response. In addition, the use of topical FK506 and PSL may be useful as a maintenance therapy for cGVHD-associated dry eye. Effectiveness of immunosuppressants also suggests the primary role of immune process in the pathogenesis of dry eye in patients with cGVHD.

3. HISTOLOGIC FINDINGS IN LACRIMAL GLAND OF CGVHD

We next examined the immunohistochemical and ultrastructural characteristics of the lacrimal gland specimens in patients who received allogeneic SCT and later developed cGVHD.[6] Lacrimal gland specimens from five patients who presented dry eye as a part of their symptoms related to cGVHD were examined by immunohistochemistry and transmission electron microscopy. Lacrimal gland specimens from five patients with SS were used as controls. Immunohistochemical analysis showed that periductal infiltration of lymphocytes were mainly T cells, whereas infiltration of lymphocytes (predominantly B cells), mainly into acinar areas, was noted in SS patients. Excessive fibrosis in the glandular interstitum and atrophy of acinar lesions was observed in cGVHD with severe dry eye, but not in patients with mild dry eye and in SS patients. Electron microscopic evaluation demonstrated fibroblasts attached to lymphocytes with primitive contact, multilayered basal lamina of blood vessels, marked thickening of basal lamina in ductal and lobular epithelium, and newly synthesized collagen fibrils in the extracellular matrix in all cGVHD patients. In contrast, these lesions were observed in SS patients infrequently. These findings suggest substantial differences in the lacrimal gland histopathology of patients with cGVHD and those with SS.

Increased numbers of CD34$^+$ fibroblasts are present at the site of acinar cell loss in the fibrotic areas around the ducts in patients with cGVHD. This finding was observed in all cGVHD but in only one SS patient. CD34$^+$ fibroblasts might play some role in the pathogenic process of the lacrimal gland in cGVHD by interacting with various inflammatory cells, and by producing excessive extracellular matrix. The CD34 molecule, also referred to as L-selectin ligand Sgp90, was originally found to be expressed on hematopoietic stem cells. CD34 expression is also found in vascular and lymphatic endothelial cells, as well as in other mesenchymal cells such as dermal dendritic cells. Because Yamazaki and Eyden have recently documented that CD34 stains intra- and inter-lobular stromal fibroblasts in normal mammary and submandibular glands,[7] it is likely that the increased number of CD34$^+$ fibroblasts in the lacrimal gland of patients with cGVHD are stromal fibroblasts residing in the intra- and inter-lobular areas. Another possibility is that CD34$^+$ fibroblasts in the lacrimal glands originated from donor fibrocytes, a novel population of blood-borne cells with a spindle-like morphology and a distinct cell-surface phenotype (collagen$^+$/CD13$^+$/CD34$^+$/CD45$^+$).[8]

To examine whether stromal fibroblasts are involved in the immune process in patients with cGVHD, freshly prepared frozen sections of the lacrimal gland from 7 patients with cGVHD and 5 patients with SS were immunostained with a variety of monoclonal antibodies. CD34$^+$ fibroblasts in periductal and inter-lobular areas strongly expressed HLA-DR and CD54 in patients with cGVHD but not in SS patients, while HLA-DR expression on ductal and acinar epithelia was observed equally in patients with cGVHD and SS patients. Stromal fibroblasts expressing CD40, CD80, and/or CD86 were detected exclusively in patients with cGVHD, but these co-stimulatory molecules were expressed by ductal and acinar epithelia in both cGVHD and SS patients. T cells infiltrated into the lacrimal gland were predominantly CD8$^+$ T cells in cGVHD, whereas the proportion of CD4$^+$ and CD8$^+$ T cells was almost equal in SS patients. In patients with cGVHD, CD4$^+$ and CD8$^+$ T cells were co-localized with stromal fibroblasts in periductal area, whereas only CD8$^+$ T cells infiltrated into the acinar and ductal epithelia.

4. A POSSIBLE PATHOGENIC PROCESS IN LACRIMAL GLAND CGVHD

Based on our histological analysis, we propose a possible pathogenic process in the lacrimal glands of patients with cGVHD (Fig. 1). Specifically, a breakdown of the blood vessel basal laminae induces the migration of donor T cells into the lacrimal gland tissues. These donor T cells are then activated and migrate into the periductal areas, and contribute to the destruction of the ductal epithelia. At the same time, CD34$^+$ stromal fibroblasts are activated by cytokines released by inflammatory cells. In addition, stromal fibroblasts expressing HLA-DR, as well as adhesion and co-stimulatory molecules, are involved in the activation of both CD4$^+$ and CD8$^+$ T cells in periductal area. These activated fibroblasts synthesize an excessive amount of extracellular matrix, resulting in rapid interstitial fibrosis. In this scenario, the suppression of activated donor T cells and/or fibroblasts early in the course of the disease may lead to the prevention of progressive dry eye.

Further investigation examining target antigens recognized by donor T cells and functional properties of stromal fibroblasts is required to define the pathogenic process in lacrimal gland cGVHD.

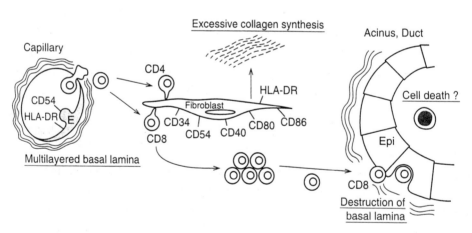

Figure 1. A scheme representing a possible pathogenic process in lacrimal gland cGVHD. E, endothelium; Epi, epithelium.

REFERENCES

1.   E.D. Thomas, R. Storbe. The development of the scientific foundation of hematopoietic cell transplantation based on animal and human studies. In: *Hematopoietic cell transplantation.* E.D. Thomas, K.G. Blume, S.J. Forman. 2nd ed. 1–11. Blackwell, Oxford (1998).

2.   K.M. Sullivan. Graft-versus-host disease. In: *Hematopoietic Cell Transplantation.* E.D. Thomas, K.G.
      Blume, S.J. Forman SJ. 2nd ed. 515–536, Blackwell, Oxford (1998).
3.   Y. Ogawa, S. Okamoto, M. Wakui, et al. Dry eye after haematopoietic stem cell transplantation. *Br J
      Ophthalmol.* 83:1125–1130 (1999).
4.   M. K. Jack, G.M. Jack M, G.E. Sale, H.M. Shulman, K.M. Sullivan. Ocular manifestations of graft-v-
      host disease. *Arch Ophthalmol.* 101:1080–1084(1983).
5.   Y. Ogawa Y, S. Okamoto, M. Kuwana, et al. Successful treatment of dry eye in two patients with
      chronic graft-versus-host disease with systemic administration of FK506 and corticosteroids; Case
      reports. *Cornea.* 20:430–434 (2001)
6.   Y. Ogawa . K. Yamazaki, M. Kuwana, et al. A significant role of stromal fibroblasts in rapidly
      progressive dry eye in patients with cGVHD. *Invest Ophthalmol Vis Sci.* 42:111–119 (2001)
7.   K. Yamazaki, B.P. Eyden. Ultrastructural and immunohistochemical studies of intralobular fibroblasts
      in human submandibular gland: the recognition of a 'CD34 positive reticular network' connected by gap
      junctions. *J Submicrosc Cytol Pathol.* 28:471–483 (1996).
8.   J. Chesney, C. Metz, A.B. Stavitsky, M. Bacher, R. Bucala. Regulated production of type I collagen
      and inflammatory cytokines by peripheral blood fibrocytes. *J Immunol.* 160:419–425 (1998).

CONTRIBUTION OF NEUROGENIC INFLAMMATION TO IRRITABLE EYE SYNDROME

Janos Feher

Ophthalmic Neuroscience Program
Institute of Ophthalmology
University of Rome "La Sapienza"
Rome, Italy

1. INTRODUCTION

Migraine is frequently associated with photophobia (with or without visual aura). Our previous clinical experience showed that ocular irritation symptoms ("dry eye symptoms") are similarly associated with photophobia (Feher, 1995), and collectively, these symptoms and conditions may be referred to as *irritable eye syndrome*. The aims of the present clinical studies were to learn more about the association between these diseases and to reveal an eventual pathophysiologic correlation between ocular irritation symptoms, headache and photophobia.

2. MATERIAL AND METHODS

Ocular irritation symptoms were established in 72 out patients by the use of previously described methods (Feher, 1998). For comparison, equal number of age and sex matched control cases were also selected. The prevalence of headache and photophobia was evaluated by questionnaire. Histologic studies were performed in 12 affected and 9 control cases: conjunctiva biopsy-specimens were used for light, polarization and electron microscopy (TEM, SEM). Headache evaluation was based on the diagnostic criteria of the International Headache Society (Marks and Rapaport, 1997).

Lacrimal Gland, Tear Film, and Dry Eye Syndromes 3
Edited by D. Sullivan *et al.*, Kluwer Academic/Plenum Publishers, 2002

1047

3. RESULTS

Table 1. Ocular irritation in association with photophobia

	Photophobia	No Photophobia
Ocular Irritation (n = 72)	58 (80%)	14 (20%)
Control (n = 72)	2 (5%)	70 (95%)

Table 2. Ocular Irritation in association with headache

	Migraine	Tension-type	Cluster-type	No headache
Ocular Irritation (n = 72)	27 (37%)	32 (44%)	4 (5%)	9 (14%)
Control (n = 72)	13 (19%)	29 (40%)	3 (4%)	27 (37%)

Table 3. Headache in association with photophobia

	Photophobia	No Photophobia
Headache (n = 63)	57	7
Migraine type headache (27)	27	0
Tension type headache (32)	26	6
Cluster type headache (4)	3	1
No headache (n = 9)	2	7
Control (n = 72)	2	70

4. DISCUSSION

The current concepts on keratoconjunctivitis sicca presume that sub-clinical ocular surface inflammation plays a key role in the pathogenesis of ocular surface irritation symptoms (Pflugfelder et al., 2000). Components of the ocular surface, main lacrimal gland and their innervation represent a functional unit. Thus, when any of them is compromised, normal lacrimal support to the ocular surface is impaired, and the subsequent immunogenic inflammation may further compromise tear flow (Stern et al., 1998)

Ocular irritation symptoms were correlated with abnormal mucus and lipid secretion, and decreased tear fluorescein clearance was a risk factor for ocular irritation, even in subject with normal Schirmer scores. Another risk factor was the presence of meibomian gland pathology. which was found to correlate with the severity of ocular irritation symptoms (Afonso et al. 1999). These observations confirmed our histological studies that ocular irritation is associated with increased mucus secretion. Furthermore, the surface epithelial cells (so called "secondary secretory cells") may also contribute to mucus secretion in ocular irritation (Feher, 1993).

An increased sensitivity to light is a constant finding in inflammatory corneal and uveal diseases (keratitis, iritis, iridocyclitis). Several experimental studies showed that corneal trauma lead to a neurogenic inflammation of the whole anterior segment of the eye, characterized by vasodilatation, exudation, and miosis. Release of Substance P (SP) from sensory nerve endings plays a crucial role in the development of these symptoms. The photophobia is an interference of ciliary nerve hypersensitivity and light stimuli. One contributing mechanism may be that any inflammation of the anterior uvea sensitizes ciliary nerves, and light induced miosis and ciliary muscle spasm evokes pain. This hypothesis is supported by the fact that atropine significantly reduces ocular pain in uveal inflammation. Another contributing mechanism may be the involvement of the retinal pigment epithelium and its sensory nerves. The choroidal involvement is supported by well known clinical experiences where chorioretinal dystrophies (retinitis pigmentosa, Stragardt dystrophy, AMD, myopic dystrophy) are almost always associated with photophobia.

Current interpretation of the headache/migraine suggests that trigeminal sensory nerve stimulation triggers the pain center. The pathophysiology of migraine and tension-type headache encompasses dilatation of dural vessels and extravasation of plasma proteins (Ghabriel et al., 1999). The neurogenic dural inflammation theory of migraine supposed that the dural membrane surrounding the brain becomes inflamed and hypersensitive due to release of neuropeptides from primary sensory nerve terminals. Substance P (SP), calcitonin gene related peptide (CGRP) and nitric oxide are all thought to play a role in dural inflammatory cascade.

Photophobia is commonly associated with migraine attack, and it may be shortened and/or attenuated if the affected patient stays in dark room. These experiences were confirmed by a double-blind, placebo-controlled study (Hyson, 1998). The threshold for both light and sound were lower in the migraine patient than controls between attacks (Main et al., 1997).

In conclusion our clinical experiences showed the ocular irritation symptoms, photophobia and headache (usually migraine or tension types), occurred together forming a syndrome "irritable eye syndrome." Trigeminal nerve hypersensitivity (neurogenic inflammation) seems to be the most likely pathogenic process explaining the co-morbidity of ocular irritation, photophobia and headache. Finally, the clinical overlap between ocular irritation and headache is the photophobia, while the pathophysiological one is the neurogenic inflammation.

REFERENCES

Afonso, A.A., Monroy, D., Stern, M.E., Feuer, W.J., Tseng, S.C.G., Pflugfelder, S.C. Correlation of tear fluorescein clearance and Schirmer test scores with ocular irritation symptoms. Ophthalmology, 106:803 (1999)

Feher, J. Pathophysiology of the Eye 1: The Preocular Tear Film. Akademiai Kiado, Budapest (1993)

Feher, J. Tear film abnormalities and mucous membrane diseases associated with neurohormonal dysfunctions. In Hurwitz, J. et all (eds) The Lacrimal System 1994. Kugler & Ghedini Publications. Milano 1995 pp. 65 (1994)

Feher, J. Diagnostic value of tear film abnormalities in a new syndrome affecting the neuroendocrine and
    immune systems. In: Sullivan, D et all. (eds): Lacrimal Gland, Tear Film, and Dry Eye Syndromes 2.
    Plenum Press, New York, pp. 839 (1998)

Ghabriel, M.N, Lu, M.X., Leigh, C., Cheung, W.C., Allt, G. Substance P-induced enhanced permeability of
    dura mater microvessels is accompanied by pronounced ultrastructural changes, but not dependent on
    the density of endothelial cell anionic sites. Acta Neuropathol (Berl.) 97:297 (1999)

Hyson, M.I. Anticephalic photoprotective premedicated mask. A report of a successful double-blind placebo-
    controlled study of a new treatment for headache with associated frontal pain and photophobia.
    Headache 38:475 (1998)

Main, A., Dowson, A., Gross, M. Photophobia and phonophobia in migraineous between attacks. Headache
    37:492 (1997):

Marks, D.R., Rapaport, A.M. Diagnosis of migraine. Semin Neurol., 17:303 (1997)

Pflugfelder, S.C, Solomon, A., Stern, M.E. The diagnosis and management of dry eye: a twenty-five year
    review. Cornea, 19:644 (2000)

Stern M.E, Beuerman, R.W., Fox, R.I., Gao, J., Mircheff, A.K., Pflugfelder, S.C. The pathology of dry eye:
    The interaction between the ocular surface and lacrimal gland. Cornea, 17:584 (1998)

OCULAR DRYING ASSOCIATED WITH ORAL ANTIHISTAMINES (LORATADINE) IN THE NORMAL POPULATION–AN EVALUATION OF EXAGGERATED DOSE EFFECT

D. Welch,[1] G. W. Ousler III,[1] L. A. Nally,[1] M. B. Abelson,[1,2,3] and K. A. Wilcox[1]

[1]Ophthalmic Research Associates
Dry Eye Department
North Andover, Massachusetts, USA
[2]Schepens Eye Research Institute
Boston, Massachusetts, USA
[3]Harvard Medical School
Boston, Massachusetts, USA

1. INTRODUCTION

Many systemic medications adversely influence the eye by altering tear flow and production and natural tear substances, or by penetrating and combining with natural components of the tear film.[1,2] Consequently, changes in tear film stability resulting from systemic medications lead to increased ocular discomfort. Oral antihistamines, anti-hypertensives (Accupril, Dyazide), antiemetics (Compazine, Dramamine), antidepressants (Celexa, Sinequan) and diuretics (Maxzide, Dyrenium) may cause dry eye syndromes.[3] In particular, patients often complain of side effects associated with oral antihistamines such as Benadryl, Zyrtec, and Claritin®. One common complaint is their "drying" effect.

Muscarinic receptors of the M_3 subtype, which control endocrine gland function, help regulate secretion of protein and fluid in the eye.[4] The anti-muscarinic activity of antihistamines can affect the aqueous layer by decreasing basal tear production and the mucous layer by decreasing mucin output of the conjunctival goblet cells.[1] Chlorpheniramine maleate, an over-the-counter antihistamine, significantly decreases tear production, as measured by Schirmer's test in normal subjects.[5] An abnormality and/or

Lacrimal Gland, Tear Film, and Dry Eye Syndromes 3
Edited by D. Sullivan *et al.*, Kluwer Academic/Plenum Publishers, 2002

1051

deficiency in any one of the three layers composing the tear film (lipid, aqueous and mucin) can lead to tear film instability, usually resulting in ocular discomfort.

Low humidity, temperature, air flow, lighting conditions, prolonged visual tasking and certain systemic medications can exacerbate the discomfort. A controlled adverse environment (CAE), which produces the signs and symptoms associated with dry eye syndrome by regulating humidity (< 5%), temperature (76 ± 6° F), air flow (non-turbulent), lighting conditions and visual tasking (television, person computers) has been developed to examine the effects of systemic medications on ocular health and discomfort. The objective of this study was to investigate the ocular effects of a commonly prescribed oral antihistamine, Claritin® (loratadine, 10 mg) on normal subjects when exposed to a CAE for 45 min.

2. METHODS

2.1. Visit 1 (Baseline)

Consent was obtained from 14 normal subjects who had not used any systemic medications for at least 30 days, and they were queried about their medical and medication history. Baseline ophthalmic examinations consisting of a visual acuity test, slit-lamp biomicroscopy examination, keratitis and conjunctival staining evaluations (according to a standardized 0-to-4-point scale), tear film breakup time (TFBUT) measurement and Schirmer's test were conducted. Subjects were exposed to a CAE for 45 min, during which their ocular discomfort was recorded at 0, 5, 10, 15, 30, and 45 min. Immediately after the CAE challenge, all ophthalmic examinations were repeated. Subjects were randomized to receive either QD (once daily) or BID (twice daily) treatment with Claritin® for the following 4 days.

2.2. Visit 2 (Day 4)

The 7 subjects who had been dosing QD and the 7 who had been dosing BID with Claritin® were queried about any changes since baseline in their medical and medication history. Treatment compliance was reviewed and all ophthalmic examinations performed. Subjects were re-challenged in a CAE for 45 min, recording ocular discomfort at all time points. Immediately after exposure to the CAE, all ophthalmic examinations were repeated and subjects exited from the study.

2.3. Evaluation Criteria

Safety evaluations included visual acuity, slit-lamp biomicroscopy findings and adverse events query. Signs were evaluated by examining the severity of corneal and conjunctival staining (on a standardized 0-to-4 scale 5 min after instillation of 5 μl non-preserved, 2% sodium fluorescein), TFBUT (5 μl non-preserved, 2% sodium fluorescein) and a Schirmer's test.[6] Symptoms, as assessed by the subject, were evaluated on a standardized 0-to-4 ocular discomfort scale. All analyses were performed comparing data obtained during and after CAE exposure at both the baseline visit and after dosing QD or BID with Claritin® for 4 days.

3. RESULTS

Of the 14 subjects who enrolled in and completed this study, 7 were randomized to dose with Claritin® QD, and the remaining 7 dosed with Claritin® BID for 4 days after the baseline visit. No changes were observed in any safety parameters.

3.1. QD Group

After the 4-day treatment and 45 minute CAE challenge, QD-dosing subjects exhibited a mean increase of 1 unit (81%) in keratitis (Fig. 1), 1 unit (71%) in conjunctival staining (Fig. 2) and a 2.2 second (41%) decrease in TFBUT (Fig. 3) compared to (Visit 1) baseline post-CAE exposure measurements. After the 4-day treatment with Claritin®, QD-treated subjects showed a trend toward increased ocular discomfort scores at all but one time point (10 min) (Fig. 4) compared to baseline CAE exposure. This decrease in discomfort scores at the 10 minute time point may be attributed to the "Natural Compensation" phenomenon, the moment a worsening of symptoms temporarily improves during exposure to a CAE. In normals, this compensatory response mechanism occurs at about 10 min.[7] Schirmer's test was performed at baseline and in the QD treatment group only. After 4 days of dosing with Claritin®, QD-treated subjects exhibited an 8.1 mm decrease (53%) in post-CAE exposure Schirmer values (baseline = 15.4 mm, Day 4 QD = 7.3 mm) at Visit 2 compared to Visit 1.

3.2. BID Group

After the 4-day BID treatment with Claritin® and 45 minute CAE challenge, subjects randomized to dose twice a day showed a mean increase of 1.7 units (130%) in keratitis (Fig. 1), 2.5 units (175%) in conjunctival staining (Fig. 2) and a 3.6 second decrease (67%) in TFBUT (Fig. 3) compared to Visit 1. BID-treated subjects showed a trend toward increased ocular discomfort during the CAE challenge after the 4 day treatment with Claritin® (Fig. 4) compared to Visit 1 CAE exposure.

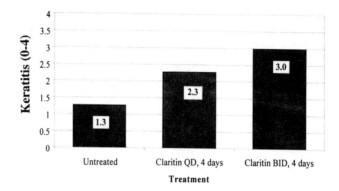

Figure 1. Mean keratitis, measured on a standardized 0-to-4-point scale, after 45 min CAE exposure. Results are based on the average of both eyes. n = 14 for the untreated group (baseline) and n = 7 for QD- and BID-treated groups.

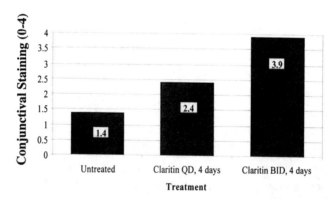

Figure 2. Mean conjunctival staining, graded on a standardized 0-to-4-point scale, after 45-min CAE exposure. Results are based on the average of both eyes. n = 14 for the untreated group (baseline) and n = 7 for QD- and BID-treated groups.

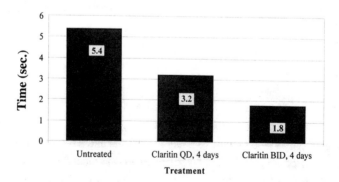

Figure 3. Mean TFBUT after 45-min CAE exposure. Results are based on the average of both eyes. n = 14 for the untreated group (baseline) and n = 7 for QD- and BID-treated groups.

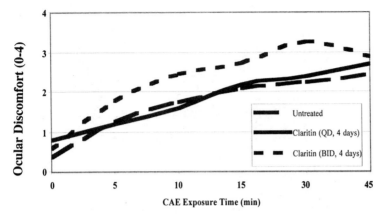

Figure 4. Mean ocular discomfort, graded on a 0-to-4 standardized scale, during 45 min CAE exposure. Results are based on the average of both eyes. n = 14 for the untreated group (baseline) and n = 7 for QD- and BID-treated groups.

4. CONCLUSIONS

Signs and symptoms consistent with ocular dryness, including an increase in corneal and conjunctival staining, a decrease in TFBUT, an increase in ocular discomfort and a decrease in the Schirmer's test, were induced by CAE exposure in normal subjects after taking Claritin® (loratadine, 10 mg) for 4 days, compared to baseline values. Mean differences in QD and BID treatment groups indicate a dose effect. Therefore, Claritin® may cause clinically meaningful damage to the ocular surface as a result of anti-muscarinic activity on the M_3 muscarinic receptors that causes a decrease in the aqueous and mucous layers of the tear film.

REFERENCES

1.   D.C. Crandall, I.H. Leopold.The influence of systemic drugs on tear film constituents. *Ophthalmology* 86: 115–25 (1979).
2.   B.C.P. Polak.Side effects of drugs and tear secretion. *Documental Ophthalmology* 67: 115–117 (1987)
3.   S.D. Jaanus.Ocular side effects of selected systemic drugs. *Opthom Clin.* 2(4): 73–96 (1992).
4.   B.G. Katzung. 1998.*Basic & Clinical Pharmacology*, Appleton & Lange, Stamford.
5.   B.H. Koffler, M.A. Lemp.The effect of an antihistamine (chlorpheniramine maleate) on tear production in humans. *Ann Ophthalmol.* 12: 217 (1980).
6.   M.B. Abelson, G.W. Ousler, L.A. Nally, K. Krenzer,Alternative reference values for tear film breakup time in normal and dry eye populations. Abstract. *Cornea: Supplement 2* 19(6): s72 (2000).
7.   G.W. Ousler., M.B. Abelson, L.A. Nally, D. WelchAn evaluation of time to "natural compensation" or improved ocular discomfort during exposure to a controlled adverse environment (CAE) in normal and dry eye populations. Abstract. *Cornea: Supplement 2* 19(6): s111 (2000).

EVALUATION OF THE TIME TO "NATURAL COMPENSATION" IN NORMAL AND DRY EYE SUBJECT POPULATIONS DURING EXPOSURE TO A CONTROLLED ADVERSE ENVIRONMENT

G. W. Ousler III,[1] M. B. Abelson,[1,2,3] L. A. Nally,[1]
D. Welch,[1] and J. S. Casavant[1]

[1]Ophthalmic Research Associates, Inc.
Dry Eye Department
North Andover, Massachusetts, USA
[2]Schepens Eye Research Institute
Boston, Massachusetts, USA
[3]Harvard Medical School
Boston, Massachusetts, USA

1. INTRODUCTION

Approximately 1 of every 5 Americans suffers from dry eye symptomatology.[1] Since the American population constantly encounters adverse environments, visual tasking, and systemic antihistamines, of which all can cause ocular drying, the high incidence of symptoms is hardly surprising. Furthermore, as the prevalence of these factors increase, the incidence of dry eye symptoms can be expected to escalate, and tolerance to symptoms of dry eye may decrease.

A controlled adverse environment (CAE) that produces and/or exacerbates the signs and symptoms associated with dry eye in a standardized manner has been developed. The CAE regulates humidity (< 5%), temperature (76 ± 6°F), airflow (non-turbulent), lighting conditions (adequate for illuminating the CAE without causing photosensitivity or minimizing the interpalpebral fissure), and visual tasking (television or personal computer use). Ocular discomfort, which is measured according to a 0–4 point standardized scale, is

Lacrimal Gland, Tear Film, and Dry Eye Syndromes 3
Edited by D. Sullivan *et al.*, Kluwer Academic/Plenum Publishers, 2002

1057

one of the subjective parameters that are measured at various time points during exposure of a subject to the CAE.

After screening hundreds of subjects using the CAE, including both normal individuals as well as those diagnosed with dry eye, we observed that a period of temporary improvement can occur in the exacerbated ocular discomfort of a subject during exposure to the CAE. We refer to this transient improvement as "natural compensation."

The objective of this study was to evaluate the time to natural compensation (TNC), which is defined as the moment after which worsening ocular discomfort temporarily improves during the exposure to the CAE.

2. METHOD

This study entailed a review of case report forms. Informed consent was obtained from 50 normal subjects, 175 subjects were diagnosed with mild to moderate dry eye, and 25 subjects were diagnosed with severe dry eye. A medical and medication history was taken for each subject prior to a baseline ophthalmic examination, which included a visual acuity test and a slit-lamp biomicroscopy. The subjects assessed their individual ocular discomfort according to a 0–4 point standardized scale. They were then exposed to the CAE for 60 min, during which they reassessed their individual ocular discomfort every 5 min. After the conclusion of each exposure to the CAE, all ophthalmic examinations were repeated, and the subjects were queried regarding any adverse events during the exposure period.

Natural compensation was defined as a 0.5 unit improvement in worsening ocular discomfort scores during exposure to the CAE. Ocular discomfort was the evaluation criterion for symptoms, and a visual acuity measurement, a slit-lamp biomicroscopy examination, and an adverse event query were the evaluation criteria for safety.

3. RESULTS

No changes were observed in any of the safety parameters. The mean TNC varied among the three evaluation groups (Fig. 1). The mean TNC for the normal subjects (n = 50) was 10.24 min. Subjects with mild to moderate dry eye (n = 175) had a mean TNC of 20.47 min. On average, the severe dry eye population did not exhibit natural compensation. However, by excluding the 18 subjects diagnosed with severe dry eye that did not experience natural compensation, the mean TNC for the remaining severe dry eye subjects (n = 7) was 48.75 min.

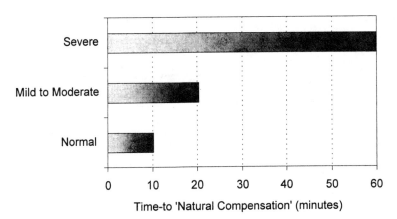

Figure 1. Mean natural compensation occurred in normal subjects (n = 50) at 10.24 min, in subjects with mild to moderate dry eye (n = 175) at 20.47 min, and in the 7 of 25 severe dry eye subjects who experienced natural compensation, the mean was 48.25 min.

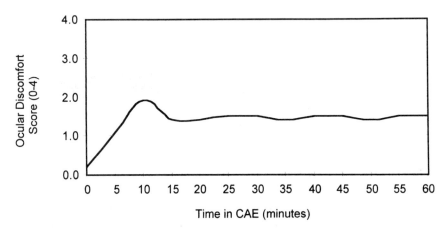

Figure 2. Normal subject profile of ocular discomfort during exposure to a CAE. Normal subjects (n = 50) experienced natural compensation after 10 min of exposure to the CAE, and subsequently there was a slight fluctuation in ocular discomfort for the duration of the exposure period. Mean ocular discomfort at 0 min = 0.2 unit, 5 min = 1.1 units, 10 min = 1.9 units, 15 min = 1.4 units, 20 min = 1.4 units, 25 min = 1.5 units, 30 min = 1.5 units, 35 min = 1.4 units, 40 min = 1.5 units, 45 min = 1.5 units, 50 min = 1.4 units, 55 min = 1.5 units, 60 min = 1.5 units.

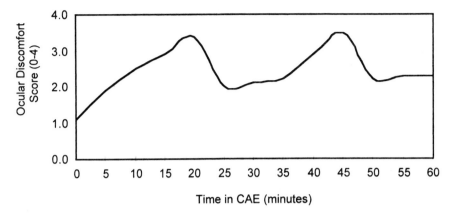

Figure 3. Mild to moderate dry eye subject profile of ocular discomfort during exposure to a CAE. Subjects (n = 175) with mild to moderate dry eye experienced natural compensation at 20 min, and then again at 45 min following another worsening in ocular discomfort. Mean ocular discomfort at 0 min = 1.1 units, 5 min = 1.9 units, 10 min = 2.5 units, 15 min = 2.9 units, 20 min = 3.4 units, 25 min = 2.0 units, 30 min = 2.1 units, 35 min = 2.2 units, 40 min = 2.9 units, 45 min = 3.5 units, 50 min = 2.2 units, 55 min = 2.3 units, 60 min = 2.3 units.

Normal subjects exhibited a mean increase of 1.7 units of discomfort over the first 10 min of exposure to the CAE (Fig. 2). At 15 min, discomfort had improved by 0.5 units, and varied slightly thereafter (0.1 units).

Subjects with mild to moderate dry eye exhibited a mean increase of 2.3 units of discomfort over the first 20 min of CAE exposure (Fig. 3). At 25 min, discomfort improved in by 1.4 units. From 25 – 45 min, ocular discomfort increased by 1.5 units. At 50 min, discomfort had improved by 1.3 units.

Most subjects with severe dry eye (n = 18) did not experience natural compensation during their exposure to the CAE (Fig. 4). In the severe dry eye population, ocular discomfort increased steadily, reaching a final plateau of 3.9 units by 20 min of CAE exposure, and then remained constant thereafter. However, when the sub-population of subjects with severe dry eye that experienced a symptomatic improvement was evaluated, a mild compensatory improvement of less than 0.5 units was observed at 40 min.

4. DISCUSSION

Ocular discomfort data on subjects exposed to a CAE indicate that the ability of a subject to naturally compensate, and thereby reduce their ocular discomfort, varies between normal subjects and dry eye patients. These data also demonstrate that the time to natural compensation, the magnitude of the response, and the frequency of natural compensation events all vary among normal, moderate dry eye, and severe dry eye subjects.

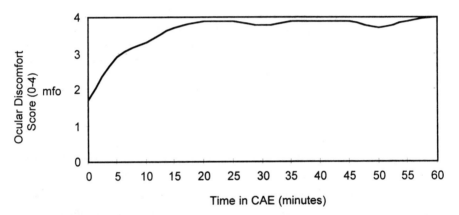

Figure 4. Severe dry eye subject profile of ocular discomfort during exposure to a CAE. Subjects (n = 25) with severe dry eye did not experience natural compensation during a 60 min exposure to the CAE. However, in those who did experience an improvement (n = 7), a mild compensatory effect of < 0.5 unit was observed after 40 min. Mean ocular discomfort at 0 min = 1.7 units, 5 min = 2.9 units, 10 min = 3.3 units, 15 min = 3.7 units, 20 min = 3.9 units, 25 min = 3.9 units, 30 min = 3.8 units, 35 min = 3.9 units, 40 min = 3.9 units, 45 min = 3.9 units, 50 min = 3.7 units, 55 min = 3.9 units, 60 min = 4.0 units.

The concept of natural compensation enhances our understanding of dry eye, and by understanding how natural compensation changes with the severity of dry eye, clinicians can classify dry eye sufferers with greater accuracy, identify more vulnerable subpopulations of dry eye sufferers, and target appropriate remedies more precisely. The role that natural compensation plays in clinical studies that evaluate treatment efficacy also warrants attention.

Normal subjects possess compensatory mechanisms such as blinking, tearing, leaving the irritating environment, that alleviate ocular discomfort and prevent surface damage when an intolerable level of discomfort is experienced.[2,3] However, some mechanisms such as blinking and removal of the irritant, can be controlled.[4] Studies conducted by our group have measured the time to compensation of another compensatory mechanism, reflex tearing, by using various dilutions of ammonia as a nasal stimulation. That study demonstrated that dry eye subjects have a delayed response to nasal stimulation compared to normal subjects (unpublished observations). Therefore, it is likely that variations in reflex tearing capacities among the 3 subject groups in the current study are responsible for the variations in observed natural compensation responses. Compared to the normal subject profile, the profiles for the mild to moderate dry eye subject population and the severe dry eye subject population indicate defective compensatory mechanisms. The response in the normal subject population was rapid, and maintained a consistent level of protection from the ocular irritation induced by the CAE. In mild to moderate dry eye subjects, natural compensation follows a second wave of discomfort, suggesting that these subjects do have a reflex tearing response to the discomfort, but that the compensatory system cannot maintain a consistent level of protection against the irritation. In the severe dry eye subject population, natural compensation was observed late in the exposure period, or, for most, not at all. This indicates that the reflex tearing response is either abnormally delayed, absent, or is inadequate to overcome the deficiencies of the compensatory system. Therefore, natural compensation is an important response that should be considered when evaluating the severity of dry eye, as well as the efficacy of dry eye treatments.

REFERENCES

1. 1997 Eagle Vision and Yankelovich Partners Survey. Cited by: Segrè L. Advanced Technology: Reshaping the Future of Vision Care. Available at: http://www.optistock.com/iis99.htm. Accessed January 18, 2001.
2. Walsh FB and Hoyt WF. *Clinical Neuro-Ophthalmology.* Vol 1. 3rd ed. Baltimore, MD: The Williams & Wilkins Company; 1969: 321–30, 554–5.
3. Tsubota K. Tear dynamics and dry eye. *Prog Retin Eye Res* 1998 Oct; 17(4): 565–96.
4. Tsubota K, Nakamori K. Effects of ocular surface area and blink rate on tear dynamics. *Arch Ophthalmol* 1995; 113: 155–158.

Dry Eye Syndromes: Diagnosis, Clinical Trials and Pharmaceutical Treatment

DIAGNOSIS AND TREATMENT OF THE DRY EYE: A CLINICAL PERSPECTIVE

J. Daniel Nelson

Department of Ophthalmology
HealthPartners Medical Group & Clinics
HealthPartners Research Foundation
Department of Ophthalmology
University of Minnesota
Minneapolis, Minnesota, USA

1. INTRODUCTION

The assumption that dry eye implies a lack of tears has caused clinicians to focus on wetting the dry eye and industry to develop artificial tear products that are retained longer and coat the ocular surface better. Simply categorizing the dry eye as being "dry" has not been very helpful in designing treatments. Thus wetting the dry eye has not proved universally effective. The term keratoconjunctivitis sicca (KCS) implies inflammation. I believe that ignoring the inflammation component of dry eye has created much of the difficulty in the diagnosis and treatment of KCS. I suggest that the key is not the lack of tears or tear production, but the presence or absence of inflammation that is the key to effective diagnosis and treatment. Developing ways to measure, identify, and quantify ocular surface, tear film, and lacrimal gland inflammation will be quite useful, as will new therapies to treat and prevent inflammation.

2. KEY CONCEPTS

Two key concepts should govern the diagnosis and treatment of dry eye. First, the lacrimal gland, meibomian glands, and ocular surface should be considered as a single functional unit.[1] All have androgen receptors and all are innervated. The concept of the lacrimal gland, meibomian glands, and ocular surface forming a single servo-unit requires that each function normally to provide and maintain a normal tear film and ocular surface. In addition, this single functional unit must have some reserve. It must be able to tolerate changes, sometimes extreme, in local and external environmental conditions. When the lacrimal gland secretion diminishes or its composition changes, it affects the ocular surface and meibomian glands. The ocular surface and meibomian glands compensate by increasing mucin and lipid production. Similarly, when the function of the meibomian glands or ocular surface is impaired, the lacrimal gland responds with increased secretion and changes in fluid composition.

There have been many breakthroughs in the treatment of dry eye, including the introduction of sterile lubricants with increased viscosity, non-preserved artificial lubricants, and artificial tears containing electrolytes and bicarbonate.[2,3] The next major breakthrough in present and future therapies will be treating the inflammation that accompanies dry eye. I believe that it is the inflammation that ties symptoms and signs of dry eye together. Thus the second key concept is that therapeutic goals should be directed toward diagnosing and dealing with the effects of inflammation of the ocular surface and adnexia. This also includes treating the effects after inflammation has done its damage.

3. PHYSIOLOGY

3.1. Lacrimal Gland

In response to conjunctival or corneal irritation, reflex lacrimal gland fluid secretion provides irrigation and flushing of the ocular surface. It also provides emergency delivery of growth factors, antimicrobial agents such as IgA, lactoferrin, lysozyme, and vitamin A. Gland function is altered by its destruction due to autoimmune disease.[4] With decreased corneal sensation, there is decreased lacrimal gland secretion.[5] Aging is associated decreasing tear secretion due to periductal fibrosis and stenosis of excretory ducts.[6] Ocular surface scarring, in ocular cicatricial pemphigoid for example, blocks lacrimal gland fluid from reaching the eye. Lacrimal gland denervation causes decrease tear and protein secretion. This suggests that protein secretion is neuronally regulated.[7] Finally, it is suggested that hormonal deficiency causes decreased lacrimal gland secretion.[8]

3.2. Accessory Lacrimal Glands

In normal squirrel monkeys fluid from the accessory lacrimal glands is sufficient to maintain a stable tear film.[9] Absence of the main lacrimal gland does not appear in itself to lead to keratoconjunctivitis sicca. Accordingly, the accessory lacrimal glands may maintain the normal homeostasis of the ocular surface, and the main lacrimal gland may provide the "emergency response" to challenges of ocular surface integrity. Accessory lacrimal gland secretion is altered in cases of destruction due to autoimmune disease, with decreased corneal sensation,[10] and with ocular surface scarring.

3.3. Meibomian Glands

Meibomian glands secrete lipid, which floats on the surface of the tear film/gel and decreases evaporation.[11] Meibomian lipids stabilize the tear film, and the lipid layer keeps external environmental irritants from contaminating the ocular surface. Androgen[12] and estrogen[13] deficiencies can alter meibomian gland secretion. In meibomian gland disease, lipid levels are decresed.[14] Altered lipid level creates an unstable tear film that can cause a four-fold increase in evaporation rate.[15] Blinking is one of the factors responsible for meibomian gland secretion, and disorders that effect blinking cause decreased secretion of lipid. The loss of meibomian glands (meibomian gland drop out) is associated with increased evaporation and increased discomfort.[16] Blepharitis is often found in association with an aqueous tear deficiency. The question arises, "Is blepharitis responsible for the increased inflammation seen in keratoconjunctivitis sicca or does keratoconjunctivitis sicca cause blepharitis?" In at least one report, meibomian gland dysfunction may have contributed to the cause of ocular surface disease seen in Sjogren's syndrome.[16]

3.4. Goblet Cells

The goblet cells are a major source of mucin, which, along with bicarbonate, forms a mucin-bicarbonate gel that covers the ocular surface. Mucin secretion by goblet cells is altered in aqueous tear deficiency due to keratoconjunctivitis sicca[17] and in cicatricial ocular surface disease where there is loss of goblet cells.[18] Epithelial cells also secrete mucin to form the glycocalyx.

3.5. Corneal Epithelium

The corneal epithelium forms a protective barrier between the environment and underlying ocular structures. More importantly, the interaction of the ocular tear film and epithelium forms the major refractive surface of the eye. The function of the epithelial surface is altered by mechanical trauma or inflammation, leading to increased corneal epithelial permeability and epithelial breakdown.

3.6. Blinking

The function of the eyelids in dry eye is often overlooked. Blinking and palpebral fissure width are important in maintaining a healthy ocular surface. Blinking creates a mechanical force, which causes lipid secretion by the meibomian glands and clears debris from the ocular surface and tear film. Compression of the nasal lacrimal sac by eyelid closure pumps tears out of the eye.[19] Without blinking there is no drainage of tears from the eyes.[20] Finally blinking serves to mix and spread tears across the ocular surface.[21]

With increased palpebral fissure width, there is increased exposed ocular surface area that can result in increased evaporation. Decreased blinking has a similar effect,[22] as is seen in thyroid disease, keratoconjunctivitis sicca,[23] and with decreased corneal sensation.

Lagophthalmos, due to seventh nerve paralysis, decreases tear drainage from the eye.[24] Various ocular surface conditions also can affect evaporation rates.[25]

3.7. Neuronal Innervation

Loss of innervation results in decreased secretion of lacrimal gland fluid[7,26] and meibomian gland lipid. In Sjogren's syndrome this is likely related to neuro-decoupling due to focal lymphocytic infiltration in the lacrimal gland. Decreased corneal sensation also results in decreased reflex tearing[1] and meibomian gland secretion. Neuronal innervation is also critical for corneal epithelial wound healing.[27] Inflammation, trigeminal nerve disorders, herpes infections and LASIK[28,29] are associated with decreased corneal sensation.

3.8. Pre-corneal Tear Layer

The pre-corneal tear layer serves as a permeability barrier and is essential for maintaining a smooth, regular optical surface.[30] Some suggest that the tear film is really a mucin-bicarbonate gel. This protective gel is likely similar to that which coats the gastric mucosa. A plausible hypothesis is that bicarbonate, secreted by the corneal epithelium, combines with the mucins secreted by goblet cells to form a protective gel. This mucin bicarbonate gel may measure up to 40 microns in thickness.[31,32] Basal tear secretion and reflex tear secretion likely serve to hydrate the mucin-bicarbonate gel. Disorders involving epithelial cells, goblet cell secretion, meibomian gland secretion, lacrimal gland secretion, and neuronal innervation all effect the integrity of the mucin-bicarbonate gel.

4. ASSESSMENT OF MEASURE FUNCTION

4.1. Lacrimal Gland

The presence of irritant or reflex tearing and emotional tearing suggest that the lacrimal gland is functional. One of the ways to distinguish between loss of lacrimal gland innervation and decreased corneal sensation is by comparing the Schirmer I test with the Schirmer I test *with* nasal stimulation. In KCS, there is decreased corneal sensation associated with decreased Schirmer I test values. However, if the lacrimal gland is not involved, the Schirmer test with nasal stimulation shows relatively normal values. With lacrimal gland inflammation and resultant neuro-decoupling within the gland, the Schirmer test with nasal stimulation values are low or absent. Thus, low values for both Schirmer I test and Schirmer I test with nasal stimulation suggests lacrimal gland inflammation and/or destruction. Tear clearance[33,34] and tear turnover[35,36] are indirect measures of lacrimal gland secretion. The tear film concentrations of lysozyme and lactoferrin are also measures of lacrimal gland secretory function.

4.2. Accessory Lacrimal Glands

Although there is controversy over whether or not a basal tear secretion truly exists,[10] the functional status of the accessory lacrimal glands can be assessed by measurement of basal secretion with a "Schirmer test Test with anesthesia" or fluorophotometry[5] allows assessment of the functional status of these glands. It is likely that normal lacrimation contains some reflex tears.[37]

4.3. Meibomian Glands

Diagnosis of subtle meibomian gland disease and mild blepharitis is difficult. Clinically meibomian gland dropout, orifice plugging, oil deposition, foamy tears, and lid margin telangectasis all suggest blepharitis. Measurement of tear breakup time, which correlates with overall tear film stability, and tear evaporation give indirect assessments of lipid layer function.

4.4. Goblet Cells

The status of goblet cells cannot be determined by a clinical exam. However, the Xeroscope gives an indirect measure of mucin function and goblet cells,[17] while impression cytology gives a direct objective measure of goblet cell density.[18]

4.5. Corneal Epithelium

Clinically, fluorescein staining of the corneal epithelium demonstrates the loss of epithelial cell membrane integrity and epithelial cells. The barrier function of the epithelium is measured by fluorophotometry. Morphological appearance of the epithelium can be evaluated by impression cytology and confocal microscopic analysis.

4.6. Tear Film/Gel

Rose bengal[38–41] and lissamine green[38,42] can demonstrate the intactness of the mucin-bicarbonate gel. When the protective affect of the mucin-bicarbonate gel is absent, staining of ocular surface epithelium results. The Xeroscope[17] and tear breakup time[43] are clinical tests that give indirect measures of tear film stability.

4.7. Neuronal Innervation

The degree of corneal sensation gives a direct, and the Schirmer I test is an indirect measure of ocular surface innervation. The Schirmer I test with nasal stimulation gives an indirect measure of neuronal innervation of the lacrimal gland.

4.8. Eyelid Blinking

The blink rate and maximum blink interval[23] can be measured easily. A decreased blink rate and increased blink interval are seen in keratoconjunctivitis sicca. Tear evaporation rates are indirect measures of the effects of decreased blinking and prolonged blink intervals.

4.9. Tear Clearance

Tear secretion, stability, and evaporation affect tear clearance. As such, this is probably the best overall measure of the lacrimal gland, meibomian gland, and ocular surface functional unit. Measurement of tear clearance is critical in the assessment and the followup of patients with keratoconjunctivitis sicca. Importantly, many factors decrease tear clearance. When suboptimal tear clearance is accompanied by inflammation, a vicious cycle results with increasing inflammation leading to further decreases in tear clearance and the retention of inflammatory cytokines on the ocular surface. Presently, tear clearance is measured by the change in the concentration of fluorescein in the tear film. Decreased tear clearance occurs when there is decreased tear secretion or blinking, increased evaporation, or following punctal occlusion. It can be measured in vivo using fluorophotometric techniques,[35,36] or by using Schirmer strips to collect fluorescein containing tears.[33] More recent innovations involve the comparison the color of fluorescein in the lateral one-third of the lower lid tear film to a standard visual scale.[44] Fluorophotometry, however, is cumbersome to perform. Further, comparison of the colored Schirmer strips to colored standards is difficult, and the amount of staining is dependent on the length of wetting along the Schirmer strip.

5. EFFECTS OF INFLAMMATION

5.1. Main and Accessory Lacrimal Glands

Inflammation of the main and accessory lacrimal glands results in decreased tear secretion due to neuro-decoupling, damage to acinar units and ducts and loss of acinar tissue. With reduced tear secretion, there is decreased tear clearance[34] that results in increased retention of inflammatory cytokines, which are released into the tear film either from the ocular surface epithelium or into lacrimal gland fluid.[45] Finally, with decreased lacrimal gland secretion there is a decreased secretion of regulatory and growth hormones, which are necessary to maintain the health of the ocular surface.[46]

5.2. Meibomian Glands

With inflammation there is abnormal lipid production or a reduction in the output of lipid[13] that results in decreased tear film stability. This leads to increased evaporation[47,12] and increased inflammation of the ocular surface and a resultant diminished tear clearance and retention of inflammatory products. In addition, blepharitis itself is likely to increase ocular surface inflammation.

5.3. Goblet Cells

Inflammation of the ocular surface initially increases the production of mucus. However, this is followed by a decrease in mucin output that results in the formation of ropy mucus and filaments.

5.4. Corneal Epithelium

Inflammation causes an increased permeability of the epithelial cells, which result in increased stimulation of the corneal nerves. The result is symptoms of pain and irritation and eventually fluorescein staining of the cornea.

5.5. Eyelids

With increased ocular surface inflammation, there is reduced blinking due to decreased corneal sensation. As a consequence, tear stability is jeopardized by diminished meibomian gland secretion, and the prolonged blink interval results in greater evaporation rates. Tear clearance is further reduced due to the absence of tear pump action.

5.6. Neuronal Innervation

Inhibited corneal sensation consequent to ocular surface inflammation results in inhibited lacrimal gland fluid and meibomian gland secretion. This reduces tear clearance, exacerbating inflammation.

5.7. Pain

Subjective and objective findings are not well correlated except perhaps in severe aqueous tear deficiency.[48,33,49] Studies show that there is poor correlation between Schirmer test and symptoms and only modest correlation with ocular surface staining and symptoms.[33] I suspect that it is inflammation that gives rise to pain. Early on, there is a loss of epithelial cell membrane integrity and cell to cell attachments which cause increased corneal permeability. Eventually punctate staining of the cornea is seen. Increasing inflammation decouples corneal sensation resulting in reduced tear secretion, decreased blinking, lowered meibomian gland secretion, and increased evaporation. This effectively decreases tear clearance and increases retention of inflammatory cytokines. Tear clearance does correlate with the patient's symptoms.[49,44]

In summary, inflammation incites a cascade of events, which result ultimately in decreased tear clearance and further increase in inflammation associated with the increased retention of inflammatory cytokines. This causes patient symptoms of pain and clinical findings of inflammation. Control and elimination of inflammation in the lacrimal gland, on the ocular surface, and in meibomian glands is a key step in treating the dry eye.

6. DIAGNOSIS

With the above as background, I want to discuss the steps in diagnosing the dry eye. The first step is to ask if inflammation is present. The answer to this question can be obtained through inquiring about patient's symptoms such as burning, foreign body sensation, and excess mucus production. Clinically, injection of the ocular surface and the presence of increased mucin, filaments, and papilla are evidence of an inflammatory process. Punctate erosions or punctate staining of cornea with fluorescein also support the presence of inflammation, as is decreased corneal sensation.

The second step is to determine whether or not the lacrimal and accessory lacrimal glands are functioning properly. If the patient is able to generate irritant or reflex tears and emotional tears, it provides some assurance that the lacrimal gland still has some residual function. The absence of reflex tears but the presence of emotional tears suggests that there may be a neuro-decoupling between the ocular surface and lacrimal gland. The best tools for measuring lacrimal gland function are the Schirmer I test and the Schirmer I test with nasal stimulation. Measurement of tear clearance is also helpful but it is not a direct measurement of lacrimal gland function.

The third step is to determine if there is adequate lacrimal gland secretion to maintain a normal ocular surface and pre-corneal tear layer. The presence of rose bengal or lissamine green staining of the conjunctiva demonstrates an inadequate mucin-bicarbonate layer. The Xeroscope and tear breakup time, give an assessment of the stability of the pre-ocular tear film, and impression cytology gives a morphologic appraisal of the ocular surface. Finally, tear turnover or tear clearance gives an assessment of the overall function of the lacrimal gland ocular surface functional unit.

The fourth step is to determine whether or not blepharitis is present. There are many different techniques to assess lipid and meibomian gland function, but most practical way is to assume that blepharitis is always present. However, this yields no insight as to whether blepharitis the primary cause of inflammation or secondary to it.

In summary, the key to diagnosis is to determine 1) whether or not inflammation is present, 2) whether the lacrimal and accessory lacrimal glands functioning properly, 3) whether there is adequate lacrimal gland and accessory lacrimal gland secretion to maintain a normal ocular surface, and 4) whether blepharitis is present. With these key diagnostic steps in mind, the following new tests would aid clinicians in the diagnosis of these symptoms:

- Quantifying the amount of inflammation and the development of specific inflammatory profiles for each dry eye disease
- More convenient and practical methods for measuring corneal sensation and tear clearance.
- Specific content profiles of lacrimal gland fluid to determine absence or presence of specific components in various ocular diseases.
- Additional tests for measuring the integrity and the thickness of the pre-corneal tear layer.

- Better methods for assessing and quantifying the degree of lipid abnormality, blepharitis, and meibomian gland dysfunction.

7. MANAGEMENT

The priority in management of the dry eye is to control ocular surface inflammation and to increase tear clearance. Inflammation may be treated topically with steroids,[50] cyclosporin A,[51] or other anti-inflammatory agents. Treating blepharitis with tetracycline and local application of heat is helpful. Topical androgen drops may prove helpful in the future.

Increasing lacrimal gland fluid secretion is accomplished by treating lacrimal gland inflammation with systemic immunomodulatory agents such as hydroxychloraquine, cyclophosphamide, and methotrexate. Treating inflammation will also help re-establish neuronal innervation of the lacrimal gland and result in increased lacrimal gland fluid secretion.

Normalizing the tear film can be accomplished by replacing missing growth factors with autologous serum,[46] by replacing missing electrolytes and bicarbonate, and by hydrating the mucin layer. Punctal occlusion can increase the retention of secreted growth factors, but this technique can be detrimental if it increases the retention of inflammatory products. Tear evaporation may be decreased with the application of calcium ointment to the lower lids[53,54] and by tarsorraphy. Restoring the protective mucin bicarbonate gel by increasing mucin secretion may also prove helpful.[2,3] PGY2 stimulators[55] and gefarnate[56] may prove helpful in increasing mucin secretion. Future therapies are likely to focus on topical application of specific missing cytokines, growth and regulatory factors, and on agents that block specific factors that cause or support inflammation.

In summary, the key to diagnosis and management of dry eye is inflammation — recognizing its presence and then treating it. It is likely that our difficulty in diagnosing the more mild forms of KCS is due to the inability to determine the presence of inflammation. It is also likely that our difficulty in treating more severe forms of dry eyes is due to our failure to treat inflammation. Future diagnostic testing and therapy for the dry eye should be aimed at developing methods to determine the presence or absence of inflammation and its underlying causes. With this, specific treatments can be developed that are safer and more effective for our patients.

REFERENCES

1. M.E. Stern, R.W. Beuerman, R.I. Fox, *et al.*, A unified theory of the role of the ocular surface in dry eye, *Adv Exp Med Biol.* 438:643 (1998).
2. D. Lopez Bernal and J.L. Ubels, Artificial tear composition and promotion of recovery of the damaged corneal epithelium, *Cornea.* 12:115 (1993).
3. P.C. Donshik, J.D. Nelson, M. Abelson, *et al.*, Effectiveness of BION tears, Cellufresh, Aquasite, and Refresh Plus for moderate to severe dry eye, *Adv Exp Med Biol.* 438:753 (1998).

4.   D.A. Sullivan, L.A. Wickham, E.M. Rocha, et al., Androgens and dry eye in Sjogren's syndrome, Ann N Y Acad Sci. 876:312 (1999).

5.   M. Goebbels, Tear secretion and tear film function in insulin dependent diabetics, Br J Ophthalmol. 84:19 (2000).

6.   H. Obata, S. Yamamoto, H. Horiuchi, et al., Histopathologic study of human lacrimal gland. Statistical analysis with special reference to aging, Ophthalmology. 102:678 (1995).

7.   M.A. Meneray, D.J. Bennett, D.H. Nguyen, et al., Effect of sensory denervation on the structure and physiologic responsiveness of rabbit lacrimal gland, Cornea. 17:99 (1998).

8.   W.D. Mathers, D. Stovall, J.A. Lane, et al., Menopause and tear function: the influence of prolactin and sex hormones on human tear production, Cornea. 17:353 (1998).

9.   D.Y. Maitchouk, R.W. Beuerman, T. Ohta, et al., Tear production after unilateral removal of the main lacrimal gland in squirrel monkeys, Arch Ophthalmol. 118:246 (2000).

10. A. Jordan and J. Baum, Basic year flow: Does it exist, Ophthalmol. 87:920 (1980).

11. W.E. Shine and J.P. McCulley, Keratoconjunctivitis sicca associated with meibomian secretion polar lipid abnormality, Arch Ophthalmol. 116:849 (1998).

12. E.M. Rocha, L.A. Wickham, L.A. da Silveira, et al., Identification of androgen receptor protein and 5alpha-reductase mRNA in human ocular tissues, Br J Ophthalmol. 84:76 (2000).

13. B. Esmaeli, J.T. Harvey and B. Hewlett, Immunohistochemical evidence for estrogen receptors in meibomian glands, Ophthalmology. 107:180 (2000).

14. N. Yokoi, F. Mossa, J.M. Tiffany, et al., Assessment of meibomian gland function in dry eye using meibometry, Arch-Ophthalmol. 117:723 (1999).

15. J.P. Craig and A. Tomlinson, Importance of the lipid layer in human tear film stability and evaporation, Optom-Vis-Sci. 74:8 (1997).

16. J. Shimazaki, E. Goto, M. Ono, et al., Meibomian gland dysfunction in patients with Sjogren syndrome, Ophthalmology. 105:1485 (1998).

17. S.C. Pflugfelder, S.C. Tseng, K. Yoshino, et al., Correlation of goblet cell density and mucosal epithelial membrane mucin expression with rose bengal staining in patients with ocular irritation, Ophthalmology. 104:223 (1997).

18. J.D. Nelson, V.R. Havener and J.D. Cameron, Cellulose acetate impressions of the ocular surface. Dry eye states, Arch Ophthalmol. 101:1869 (1983).

19. M.G. Doane, Blinking and the mechanics of the lacrimal drainage system, Ophthalmology. 88:844 (1981).

20. W.L. White, A.T. Glover and A.B. Buckner, Effect of blinking on tear elimination as evaluated by dacryoscintigraphy, Ophthalmology. 98:367 (1991).

21. K. Tsubota, Tear dynamics and dry eye, Prog Retin Eye Res. 17:565 (1998).

22. K. Nakamori, M. Odawara, T. Nakajima, et al., Blinking is controlled primarily by ocular surface conditions, Am J Ophthalmol. 124:24 (1997).

23. K. Tsubota and K. Nakamori, Effects of ocular surface area and blink rate on tear dynamics, Arch Ophthalmol. 113:155 (1995).

24. P. Arrigg and D. Miller, A new lid sign in seventh nerve palsy, Ann Ophthalmol. 17:43 (1985).

25. K. Tsubota, S. Hata, Y. Okusawa, et al., Quantitative videographic analysis of blinking in normal subjects and patients with dry eye, Arch Ophthalmol. 114:715 (1996).

26. M.F. Salvatore, L. Pedroza and R.W. Beuerman, Denervation of rabbit lacrimal gland increases levels of transferrin and unidentified tear proteins of 44 and 36 kDa, Curr Eye Res. 18:455 (1999).

27. K. Araki, S. Kinoshita, Y. Kuwayama, et al., [Corneal epithelial wound healing in the denervated eye], Nippon Ganka Gakkai Zasshi. 97:906 (1993).

28. A.J. Kanellopoulos, I.G. Pallikaris, E.D. Donnenfeld, et al., Comparison of corneal sensation following photorefractive keratectomy and laser in situ keratomileusis, J Cataract Refract Surg. 23:34 (1997).

29. R.S. Chuck, P.A. Quiros, A.C. Perez, et al., Corneal sensation after laser in situ keratomileusis, J Cataract Refract Surg. 26:337 (2000).

30. D. Dursan, D. Monroy, R. Knighton, *et al.*, The effects of experimental tear film removal on corneal surface regularity and barrier function, *Ophthalmol.* 197:17541760 (2000).

31. J.I. Prydal, P. Artal, H. Woon, *et al.*, Study of human precorneal tear film thickness and structure using laser interferometry, *Invest Ophthalmol Vis Sci.* 33:2006 (1992).

32. J.I. Prydal and F.W. Campbell, Study of precorneal tear film thickness and structure by interferometry and confocal microscopy, *Invest Ophthalmol Vis Sci.* 33:1996 (1992).

33. S.C. Pflugfelder, S.C. Tseng, O. Sanabria, *et al.*, Evaluation of subjective assessments and objective diagnostic tests for diagnosing tear-film disorders known to cause ocular irritation, *Cornea.* 17:38 (1998).

34. P. Prabhasawat and S.C. Tseng, Frequent association of delayed tear clearance in ocular irritation, *Br J Ophthalmol.* 82:666 (1998).

35. J.D. Nelson, Simultaneous evaluation of tear turnover and corneal epithelial permeability by fluorophotometry in normal subjects and patients with keratoconjunctivitis sicca (KCS), *Trans Am Ophthalmol Soc.* 93:709 (1995).

36. A. Joshi, D. Maurice and J.R. Paugh, A new method for determining corneal epithelial barrier to fluorescein in humans, *Invest Ophthalmol Vis Sci.* 37:1008 (1996).

37. M. Tang, P. Zuure, R. Pardo, *et al.*, Reflex lacrimation in patiets with glaucoma and healthy control subjects by fluorophotometry, *Invest Ophthalmol Vis Sci.* 41:709 (2000).

38. M. Norn, Lissamine green vital staining of cornea and conjunctiva, *Acta Ophthalmol.* 483 (1973).

39. R.P. Feenstra and S.C. Tseng, Comparison of fluorescein and rose bengal staining, *Ophthalmology.* 99:605 (1992).

40. R.P. Feenstra and S.C. Tseng, What is actually stained by rose bengal?, *Arch Ophthalmol.* 110:984 (1992).

41. J. Chodosh, R.D. Dix, R.C. Howell, *et al.*, Staining characteristics and antiviral activity of sulforhodamine B and lissamine green B, *Invest Ophthalmol Vis Sci.* 35:1046 (1994).

42. F.J. Manning, S.R. Wehrly and G.N. Foulks, Patient tolerance and ocular surface staining characteristics of lissamine green versus rose bengal, *Ophthalmology.* 102:1953 (1995).

43. M. Lemp and J. Hamill, Factors affecting tear film breakup in normal eyes, *Arch Ophthalmol.* 89:103 (1973).

44. A. Macri, M. Rolando and S. Pflugfelder, A standardized visual scale for evaluation of tear fluorescein clearance, *Ophthalmol.* 107:1338 (2000).

45. S.C. Pflugfelder, D. Jones, Z. Ji, *et al.*, Altered cytokine balance in the tear fluid and conjunctiva of patients with Sjogren's syndrome keratoconjunctivitis sicca, *Curr Eye Res.* 19:201 (1999).

46. K. Tsubota, E. Goto, H. Fujita, *et al.*, Treatment of dry eye by autologous serum application in Sjogren's syndrome [see comments], *Br J Ophthalmol.* 83:390 (1999).

47. M. Rolando, M.F. Refojo and K.R. Kenyon, Increased tear evaporation in eyes with keratoconjunctivitis sicca, *Arch Ophthalmol.* 101:557 (1983).

48. O.D. Schein, J.M. Tielsch, B. Munoz, *et al.*, Relation between signs and symptoms of dry eye in the elderly. A population-based perspective, *Ophthalmology.* 104:1395 (1997).

49. A.A. Afonso, D. Monroy, M.E. Stern, *et al.*, Correlation of tear fluorescein clearance and Schirmer test scores with ocular irritation symptoms, *Ophthalmology.* 106:803 (1999).

50. P. Marsh and S.C. Pflugfelder, Topical nonpreserved methylprednisolone therapy for keratoconjunctivitis sicca in Sjogren syndrome, *Ophthalmology.* 106:811 (1999).

51. K. Sall, O.D. Stevenson, T.K. Mundorf, *et al.*, Two multicenter, randomized studies of the efficacy and safety of cyclosporine ophthalmic emulsion in moderate to severe dry eye disease. CsA Phase 3 Study Group, *Ophthalmology.* 107:631 (2000).

52. A. Solomon, M. Rosenblatt, L. De-Quan, *et al.*, Doxycycline inhibition of Interleukin-1 in the corneal epithelium, *Invest Ophthalmol Vis Sci.* 41:2544 (2000).

53. D.L. MacKeen, H.W. Roth, M.G. Doane, *et al.*, Supracutaneous treatment of dry eye patients with calcium carbonate, *Adv Exp Med Biol.* 438:985 (1998).

54. K. Tsubota, Y. Monden, Y. Yagi, *et al.*, New treatment of dry eye: the effect of calcium ointment
     through eyelid skin delivery, *Br J Ophthalmol*. 83:767 (1999).
55. J.E. Jumblatt and M.M. Jumblatt, Regulation of ocular mucin secretion by P2Y2 nucleotide receptors
     in rabbit and human conjunctiva, *Exp Eye Res*. 67:341 (1998).
56. M. Nakamura, K. Endo, K. Nakata, *et al.*, Gefarnate stimulates secretion of mucin-like glycoproteins
     by corneal epithelium in vitro and protects corneal epithelium from desiccation in vivo, *Exp Eye Res*.
     65:569 (1997).

DRY EYE SYNDROMES: DIAGNOSIS, CLINICAL TRIALS AND PHARMACEUTICAL TREATMENT–'IMPROVING CLINICAL TRIALS'

M. B. Abelson,[1,2,3] G. W. Ousler III,[3] L. A. Nally,[3] and
T. B. Emory[3]

[1]Harvard Medical School
Boston, Massachusetts, USA
[2]Schepens Eye Research Institute
Boston, Massachusetts, USA
[3] Ophthalmic Research Associates, Inc.
Dry Eye Department
North Andover, Massachusetts, USA

1. GENERAL INTRODUCTION

There is a shift occurring in the field of dry eye bringing us from a time when the condition could only be managed quantitatively to anticipating being able to treat it in the near future qualitatively. Our improved understanding of the underlying etiologies of dry eye syndrome is driving this new therapeutic focus. Research showing that dry eye is not just a factor of decreased tear volume, but rather may consist of deficiencies of various tear components is at the forefront of the field. Thus, we now understand that dry eye, in fact, comprises multiple different disease states. Any two dry eye patients may suffer from different basic conditions. Diagnostic measures that can differentiate between dry eye patients are becoming increasingly important, and will be necessary to properly match the right treatment to the patient. Further, as drug development in dry eye is proceeding in several different pharmacological categories, it will be necessary to design standardized clinical trials according to the mechanism of action of the drug and select for the appropriate patients to include in these studies.

Agents that are capable of reducing ocular discomfort with   greater   efficacy   and

Lacrimal Gland, Tear Film, and Dry Eye Syndromes 3
Edited by D. Sullivan *et al.*, Kluwer Academic/Plenum Publishers, 2002

1079

longer duration compared with existing FDA monograph-based tear substitutes, while promoting healing or preventing damage to the ocular surface, are at hand. This paper will summarize our thoughts on the integration of some of the phenomenon behind dry eye, factors driving future therapies, and specific improvements in clinical trials that will most effectively measure efficacy of these drugs.

2. CURRENT FDA REGULATIONS OF DRY EYE PRODUCTS

No clinical trial has ever demonstrated clinical significance in the relief or treatment of signs and symptoms associated with dry eye. Current therapies are only able to reduce the symptoms (e.g. burning, discomfort) but not the signs (e.g. reduce keratitis, lengthen TFBUT, decrease fluoroscein staining, etc.) associated with dryness of the eye, and do not address any of the underlying causes.

Since 1964 the FDA monograph has recognized the following agents as being safe and effective in the temporary relief of symptoms associated with dryness of the eye:
astringents: locally acting pharmacological agents that, by precipitating protein, help clear mucus from the outer surface of the eye;
demulcents: agents, usually water-soluble polymers, applied topically to the eye to protect and lubricate mucous membrane surfaces and relieve dryness and irritation;
emollients: agents, usually a fat or oil, applied locally to eyelids to protect or soften tissues and to prevent drying and cracking.

These agents, or any combination, may be used in medications for dry eye. Currently, all dry eye products are over-the-counter tear substitutes.
The active ingredients that researchers can use are limited. Some examples include cellulose, gelatin, polyethylene, polysorbate, polyvinyl alcohol, and mineral oil. These ingredients may be used in tear substitutes to reduce bothersome symptoms, but they do not heal damaged tissue or focus on tear quality. Agents that reduce ocular discomfort more effectively for a longer duration and promote healing or prevent damage by virtue of replacing or substituting for the deficient compound of the tear film, such as mucomimetics, anti-evaporatives, anti-inflammatories, secretagogues and improved polymers are within our grasps. But the correct clinical trial must be run to prove their efficacy and safety to the FDA.

3. CLINICAL TRIAL DESIGNS

Environmental trials have been the standard design for dry eye studies. There are several improvements that should be made in order to appropriately study the efficacy of new drugs for dry eye. Variables introduced and need to be accounted for in current environmental designs include: the environment, patient population, clinical endpoints, trial duration, and matching the compound's mode of action to trial design. It is also important to keep in mind that evidence does not suggest the correlation of subjective symptoms and objective signs of dry eye. A successful clinical trial needs to evaluate subjective symptoms as well as objective clinical test while accounting for the aforementioned variables.[1]

3.1. Controlling Environmental Factors

Dry eye syndrome and exposure to an adverse environment are not mutually exclusive. It has been estimated that only 5% of the population is diagnosed with dry eye; however, the percent of individuals who experience ocular discomfort as a result of their environment at one time or another is much greater. People with normal ocular health experience dry eye symptom while in airplanes and dry environments, and while performing a visual task (e.g. personal computer use, reading, or watching television). Adverse environments cause normals to experience temporary periods of dry eye, while people who have mild to moderate dry eye experience a synergistic effect and can manifest signs and symptoms of severe dry eye when exposed to such conditions. Environmental conditions such as climate, time of year, life style, and extent of exposure to the above situations all influence the signs and symptoms a specific patient may experience at any one time. Therefore, when these issues are not controlled for, variation is introduced to environmental trial designs.

However, a new controlled model for studying ocular surface disease has been developed, the Controlled Adverse Environment (CAE). The CAE is an environmental chamber that exacerbates the signs and symptoms of dry eye by very precisely regulating humidity (<5%), temperature (76 ± 6°F), airflow (non-turbulent), lighting conditions (adequate to illuminate environment without causing photosensitivity or minimizing the interpalpebral fissure), and visual tasking (television or PC use). By integrating specific diagnostic equipment, such as digital imaging, fluorophotometry, and eye tracking, the model is able to measure both objective and subjective parameters according to standardized scales. The CAE produces and/or exacerbates the signs and symptoms associated with dry eye in a standardized manner,[2] and allows researchers to standardize trials making them more precise, reliable, predictable, reproducible, and sensible, as well as allowing further investigations into underlying mechanisms of dry eye. In the CAE model, examining patients at baseline and monitoring their response over time while exposed to the CAE, controls for these variables. Such a model is essential to properly determine treatment effects of possible therapies.

3.2. Target Trial Population

When selecting subjects for a study, one must be sure to choose the proper target population. Based on the mode of action of an investigational drug, patients should be enrolled into dry eye trials according to specific underlying mechanisms, such as their specific tear deficiency (lipid, aqueous, mucous, and combinations) and severity of their condition. To accomplish this, methods for properly screening, characterizing, and categorizing patients needs to be identified and implemented in trials. In addition, there are other concomitant conditions that cause dry eye, which should also be recognized and controlled for. Some of these conditions include Sjogren's syndrome, lupus, Stevens-Johnson's Syndrome, pemphigoid, inflammatory disease, other ocular surface disease or lid margin disease.

For example, lid margin metaplasia is characterized by keratinization and breakdown of the lid margin epithelium and gray lines, altering the lipid secretions to drip forward rather than onto the ocular surface. This condition leads to decreased oil release, enhanced tear film evaporation and, consequently, the exacerbation of dry eye.[3] If a patient suffering from metaplasia were enrolled into a trial examining the effects of a secretagogue that increases lipid secretions, it would appear as though the drug was ineffective. Although, in reality, the therapy could be very efficacious and the patient's condition itself produced erroneous results.

Similarly, it has been observed that when patients are exposed to a controlled adverse environment a period of temporary relief in their ocular discomfort may occur which correlates with the severity of their dry eye. This occurrence has been termed "natural compensation." The degree and frequency of natural compensation is greater in both normals and mild to moderate dry eye patients when compared to those patients with moderate to severe dry eye (normals occurred in 10 min, mild to moderate dry eye in 20 min, moderate to severe dry eye >40 min or not at all).[4] This suggests that natural compensation may occur based on a patient's ability to reflex tear, alleviating the ocular surface of discomfort. When screening patients for a clinical trail, it is important to keep this finding in mind. Enrolling a patient who is able to compensate for their ocular discomfort without treatment would adversely influence the clinical evaluation and efficacy of a treatment could not be shown.

Choosing the right target population is a crucial step in designing a clinical trial and must not be overlooked in order to prove a drug efficacious.

3.3. Clinical Endpoints

Researchers in dry eye clinical trials can evaluate many endpoints. These include signs of dry eye such as keratitis and conjunctival staining, rose bengal staining, lissamine green B staining, tear film breakup time (TFBUT), blink rate, Schirmer's test, Zone Quick test, tear meniscus height, impression cytology, osmolality, visual function, Fern testing, and Newtonian rings, as well as symptoms of dry eye including ocular discomfort, dryness, burning, stinging, foreign body sensation, grittiness, and irritation.

By understanding a potential drug's mode of action, one can appropriately choose a clinical endpoint. It is important to have the proper endpoint; a study can be inconclusive and lack significance if the wrong endpoint is being evaluated. A clinical endpoint must be relevant, reproducible, and have an appropriate scale to detect changes the drug will cause. The endpoint(s) chosen must reflect the drug's mode of action. For example measuring changes in tear secretion (e.g. using a fluorophotometer) when evaluating a tear substitute would provide irrelevant data and no changes would be detected. Ocular discomfort, fluorescein staining and TFBUT are endpoints, which can be used to evaluate the effectiveness of tear substitutes, secretagogues, anti-inflammatory agents and tear replacement therapies (e.g. mucomimetics). Tear meniscus height, Schirmer's test and fluorophotometry can be used to show the efficacy of secretagogues.

Sometimes it is necessary to refine a clinical endpoint. Rose bengal has been a standard tool to selectively stain damaged cells on the cornea and conjunctiva. Studies have demonstrated that lissamine green B stains the same cells as rose bengal and that these stained cells are more clearly defined under a slit lamp. Lissamine green B is also non-toxic to living cells and causes less irritation upon instilled compared to rose bengal. Therefore, it has been proposed that lissamine green B is a more tolerable alternative for clinical evaluation of ocular surface damage in the assessment of dry eye. In addition, lissamine green B showed a greater change in staining post-CAE exposure providing valuable data for the assessment of time and condition dependant changes of the ocular surface.[5]

Thus we can see that it is important to choose a clinically significant endpoint, but it is also crucial that these endpoints continue to be evaluated. A researcher must constantly strive to find the very best endpoint, one that is clinically significant and does not alter results due to unwanted side effects or outside influences.

3.4. Standardizing Clinical Evaluations

When evaluating clinical signs and symptoms, many variables are introduced. For instance, symptoms of dry eye are very subjective. Each dry eye patient may use a different descriptive term to express his/her ocular discomfort. Further, physicians have differing methodology for examining the signs of dry eye, thus results can be highly variable. It is therefore imperative that the evaluations of both signs and symptoms be standardized Efforts are presently being pursued to standardize diagnostic tools used in clinical trials as well as controlled models that exacerbate the signs and symptoms of dry eye such as the CAE.

For example, a clinically significant evaluation that, in the past, has been significantly variable is TFBUT. The interval between the last complete blink and the first appearance of a micelle (random dry spot on the ocular surface) is measured as the TFBUT. The TFBUT in patients diagnosed with dry eye is very rapid due to decreased quantities of aqueous tears and conjunctival mucin. Traditionally large quantities of fluorescein have been used to measure TFBUT. This influences the thickness of the tear film, increasing the measured value for TFBUT and its variability. With this traditional technique, TFBUT was determined to be greater that 10 sec in normals and less than 10 sec in dry eye patients. It has recently been shown that well controlled, micro-quantities of fluorescein (5 μl) directly influence the accuracy and reproducibility of TFBUT measurements.[6,7] By using 5 μl of fluorescein new reference values have been established. TFBUT was determined to be greater than 5 sec in normals and less that 5 sec in dry eye patients.[8] In standardizing the procedure of measuring TFBUT, an accurate, reproducible, and precise evaluation is reached.

This precise method of measuring TFBUT led to the discovery of a correlation between ocular discomfort and TFBUT. In a study evaluating 33 subjects diagnosed with dry eye, 73% of the population experienced ocular discomfort within 1 second of TFBUT reported by the examiner. This suggests the possibility of an additional method of measuring tear film stability that would be simple and non-invasive.[9]

Several pieces of new technology have also been employed to capture the proper data needed to standardize the evaluation dry eye conditions. These include, but are not limited to digital imaging, fluorophotometry, and eye tracking.

Digital imaging is a very useful tool for looking at TFBUT and fluorescein staining. By capturing the images digitally and cataloging them on a computer database, one can efficiently record and review the information. High-resolution digital images can be scrutinized at a later date to ensure accuracy and add validity.

Using new technologies in combination also leads to the recognition and validation of new clinical evaluations and endpoints. For example, the use of digital imaging to measure TFBUT and eye tracking to measure blink rate allows us to consider the relationship between blink rate and TFBUT. In an ideal system in which the ocular surface is continually protected, the time to tear film destabilization (TFBUT) should be equal to or greater than the time to the next complete blink. In dry eye states, where there is a decreased TFBUT, the ocular surface is temporarily unprotected resulting in ocular discomfort and the development of superficial punctate keratitis (SPK). We can thus evaluate how well a new agent protects the ocular surface by measuring increases in TFBUT and its correlation to the blink rate interval.[9] When defining a clinically significant increase in TFBUT after a treatment with a dry eye agent, it is important to consider this relationship. With this new, concept protection of the ocular surface can be evaluated.

3.5. Appropriate Trial Duration

When designing a clinical trial, it is crucial to keep in mind that the trial design, including the length of the trial, must match the component's mode of action. Without matching a trial design and duration to the mode of action of the component being evaluated, the results will be inconclusive. When looking at the efficacy of tear substitutes, or polymers, a trial duration would be short term, one to a few days in length. This is because tear substitutes provide relief instantly and their effect is not long lasting. On the other hand, if a secretagogue or anti-inflammatory agent was being evaluated the trial would have to last much longer, possibly weeks, in order to provide sufficient time for the therapy to become efficacious. The mode of action for such drugs is a physiological change and it takes time for the drug to take effect. When designing a clinical trial for tear replacement therapies (e.g. mucomimetics), you can consider implementing trials of both long and short duration.

4. CONCLUSIONS

Further understanding of the underlying physiological causes of dry eye syndrome will direct us towards an era where this condition is treated, rather than only managed. We need to look beyond the current FDA monographed agents and artificial tear substitutes, to agents capable of reducing ocular discomfort with greater efficacy and duration, while promoting healing and preventing damage to the ocular surface. Deficient components of the tear film need to be replaced or mimicked to protect the ocular surface. With all the new compounds being tested, it is paramount that we improve clinical trials. This will lead to proper evaluation of newer agents and the development of products that will provide the necessary protection to the ocular surface. If the ocular surface is properly protected, whether it is by increased aqueous secretion or lengthened TFBUT, then it will allow for its healing. With healing will come the reduction of inflammation associated with the damaged tissue. Consequently, if there is less inflammation, then it will reduce the irritation of the ocular surface, which will make for an improved quality of life.

The improvements that must be accomplished are not limited to the suggestions made by this paper. If the right target population is chosen, with the proper trial design, including duration and clinically significant endpoints, while keeping the drugs mode of action in mind, then clinical trials have a better chance of being successful. They will also accurately, reliably and reproducibly assess the efficacy of new agents.

The CAE is a model that has allowed for a standardized way to determine a trial population and evaluate future treatments in a controlled manner. It has also allowed for investigations into mechanisms underlying the dry eye condition. Continuing to learn about the ocular surface and the physiology behind dry eye while, we also improve upon treatment agents and clinical trials, will lead us towards our common goals: finding a cure for dry eye.

REFERENCES

1. M.A. Lemp, Report of the national eye institute/industry workshop on clinical trials in dry eyes, *The CLAO Journal*. 21(4):221–232 (1995).
2. A. Giovanoni, M.B. Abelson, S. Rosen, and D. Welch, A controlled environment approach to the study of dry eye, *ARVO Abstract Issue*. 39(4):S537 (1998).
3. K. Grant, M.B. Abelson, M.A. George, C. Connell, and A. Giovanoni, Metaplastic changes to the lid margin epithelium–a contributing factor in Dry Eye Syndrome, *ARVO Abstract Issue*. 39(4):S538 (1998).
4. G.W. Ousler III, M.B. Abelson, L.A. Nally, and D. Welch, An evaluation of time to 'natural compensation' or improved ocular discomfort during exposure to a controlled adverse environment (CAE) in normal and dry eye populations, *Cornea*. 19(2):S111 (2000).
5. K.L. Krenzer, G.W. Ousler III, A. Slugg, and A. Nau, A comparison of the staining characteristics of Rose Bengal and Lissamine Green in their evaluation of the clinical signs of dry eye, *ARVO Abstract Issue*. 14(4):S928 (2000).

6.   A.M. Abdul-Fattah, H.N. Bhargva, D.R. Korb, T. Glonek, and J.V. Greiner, Quantitative in vitro comparison of fluorescein delivery to the eye via impregnated strip and volumetric techniques, *ARVO Abstract Issue*. 40(4):S544 (1999).

7.   R. Marquardt, R. Stodtmeiser, and T. Christ, Modification of tear film breakup time test for increased reliability, In: *The Preocular Tear Film in Health, Disease and Contact Lens Wear,* F.J. Holly, ed., Dry Eye Institute. Lubbock, TX 1986.

8.   M.B. Abelson, G.W. Ousler III, L.A. Nally, and K. Krenzer, Alternative reference values for tear film breakup time in normal and dry eye populations, *Cornea.* 19(2):S72 (2000).

9.   M. B. Abelson, K.A. Wilcox, G.W. Ousler III, The coming revolution in dry eye treatment, *Ophthalmology Management.* 5(5): 53 – 60 (2000).

USING OSMOLARITY TO DIAGNOSE DRY EYE: A COMPARTMENTAL HYPOTHESIS AND REVIEW OF OUR ASSUMPTIONS

Anthony J. Bron,[1] John M. Tiffany,[1] Norihiko Yokoi,[2] and Scott M. Gouveia[1]

[1]Nuffield Laboratory of Ophthalmology
University of Oxford, United Kingdom
[2]Prefectural University
Kyoto, Japan

1. INTRODUCTION

1.1. Dry Eye

Two major forms of dry eye exist, aqueous-deficient (ADDE) and evaporative (EDE).[1] A major inflammatory component is also recognised.[2-4] The two forms have global features in common that identify them as dry eye, but do not specify cause. Of these, tear hyperosmolarity arises either as a result of evaporation in the presence of reduced tear flow, or excessive evaporation in the presence of a normal flow rate. Hyperosmolarity m a y contribute to inflammatory processes at the ocular surface by causing epithelial damage, possibly inducing cytokine release from epithelial cells.[5] Inflammation in dry eye is otherwise due to a complex process including autoimmune lacrimal gland and duct destruction, and release of inflammatory mediators from the gland and conjunctiva into tears. Tear instability results from a loss of stabilizing components of tears, such as mucin and lipocalin.[6] Ocular surface damage may result from tear hyperosmolarity, proinflammatory mediators in the tears, and decreased lubrication between the lids and globe. These factors are probably responsible for many symptoms of dry eye.

Lacrimal Gland, Tear Film, and Dry Eye Syndromes 3
Edited by D. Sullivan *et al.*, Kluwer Academic/Plenum Publishers, 2002

1087

1.2. Background Tear Physiology

1.2.1. Volume and Distribution of the Tears. With the eyes open, the tears are distributed in three compartments: the conjunctival sac, preocular tear film and tear menisci. Table 1 gives the volumes calculated by Mishima et al. on the basis of fluorimetry.[7] The distribution of these variables in the normal population and how they change in different positions of gaze are not known.

Table 1. Tear film and meniscus volumes, and area/volume ratios for normal and dry eyes

	Normal			Dry Eye		
	Volume µl	Area mm²	Area/vol. ratio	Volume µl	Area mm²	Area/vol. ratio
Tear Film	1[a]	260[b]	260	NA	260	NA
Meniscus	2.9[a]	29	10	0.67	19.6	29.2

[a]Mishima, 1966; [b]Tiffany et al., 1998; NA, not available.

In the primary position in the normal adult, the width of the palpebral aperture is about 9–11 mm, and the surface area of the exposed globe is about 220–260 mm².[8,9] In full downgaze, the vertical height of the aperture may be reduced to the width of 2 or 3 mm, equivalent to a surface area of 40 mm², while in full vertical upgaze, the palpebral aperture widens to a width of about 15 mm, equivalent to an evaporative surface area of 376 mm².[9] The evaporative surface of the eye may also be increased in the presence of proptosis. Widening of the palpebral aperture, even if the evaporative loss per square millimeter remains stable, leads to an increased total water loss from the eye in unit time.[8]

It is not known what happens to the tear distribution during vertical eye movements, but in the simplest model the upper lid "slides" over the tear layers with limited mixing. Considering a movement from the primary position to one of upgaze, fluid that had been in the sac compartment would be "revealed" in the upper part of the tear film, probably with a delivery of fluid from the upper meniscus into the tear film, which in turn could receive fluid from the sac. The situation would be reversed in downgaze. For the lower lid, such a description would be less appropriate, since vertical movement of the lower lid is limited. This could account for the delayed appearance of dyes in the tear film, after instillation into the lower conjunctival sac.[10]

1.2.2. Flow and Volume. Mishima et al. estimated resting tear flow to be as high as 1.2 µl min⁻¹ [7] or as low as 0.15 µl min⁻¹ [11] in more basal conditions. Nocturnal flow is negligible.[12] A positive relationship exists between tear flow and tear volume,[7] i.e., the lower the flow, the lower the total volume and, as a consequence, the lower the volume of the menisci and tear film. The corollary is that meniscus and film volume would be positively related to total tear volume (Fig. 1A) and in a population meniscus volume in the steady state might be a surrogate for tear volume.

1.2.3. Tear Film Thickness. In the same way, different measurements of tear film thickness have been presented, ranging from 10 μm reported by Mishima,[13] 40 μm by Prydal et al.[14] and 3 μm by King-Smith.[15] Prydal et al.[14] determined the aqueous layer occupied 10 μm of the overall film thickness. Population figures in a large data set are not available.

Wong et al. have proposed a mathematical relationship between meniscus curvature and tear film thickness[16,17] as follows:

$$t = R \ (2.132\{\mu \ U/\sigma\}^{2/3})$$

where t = tear film thickness, R = radius of meniscus curvature, μ = tear viscosity, U = lid velocity and σ = surface tension. From this, the tear film and meniscus volumes can be calculated to be about 12 μm. Tear film thickness would be expected to be a critical determinant of tear molarity at a given evaporation rate from the ocular surface.

1.2.4. Meniscus Volume. Meniscus volume can be estimated from its anterior radius of curvature by multiplying its cross-sectional area by the total lid length (e.g., 50 mm). Meniscus curvature can be measured reproducibly by the non-invasive technique of meniscometry, in which the radius is gauged from the size of the specular image of an illuminated target.[18] Using this technique in a normal adult population of 45.6 ± 21.0 years (mean age ± SD), the radius of curvature of the lower meniscus was estimated at 0.365 mm ± 0.153.[18] If the assumption is made that the exposed surface of the meniscus represents one quarter of a circle in cross-section, then its volume would range from 0.48 μl to 2.88 μl. A higher value for R is reported using slit-image photography, giving 3.19 μl.[19]

2. MEASUREMENT OF TEAR OSMOLARITY

Osmolarity is generally measured using a depression of freezing point method with the Clifton osmometer. Measurement is made on a 10-nl aliquot taken from a 200-nl tear sample from the lower meniscus. The Clifton osmometer has been accepted as the "gold standard" in the diagnosis of dry eye.[20,21] Clinically, a cutoff of 312 mOsm differentiates ADDE and EDE from the normal.[21] Experimentally, hyperosmolarity is associated with surface damage in animal models of ADDE and EDE.[22,23] Corneal epithelial cells are damaged by hyperosmolarity in vitro[23,24] and in vivo.[25]

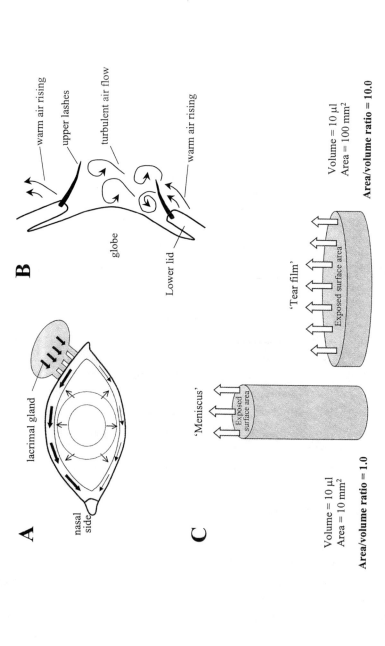

Figure 1. Schematic representations: **(A)** Direction of tear flow in the menisci and tear film; arrow width is proportional to flow rate. **(B)** Turbulent air flow over the ocular surface due to the lashes. **(C)** Differences in area/volume ratio of an idealized tear film and tear meniscus of the same total volume but different exposed surface area. Assuming an equal rate of evaporation unit area (shown by vertical arrows), the concentrating effect of evaporation will be greater where the area/volume ration is higher.

2.1. Factors Influencing Tear Osmolarity

The tears are secreted as an isotonic fluid[26] whose tonicity is increased slightly in the waking state by evaporation. This rise in tonicity accounts for the slight reduction in corneal thickness that occurs following waking due to the osmotic removal of water from the cornea.[27] In general, it may be assumed the hypertonic tears of the preocular film mix with the isotonic lacrimal fluid and are diluted by it. However, the degree of mixing is imperfect and may vary from individual to individual. Some evidence points to a partial segregation of the menisci from the preocular tear film[28] and a suggestion that flow may be greater in the menisci than in the film (see section **3.2** below), with a higher rate of flow in the upper than lower meniscus.

Several factors may contribute to an increase in tear osmolarity in the waking state. Total water loss from the exposed ocular surface increases with increased surface area (as in upgaze or with proptosis). In the same way, a lengthening of the blink interval increases water loss. Since evaporative loss is temperature-dependent, it may vary regionally across the ocular surface, as surface temperature of the open eye is highest at the limbus and lowest over the avascular cornea.[29] Water loss will be influenced by air flow across the ocular surface, being increased per unit area by high air flow, and conversely, will fall with increasing ambient humidity. It is likely that even in the absence of a high air flow over the eye surface, subtle variations in air movement will occur across it due to convective factors (mentioned above) and turbulence induced by the presence of the lashes (Fig. 1B). A faster upward flow of air would be expected from the warmer, lower lid margin than from the cooler corneal surface, for instance. The effect of changing vertical gaze position was mentioned earlier, and it would also be expected a change in head posture could affect evaporation, e.g., a face-down position would tend to retain a more humid ambient air layer over the ocular surface than a face-up position. A quantitative or qualitative inadequacy of the oil layer of the tear film would increase water loss per unit area of the film and is probably the most important cause of evaporative dry eye. Whatever the factors leading to an increase in tear molarity, they will be counterbalanced to some extent by equilibration with the extraocular fluid of the ocular surface tissues. In the presence of increased permeability of the conjunctiva, such as that encountered at the ocular surface in dry eye, equilibration would be expected to be greatly enhanced.

2.2. Assumptions and Inferences of Osmometry

In performing and interpreting osmometry, certain assumptions are made whose tenability is untested. The chief of these is that osmolarity of the meniscus is identical to that of the tear film. In the past, increased tear osmolarity was assumed to relate to the entire exposed ocular surface. Because the tear film is extremely thin, it has so far not been possible to measure its osmolarity directly. Additional assumptions are as follows: evaporation is uniform across the preocular tears, and a uniform flow of the preocular tears occurs so the concentrating effect of evaporation from the ocular surface may be generalized to any point on the tear film. These assumptions imply a rapid mixing between the compartments of the tears. Those researchers measuring evaporation from the ocular surface pay attention to the standardization of environmental temperature, but ignore the possibilities of regional fluctuations in temperature over the ocular surface. Finally, apart from Mishmia and Maurice,[27,30] little attention is paid to considerations of ocular surface permeability.

3. A REVISED CONCEPT OF TEAR HYPEROSMOLARITY IN DRY EYE

3.1. A Compartmental Hypothesis

A review of the literature suggests many of the above assumptions are untrue, and as a result, regional variations exist in tear osmolarity over the ocular surface that may determine the characteristic distribution of ocular surface damage in dry eye. We hypothesise that in the waking state, tear osmolarity is influenced by a compartmentalization of tears at the ocular surface. Consequently, (1) the osmolarity of the meniscal and tear film fluids is not identical, (2) the osmolarity of the tear film always exceeds that of the meniscus, and (3) this differential is amplified in dry eye states.We believe this has important implications for understanding ocular surface damage in dry eye. At present evidence for this hypothesis is indirect.

3.2. Evidence for the Compartmental Hypothesis

MacDonald and Brubaker[28] proposed that negative hydrostatic pressure in the meniscus draws water from the preocular tear film between blinks, leading to a thinning of the tear film compartment. This accounts for the appearance of the "black line" appearing in the fluorescein-stained tear film at the junction between the two compartments shortly after a blink, suggesting they are relatively segregated when the eyes are open. The reduction in tear film thickness that occurs when the eyes are open[15] is presumably due to evaporation and transfer of fluid from the film to the meniscus. Also, although it is usually assumed an even flow or distribution of air over the ocular surface exists, this is unlikely, because differences in temperature of the lids, conjunctiva, limbus and cornea may cause turbulence in the convective flow. Turbulence is likely increased by the presence of the lashes (Fig. 1B).

Studies with instilled fluorescein have shown variable and generally poor mixing of dye.[10] Scintigraphic studies suggest relatively greater volume flow occurs in the upper, compared to the lower, meniscus after drop instillation (C.G. Wilson, personal communication).

Most important, a differential effect of evaporation of the tears from the meniscus and tear film compartments is likely, because of their different area/volume ratios, i.e., even assuming the rate of evaporation per unit area from the meniscus and tear film is identical, the greater surface of the film will ensure a greater concentration of solute under the film than the meniscus, in proportion to the ratio of the area/volume ratios of each compartment (Fig. 1C). Using available data, it is possible to calculate the area/volume ratios (Table 1).

On this basis, even neglecting aspects of differential flow between the two compartments, the concentrating effect of evaporation from the tear film compartment would be estimated about 26 times that from the meniscus. Recalculating these values using the formula of Creech et al.[17] and available data concerning either the radius of meniscus curvature or film thickness, the expected evaporative differential between the tear film and meniscus is modified to about 4-fold (3.82).

In dry eye, overall tear volume is reduced and this is accompanied by a fall in the volume and radius of curvature of the meniscus (radius = 0.25 mm, meniscal volume = 0.67 μl). The figures indicate the influence of changing radius of curvature on the area/volume ratios.

3.3. Implications for Dry Eye

These predictions have implications for dry eye:

• Meniscus samples will tend to underestimate tear hyperosmolarity at the ocular surface and hence the potential for hyperosmotic damage over the ocular surface.

• Since dilution of the meniscus tears by inflowing tears will be greatest when lacrimal function is normal, the discrepancy between meniscus osmolarity and film osmolarity will be greater in evaporative dry eye than in aqueous-deficient dry eye, i.e., for an identical meniscus osmolarity, the molarity of the tear film will be greater in EDE than ADDE.

• The corollary is that in dry eye, for a given tear film osmolarity, the radius of curvature and volume of the tear meniscus will be greater in EDE than ADDE, where reduced lacrimal secretion and the associated loss of tear volume will itself lead to a reduction in meniscus curvature.

• To test these hypotheses, an urgent need exists to develop techniques for the measurement of osmolarity directly in the tear film.

4. CONCLUSIONS

A relative segregation of the tear film and meniscus when the eyes are open, and differential flow rates and area/volume ratios in the tear meniscus and tear film, probably influence the pattern of ocular surface damage occurring in dry eye, tending to accentuate hyperosmolarity in the tear film compared to the meniscus. This is probably compounded by altered convective air flow over the ocular surface related to the presence of the lashes and regional differences in temperature of the lids and exposed ocular surfaces. A given osmolarity in tears collected from the meniscus may have different implications for the ocular surface in ADDE and EDE.

REFERENCES

1. NEI/Industry report (1995).
2. M.E. Stern, R.W. Beuerman, R.I. Fox, J.-P. Gao, A.K. Mircheff, and S.C. Pflugfelder. The pathology of dry eye: the interaction between the ocular surface and lacrimal glands. *Cornea.* 17:584–589 (1998).
3. F. Brignole, P-J. Pisella, M. Goldschild, M. De Saint Jean, A. Goguel, and C. Baudouin. Flow cytometric analysis of inflammatory markers in conjunctival epithelial cells of patients with dry eyes. *Invest Ophthalmol Vis Sci.* 41:1356–1363 (2000).
4. S.C. Pflugfelder. Advances in the diagnosis and management of keratoconjunctivitis sicca. *Current Opinion in Ophthalmology.* 9:50–53 (1998).
5. S.C. Pflugfelder, Z-G. Liu, D. Monroy, D.-Q. Li, M.E. Carvajal, S.A. Price-Schiavi, N. Idris, A. Solomon, A. Perez, and K.L. Carraway. Detection of sialomucin complex (MUC4) in human ocular surface epithelium and tear fluid. *Invest Ophthalmol Vis Sci.* 41:1316–1326 (2000).
6. J. Tiffany and B. Nagyova. The role of lipocalin in determining the physical properties of tears. *This volume.*
7. S. Mishima, A. Gasset, S.D. Klyce Jr., and J.L. Baum. Determination of tear volume and tear flow. *Invest Ophthalmol.* 5:264–276 (1966).
8. K. Tsubota, and K. Nakamori. Effects of ocular surface area and blink rate on tear dynamics. *Arch Ophthalmol.* 113:155–158 (1995).
9. J.M. Tiffany, B.S. Todd, and M.R. Baker. Calculation of the exposed area of the human eye. *Invest Ophthalmol Vis Sci.* 38:S155 (1997).
10. N.M. Sang, and D.M. Maurice. Poor mixing of microdrops with the tear fluid reduces the accuracy of tear flow estimates by fluorophotometry. *Curr Eye Res.* 14:275–280 (1995).
11. W.D. Mathers, and T.E. Daley. Tear flow and evaporation in patients with and without dry eye. *Ophthalmology.* 103:664–669 (1996).
12. R.A. Sack, K.O. Tan, and A. Tan. Diurnal tear cycle: evidence for a nocturnal inflammatory constitutive tear fluid. *Invest Ophthalmol Vis Sci.* 33:626–640 (1992).
13. S. Mishima. Some physiological aspects of the precorneal tear film. *Arch Ophthalmol.* 73:233–241 (1965).
14. J.I. Prydal, P. Artal, H. Woon, and F.W. Campbell. Study of human precorneal tear film thickness and structure using laser interferometry. *Invest Ophthalmol Vis Sci.* 33:2006–2011 (1992).

15. P.E. King-Smith, B.A. Fink, N. Fogt, K.K. Nichols, R.M. Hill, and G.S. Wilson. The thickness of the human precorneal tear film: evidence from reflection spectra. *Invest Ophthalmol Vis Sci.* 41:3348–3359 (2000).

16. H. Wong, I. Fatt, and C.J. Radke. Deposition and thinning of the human tear film. *J Colloid Interface Sci.* 184:44–51 (1996).

17. J.L. Creech, L.T. Do, I. Fatt, and C.J. Radke. *In vivo* tear-film thickness determination and implications for tear-film stability. *Curr Eye Res.* 17:1058–1066 (1998).

18. N. Yokoi, A. Bron, J. Tiffany, N. Brown, J. Hsuan, and C. Fowler. Reflective meniscometry: a non-invasive method to measure tear meniscus curvature. *Brit J Ophthalmol.* 83:92–97 (1999).

19. J.C. Mainstone, A.S. Bruce, and T.R. Golding. Tear meniscus measurement in the diagnosis of dry eye. *Curr Eye Res.* 15:653–661 (1996).

20. R.L. Farris. Tear osmolarity--a new gold standard? In: Sullivan DA, ed. *Lacrimal Gland, Tear Film and Dry Eye Syndromes.* New York: Plenum Press; 1994:495–503.

21. J.P. Gilbard, R.L. Farris, and J. Santamaria, II. Osmolarity of tear microvolumes in keratoconjunctivitis sicca. *Arch Ophthalmol.* 96:677–681 (1978).

22. J.P. Gilbard, S.R. Rossi, K.L. Gray, L.A. Hanninen, and K.R. Kenyon. Tear film osmolarity and ocular surface disease in two rabbit models for keratoconjunctivitis sicca. *Invest Ophthalmol Vis Sci.* 29:374–378 (1988).

23. J.P. Gilbard, S.R. Rossi, and K.G. Heyda. Tear film and ocular surface changes after closure of the meibomian gland orifices in the rabbit. *Ophthalmology.* 96:1180–1186 (1989).

24. J.P. Gilbard, J.B. Carter, D.N. Sang, M.F. Refojo, L.A. Hanninen, and K.R. Kenyon. Morphologic effect of hyperosmolarity on rabbit corneal epithelium. *Ophthalmology.* 91:1205–1212 (1984).

25. A.J.W. Huang, R. Belldegrün, L. Hanninen, K.R. Kenyon, S.C.G. Tseng, and M.F. Refojo. Effect of hypertonic solutions on conjunctival epithelium and mucinlike glycoprotein discharge. *Cornea.* 8:15–20 (1989).

26. J.M. Tiffany. Tears and conjunctiva. In: Harding JJ, ed. *Biochemistry of the eye.* London: Chapman and Hall; 1997:1–15.

27. S. Mishima, and D.M. Maurice. The effect of normal evaporation on the eye. *Exp Eye Res.* 1:46–52 (1961).

28. J.E. McDonald, and S. Brubaker. Meniscus-induced thinning of tear films. *Amer J Ophthalmol.* 72:139–146 (1971).

29. P.B. Morgan, M.P. Soh, N. Efron, and A.B. Tullo. Potential applications of ocular thermography. *Optom Vis Sci.* 70:568–576 (1993).

30. S. Mishima, and D.M. Maurice. The oily layer of the tear film and evaporation from the corneal surface. *Exp Eye Res.* 1:39–45 (1961).

RELIABILITY OF MEASUREMENTS OF TEAR PHYSIOLOGY

Alan Tomlinson,[1] Lee Choon Thai,[1] Marshall G.Doane,[2] and
Angus McFadyen[3]

[1]Department of Vision Sciences
Glasgow Caledonian University
Glasgow, Scotland
[2]Schepens Eye Research Institute
Harvard Medical School
Boston, Massachusetts, USA
[3]Department of Mathematics
Glasgow Caledonian University
Glasgow, Scotland

1. INTRODUCTION

Evaluation of the effects of various interventions on tear physiology, such as contact lens wear or eye drop instillation, often require multiple measurements on different occasions. The effect of the intervention is defined as the change in the physiological parameter from a "baseline" level. Therefore, the measurement of the baseline value is important, and the repeatability of the measurement must be reliable. The repeatability of baseline levels may be affected by changes in the physiological state of the individual, inconsistency in methods used for obtaining measurement, and intra- and inter-diagnostician variability in performing measurements.

The repeatability of tear physiology measurements has been considered by a number of researchers. Most attention has previously focused on repeatability of standard clinical tests for evaluating tear production (the Schirmer strip[1-3] and phenol-red thread test[4,5]), and tear breakup (fluorescein staining[3,6,7]). Statistical analyses of repeatability have also varied, ranging from simple tests such as the application of repeated 't' tests[1] and correlation statistics,[4] to analyses of variance.[2,5] Recently, the reliability and validity of a new subjective questionnaire for the assessment of symptoms of ocular irritation, "The Ocular

Lacrimal Gland, Tear Film, and Dry Eye Syndromes 3
Edited by D. Sullivan *et al.*, Kluwer Academic/Plenum Publishers, 2002

1097

Surface Disease Index" (OSDI)[8] has been evaluated using a variety of sophisticated, statistical analyses.[9] Therefore, the determination of repeatability of the various diagnostic tests depends not only on the factors described above, but also on the methods of statistical analyses used to determine repeatability. Thus, the statistical test employed, as well as its appropriateness for the protocol design is critical.

In this study, three measures of tear physiology, standard clinical tests for tear stability, are studied for their repeatability: a non-invasive tear breakup time measure, a laboratory technique for the measurement of tear film evaporation, and a dynamic method for the observation of tear film structure.

2. METHODS

2.1. Subjects

Twenty subjects, 3 males and 17 females, aged 16 to 45 years, were asymptomatic for dry eye or anterior surface disease, and were not taking medications likely to affect the tear film. Some were contact lens wearers, but on the days that measurements were taken had not been wearing contact lenses for at least 12 h. Measurements of the pre-corneal tear film were performed on all subjects on each of five occasions separated by at least 24 h. Data were collated and subjected to a variety of statistical analyses to determine the repeatability of these "baseline" measurements of pre-corneal tear film physiology.

2.2. Determination of Tear Stability

The time to tear breakup was measured with the HirCal grid.[10] Using this method, three measurements of tear film pre-rupture phase time (TP-RPT) were taken for each subject in a given session, and the average was used as the raw data point.

2.3. Determination of Tear Evaporation

Tear evaporation rate was measured with a modified Servo-Med Evaporimeter (EP-3) (Servo-Med, Kinna, Sweden), which measures the relative humidity and temperature at two points above the corneal surface. The evaporation rate was determined from the vapour pressure gradient between these two points following the method described previously.[11]

2.4. Evaluation of Tear Film Structure

Pre-corneal tear film structure was evaluated dynamically with a thin film interferometer linked to a video camera as described previously.[12,13] The instrament has since been modified such that the thickness of the superficial lipid layer of the pre-corneal tear film can be determined with white light inspection using an incandescent 30 watt white light source with a 1.0 neutral density filter to balance intensity with the green filter. Interference fringe patterns from the pre-contact lens tear film can be viewed with a monochromatic green filter (narrow-bandpass, with 8 nm transmission bandwith at half intensity, centred at a wavelength of 546 nm). The pre-corneal tear film structure was described according to a new modification of a previously reported[12] system of grading, specifically designed for this study (Table 1). In the modified system of grading, the pre-corneal tear film can be described by one of five grades of pre-corneal tear film lipid quality. It is difficult to compare this new grading system with previous systems that used visual observation of interference fringes with slit lamp mounted instrumentation because of differences in the wavelengths of illumination. However the grades 2, 4 and 5 in our system roughly correspond to open meshworks, wave and amorphous, respectively, as described by Guillon.[14]

Table 1. Grading of dynamic, tear film interferometry recordings of pre-corneal tear film lipid quality

Grade	Description
1	"islands" of non-mixing lipid
2	mixed lipid layer with some clusters
3	stable pattern with the appearance of micelles
4	mixed lipid layer with vague pattern
5	mixed lipid layer with no pattern

2.5. Statistical Analyses

The continuous variables, evaporation and tear breakup, were analysed using descriptive statistics, parametric ANOVA, and product moment correlations. The pre-corneal lipid quality scores were analysed using summary statistics, Friedman ANOVA and Spearman's rank correlations. Significance was determined at p values ≤ 0.05 for all inferential tests, and all analyses were performed on either Minitab[15] v12 or SPSS[16] v10.

The overall reliability of data was determined using the Intra-Class Correlation (ICC) theory.[17] Correlation methods simply assess the degree of linearity between two variables, but do not measure agreement in a true sense. Fig. 1 is an illustration of the comparison of two measurements using both correlation and ICC analyses. The correlation of 0.95 is highly significant ($p < 0.001$), and on this basis reliability would be assumed to be very high, but the line does not pass through the origin. For the same data, the ICC calculation yielded a value of 0.42, which is clearly not acceptable at the 0.7 threshold[18] required for excellent reliability (some degree of reliability is indicated by an ICC ≥ 0.4). When assessing pairwise reliability, a simple plot may yield better information than a correlation coefficient.

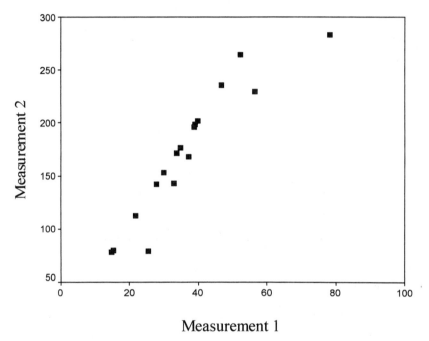

Figure 1. Illustrates the comparison of two measurements by simple correlation (r = 0.95) and Intra-Class Correlation (ICC = 0.42) analyses.

3. RESULTS

This study yielded quantitative (tear evaporation and breakup) and qualitative (pre-corneal lipid quality) data that was analysed by ANOVA and the Intra-Class Correlations techniques.[17] The mean and standard deviations, and ranges of the values obtained for evaporation rate, tear breakup time, and tear film structure for the 20 subjects in this study over five visits are shown in Table 2. The within-subjects variation over 5 visits for these parameters are illustrated in Figs.s 2–4. However, no significant variability between measurements of evaporation and tear breakup time was found between visits by the ANOVA technique (p = 0.084, p = 0.853, respectively). The qualitative data, obtained for pre-corneal tear film lipid quality, was analysed by a Freidman ANOVA, and was also non-significant (p = 0.061). The Intra-Class Correlation analysis indicated some lack of reliability for evaporation (ICC = 0.4) and tear quality (ICC = 0.3) when considering all 5 visits. Comparisons of specific pairs of visits gave ICC of > 0.7 for 2 out of 10 measurements for evaporation and 0 out of 10 for tear quality. Tear breakup time was more reliable (ICC = 0.6) overall, but only 3 out of 10 specific visit comparisons were above 0.7.

The limited reliability found by the ICC test would not have been detected by a frequently applied correlation test, e.g. a product moment correlation test gave r values of up to 0.78 for evaporation, up to 0.87 for TP-RPT and up to 0.55 for tear lipid quality (in the latter case by the Spearman Rank test). A correlation co-efficient greater than 0.44 is significant at the p = 0.05 level for the sample size of this study.

Table 2. Measurements of evaporation rate, tear thinning time (TTT in sec) and pre-corneal tear film lipid quality (PCTFLQ) grading by dynamic interferometry obtained on five separate occasions

Subject[a]	Evaporation rate[a]	TTT[a]	PCTFLQ[a]	PCTFLQ[b]
1	43.27 ± 13.15	13.60 ± 2.90	1.9 ± 0.2	1.5 –2.0
2	20.44 ± 14.83	23.40 ± 8.74	2.5 ± 0.8	1.5 –3.5
3	56.29 ± 17.67	49.67 ± 17.01	3.0 ± 0.6	2.5 –4.0
4	20.54 ± 5.61	8.47 ± 2.67	1.9 ± 0.4	1.5 –2.5
5	24.71 ± 11.52	22.27 ± 11.70	2.0 ± 0.4	1.5 –2.5
6	38.43 ± 22.14	12.73 ± 1.98	2.8 ± 0.9	1.5 –4.0
7	46.28 ± 9.01	13.47 ± 2.61	2.0 ± 0.0	2.0
8	47.65 ± 15.41	18.67 ± 3.67	3.2 ± 0.6	2.5 –4.0
9	35.42 ± 15.05	5.73 ± 1.23	2.8 ± 0.6	2.0 –3.5
10	53.56 ± 18.23	13.53 ± 4.18	2.4 ± 1.1	1.5 –4.0
11	23.01 ± 9.30	66.93 ± 23.20	1.9 ± 0.2	1.5 –2.0
12	61.46 ± 14.70	8.13 ± 0.99	2.0 ± 0.0	2.0
13	53.43 ± 12.59	9.93 ± 2.44	2.8 ± 0.9	2.0 –4.0
14	44.27 ± 13.37	17.27 ± 9.30	3.0 ± 0.8	2.0 –4.0
15	47.24 ± 17.31	43.93 ± 31.35	1.7 ± 0.6	1.0 –2.0
16	44.78 ± 22.25	12.93 ± 5.00	2.4 ± 0.7	2.0 –3.5
17	43.55 ± 20.22	13.60 ± 2.44	2.1 ± 0.4	1.5 –2.5
18	31.07 ± 13.97	30.13 ± 27.82	2.8 ± 0.8	2.0 –3.5
19	35.62 ± 11.03	5.40 ± 1.12	3.4 ± 0.9	2.5 –4.5
20	9.88 ± 7.91	39.00 ± 3.23	2.8 ± 0.9	1.5 –4.0

[a]mean ± sd units; [b]range

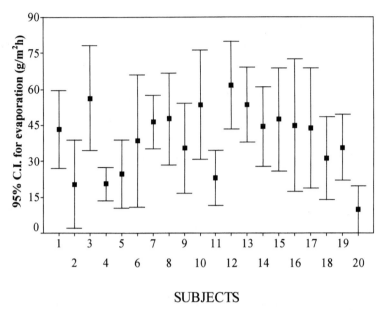

Figure 2. Within-subject variation in measurements of tear evaporation over 5 visits.

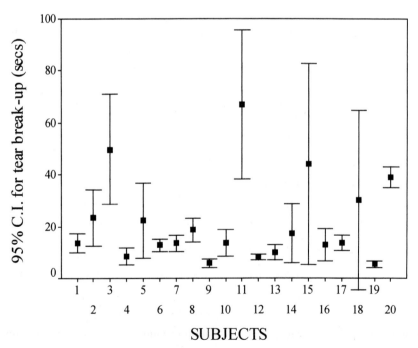

Figure 3. Within-subject variation in measurements of tear breakup time over 5 visits.

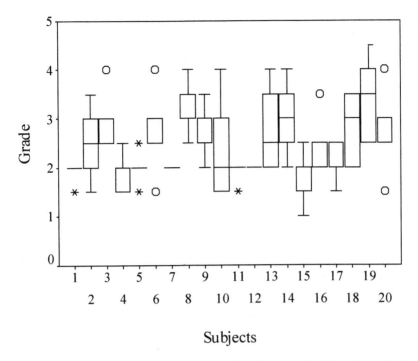

Figure 4. Box plots of within-subject variation in grading of pre-corneal lipid quality over 5 visits.

The intra-observer reliability of our new tear grading system was evaluated in this study. Two separate intra-observer comparisons of interferometry video recordings of pre-corneal tear film quality were made on all five visits. ICC scores for these comparisons were between 0.61 and 0.79. No inter-observer reliability of the new system was possible in this protocol.

4. DISCUSSION

The measurement of tear breakup time was found to be repeatable in this study when data were analysed by an ANOVA and Intra-Class Correlation analyses. This is consistent with the results of a similar analysis carried out by Cho et al.[6] It is also consistent with the less rigorous statistical analysis of this measurement carried out by Little and Bruce.[4] Others have also reported an acceptable reliability in breakup time measurements when a well-defined technique of examination was employed.[3,7,19]

Tear evaporation and pre-corneal tear film lipid quality assessment did not vary significantly in this study when data were analysed by the ANOVA and correlation methods. However, there is some indication that limited reliability exists between these measurements when the Intra-Class Correlation technique is applied. To our knowledge, no previous measurements exist in the literature of the repeatability of either tear evaporation or pre-corneal tear film lipid quality. Whether or not the measurements of evaporation or tear lipid quality are repeatable depends on the method of statistical analysis employed. Our view is that while a degree of reliability is indicated by the correlation and ANOVA techniques, the strictest criterion for repeatability, the Intra-Class Correlation, indicates a limited repeatability. The conclusion that these measurements have limited repeatability, may in part explain why there have been relatively few reports in the literature that have found correlations between objective measurements of tear physiology.[20-22] Poor repeatability may be the basis for the difficulties in finding correlations between subjective reports of symptoms of dry eye.[23] Perhaps the limited repeatability of the three measurements of tear physiology in this study is reassuring to the researcher given the physiological potential for great volatility in the tear system,[24] and instrument and observer variabilities in measurement.

There is a requirement in tear research to take repeated measurements of tear parameters on several occasions in order to study the effects of different interventions, e.g., contact lenses[25-27] and eye lubricants.[28,29] A "wash-out" period between measurements is also required, which prevents measurements of several aspects of tear physiology from being taken on the same occasion. Given the limited repeatability of baseline values for some parameters, the effect of any intervention on a tear physiology measurement must be determined by comparison to a baseline value obtained on the same measurement occasion. The use of such comparative measurements avoids many of the problems of variability inherent in the consideration of absolute values. This is because any variability in the initial baseline value for a given tear measurement at each occasion may affect the recorded value in the presence of an intervention. It is surprising that the tear literature does not contain more reports of repeatability for measurement of tear physiology. Certainly any new techniques developed for the description of physical or chemical parameters of the tear film must be evaluated for their repeatability.

REFERENCES

1. S. Patel, J. Farrell, H. Bevan, Repeatability and variability of Schirmer Test, *Optician.* 194(5122):12 (1987).
2. P. Cho, M. Yap, Schirmer Test II: A clinical study of its repeatability, *Optom. Vis. Sc.* 70:157 (1993).
3. A. Shapero, S. Merin, Schirmer test and breakup time of the tear film in normal subjects, *Am J. Ophthalmol.* 88:752 (1979).
4. S.A. Little, A.S. Bruce, Repeatability of the phenol-red thread and tear thinning time tests for tear film function, *Clin. Exper. Optom.* 77:64 (1994).
5. P. Cho, The cotton thread test: brief review and a clinical study of its reliability on Hong Kong/Chinese, *Optom. Vis. Sci.* 70:804 (1993).
6. P. Cho, B. Brown, I. Chan, R. Conway, M. Yap, Reliability of the tear breakup time technique of assessing tear stability and the locations of tear breakup in Hong Kong/Chinese, *Optom. Vis. Sci.* 69:879 (1992).
7. M.A. Lemp, J.R. Hamill, Factors affecting tear film breakup in normal eyes, *Arch. Ophthalmol.* 89:103 (1973).

8.   J.G. Walt, M.M. Rowe, K.L.Stern, Evaluating the functional impact of dry eye: the Ocular Surface Disease Index, *Drug Inf. J.* 31:1436 (1997).

9.   R.M. Schiffman, M.D. Christianson, G. Jacobsen, J.D. Hirsch, B.L. Reis, Reliability and validity of the Ocular Surface Disease Index, *Arch. Opthalmol.* 118:615 (2000).

10. N. Hirji, S. Patel, M. Callendar, Human tear film (pre-rupture phase time) (TP-RPT) a non-evasive technique for evaluating the pre-corneal tear film using a Noval keratometer mire, *Ophthal. Physiol. Opt.* 9:139 (1989).

11. G.R. Trees, A. Tomlinson, Effect of artificial tear solutions and saline on tear film evaporation, *Optom. Vis. Sci.* 67:886 (1990).

12. M.G. Doane, M.E. Lee, Tear film interferometry as a diagnostic tool for evaluating normal and dry eye tear film, *Ad. Exp. Med. Biol.* 438:297 (1998).

13. M.G. Doane, An instrument for invivo tear film interferometry, *Optom. Vis. Sci.* 66:383 (1989).

14. J.P. Guillon, Abnormal Lipid Layers, *Adv. Exp. Med. Biol.* 438:309 (1998).

15. Minitab Inc., 3081 Enterprise Drive, State College, PA 16801–3008, USA.

16. SPSS Inc., 233 South Wacker Drive, Chicago, II 60606–6307, USA.

17. P.E. Shrout, J.L. Fleiss, Intraclass correlations: uses in assessing rater reliability, *Psychol. Bull.* 86:420 (1979).

18. J. Nunnally, *Psychometric Theory*, McGraw Hill Book Company, NY, (1978).

19. R. Marquardt, R. Stodtmeister, T. Christ, Modification of tear film breakup time test for increased reliability, in: *The Pre-Ocular Tear Film in Health, Disease, and Contact Lens Wear*, Dry Eye Institute, Lubbock, (1986).

20. J.P. Craig, Tear physiology in the normal and dry eye, PhD Thesis, Glasgow Caledonian University, Glasgow, (1995).

21. J.P. Craig, A. Tomlinson, Importance of the lipid layer in human tear film stability and evaporation, *Optom. Vis. Sci.* 74:8 (1997).

22. W.D. Mathers, Model of ocular tear film function, *Cornea.* 15:110 (1996).

23. M.A. Lemp, Report of the National Eye Research/Industry Workshop on clinical trials in dry eye, *CLAO J.* 21:221 (1995).

24. R.E. Records, Tear film, in: *Physiology of the Eye and Visual System*, R.E. Records, ed., Harper and Row, Hagerstown MD, (1997).

25. F.J. Holly, Tear film physiology and contact lens wear II. Contact lens tear film interaction. *Am J. Optom. Physiol. Opt.* 58:331 (1981).

26. A. Sharma, E. Ruckenstein, Mechanism of tear film rupture and its implications for contact lens tolerance, *Am J. Optom. Physiol. Opt.* 62:246 (1985).

27. A. Tomlinson, T.H. Cedarstaff, Tear evaporation and the human eye: the effects of contact lens wear, *J. Brit. Contact Lens Assn.* 5:141 (1982).

28. E.I. Pearce, A.Tomlinson, K.J. Blades, H.K. Falkenberg, B. Lindsay, C.G. Wilson, Effects of an oil and water emulsion on tear evaporation rate, *Adv. Exp. Med. Biol.* (in press 2001)

29. G.R. Trees, A. Tomlinson, *Optom. Vis. Sc.* Effect of artificial tear solutions and saline on tear film evaporation, *Optom. Vis. Sci.* 67:886 (1990).

USE OF CEVIMELINE, A MUSCARINIC M1 AND M3 AGONIST, IN THE TREATMENT OF SJÖGREN'S SYNDROME

Robert I. Fox

Allergy and Rheumatology Clinic
Scripps Memorial Hospital and Research Foundation
La Jolla, California, USA

1. INTRODUCTION

Sjögren's syndrome (SS) is characterized by symptoms of dryness of the eyes and mouth associated with lymphoid infiltrates of the lacrimal and salivary glands. Confusion exists about the criteria for diagnosis of this disease,[1] but the frequency, according to a newly proposed international criterion that requires either a characteristic minor salivary gland biopsy or the presence of an antibody against SS-A/SS-B antigen,[2] is about 0.5% of adult women. SS may exist with other autoimmune disorders (secondary SS associated with rheumatoid arthritis, systemic lupus erythematosus or scleroderma), or as a primary condition. Primary and secondary SS are systemic autoimmune disorders, although the most recognizable manifestations are the ocular and oral components of the disease.

The symptoms of ocular dryness result from increased friction as the upper lid transverses the surface of the orbital globe.[3] The epithelial cells lining the ocular surface and the inner side of the eyelid contain transmembrane mucins that facilitate a low friction gliding motion.[4] The sensation of dryness reflects increased friction as the eyelid traverses the globe, or as the tongue moves over the buccal mucosa.[5] The tear film is a bilipid layer containing water, lipids and nutrients including proteins, hormones and growth factors partly derived from the plasma.[6] In a similar manner, the symptoms of dry mouth are derived from the need for the mucous membranes of the buccal mucosa and tongue to slide with low friction during deglutition and speech.

The perceptions of "dry eyes" and "dry mouth" in SS are derived from signals originating from a neural functional unit.[7]. The ocular surface is heavily innervated with afferent nerves leading to an area of the midbrain called the "lacrimatory" nucleus.[8] This nucleus also receives neural signals from higher cortical centers, and is the presumed target for the reversible dryness associated with centrally acting drugs that have "anti-cholinergic side effects," such as tricyclic antidepressants, or blood pressure medications such as clonidine. Similarly, afferent nerves from the buccal mucosa travel to an adjacent area in the midbrain called the salivatory nucleus, which also receives inputs from higher cortical centers. After the signals from the periphery and the higher cortical center are "integrated" in the midbrain, efferent signals are sent to the blood vessels, primarily via adrenergic nerves, and the lacrimal/salivary glands via cholinergic nerves. The major neurotransmitter of the cholineric signals is acetylcholine, which binds to a family of different receptors (described below), while adrenergic efferent pathways use norepinephrine as their neurotransmitter.

In SS, the secretory responses of lacrimal and salivary glands in response to symptoms of dryness are deficient.[9] The reasons for these deficiencies are likely multifactorial, but do involve a decrease in the number of secretory units, i.e., acini and ducts. Lymphocyte participation in the death of acinar/ductal cells via perforin/granzyme[18] and Fas/Fas L pathways[19] has been demonstrated. However, a common misconception about SS is that the glands are totally destroyed; therefore, the disease is commonly viewed as analogous to insulin deficiency after destruction of pancreatic islet cells in type I diabetes. In fact, only about half the acinar or ductal structures are destroyed in patients with long-standing symptoms of severe dryness.[10,11] This raises the question of why the residual acinar/ductal cells are dysfunctional. Since the characteristic laboratory finding in SS is the presence of focal lymphocytes in the salivary gland,[11] it is likely the presence of lymphocyte-derived cytokines, including interleukin-1 and TNF-α, plays a role in inhibiting the release of secretions in response to neurotransmitters.[12] This inhibition of secretion has been demonstrated by exposing cholinergic ganglia to these cytokines in vitro.[13,14] Recently, the potential role of antibodies against muscarinic M3 receptors and the action of cytokines such as TNF-a to diminish the response of the muscarinic receptor to stimulation by agonists.[15–17,20] Another mechanism of glandular dysfunction in SS may involve the release metaloproteinases[21–23] that digest cell matrices required for optimal cell function.[24] As a result of diminished tear and saliva secretion, the ocular and oral mucosal surfaces develop the characteristics of a chronic wound, resulting in further release of pro-inflammatory cytokines that perpetuate the problem.[25–28]

Thus, since SS involves an interesting pathologic combination of immune, exocrine and neurochemical processes, a better understanding of the pathogenesis of the aberrant immune responses resulting in the release of cytokines and metalloproteinases could lead to improved therapeutic approaches. However, improvement of SS symptoms can also be achieved by functional optimization of the residual secretory units. This review explores the rationale for using M1 and M3 muscarinic receptor agonists in the treatment of SS, with specific emphasis on two agents, pilocarpine and cevimeline, recently approved by the U.S. Food and Drug administration.

2. MUSCARINIC RECEPTORS IN SS

The expression of muscarinic receptors (mACR) of the m1 and m3 subtypes in lacrimal and salivary glands is of particular interest in SS. Although the m5 receptor has been cloned and is structurally similar to m1/m3, its function remains unclear, and expression in the secretory glands has not been detected.[29,30] Therefore, since little expression of m2 or m4 receptors exists in salivary or lacrimal glands, it is likely the symptoms of SS are mediated through m1 and m3 receptors.

Activation of m1 receptors may alter the apoptotic response of glandular cells. This is important in SS because increased apoptosis is induced by either the granzyme/perforin[18] or Fas/Fas ligand mediated pathways.[19] Even in the absence of apoptotic death, activation of cell surface Fas (CD95) interferes with m3- or growth factor-induced responses such as calcium ion flux.[20] Conversely, the stimulation of m1 receptors interferes with apoptotic pathways by influencing the levels of bcl2/bcl-x.[20] Thus, treatment with m1 and m3 agonists may help preserve glandular survival in an inflammatory micro-environment.

IgG obtained from sera of primary Sjögren's syndrome patients (pSS-IgG) binds to M3 ACR in rat exorbital lacrimal glands,[15] and this binding is attenuated by pre-incubation of the pSS-IgG with a synthetic peptide corresponding to the second extracellular loop of M3 ACR. pSS-IgG also binds irreversibly to muscarinic acetylcholine receptors, displacing the specific cholinergic antagonist QNB. Moreover, these antibodies trigger intracellular signals coupled to M3 muscaric cholinoceptors, such as nitric oxide synthase activation and cGMP production. Both these pSS-IgG effects mimic carbachol action, and are abrogated by specific muscarinic antagonists, such as 4-DAMP. Chronic interaction between these autoantibodies and lacrimal gland muscarinic acetylcholine receptors could lead to tissue damage through nitric oxide release. Antibodies from SS patients also inhibit contraction in the smooth muscle contraction assay, demonstrating they are M3-muscarinic receptor antagonists.[17] In addition, NOD mice, a model for human SS, lose secretory function after infusion of pSS-IgG.[38,39] Furthermore, when submandibular glands are exposed to pSS-IgG with cytokines in vitro, only a small cytokine effect is observed, suggesting pSS-IgG inhibits the actions of acetylcholine.[40] However, the reactivity of SS sera with human M3 ACR or with biopsies of SS lacrimal or salivary gland acinar cells has not been studied.

The generation of second messengers in response to M1/M3 stimulation may involve activation of MAP kinases,[41] which may activate the same pathways as those activated by cytokines and growth factors.[42] Thus, inflammation in SS glands may result in "cross talk" between second messengers at multiple levels that dampen or prevent efficient glandular secretion responses.

3. CEVIMELINE FOR TREATMENT OF SS

Cevimeline is a muscarinic agonist and a structurally constrained analog of acetylcholine.[43,44] It increases salivary secretion in the murine MRL/lpr model of SS[45] and in a 6-week, randomized, double-blind, placebo-controlled trial involving 75 SS patients with a mean age of 53.6 years, 30 mg tid. cevimeline produced statistically significant global improvements in salivary symptoms and flow rates (Fig. 1 and Table 1).[46] Indeed, 76% of the patients in the 30-mg tid group reported global improvement in their dry mouth symptoms compared to 35% in the placebo group (Table 1 and Fig. 1A). Increasing the dose to 60 mg tid produced no additional benefit. Regular use of the drug at 30 mg significantly reduced associated oral discomfort, nocturnal fluid ingestion and the need for saliva substitutes. No serious adverse effects were noted among the study population, although one commonly observed adverse effect was excessive sweating. It should be noted patients with significant cardiovascular, pulmonary or gastrointestinal disease, and the patient with narrow-angle glaucoma, were excluded from these cevimeline studies. Although not approved by the FDA for the indication of dry eyes, symptomatic improvement of ocular symptoms was observed (Fig. 2 and Tables 2, 3). Currently, exfoliative cytologic studies of the conjunctiva are in progress to obtain support for the clinical symptoms of improvement to confirm an initial study performed by Tsubota in Japan (Figs. 2, 3). In comparison, Table 4 shows previous studies on the benefit of pilocarpine in improving symptoms of oral dryness.

Table 1. Statistical significance of treatment comparisons for salivary flow. Change from baseline to endpoint

Comparison	SB95US01	SB96US02	SB96US04
Overall			
Placebo, 30 mg, 60 mg	Significant		
Placebo, 15 mg, 30 mg		Significant	Significant
Between Treatments			
15 mg vs Placebo		Not Significant[1]	Not Significant
30 mg vs Placebo	Significant	Significant	Significant
60 mg vs Placebo	Significant		

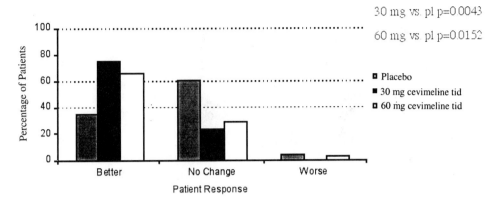

Figure 1. Cevimeline in Sjogren's Syndrome: Improve in oral symptoms.

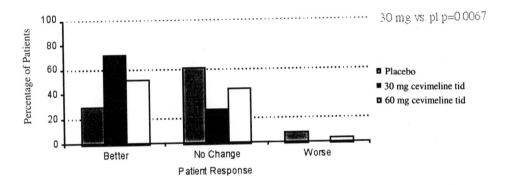

Figure 2. Cevimeline dry eye data.

Table 2. Summary of Ocular Dryness VAS

Comparison	SB95US01	SB96US02
Overall		
Placebo, 30 mg, 60 mg	Significant	
Placebo, 15 mg, 30 mg		Significant
Between Treatments		
15 mg vs Placebo		Not Significant
30 mg vs Placebo	Significant	Significant
60 mg vs Placebo	Significant	

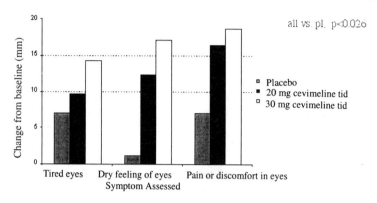

Figure 3. Cevimeline: Dry eye data.

Table 3. Cevimeline: Summary of measures

Efficacy Parameter	Cevimeline tid		
	15 mg	30 mg	60 mg
Subjective Measures			
Patient Global Evaluation of Dry Mouth		Significant	Significant
Patient Global Evaluation of Dry Eyes		Significant	
Patient Global Evaluation of Overall Dryness		Significant	Significant
Individual Symptoms of Dry Mouth	Significant[1]	Significant[2]	Significant[3]
Individual Symptoms of Dry Eyes	Significant[4]	Significant[5]	Significant[6]
Objective Measures			
Salivary Flow	Significant[7]	Significant	Significant
Lacrimal Flow	Significant	Significant[8]	
Use of Artificial Saliva			
Use of Artifical Tears	Significant	Significant	
Fluid Intake			

Table 4. Summary of randomized double-blind placebo-controlled studies demonstrating the clinical efficacy of oral muscarinic agonists on salivary gland secretion and symptomatic relief of xerostomia

Reference	N[a]	Diagnosis	Treatment Regimen	Salivary Flow Rate	Improvement in Symptoms
Fox et al. (547)	6	SS	PC 5 mg bid	increased	improved 5/6
Fox et al. (48)	31	SS	PC 5 mg tid	increased	improved 27/31
Greenspan (49)	12	XRT	PC 5 mg qid	increased	improved 9/12
Johnson (50)	191	XRT	PC 5-10 mg tid	increased	improved 102/191
Vivino (51)	373	SS	PC 5 mg qid	increased	improved
Fox (52)	196	SS	CEV 30 mg tid	increased	improved

[a]number of patients

4. COMPARISON OF PILOCARPINE AND CEVIMELINE

Direct double-blind clinical comparisons of pilocarpine and cevimeline in SS patients are not yet available. However, a comparison of salivation induced by both agents in rats and dogs demonstrated intravenously infused pilocarine induced a marked, though short-lasting, effect, whereas cevimeline-induced salivation lasted about 2-fold longer than pilocarpine. The cevimeline-induced plasma level that resulted in a salivation increase of 0.4 ml/min or greater was maintained for over 90 min compared to less than 20 min for pilocarpine.[47] In normal rats, stimulated salivary flow rates predominantly involved muscarinic M3 receptors, although M1 receptor stimulation played a significant role.[48] A cevimeline-induced increase in salivary secretory rate was also noted in the MRL/lpr mouse model of SS by Iwabuchi.[49] Interestingly, the glandular atrophy that occurs in untreated mice was partially reversed by cevimeline treatment. This protection of atrophy has not been reported in pilocarpine-treated murine models of SS.[50]

A comparison of the capacity of cevimeline and pilocarpine to bind to rat salivary gland muscarinic receptors has been reported (Fig. 3).[47] Specific uquinnuclididinyl benzilate (^{3}H-QNB) binding to purified membrane fractions was concentration dependent, saturable and inhibited by atropine. Scatchard plot analysis of the saturation curve revealed a single ^{3}H-QNB binding component with a Kd of 22 pM and a maximal binding capacity of 60 fmol/mg protein. The competitive inhibition curve indicated a Ki (μM) of 1.2 for cevimeline and 4.6 for pilocarpine. In an extension of these studies, CHO cells were transfected with human M1 to M4 receptors (Human Biology, Inc), and the response to PI and cAMP response was measured. Results showed a 10-fold preferential binding of cevimeline or pilocarpine to M3 receptors (0.003 mM) compared to M2 receptors (0.01 mM).[46] Cevimeline demonstrated greater M1 agonist activity than pilocarpine by various assay methods.[51] This may be important since cevimeline-induced signaling through the M1 receptor could potentially prevent apoptosis (or glandular dysfunction) by blocking caspase activation through phosphoinositide 3-kinase- and MAPK/ERK-independent pathways.[52] Also, cevimeline could still deliver stimulatory signals through the M1 receptor, even in the presence of putative antibodies that block M3 receptors.

5. CONCLUSION

The production of saliva or tears requires the integration of neural, vascular and exocrine function, so it is not surprising the disturbance in salivary and lacrimal production seen in Sjögren's syndrome is now believed to involve aberrant immune-mediated neuro-hormonal events. The use of muscarinic agonists such as pilocarpine and cevimeline , which act through stimulation of M3 and M1 receptors, has been a valuable therapeutic option for the relief of the dry mouth and eyes accompanying SS. In addition to stimulation of secretory function, stimulation of these receptors may help prevent apoptosis and optimization of function of the residual glandular cells.

REFERENCES

1. Fox, R., *Classification Criteria for Sjogren's syndrome.* Rheumatic Disease Clinics of North America: Current Controversies in Rheumatology, 1994. 20: p. 391–407.
2. Vitali, C. and S. Bombardieri, *The Europian classification criteria for Sjogren's syndrome (SS): proposal for modification of the rules for classification suggested by the analysis of the receiver operating characteristic (ROCS) curve of the criteria performance.* J. Rheum, 1997. 24 (supplement): p. S18.
3. Nelson, J.D., *Diagnosis of keratoconjunctivitis sicca.* Int Ophthalmol Clin, 1994. 34(1): p. 37–56.
4. Watanabe, H., *et al., Human corneal and conjunctival epithelia produce a mucin-like glycoprotein for the apical surface.* Invest Ophthalmol Vis Sci, 1995. 36(2): p. 337–44.
5. Fox, R.I., M. Stern, and P. Michelson, *Update in Sjogren syndrome [In Process Citation].* Curr Opin Rheumatol, 2000. 12(5): p. 391–8.
6. Lemp, M.A., *Tear film: new concepts and implications for the management of the dry eye.* Trans New Orleans Acad Ophthalmol, 1987. 35: p. 53–64.
7. Fox, R.I., *et al., Evolving concepts of diagnosis, pathogenesis, and therapy of Sjogren's syndrome.* Curr Opin Rheumatol, 1998. 10(5): p. 446–56.
8. Stern, M.E., *et al., A unified theory of the role of the ocular surface in dry eye.* Adv Exp Med Biol, 1998. 438: p. 643–51.
9. Andoh, Y., *et al., Morphometric analysis of secretory glands in Sjögren's syndrome.* Am Rev Respir Dis, 1993. 148(5): p. 1358–62.
10. Daniels, T.E. and J.P. Whitcher, *Association of patterns of labial salivary gland inflammation with keratoconjunctivitis sicca. Analysis of 618 patients with suspected Sjogren's syndrome.* Arthritis Rheum, 1994. 37(6): p. 869–77.
11. Kovacs, L., *et al., Impaired microvascular response to cholinergic stimuli in primary Sjogren's syndrome.* Ann Rheum Dis, 2000. 59(1): p. 48–53.
12. Main, C., P. Blennerhassett, and S.M. Collins, *Human recombinant interleukin 1 beta suppresses acetylcholine release from rat myenteric plexus.* Gastroenterology, 1993. 104(6): p. 1648–54.
13. Quirion, R., *et al., Growth factors and lymphokines: modulators of cholinergic neuronal activity.* Can J Neurol Sci, 1991. 18(3 Suppl): p. 390–3.
14. Bacman, S.R., *et al., Human primary Sjogren's syndrome autoantibodies as mediators of nitric oxide release coupled to lacrimal gland muscarinic acetylcholine receptors.* Curr Eye Res, 1998. 17(12): p. 1135–42.
15. Waterman, S.A., T.P. Gordon, and M. Rischmueller, *Inhibitory effects of muscarinic receptor autoantibodies on parasympathetic neurotransmission in Sjogren's syndrome.* Arthritis Rheum, 2000. 43(7): p. 1647–54.
16. Kong, L., *et al., Fas and Fas ligand expression in the salivary glands of patients with primary Sjogren's syndrome.* Arthritis Rheum, 1997. 40(1): p. 87–97.
17. Liu, X.B., *et al., G-protein signaling abnormalities mediated by CD95 in salivary epithelial cells.* Cell Death Differ, 2000. 7(11): p. 1119–1126.
18. Tominaga, M., *et al., Expression of metalloproteinase-2 (gelatinase A) in labial salivary glands of patients with Sjogren's syndrome with HTLV-I infection.* Clin Exp Rheumatol, 1999. 17(4): p. 463–6.
19. Azuma, M., *et al., Suppression of tumor necrosis factor alpha-induced matrix metalloproteinase 9 production by the introduction of a super-repressor form of inhibitor of nuclear factor kappaBalpha complementary DNA into immortalized human salivary gland acinar cells. Prevention of the destruction of the acinar structure in Sjogren's syndrome salivary glands.* Arthritis Rheum, 2000. 43(8): p. 1756–67.

20. Laurie, G.W., *et al.*, *Immunological and partial sequence identity of mouse BM180 with wheat alpha-gliadin.* Biochem Biophys Res Commun, 1995. 217(1): p. 10–5.

21. Pflugfelder, S.C., *et al.*, *Conjunctival cytologic features of primary Sjogren's syndrome.* Ophthalmology, 1990. 97(8): p. 985–91.

22. Jones, D.T., *et al.*, *Alterations of ocular surface gene expression in Sjogren's syndrome.* Adv Exp Med Biol, 1998. 438: p. 533–6.

23. Marsh, P. and S.C. Pflugfelder, *Topical nonpreserved methylprednisolone therapy for keratoconjunctivitis sicca in Sjogren syndrome.* Ophthalmology, 1999. 106(4): p. 811–6.

24. Liu, Z., *et al.*, *Increased expression of the type 1 growth factor receptor family in the conjunctival epithelium of patients with keratoconjunctivitis sicca.* Am J Ophthalmol, 2000. 129(4): p. 472–480.

25. Baumgold, J., *et al.*, *Comparison of second-messenger responses to muscarinic receptor stimulation in M1-transfected A9 L cells.* Cell Signal, 1995. 7(1): p. 39–43.

26. Laurie, G., *et al.*, *BM180: a novel basement membrane protein with a role in stimulus secretion coupling by lacrimal acinar cells.* Am. J. Physiol, 1996. 39: p. C1743–C1750.

27. Wu, A., *et al.*, *Effect of tumor necrosis factor alpha and interferon gamma on the growth of a human salivary gland cell line.* J. Cell Physiol, 1994. 161: p. 217–226.

28. Wu, A.J., *et al.*, *Modulation of MMP-2 (gelatinase A) and MMP-9 (gelatinase B) by interferon-gamma in a human salivary gland cell line.* J Cell Physiol, 1997. 171(2): p. 117–24.

29. Cheresh, D.A., J. Leng, and R.L. Klemke, *Regulation of cell contraction and membrane ruffling by distinct signals in migratory cells.* J Cell Biol, 1999. 146(5): p. 1107–16.

30. Eva, C., J.L. Meek, and E. Costa, *Vasoactive intestinal peptide which coexists with acetylcholine decreases acetylcholine turnover in mouse salivary glands.* J Pharmacol Exp Ther, 1985. 232(3): p. 670–4.

31. Yamamoto, H., *et al.*, *Detection of alterations in the levels of neuropeptides and salivary gland responses in the non-obese diabetic mouse model for autoimmune sialoadenitis.* Scand J Immunol, 1997. 45(1): p. 55–61.

32. Humphreys-Beher, M.G. and A.B. Peck, *New concepts for the development of autoimmune exocrinopathy derived from studies with the NOD mouse model.* Arch Oral Biol, 1999. 44 Suppl 1: p. S21–5.

33. Robinson, C.P., *et al.*, *Transfer of human serum IgG to nonobese diabetic Igmu null mice reveals a role for autoantibodies in the loss of secretory function of exocrine tissues in Sjogren's syndrome.* Proc Natl Acad Sci U S A, 1998. 95(13): p. 7538–43.

34. Dawson, L.J., S.E. Christmas, and P.M. Smith, *An investigation of interactions between the immune system and stimulus- secretion coupling in mouse submandibular acinar cells. A possible mechanism to account for reduced salivary flow rates associated with the onset of Sjogren's syndrome.* Rheumatology (Oxford), 2000. 39(11): p. 1226–33.

35. Wotta, D.R., *et al.*, *M1, M3 and M5 muscarinic receptors stimulate mitogen-activated protein kinase.* Pharmacology, 1998. 56(4): p. 175–86.

36. Lu, G., *et al.*, *Tumor necrosis factor-alpha and interleukin-1 induce activation of MAP kinase and SAP kinase in human neuroma fibroblasts.* Neurochem Int, 1997. 30(4–5): p. 401–10.

37. Fisher, A., *et al.*, *(+-)-cis-2-methyl-spiro(1,3-oxathiolane-5,3') quinuclidine (AF102B): a new M1 agonist attenuates cognitive dysfunctions in AF64A-treated rats.* Neurosci Lett, 1989. 102(2–3): p. 325–31.

38. Fisher, A., *Therapeutic strategies in Alzheimer's disease: M1 muscarinic agonists.* Jpn J Pharmacol, 2000. 84(2): p. 101–12.

39. Iga, Y., *et al.*, *(+/-)-cis-2-methylspiro[1,3-oxathiolane-5,3'-quinuclidine] hydrochloride, hemihydrate (SNI-2011, cevimeline hydrochloride) induces saliva and tear secretions in rats and mice: the role of muscarinic acetylcholine receptors.* Jpn J Pharmacol, 1998. 78(3): p. 373–80.

40. Fox, R., *et al.*, *Randomized, placebo controlled trial of SNI-2011, a novel M3 muscarinic receptor agonist, for the treatment of Sjogrenn's syndrome.* Arth Rheum, 1998. 41 (supplement): p. S288.
41. Masunaga, H., *et al.*, *Long-lasting salivation induced by a novel muscarinic receptor agonist SNI-2011 in rats and dogs.* Eur J Pharmacol, 1997. 339(1): p. 1–9.
42. Iwabuchi, Y. and T. Masuhara, *Sialogogic activities of SNI-2011 compared with those of pilocarpine and McN-A-343 in rat salivary glands: identification of a potential therapeutic agent for treatment of Sjorgen's syndrome.* Gen Pharmacol, 1994. 25(1): p. 123–9.
43. Iwabuchi, Y., M. Katagiri, and T. Masuhara, *Salivary secretion and histopathological effects after single administration of the muscarinic agonist SNI-2011 in MRL/lpr mice.* Arch Int Pharmacodyn Ther, 1994. 328(3): p. 315–25.
44. Nagler, R.M., D. Laufer, and A. Nagler, *Parotid gland dysfunction in an animal model of chronic graft-vs-host disease.* Arch Otolaryngol Head Neck Surg, 1996. 122(10): p. 1057–60.
45. Fisher, A., *et al.*, *Selective signaling via unique M1 muscarinic agonists.* Ann N Y Acad Sci, 1993. 695: p. 300–3.
46. Leloup, C., *et al.*, *M1 muscarinic receptors block caspase activation by phosphoinositide 3-kinase- and MAPK/ERK-independent pathways.* Cell Death Differ, 2000. 7(9): p. 825–33.
47. Fox, P.C., *et al.*, *Pilocarpine treatment of salivary gland hypofunction and dry mouth (xerostomia).* Arch Intern Med, 1991. 151(6): p. 1149–52.
48. Greenspan, D. and T.E. Daniels, *Effectiveness of pilocarpine in postradiation xerostomia.* Cancer, 1987. 59: p. 1123–1125.
49. Johnson, J.T., *et al.*, *Oral pilocarpine for post-irradiation xerostomia in patients with head and neck cancer [see comments].* N Engl J Med, 1993. 329(6): p. 390–5.
50. Vivino, F.B., *et al.*, *Pilocarpine tablets for the treatment of dry mouth and dry eye symptoms in patients with Sjogren syndrome: a randomized, placebo-controlled, fixed-dose, multicenter trial. P92-01 Study Group [In Process Citation].* Arch Intern Med, 1999. 159(2): p. 174–81.
51. Fox, R., *et al.*, *Randomized, placebo controlled trial of SNI-2011, a novel M3 muscarinic receptor agonist, for the treatment of Sjogrenn's syndrome.* Arth Rheum, 1998. 41 (supplement): p. S288.

DRUG DEVELOPMENT ISSUES IN PHARMACOLOGICAL TREATMENTS FOR DRY EYE

Gary D. Novack [1]

Pharma•Logic Development, Inc.
San Rafael, California, USA

1. INTRODUCTION

Several pharmacological approaches for the treatment of dry eye are currently being developed by pharmaceutical firms, including attenuation of the immune response, hormonal modulation, and enhanced mucin release. Since an overview of the procedural aspects of ophthalmic drug development, i.e. Investigational New Drug Exemption (IND), Clinical Phases 1 through 3, New Drug Application (NDA), etc.. [1-2], as well as the financial implications of this development for glaucoma drugs[3] were presented previously, this review focuses on investment issues in the development of pharmacological treatments for dry eye. Development cost and timing, and the market potential of a putative pharmacological treatment for dry eye, were evaluated.

2. METHODS AND RESULTS

As of mid-October 2000, there has been no approval of topical pharmacological treatments for dry eye in the United States, Europe or Japan. Thus, currently there is no precedent for a development model. However, there are at least two drugs in late phase development for dry eye.[4-6] I assumed sponsors of these drugs conducted their trials in conjunction with discussions with regulatory agencies, and the published studies were used in preparation of this estimate.

---

[1] The author serves a consultant to a number of firms developing dry eye treatments. He also owns stock in two companies developing or marketing ophthalmic pharmaceuticals.

Shown in **Table 1** are the estimates of the sample sizes for each phase of clinical research. Early stage trials were assumed to be double-masked, paired comparison trials (randomized on eye), with the latter trials being double-masked, vehicle-controlled trials (randomized on subject), using the benefits of unequal randomization to increase the population of patients exposed to the investigational medication. Efficacy sample sizes were based upon power of 80% or more to detect a 0.5 to 1.0 grade change (0 to 3 scale) with a standard deviation of 1.0 gradewith $\alpha = 0.05$. One interpretation of International Conference on Harmonisation (ICH) requirements is that there must be 1500 patients exposed to the drug in the anticipated dosage regimen. Thus, a supplemental sample is listed for this requirement.

Table 1. Clinical study plan for developing cost estimates

Trial	Pts[a]	# on Rx[b]	$/pt[c]	Extended[d]	120% Admin.[e]	Total	150%[f]
				$US K			
P1	50	35	2	100	120	220	330
P2	250	125	2.5	625	750	1,375	2,063
P3	600	402	3.5	2,100	2,520	4,620	6,930
ICH Plus	1500	1,005	3.5	5,250	6,300	11,550	17,325
Total		1,567					26,647

[a]Number patients in trial; [b]Number taking developmental drug; [c]Cost per patient; [d]$ X number of patients; [e]Administrative costs at 120% of extended; [f]Additional administrative costs of 50%

Patient costs are based upon the author's experience in the payments made to investigators for studies of similar type. The cost to administer the study (plan, monitor, quality assurance, data manger, analyze write up) is estimated at 150%, again, based upon the author's experience. An additional 50% was estimated to cover costs of clinical supplies and manufacturing scale-up, additional toxicology studies, and other project costs. Using these estimates, the cost of the development program would be approximately $27 million (US year 2000 dollars).

The time for each phase of development may be highly variable.[3] The estimates for each clinical phase used for this analysis are shown in Table 2. The total time from investigational new drug (IND) filing to new drug application (NDA) filing was estimated at 43 months. The total time required for review of a NDA was estimated at 12 months. This is based upon a review of the past decade of intervals for approved ophthalmic drugs at the U.S. Food and Drug Administration.[7,8]

The value of the lubricant market in key worldwide markets in 1996 was $236 million (IMS Americas, Plymouth Meeting, PA). However, in the same way as there is no approved product to use for estimate of development costs and timing, there is no current market from which to estimate the value of a new product. Thus, market potential was estimated based upon the size of the patient population who might use this drug, the penetration of the new product into this population, and the price.

Table 2. Development times used for estimate

Phase	Time (months)
IND	1
Phase 1a	3
Phase 1b	3
Phase 2	12
Phase 3	18
NDA preparation	6
NDA Review	12
Total	55

The prevalence of dry eye disease has been evaluated in epidemiological studies.[9,10] Using the figures from these studies and the U.S. population, it is assumed that approximately 10 million patients in the U.S. with mild to moderate dry eye would be candidates for a new pharmacological treatment for dry eye. Based upon input from experienced marketing colleagues, it was assumed that the total penetration would be 30%, and that it would take four years after product launch to reach this maximum penetration. For the revenue assumption, it was assumed the price would be $1.50 per day from the manufacturer . For return on investment calculations, five years of sales were assumed, with a residual value at year 6 of 3-times sales. There was no adjustment for inflation. When evaluating the market potential based upon these estimates, it was estimated that the total market for pharmacological treatment of mild to moderate dry eye could be approximately $542 million per year in the U.S. at year 5.

Two types of financial measures were calculated: net present value (NPV) and internal rate of return (IRR). NPV was sufficiently positive at 5%, 8%, 10% and 40% cost of financing. IRR was 188%.

3. DISCUSSION

This analysis showed a very positive financial reward for the development of a pharmacological treatment for dry eye. For relatively risky investments such as novel technology, the NPV at the high rate of 40% is used as a "hurdle". The investment in developing a pharmacological treatment for dry eye would be positive even at this high hurdle rate. The analysis was positive enough that even with a more pessimistic estimate (e.g., 25% fewer patients, more competition, slower market penetration, etc.), it would still be very positive.

There are several caveats to consider in this analysis. First, this analysis does not include the discovery effort. It assumes that innovative scientists have researched the mechanisms of dry eye, synthesized and evaluated the novel compound in preclinical studies. Second, this analysis does not include the time and resources in getting the drug to IND stage, nor does it include production (typically 10% of product price) or marketing costs. Third, there are a host of variables in clinical research and the regulatory review process that may substantially affect the development time, costs, and especially risks. For example, enrollment rates may be slower than planned due to an intense allergy season, or a general issue with the Institutional Review Board unrelated to your product may put your study on hold. Fourth, this assumes that all costs would be balanced against a U.S. market. It may be that for relatively small incremental costs, the drug may be approved for marketing in Japan and Europe, where substantial additional revenues might be incurred. Finally, this is a preliminary analysis, and many more factors should be addressed.

In summary, if there was a successful pharmacological treatment for dry eye, its high market potential approaches those of other chronic treatment ophthalmic drugs (e.g., glaucoma), and assuming a firm could manage cash flow during the development, it would be a worthwhile investment.

REFERENCES

1. G. D. Novack, The development of new drugs for ophthalmology Am J Ophthalmol 114, 357 (1992).
2. G. D. Novack, Ophthalmic drug development: Procedural considerations J Glaucoma 7, 202 (1998).
3. G. D. Novack, Financing new drug development in ophthalmology J Glaucoma 9, 195 (2000).
4. K. Sall, O. D. Stevenson, T. K. Mundorf, and B. L. Reis, Two multicenter, randomized studies of the efficacy and safety of cyclosporine ophthalmic emulsion in moderate to severe dry eye disease. CsA Phase 3 Study Group Ophthalmology 107, 631 (2000).
5. D. Stevenson, J. Tauber, B. L. Reis, and The Cyclosporin A Phase 2 Study Group, Efficacy and safety of cyclosporin A ophthalmic emulsion in the treatment of moderate to severe dry eye disease: A dose-ranging, randomized trial Ophthalmology 107, 967 (2000).
6. M. V. Mundasad et al., Ocular Safety of INS365 Ophthalmic Solution: A P2Y$_2$ Agonist, In Healthy Subjects J Ocul Pharmacol Ther 17, 173 (2001).
7. G. D. Novack, Review times for ophthalmic new drug applications (NDAs) at the U.S. Food and Drug Administration (FDA) Am J Ophthalmol 126, 122 (1998).
8. G. D. Novack, Update on regulatory review intervals for ophthalmic New Drug Applications at the U.S. Food and Drug Administration Am J Ophthalmol 130, 664 (2000).
9. O. D. Schein et al., Prevalence of dry eye among the elderly Am J Ophthalmol 124, 723 (1997).
10. S. E. Moss, R. Klein, and B. E. K. Klein, Prevalence of and Risk Factors for Dry Eye Syndrome Arch Ophthalmol 118, 1264 (2000).

ALTERNATIVE REFERENCE VALUES FOR TEAR FILM BREAK UP TIME IN NORMAL AND DRY EYE POPULATIONS

Mark B. Abelson,[1,2,3] George W. Ousler III,[1] Lauren A. Nally,[1] Donna Welch,[1] and Kathleen Krenzer[1]

[1]Ophthalmic Research Associates
Dry Eye Department
North Andover, Massachusetts, USA
[2]Schepens Eye Research Institute
Boston, Massachusetts, USA
[3]Harvard Medical School
Boston, Massachusetts, USA

1. INTRODUCTION

Over the past decade our understanding of dry eye has greatly increased, but the number affected by the condition has also risen. It is estimated 1 of 5 Americans suffers from dry eye[1] and about 7–10 million currently require artificial tear preparations to treat its symptoms.[2] The rise in dry eye may be due to several factors, including adverse environments, increased visual tasking (e.g., personal computers and television) and the more frequent use of systemic antihistamines. The increased prevalence has led to the need for an even better understanding of the disease and its therapeutic approaches. Consequently, tremendous improvements have occurred in the accuracy and preciseness of the diagnostic tools used to evaluate the condition.

One of the most widely used clinical tests is tear film breakup time (TFBUT). The study of tear film stability began in 1876 with Decker's work.[3] We have since learned from Lemp's studies that the precorneal tear film forms a continuous, protective layer over the corneal surface between blinks. As blinking is prevented, this film begins to break up, creating random dry eye spots (micelles). A weakened or absent tear film results in ocular discomfort, photophobia, forced blinks and ultimately keratitis.[4] The interval between the last complete blink and the first appearance of a micelle is measured as TFBUT and is

Lacrimal Gland, Tear Film, and Dry Eye Syndromes 3
Edited by D. Sullivan *et al.*, Kluwer Academic/Plenum Publishers, 2002

1121

rapid in dry eye patients. This decrease may be due to decreased quantities of aqueous tears and conjunctival mucin. Reduced conjunctival mucin, reflected by a decrease in conjunctival goblet cell density or an increase in tear surface tension, will cause the decreased TFBUT seen in many dry eye patients.[5]

TFBUT is a global criterion of dry eye and currently the only routine clinical test available to measure tear film stability despite controversy over its validity.[5] The differences in TFBUT values may in part be due to variations in the procedures for measuring TFBUT, including the quantity of fluorescein used when performing the test.[3] The amount of fluorescein directly influences the accuracy and reproducibility of TFBUT measurements and microquantities increase accuracy and reproducibility.[6-8] This study investigates whether the use of 5 μl of fluorescein to evaluate TFBUT will produce results different from the standard indices for normal patients (> 10 sec) and those diagnosed with dry eye (< 10 sec), as reported by the 1995 NEI/Industry Workshop on Clinical Trials in Dry Eyes.[5]

2. METHODS

Participants recruited included 100 patients with normal ocular health and 100 with a reported history of dry eye. They consented to the study and were queried about their medical and ocular history as well as concomitant medications. Baseline ophthalmic examinations were conducted on all subjects, including a visual acuity assessment and slit-lamp biomicroscopy.

TFBUT was then performed according to the following procedures: 5 μl of non-preserved fluorescein sodium 2% were instilled with a micropipette into the inferior temporal bulbar conjunctiva of each eye. Patients were instructed to blink several times without squeezing their lids together, ensuring an even distribution of the fluorescein, and then stare directly ahead without blinking. The patient's lids were not held open or manipulated. The tear film was examined through a cobalt blue light at a magnification of X16 while being recorded by a digital video imaging system equipped with a Wratten No. 12 barrier filter. A stopwatch was used to measure the interval between the last complete blink and the first appearance of an expanding micelle. Three consecutive measurements of TFBUT were recorded per eye and an average was taken. In cases in which one time was an outlier, the measurement was discarded and a fourth taken. TFBUT, visual acuity assessment and slit-lamp biomicroscopy were repeated after the measurements of tear film breakup were taken.

3. RESULTS

All participants completed the study, and no changes were observed in any of the safety parameters (visual acuity assessment and slit lamp-biomicroscopy examination) for

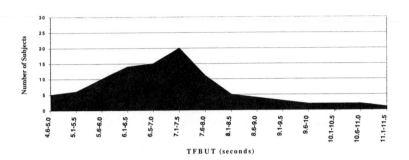

Figure 1. TFBUT in 100 normal subjects after application of 5 μl of 2% fluorescein.

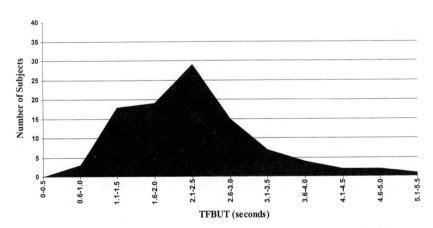

Figure 2. TFBUT in 100 subjects with a history of dry eye after application of 5 μl of 2% fluorescein.

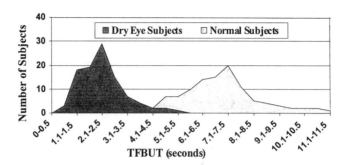

Figure 3. TFBUT in 100 normal subjects and 100 patients with a reported history of dry eye using 5 μl of fluorescein.

either the normal or the dry eye group. The mean TFBUT in patients with normal ocular health was 7.1 sec. The longest measurement was 11.4 sec and the shortest 4.7 sec. Fig. 1 shows TFBUT distribution.

For patients with a history of dry eye, the mean TFBUT was 2.2 sec with a range of 0.9 – 5.2 sec. Fig. 2 shows TFBUT distribution for this group.

On average, patients with normal ocular health have TFBUT > 5 sec (mean = 7.1 sec) while those diagnosed with dry eye have TFBUT < 5 sec (mean = 2.2 sec) (P < 0.0001) (Fig. 3).

4. DISCUSSION

The data suggest that traditional reference values of TFBUT for patients with normal ocular health (> 10 sec) and for those diagnosed with dry eye (< 10 sec) may not apply when using well-controlled microquantities of fluorescein. The accuracy and reproducibility of TFBUT depend on the amount of fluorescein instilled in the eye. The addition of fluorescein can create tear film instability that may result in an inaccurate measurement of TFBUT.[9] Therefore, the technique used in measuring TFBUT must be standardized to compare results. Varying amounts of fluorescein (ranging from 20 to 40 _l), which have previously been used, may lengthen TFBUT, affecting clinical assessments of tear film stability. The revised reference values for TFBUT could provide a more physiological assessment of tear film breakup, therefore warranting consideration and further validation.

REFERENCES

1.   1997 Eagle Vision and Yankelovich Partners Survey. Cited by: Segrè L. Advanced Technology: Reshaping the Future of Vision Care. Available at: http://www.optistock.com /iis99.htm. Accessed January 18, 2001.

2.   Kaswan R. Cyclosporin drops: a potential breakthrough for dry eyes. In: *Research to Prevent Blindness Science Writers Seminar*. New York: Research to Prevent Blindness; 1989:18–20. Cited by: Oden NL, Lilienfeld DE, Lemp MA, Nelson JD, Ederer F. Sensitivity and specificity or a screening questionnaire for dry eye. *Lacrimal Gland, Tear Film, and Dry Eye Syndromes 2*. ed Sullivan DA, Dart DA, Meneray MA. New York: Plenum Press; 1998:807–820.

3.   Cho P, Brown B. Review of the tear break up time and a closer look at the tear break up time of Hong Kong Chinese. *Optom Vis Sci*. 1993;70(1):30–8.

4.   Lemp MA, Dohlman CH, Holly FJ. Corneal desiccation despite normal tear volume. *Annals of Ophthalmology*. 1970; 284: 258–261.

5.   Lemp MA. Report of National Eye Institute/Industry Workshop on clinical trials in dry eyes. *CLAO J*. 1995;21:221–232.

6.   Marquardt R, Stodtmeiser R, Christ T. Modification of tear film breakup time test for increased reliability. In: Holly FJ, ed. *The Preocular Tear Film in Health, Disease and Contact Lens Wear*. Lubbock, Texas: Dry Eye Institute, 1986:57–63.

7.   Korb DR, Finnemore VM, Herman JP, et al. A new method for the fluorescein breakup time test. ARVO Abstracts. *Invest Ophthalmol*. 1998;40:B748.

8.   Korb DR, Finnemore VM, Herman JP, et al. The DET™ test–a new method for dry eye/tear film evaluation. ARVO Abstracts. *Invest Ophthalmol*. 1999;41:B809.

9.   Abdul-Fattah AM, Bhargava HN, Korb DR, et al. Quantitative *in vitro* comparison of fluorescein delivery to the eye via impregnated strip and volumetric techniques. ARVO Abstracts. *Invest Ophthalmol*. 1998;40:B747.

OCULAR SENSATIONS AND SYMPTOMS ASSOCIATED WITH TEAR BREAK UP

Carolyn G. Begley,[1] Debra Renner,[1] Graeme Wilson,
Salih Al-Oliky,[1] and Trefford Simpson[2]

[1]Indiana University
School of Optometry
Bloomington, Indiana, USA
[2]University of Waterloo
School of Optometry
Waterloo, Ontario, Canada

1. INTRODUCTION

The disorder dry eye presents a diagnostic dilemma because dry eye symptoms show only a weak association with clinical test results.[1-3] The reason for this is unknown, but may be due to a poor understanding of the cause of dry eye symptoms. It is the intent of this study to investigate the symptoms and sensations associated with tear breakup to provide a link between symptoms and objective measurements of local surface drying.

There is no "gold standard" clinical test for dry eye, leaving symptoms as an important diagnostic tool.[3] However, typical dry eye symptoms, such as eye soreness, irritation, and discomfort, often fall in the category of general symptoms of ocular irritation,[3,4] and thus are not specific to dry eye. In addition, dry eye symptoms may not be connected with measurable ocular surface damage in the form of surface staining with rose bengal or lissamine green.[3,5] Therefore, an investigation into the cause of dry eye symptoms appears warranted.

The name "dry eye" implies that drying of the ocular surface is implicated in the etiology of the disease. It is well known that the tear film of dry eye patients is relatively unstable and disrupts more quickly than in most normal patients.[5-8] This process of tear disruption or breakup can be visualized by instilling sodium fluorescein dye into the tear

Lacrimal Gland, Tear Film, and Dry Eye Syndromes 3
Edited by D. Sullivan *et al.*, Kluwer Academic/Plenum Publishers, 2002

1127

film, and watching for the appearance of dark areas that lack fluorescence. The clinical assumption is that these dark areas represent dry spots, in which the fluid portion of the tear film is absent or minimal. Sodium fluorescein is a water-soluble dye, and thus distributes uniformly in a fluid medium. Therefore, a lack of fluorescence in the tear film should indicate an absence of the fluid component of the tears. Based on this idea, we developed a novel method for objectively determining whether dry spots have occurred in the pre-corneal tear film and studied their association with ocular sensations and dry eye symptoms.

2. METHODS

Nineteen normal and 11 dry eye subjects participated in this study, which was conducted at the Indiana University School of Optometry. Informed consent was obtained from all subjects. Dry eye subjects were recruited from patients with a dry eye diagnosis at the Indiana University School of Optometry Eye Clinics.

All subjects first completed a pre-test symptom questionnaire that queried the intensity of nine ocular symptoms: discomfort, dryness, visual changes, soreness and irritation, grittiness and scratchiness, foreign body sensation, light sensitivity, and itching. The symptoms were rated on a scale from 0 (no symptom) to 5 (very intense). In this questionnaire, subjects were asked to report how their eyes felt at the time they filled out the questionnaire.

Experimental trials consisted of the following events. Fluorescein was instilled into one eye (randomly chosen) using DET (Akorn) strips. Subjects were seated behind the slit lamp biomicroscope and instructed to blink several times before opening the tested eye for as long as possible. The untested eye was held shut with the subject's fingers. The fluorescent image of the tested eye was videotaped using a slit lamp biomicroscope equipped with a video camera. When subjects could no longer keep the eye open, and blinked, videotaping was stopped. Subjects then filled out a McGill Pain Questionnaire (MPQ), which listed 78 adjectives in 20 groupings that describe pain or discomfort.[9] In this experiment, we allowed subjects to check any of the 78 words that described their ocular sensations during the experimental conditions of keeping one eye open as long as possible. The MPQ also contained a Present Pain Intensity (PPI) scale, in which subjects recorded the level of pain involved on a 0 (no pain) to 5 (excruciating pain) scale. Finally, subjects were asked to draw a picture depicting the ocular locale of their sensations.

Experimental trials were repeated 2 more times on the same eye, and then 3 times for the other eye. Following testing, a post-test symptom questionnaire identical to the pre-test symptom questionnaire, was completed.

Videotapes of fluorescein images of the eyes were digitized, individual frames converted to grayscale images, and maps of relative tear film fluorescence generated using a program written in MATLAB (The MathWorks, Inc., Natick, MA). An assumption was made that fluorescein dye distributed uniformly throughout the tears initially and at all stages in the formation of dry spots. Therefore, changes in tear film fluorescence during trial periods of holding the eyes open represented relative thickness differences in the tears. Tear breakup was assessed using MATLAB generated grayscale maps of the corneal fluorescein image.

3. RESULTS

During experimental trials, subjects held their eyes open for variable amounts of time, ranging from 9 to 35 seconds. Tear breakup occurred in 93% of trials, and all subjects showed tear breakup in at least one trial. Fig. 1 shows fluorescein images and resultant maps for a left eye trial of one subject. In Fig. 1a, the subject has just begun to hold the eye open. Fluorescence is relatively uniform over the surface of the eye, as illustrated by the map of relative fluorescence (Fig. 1b). Thirteen seconds later, changes have occurred in the fluorescein image (Fig. 1c), which are further illustrated by the map of relative fluorescence (Fig. 1d). The black areas in map (Fig. 1d) are defined as areas of tear breakup because their fluorescence matched background levels.

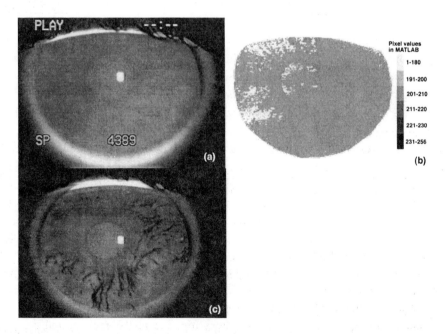

Figure 1. Data from the left eye of one subject: (a) fluorescein image immediately after the subject began to hold the eye open and (b) corresponding relative tear thickness map; (c) fluorescein image 13 sec later and (d) corresponding relative tear thickness map.

Results from the MPQ showed that some adjectives were chosen more often than others to describe the experience of keeping the eyes open as long as possible. Fig. 2a illustrates the average percentage (across all 6 trials) of dry eye and normal subjects who chose each descriptive adjective. The adjective "stinging" was most frequently chosen to describe the sensation of holding the eyes open as long as possible (Fig. 2a). "Burning" was also frequently chosen. Other adjectives were less frequently chosen, but many showed similar frequencies between dry eye and normal subjects. However, some adjectives such as "pricking" and "tingling" were chosen frequently by normal subjects, whereas only a few dry eye subjects chose them. The MPQ lists 78 adjectives, most of which were not chosen by any of the subjects in any trial. Most adjectives chosen at least 10% of the time are pictured in Fig. 2a. Fig. 2b illustrates the level of discomfort (PPI) subjects reported for one trial (second trial on the right eye). Most subjects chose 1 (mild) or 2 (discomforting) to describe the level of pain during trials in which one eye was held open as long as possible. There was no significant difference in the distribution of PPI scores between experimental trials (Kruskal-Wallis test; $P = 0.9952$).

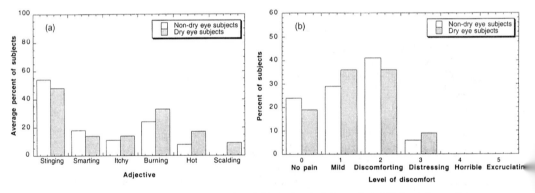

Figure 2. (a) Adjectives chosen from the MPQ and (b) the level of discomfort (PPI from the MPQ) used by dry eye and normal subjects to describe ocular sensations while holding the eyes open.

We also asked subjects to draw the ocular location(s) of their sensations while holding the eyes open. Most drawings by subjects did not correspond to the ocular location of the tear breakup during the trial. Only 30% of normals and 33% of dry eye subjects drew ocular locations close to the areas of tear breakup during the trial.

We found a shift in symptoms between those reported at the pre-test and post-test symptom questionnaires. Fig. 3a shows the intensity of discomfort reported by normal subjects at the pre- and post-test questionnaires. Before testing, 80% of normal subjects reported feeling no discomfort, but after testing the majority had shifted to "not at all intense" ocular discomfort. Fig. 3b shows the same information for dry eye subjects, many of whom began the experiment with more intense ocular discomfort than the normal subjects. Most dry eye subjects also shifted toward more intense ocular discomfort after testing. Although dry eye subjects began the experiment with more ocular discomfort, both groups shifted about the same amount on our scale (Fig. 3c).

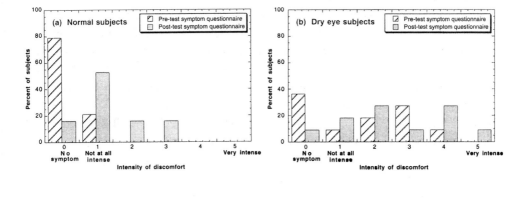

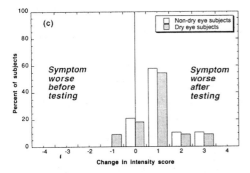

Figure 3. Intensity of discomfort in the pre- and post-test questionnaires reported by normal (**a**) and dry eye (**b**) subjects. (**c**) Difference in discomfort between pre- and post-testing discomfort.

Although all nine symptoms are not shown from the pre- and post-test questionnaire, dry eye subjects showed significantly more intense ocular symptoms at pre-testing than normal subjects ($P < 0.05$; Mann-Whitney U). When pre- and post-test symptom scores were compared, all subjects significantly shifted toward more intense ocular symptoms post-testing. The amount of the shift on our intensity scale was not significantly different between groups ($P > 0.05$; Wilcoxin signed rank test).

4. DISCUSSION

Holding the eyes open as long as possible usually results in tear breakup, and is associated with ocular sensations that fit into the general categories of stinging and burning, sharp, pricking pain, and temperature. These sensations are consistent with the innervation of the corneal surface, and suggest that tear breakup is capable of stimulating surface nociceptors.[10]

The events that occur during tear breakup over the corneal surface are largely unknown. We have used the novel technique of creating relative tear thickness maps by the intensity of fluorescence to objectively determine whether tear breakup occurred. This technique assumes that sodium fluorescein is uniformly distributed in the fluid component of the tears, and that a lack of fluorescence (similar to background levels) represents an absence of the fluid portion of the tear layer. However, this technique does not provide information about the sequence of events within a dark or non-fluorescent spot of tear breakup.

We found that the most commonly reported sensations associated with holding the eyes open as long as possible were stinging and burning pain. If we assume that the stinging and burning pain occurred due to tear breakup, then it appears likely that chemical stimulation of corneal surface nerves has occurred. A number of chemical agents, including hyptertonic saline, are known to stimulate surface polymodal and/or chemical nociceptors and cause a sensation of stinging pain.[10-14] During tear breakup, it is probable that local areas of drying become hypertonic as the fluid portion of the tear film either evaporates or recedes. This local area of hypertonicity may stimulate corneal nociceptors, which are sensitive to chemical and mechanical stimuli.[10,11,14] Sensations of pricking pain may also be due to stimulation of these fibers as the area of drying locally deforms corneal tissue. In this experiment, the levels of pain were generally reported as mild, but we allowed subjects to blink as soon as it was really necessary. A blink should rewet local dry spots, and quickly re-mix the fluid component of the tears, thus removing irritant stimuli.

We found that dry eye symptoms increased after only a few trials of holding the eyes open as long as possible. This suggests that tear breakup is implicated in causing dry eye symptoms. It is imperative that we understand the cause of dry eye symptoms because no "gold standard" clinical test for the diagnosis or treatment of dry eye exists. Holding the eyes open as long as possible may provide a model for discovering the etiology of dry eye symptoms associated with tear breakup.

REFERENCES

1. A.B. Bjerrum. Test and symptoms in keratoconjunctivitis sicca and their correlation. Acta Ophthalmol Scand, 74:436–441 (1996).
2. E.M. Hay, T.B. Pal, A. Hajeer, H. Chambers, A.J. Silman. Weak association between subjective symptoms of and objective testing for dry eyes and dry mouth: results from a population based study. Ann Rheum Dis, 57:20–24 (1998).
3. O.D. Schein, B. Munoz, J.M. Tielsch, B. Bandeen-Roche, West S. Prevalence of dry eye among the elderly. Am J Ophthalmol, 124:723–28 (1997).
4. K.K. Nichols, C.G. Begley, B. Caffery, L.A. Jones. Symptoms of ocular irritation in patients diagnosed with dry eye. Optom Vis Sci, 76(12):838–844 (1999).
5. S.H. Lee, S.C.G. Tseng. Rose bengal staining and cytologic characteristics associated with lipid tear deficiency. Am J Ophthalmol, 124:736–50 (1997).

6.   C.A. Paschides, G. Kitsios, K.X. Karakostas, C. Psillas, H.M. Moutsopoulos. Evaluation of tear breakup time, Schirmer's I test and rose bengal staining as confirmatory tests for keratoconjunctivitis sicca. Clin Exp Rheum, 7:155–157 (1989).

7.   C. Vitali, H.M. Moutsopoulos, S. Bombardieri. The European Community Study Group on Diagnostic Criteria for Sjögren's Syndrome. Sensitivity and specificity of tests for ocular and oral involvement in Sjögren's Syndrome. Ann Rheum Dis, 53:637–647 (1994).

8.   C. Snyder, R. Fullard. Clinical profiles of non dry eye patients and correlations with tear protein levels. Int Ophthalmol, 15:383–389 (1991).

9.   R. Melzack, W.S. Torgerson. On the language of pain. Anesthesiol, 34(1):50–59 (1971).

10. C. Belmonte, J. Garcia-Hirschfeld, J. Gallar. Neurobiology Ocular Pain. Progress in Retinal and Eye Res, 16(1):117–156 (1996).

11. C. Belmonte, J. Gallar, M.A. Pozo, I. Rebollo. Excitation by irritant chemical substances of sensory afferent units in the cat's cornea. J Physiol, 437(709–725)(1991).

12. A. Mandahl. Hypertonic saline test for ophthalmic nerve impairment. Acta Ophthalmol, 71:556–559 (1993).

13. M. Vesaluoma, L. Müller, J. Gallar, A. Lambiase, J. Moilanen, T. Hack, C. Belmonte, I. Tervo. Effects of oleoresin capsicum pepper spray on human corneal morphology land sensitivity. Invest Ophthalmol Vis Sci, 41(8):2138–2147 (2000).

14. M.B. MacIver, D.L. Tanelian. Structural and functional specialization of A delta and C fiber free nerve endings innervating rabbit corneal epithelium. J Neurosci, 13(10):4511–4524 (1993).

EVALUATION OF THE AKORN DRY EYE TEST (DET™) AS A PREDICTOR OF CONTACT LENS COMFORT

David Higgins,[1] Kevin Webb,[1] Stanley Shapleigh,[1] David Huebner,[2] Terry Carmolli,[2] and Brenda Hall[3]

[1]Kittery Optometric Associates
Kittery, Maine, USA
[2]Eyesight Ophthalmic Services, P.A.
Exeter, New Hampshire, USA
[3]Biocompatibles Eyecare Inc
Tewksbury, Massachusetts, USA

1. INTRODUCTION

Tear Breakup Time (BUT), the only clinical method of measuring tear film stability, is considered to be the primary and "first test" for the diagnosis of dry eyes.[1,2] A recommendation, by the National Eye Institute/Industry Workshop on Clinical Trials in Dry Eyes, that a test for tear stability be used as a global criterion of dry eyes substantiates the importance of the BUT measurement.[1] An unstable tear film is known to be a common denominator amongst a number of conditions causing dry eyes, ocular irritation, and contact lens intolerance[1,3]. BUT is defined as the interval following a blink to the occurrence of gaps or breaks in the tear film. Lemp et al.[4,5] described a normal tear BUT to be > 10 s. Marginal dry eyes show BUTs of 5-9 s and a BUT of <5 s is indicative of dry eye disorder.

In clinical practice fluorescein is used to measure tear breakup time (FBUT). However, following its introduction over 30 years ago, the FBUT test has been criticized as invasive, inaccurate and non-reproducible[6-9]. In a recent survey, only 19% of practitioners rated the FBUT test as their first choice test for the diagnosis of tear film and dry eye disorder.[10] The main problems with current FBUT measurements are the invasive nature of the test and the lack of control over the volume and concentration of fluorescein instilled.[6-8] Synder and Paugh recently calculated that a standard rose bengal strip delivers approximately 17 µl of fluid to the eye after wetting with 2 drops of saline and allowing the

Lacrimal Gland, Tear Film, and Dry Eye Syndromes 3
Edited by D. Sullivan *et al.*, Kluwer Academic/Plenum Publishers, 2002

1135

excess to fall off by gravity.[11] The physical dimensions of the rose bengal and fluorescein strips are identical, and it can be estimated that a standard fluorescein strip delivers a similar volume to the eye. Since the total tear film volume is only 7 μl, it is clear that the excessive volume delivered by the standard strips will destabilize the tear film and compromise the measurement.

Korb and co-workers[12] found that the accuracy and reproducibility of the FBUT test could be improved by following the suggestion of Marquardt et al.[8] to use smaller volumes of fluorescein (1-2 μl). However, this method required the use of a laboratory digital micropipette, which is not practical for routine clinical practice. They designed a new strip that was optimized to deliver precisely 1 μl fluorescein to the tear film.[13] The new design introduces the same amount of fluorescence as the digital micropipette but without the disadvantages to the patient. In addition this design minimizes disruption of the tear film and ocular sensation that could induce reflex tearing. The new modified strip has recently been commercialized by Akorn Ophthalmics as the Akorn Dry Eye Test or Akorn DET™ (Fig. 1).

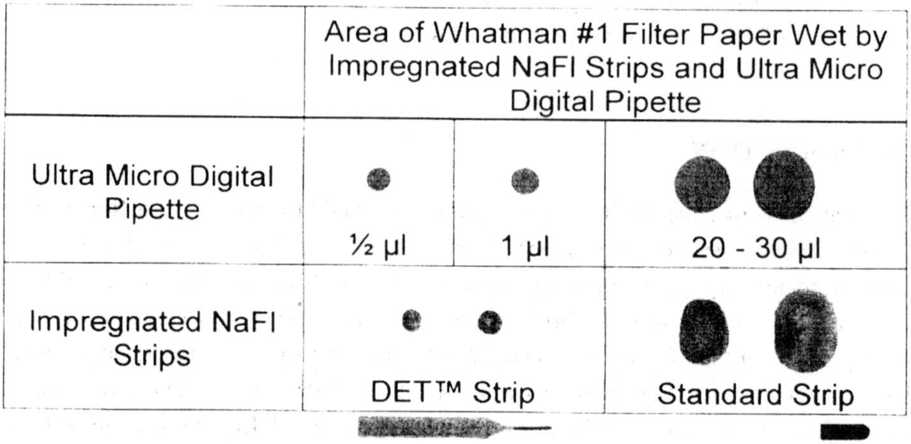

	Area of Whatman #1 Filter Paper Wet by Impregnated NaFl Strips and Ultra Micro Digital Pipette		
Ultra Micro Digital Pipette	½ μl	1 μl	20 - 30 μl
Impregnated NaFl Strips	DET™ Strip		Standard Strip

Figure 1. Comparison of volumes of liquid fluorescein delivered by DET™ and standard strips and ultra micro digital pipette.[12]

The purpose of this initial study was to determine the relationship between contact lens comfort and tear breakup time, using the Akorn Dry Eye Test (DET™). Standard methods of measuring FBUT do not demonstrate adequate accuracy and reproducibility to warrant consideration in this study. A second objective was to measure comfort with omafilcon A lenses in eyes with varying tear breakup times.

2. METHOD

This study was a two-week, open label, daily wear, comparative study at 3 US sites. Forty-seven current contact lens wearers were asked to rate overall comfort, left and right eye separately, with their existing lens on a 0 – 10 scale (0 = extremely uncomfortable; 10 = extremely comfortable). Each patient also completed a comfort questionnaire. Tear breakup time was measured within 15 minutes following lens removal using the Akorn DET™ test according to the recommended procedure..

One or two drops of non-preserved saline were applied to the impregnated paper tip. Excess fluid automatically fell off and shaking was not required or desirable. Patients were asked to look down and inward. The DET™ strip was gently touched onto the superior temporal bulbar conjunctiva for 1 to 2 s. The patient was asked to blink three times and open eyes naturally. FBUT measurement were taken immediately. Two consecutive measurements were performed. If they were not consistent, a third measurement was conducted and the results averaged. The procedure was repeated using a new strip for the second eye.

All patients were fitted bilaterally with omafilcon A lenses that were dispensed on a daily wear basis for two weeks. At the two-week follow-up visit, each patient rated comfort with omafilcon A lenses on a 0–10 scale and completed a comfort questionnaire.

3. RESULTS AND DISCUSSION

The Akorn DET™ test was used to measure FBUT. The DET™ score in seconds is equivalent to the FBUT in seconds. At the initial visit, 18 eyes (19.1%) showed a FBUT $\leq$ 5 s; 45 eyes (47.9%) showed FBUT between 5 and 10 s; 31 eyes (33%) showed FBUT > 10 s. Fig. 2 shows the number of eyes where all three measurements were precisely the same within 1 s, 2 s, 3 s, or 4 s. Ninety-seven percent of results were found to be within 3 s across the three measurements. These results compare favorably with Korb, the developer of the DET™ test, who found that 96% of results were within 3 s of each other.[12,13] Other investigators have found that variations of up to 20 s or more are common using standard strips.[6]

Comfort with their existing lenses was determined for the three groups of eyes based on the initial FBUT measurements. Average comfort ratings with their existing lens was 5.9 ± 1.97, 6.2 ± 1.75, and 6.7 ± 1.96 for the group of eyes with a FBUT of <5 s, 5 to 10 s, and >10 s respectively. There was no statistically significant difference between the three groups.

Comfort with omafilcon A after two weeks of wear on average ranged between 8 and 8.8. Paired t-test analysis, based on the FBUT groupings at the initial visit, confirms that each group of eyes show a significant ($p < 0.0001$) improvement in comfort over that found with the patients current brand (Fig. 3). In addition, the improvement in comfort in eyes with a FBUT of <5 s is significantly greater than that found by the group with FBUT of >10 s.

Pearson correlation analysis was used to determine correlations between FBUT at the initial visit and difference in comfort between the existing lens and omafilcon A lenses after two weeks of wear. Results show a significant correlation ($p < 0.001$) between FBUT and improvement in comfort with omafilcon A lenses. These results indicate that a greater

Reproducibility

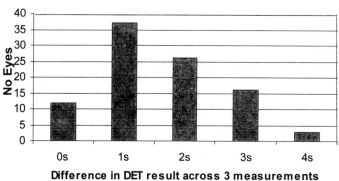

Figure 2. Difference in DET result across three measurements.

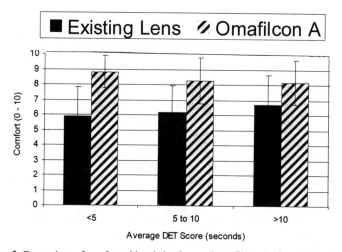

Figure 3. Comparison of comfort with existing lens and omafilcon A after two weeks wear.

improvement in comfort is experienced in the group of eyes with a FBUT of <5 s when compared to the group with a FBUT of >10 s.

Patients also responded on a number of points relating to comfort and dryness by confirming whether they agreed or disagreed with specific statements. The responses given following two weeks wear of omafilcon A lenses are compared to those given for their existing lens. The lens brands worn by the patients prior to the study were mainly low and mid-water content lenses.

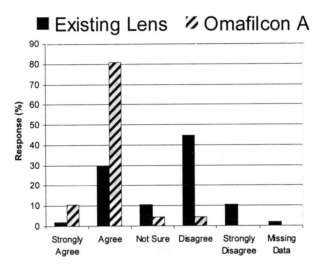

Figure 4. Percentage of patients who found their lenses to be comfortable at the end of the day.

A higher proportion of patients found omafilcon A lenses to be comfortable at the end of the day- compared with their existing lens, 91.7% versus 32% (Fig. 4). In addition, this study showed that fewer patients found omafilcon A lenses to be dry at the end of the day, 34% versus 83%. Comfort with omafilcon A lenses in dry environments was rated "better" or "much better" than their existing lens in 68% of patients. Also, 59.6% found improved wearing times with omafilcon A lenses.

4. CONCLUSION

The Akorn Dry Eye Test is a clinically practical method to accurately measure tear film stability. The results from this study show that all patients can benefit from improved comfort with omafilcon A lenses. However, eyes with poor tear stability (a low breakup time) show a greater improvement in comfort than those with a normal tear film. The Akorn DET™ test may prove to be a useful tool in helping practitioners determine which materials are best suited to their patients based on the quality of their patient's tear film.

REFERENCES

1. M.A. Lemp, Report of the National Eye Institute/Industry Workshop on clinical trials in dry eyes, *CLAO* 21 (4):221 (1995).
2. S.C. Pflugfelder, S.C. Tseng, O. Sanabria, H. Kell, C.G. Garcia, C. Felix, W. Feuer, and B.L. Reis, Evaluation of subjective assessments and objective diagnostic tests for diagnosing tear film disorders known to cause irritation, *Cornea* 17 (1):38 (1998).
3. F.J. Holly, Formation and rupture of the tear film, *Exp. Eye Res.*, 15:515 (1973).
4. M.A. Lemp, C.H. Dohlman, and F.J. Holly, Corneal dessication despite normal tear volume, *Ann Ophthalmol*, 2:258 (1970).
5. M.A. Lemp, C.H. Dohlman, T. Kuabara, F.J. Holly, and J.M. Carroll, Dry eye secondary to mucus deficiency, *Trans. Am. Acad. Ophthalmol. Otolaryngol.*, 75:1223 (1971).
6. G.T. Vanley, I.H. Leopold, and T.H. Greg, Interpretation of tear film breakup, *Arch. Ophthalmol.* 95:445 (1977).
7. L.S. Mengher, A.J. Bron, S.R. Tonge, and D.J. Gilbert, Effect of fluorescein instillation on the precorneal tear film stability, *Curr. Eye Res.* 4:9 (1985).
8. R. Marquardt, R. Stodmeister, and T. Christ, Modification of tear film breakup time test for increased reliability, in: *The Precorneal Tear Film in Health, Disease and Contact Lens Wear.* F.J. Holly, ed., Lubbock, TX: Dry Eye Institute, (1986).
9. J.R. Larke. *The Eye in Contact Lens Wear,* 2nd Edition. Butterworth-Heinemann, Oxford, (1997).
10. D.R. Korb, Survey of preferred test for diagnosis of the tear film and dry eye, *Cornea* 19:483 (2000).
11. C. Synder, and J.R. Paugh, Rose bengal dye concentration and volume delivered via dye-impregnated paper strips, *Optom. & Vis. Sci.* 75 (5):339 (1998).
12. D.R. Korb, V.M. Finnemore, J.P. Herman, J.V. Greiner, and T. Glonek, A new method for the fluorescein breakup time test, *American Academy of Optometry*, San Francisco, USA, (1998).
13. D.R. Korb, V.M. Finnemore, J.P Herman, J.V. Greiner, and T. Glonek, A new method for the fluorescein breakup time test, *Invest. Ophthalmol. Vis. Sci,.* 40:S544 (1999).

PREDICTING OPTICAL EFFECTS OF TEAR FILM BREAK UP ON RETINAL IMAGE QUALITY USING THE SHACK-HARTMANN ABERROMETER AND COMPUTATIONAL OPTICAL MODELING

Nikole L. Himebaugh, Larry N. Thibos, Carolyn G. Begley, Arthur Bradley, and Graeme Wilson

School of Optometry
Indiana University
Bloomington, Indiana, USA

1. INTRODUCTION

The pre-corneal tear film is the most anterior refracting surface of the eye. The large change in refractive index from air to tears makes it the surface with the greatest optical power. Therefore, a tear film of non-uniform thickness can potentially have a significant impact on the optical quality of the eye. Many clinical studies have shown that, during periods between blinks, the tear film is not stable and does not remain uniform on the surface of the eye.[1,2] Instead, it appears to disrupt locally, a phenomenon termed tear film breakup by clinicians, resulting in a tear film which is non-uniform in thickness. Although tear breakup is not usually regarded as an optical phenomenon, there is experimental evidence supporting the hypothesis that local changes in tear film thickness impact the optical quality of the eye.[3-5] The conclusions from these experimental studies are supported by clinical evidence of "blurry,"[6,7] "disturbed,"[8,9] and "fluctuating"[10] vision, and reduced visual acuity[11] reported by dry eye patients and contact lens wearers.

Sodium fluorescein is the standard clinical method for evaluating tear film breakup. Although this technique is utilized to assess the quality of the tears, it does not allow for quantification of the optical or visual consequences associated with tear film breakup. Thibos and colleagues have introduced a non-invasive optical method, the Shack-Hartmann (S-H) aberrometer, for monitoring tear film breakup.[3,4] In this investigation, we used the S-H aberrometer to measure the optical changes that accompany tear film breakup. By comparing the spatial and temporal characteristics of tear breakup obtained with the S-H aberrometer to those obtained with the standard fluorescein method, we are

Lacrimal Gland, Tear Film, and Dry Eye Syndromes 3
Edited by D. Sullivan *et al.*, Kluwer Academic/Plenum Publishers, 2002

1141

able to quantify the optical changes that accompany tear film breakup across the region of the cornea responsible for foveal image quality.

2. METHODS

Ten normal subjects, 7 females and 3 males, ages 24–45, participated in this study. Spatial and temporal changes in the tear film during periods of non-blinking were monitored using the traditional sodium fluorescein method. The S-H aberrometer was used to quantify changes in the optics of the eye during tear film breakup.

The S-H aberrometer system (Fig. 1) has been described in detail by Salmon et al.[12] and Thibos and Hong.[4] The S-H aberrometer creates multiple images of a single spot of light reflected from the retina. The retinal spot is produced by a narrow (~1 mm) laser beam focused onto the retina by the eye's optical system. The multiple images of this spot are created by an array of small lenslets placed one focal length from the video sensor. This micro-lenslet array divides the reflected wavefront of light exiting the eye into a large number of smaller wavefronts, each of which is focused to a small spot on the sensor.

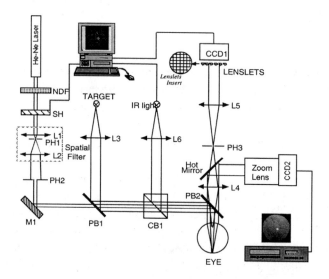

Figure 1. S-H aberrometer. Reflected light from a point source on the retina of the subject's eye is focused by a microlenslet array (lenslets) into multiple images on a CCD camera.

The principle of the S-H aberrometer is shown in Fig. 2. For a perfect (aberration-free) eye, the emerging plane wave will be focused into an array of point images with the same geometry as the micro-lenslet array. For an imperfect aberrated eye, the pattern of dot images will not match the lenslet array geometry. The spatial displacement of each spot relative to the optical axis of the corresponding lenslet is a direct measure of the local slope of the incident wavefront. Integration of these slope measurements reveals the shape of the aberrated wavefront. By this method, a single brief exposure of the sensor and subsequent computer processing of the captured image yields a detailed picture of the overall shape of the reflected wavefront and the eye that produced it.

The micro-lenslet array (Adaptive Optics Associates, Boston, MA) in our system is an array of lenslets with center-to-center spacing = 0.4 mm, which provided over 180 measurements of the wavefront slope over a 6 mm pupil. The computational method of Liang et al.[13] is used to fit the original slope data to the derivatives of Zernike circular polynomials up to the 10th order. The fitted coefficients are used to reconstruct the shape of the aberrated wavefront. The reconstructed wavefront is used to compute the modulation transfer function and simulated retinal images. All analyses and computational modeling were performed in MATLAB (Mathworks, Inc., Natik, MA).

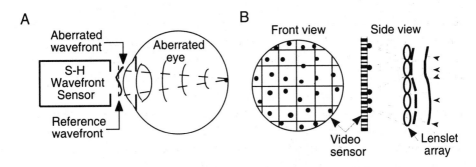

Figure 2. Principle of the Shack-Hartmann (S-H) aberrometer. **(A)** Reflected light from a point source on the retina emerges from the eye as an aberrated wavefront. Wavefront shape is measured with a S-H sensor, which is conjugate to the eye's entrance pupil. **(B)** S-H sensor subdivides the wavefront into a few hundred small beams that are focused onto a CCD video sensor by an array of small lenses. The displacement of each image relative to the grid of optical axes is determined by the local slope of the wavefront over the corresponding lenslet.

To ensure equal pupil sizes for direct comparison of the two techniques, the right eye of each subject was dilated with 0.5% tropicamide. Baseline S-H measurements were obtained prior to testing to determine the optics of the eye with an intact tear film. Both eyes were anesthetized with 0.5% proparacaine to minimize discomfort and to prevent reflexive tearing and blinking. This was necessary to observe the spatial and temporal characteristics of tear film breakup beyond the first area of disruption. To obtain the fluorescein image, sodium fluorescein was instilled into the eye, and tear film breakup was observed (wide beam method) using a Zeiss 20 SL biomicroscope (magnification of 8x) with a cobalt blue light source (approximately 30 degree beam angle) and a Wratten #8 filter. The subject's head was positioned and stabilized using a standard biomicroscope headrest. Subjects were asked to blink three to four times and hold their eyes open as long as they could do so comfortably (1–2 minutes). For each subject, the experiment began by monitoring tear film breakup with the fluorescein technique. When obvious tear film breakup was observed, the subject moved without blinking to the S-H aberrometer for one measurement. Still without blinking, the subject returned to the biomicroscope for continued monitoring of the tear film with the fluorescein technique. The spatial and temporal characteristics of tear film breakup were captured with a high resolution CCD video camera (Sony DXC-107AP), and recorded at 30 frames/second on an S-VHS video recorder (Mitsubishi HS-U69) with a video image resolution of 720 x 486 for later display and analysis.

3. RESULTS

For all subjects, spatially irregular patterns of tear breakup appeared as dark areas lacking fluorescence in the fluorescein image. Relocated, blurred and missing spots in the S-H image occurred in the same topographic locations as tear film breakup in the fluorescein image. Spatial patterns of thinning and breakup of the tear film were highly idiosyncratic with large variation between subjects. Nevertheless, both techniques revealed similar patterns with a high degree of spatial and temporal correlation for a given eye. These results clearly show that the optics of the human eye are affected by tear breakup, and increased aberrations and scatter exist in the regions lacking tear film.

Figure 3 compares tear fluorescence and optical quality for one subject immediately after a blink (Fig. 3a–d) and then after the subject had refrained from blinking for 27 to 42 sec (Fig. 3e–h). Initially, the fluorescein image (Fig. 3a) shows no tear film disruption. The corresponding S-H image (Fig. 3b) shows a regular array of clearly imaged dots, indicating low levels of optical aberration and little scatter. These impressions are confirmed quantitatively by the relatively flat wavefront deviation map (Fig. 3c). The modulation transfer function (MTF) (Fig. 4) and the computed retinal image (Fig. 3d) for this eye confirms its relatively good optical quality. All retinal image quality analysis was calculated for a 6 mm pupil. Therefore, optical quality of this eye is initially slightly worse than would be expected for a smaller, non-dilated pupil due to higher order aberrations that are introduced with an increase in pupil size.

After the subject held the eye open for 27 sec (Fig. 3e) the fluorescein image shows large areas of tear film breakup centrally and peripherally. The subsequently recorded S-H image (t = 42 sec.) is grossly distorted (Fig. 3f). Displaced and blurred dots, indicating large optical aberrations and scatter, were observed in the same corneal locations as areas of tear film breakup in the fluorescein image. Quantitative analysis of the S-H image indicates a significant increase in aberrations as demonstrated in the wavefront deviation map (Fig. 3g). The impact of these large aberrations on retinal image quality can be determined by calculating the MTF (Fig. 4) and simulated retinal images from the reconstructed wavefront. The MTF shows a marked reduction in retinal image contrast over a large range of spatial frequencies as a result of tear breakup. Simulated visual acuity shown in Fig. 3h confirms the poor optical quality of this eye during tear film breakup. The letters are blurry and appear to be doubled as a result of tear breakup. Our analysis predicts a reduction in visual acuity to approximately 20/100 for this subject as a result of tear film breakup.

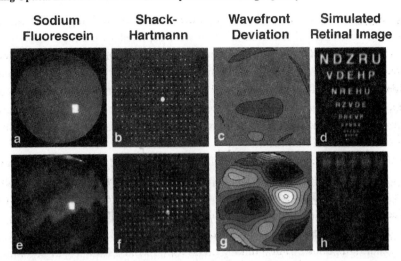

Figure 3. Optical effects of tear film breakup. The top row of images (**a–d**) was captured immediately after a blink; the bottom row of images (**e–h**) was captured during tear film breakup. Left column, sodium fluorescein images captured at baseline (**a**) and after the subject had held her eyes open for 27 sec (**d**); second column, data images captured by S-H aberrometer at baseline (**b**) and after the subject had held her eye open for 42 sec (**f**); third column, contour maps (contour interval = 0.5 _m) of aberrated wavefront emerging from the eye computed from the S-H image; fourth column, simulated retinal images computed from the re-constructed wavefront.

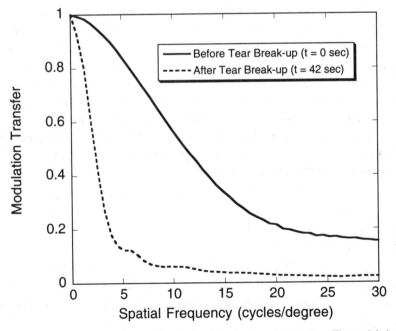

Figure 4. Quantitative description of image degradation following tear film breakup. The modulation transfer function (MTF) was calculated from the re-constructed wavefront for a 6 mm pupil before tear film breakup (solid) and after tear film breakup (dashed).

4. DISCUSSION

Due to the significant power of the eye's most anterior optical surface, small changes in the tear film can result in large changes in the optical quality of the eye. In this study, we have shown that a significant loss of optical quality occurs during extended periods of non-blinking, suggesting tear breakup as a cause of the decrement. These results support those of others,[4,5,14,15] and may explain symptoms of visual disturbances reported by dry eye patients [7-11] and contact lens wearers.[6]

In this study, the aberrations and scatter measured by the S-H aberrometer during periods of non-blinking were correlated spatially and temporally with areas of tear film breakup seen in the fluorescein image. The high degree of correlation between the two techniques supports the conclusion that the loss of optical quality of the eye during periods of non-blinking was due to tear film breakup. It is also possible that the optical changes measured during periods of non-blinking could have resulted from tear film thinning, re-distribution of the tear film, changes in the refractive index of the tear film and/or cornea, or structural changes in the corneal epithelium.

Subjects in our study held their eyes open for extended periods of time (1–2 minutes) after instillation of anesthetic. Because the S-H data were collected up to or over a minute after the subject's last blink, these results over-estimate the magnitude of the optical effects of typical tear film disruption that occurs during normal blinking. It is unlikely that the magnitude of optical degradation observed in our experiment would occur under normal viewing conditions and in normal eyes since the average blink rate is 12 blinks /min.[16] However, Tutt et al.[5] observed a reduction in image quality and contrast sensitivity immediately after the blink. Visual tasks requiring concentration, such as computer use, inhibit blinking[16-18] and thus may be expected to cause symptoms of blurry vision. In addition, patients with dry eye often demonstrate a faster tear film breakup.[9,19,20] If tear film disruption begins before the blink, then optical quality should be diminished. This hypothesis was supported by Goto et al.[11] who demonstrated a decline in visual acuity when dry eye patients held their eyes open for 10 sec. In patients with primary Sjögren's Syndrome, 42%–80% experience "disturbed vision."[8,9] Therefore, it is likely that tear film breakup causes some degree of retinal image degradation in dry eye patients, and perhaps in normal patients, depending on the visual task.

ACKNOWLEDGEMENTS

We gratefully acknowledge the support of Xin Hong and Kevin Haggerty for computer programming used in data analysis. This project was supported by NEI Grant R01-EY05109 (LNT) and a Bausch & Lomb/American Optometric Foundation Ezell Fellowship (NLH).

REFERENCES

1. P. Cho, B. Brown, I. Chan, R. Conway, and M. Yap, Reliability of the tear breakup time technique of assessing tear stability and the locations of the tear breakup in Hong Kong Chinese, *Optom Vis Sci.* 69:879–885 (1992).

2. D. Korb, D. Baron, J. Herman, V. Finnemore, J. Exford, J. Hermosa, C. Leahy, T. Glonek, and J. Greiner, Tear film lipid layer thickness as a function of blinking, *Cornea.* 13:354–359 (1994).

3. L.N. Thibos, X. Hong, A. Bradley, and C.G. Begley, Deterioration of retinal image quality due to breakup of the corneal tear film, *Invest Ophthalmol Vis Sci.* 40:S544 (1999).

4   L.N. Thibos and X. Hong, Clinical applications of the Shack-Hartmann aberrometer, *Optom Vis Sci.* 76:817–825 (1999).

5. R. Tutt, A. Bradley, C. Begley, and L. Thibos, Optical and visual impact of tear breakup in human eyes, *Invest Ophthalmol Vis Sci.* 41:4117–4123 (2000).

6. C.G. Begley, B. Caffery, K.K. Nichols, and R. Chalmers, Responses of contact lens wearers to a dry eye survey, *Optom Vis Sci.* 77:40–46 (2000).

7. I. Toda, H. Fujishima, and K. Tsubota, Ocular fatigue is the major symptom of dry eye. *Acta Ophthalmol.* 71:347–352 (1993).

8. K.B. Bjerrum, Test and symptoms in keratoconjunctivitis sicca and their correlation, *Acta Ophthalmol Scand.* 74:436–441 (1996).

9. C. Vitali, H.M. Moutsopoulos, and S. Bombardieri, The European community study group on diagnostic criteria for Sjögren's syndrome. Sensitivity and specificity of tests for ocular and oral involvement in Sjögren's syndrome, *Ann Rheum Dis.* 53:637–647 (1994).

10. S.H. Lee and S.C.G. Tseng, Rose bengal staining and cytologic characteristics associated with lipid tear deficiency, *Am J Ophthalmol.* 124:736–750 (1997).

11. E. Goto, S. Shimmura, Y. Yagi, and K. Tsubota, Decreased visual acuity in dry eye patient during gazing, *Invest Ophthalmol Vis Sci.* 39:S539 (1998).

12. T.O. Salmon, L.N. Thibos, and A. Bradley, Comparison of the eye's wavefront aberration measured psychophysically and with the Shack-Hartmann wavefront sensor, *J Opt Soc Am A.* 15:2457–2465 (1998).

13. J. Liang, B. Brimm, S. Goelz, and J.F. Bille, Objective measurement of wave aberrations of the human eye with the use of a Hartmann-Shack wave-front sensor, *J Opt Soc Am A.* 11:1949–1957 (1994).

14. C. Albarran, A.M. Pons, A. Lorente, and J.M. Artigas, Influence of the tear film on optical quality of the eye, *Contact Lens and Anterior Eye.* 20:129–135 (1997).

15. G. Timberlake, M. Doane, and J. Bertera, Short-term, low contrast visual acuity reduction associated with in vivo contact lens drying, *Optom Vis Sci.* 69:755–760 (1992).

16. M.C. Acosta, J. Gallar, and C. Belmonte, The influence of eye solutions on blinking and ocular comfort at rest and during work at video display terminals, *Exp Eye Res.* 68:663–669 (1999).

17. K. Nakamori, M. Odawara, T. Nakajima, T. Mizutani, and K. Tsubota, Blinking is controlled primarily by ocular surface conditions, *Am J Ophthalmol.* 124:24–30 (1997).

18. S. Patel, R. Henderson, L. Bradley, B. Galloway, and H. Hunter, Effect of visual display unit use on blink rate and tear stability, *Optom Vis Sci.* 68:888–892 (1991).

19. C.A. Paschides, G. Kitsios, K.X. Karakostas, C. Psillas, and H.M. Moutsopoulos, Evaluation of tear breakup time, Schirmer's-I test and rose bengal staining as confirmatory tests for keratoconjunctivitis sicca, *Clin Exp Rheumatol.* 7:155–157 (1989).

20. G. Petroutsos, C.A. Paschides, K.X. Karakostas, and K. Psilas, Diagnostic tests for dry eye disease in normals and dry eye patients with and without Sjögren's syndrome, *Ophthalmic Res.* 24:326–331 (1992).

IMPROVEMENT OF TEAR STABILITY FOLLOWING WARM COMPRESSION IN PATIENTS WITH MEIBOMIAN GLAND DYSFUNCTION

Eiki Goto,[1] Koji Endo,[2] Atsushi Suzuki,[3] Yoshiaki Fujikura,[2] and Kazuo Tsubota[1]

[1]Department of Ophthalmology Tokyo Dental College
Chiba, Japan
[2]Analytical Research Center
KAO Corporation
Tochigi, Japan
[3]Health Care Products Research Laboratories
KAO Corporation
Tokyo, Japan

1. INTRODUCTION

Meibomian gland dysfunction (MGD) has been reported as evaporative dry eye or lipid deficiency dry eye.[1-4] Warm compression has been recognized as conventional and effective therapy for MGD patients. Following 5 min of warm compression, we measured the change of tear evaporation of MGD patients using a newly developed tear evaporimeter with modern hardware and software.[5]

2. METHODS

We examined a consecutive series of 6 eyes from 6 MGD patients. All patients were classified as having noninflamed obstructive MGD.[3,6] For diagnosis of MGD, transillumination observation (meibography) using a fiber-optic device (L-3920; Inami Co., Tokyo, Japan) was performed.[3,4,7] Eyes with anterior blepharitis of more than moderate severity, infectious conjunctivitis, MGD with acute inflammation, meibomitis

Lacrimal Gland, Tear Film, and Dry Eye Syndromes 3
Edited by D. Sullivan *et al.*, Kluwer Academic/Plenum Publishers, 2002

1149

and seborrheic MGD were excluded from the study. The patients consisted of 2 males and 4 females (45.8 ± 14.8 years). Only the right eyes were used for analysis. Among them, 3 patients were diagnosed with Sjøgren's syndrome according to the criteria of Fox et al.[8] The Schirmer's test for 2 patients was less than 5-mm.[9] For these experiments, Schirmer's test was performed on another day to avoid influencing the tear evaporation test due to ocular surface damage. The average and standard deviations of the ocular surface examinations were as follows: tear interferometry grading, 3.2 ± 1.0; rose bengal staining score, 2.3 ± 1.9; fluorescein score, 1.2 ± 1.6; tear breakup time (BUT), 3.0 ± 2.1 sec; Schirmer's test, 6.0 ± 4.3 mm; meibography grading, 1.0 ± 0.0. A prospective comparison was performed using the following examinations: (1) tear evaporation test, (2) 5 min of eye warming, (3) tear evaporation test after warm compression, (4) measurement of BUT and (4) assessment of meibomian gland (MG) lipid expressibility.[3,4,10,11] A device for warm compression and steaming (KAO Corp., Tokyo, Japan) was used in this study,[12] and the evaporation of tears under normal blinking was measured by our newly developed microbalance technology.[5] Briefly, this is composed of a relative humidity sensor with high sensitivity and rapid response. It is housed within a ventilated chamber attached to an eye cap that snugly covers the eye. The data were collected and calculated using real-time presentation software run under Windows 98 (Fig. 1).

BUT was measured three times, and the measurements were averaged.[11] For the assessment of orifice obstruction and lipid expressibility, digital pressure was applied on

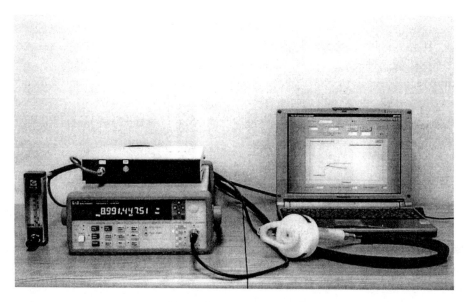

Figure 1. Tear evaporimeter. (**top right**) personal computer with real-time data acquisition software, (**bottom right**) humidity sensor probe, (**bottom left**) frequency counter, (**top left**) power supply.

the upper tarsus, and the degree of ease in expressing meibomian secretion (meibum) was evaluated semiquantitatively as follows: grade 0, clear meibum is easily expressed; grade 1, cloudy meibum is expressed with mild pressure; grade 2, cloudy meibum is expressed with more than moderate pressure and grade 3, meibum cannot be expressed even with the hard pressure.[3,4] A questionnaire was administered to patients following the warm compression procedure. All data are presented as mean ± SD. Findings at, prior to and following the treatment were compared and analyzed by Wilcoxons's signed rank test for nonparametric paired data. A level of $P < 0.05$ was accepted as statistically significant.

3. RESULTS

The tear evaporation rate (mean ± SD) was decreased significantly from 6.6 ± 0.97 to 4.7 ± 1.5 (10^{-7} g·cm^{-2}sec^{-1} ($P = 0.028$) after 5 min of warm compression (Table 1). BUT was prolonged from 3.0 ± 2.1 sec to 11 ± 2.7 sec ($P = 0.028$). Orifice obstruction score decreased from 2.2 ± 0.41 to 1.0 ± 0.0 ($P = 0.028$). No complications or complaints were reported in the patient questionnaire.

Table 1. Change of tear stability prior to and following warm compression

	Prior to[a]	Following[b]	P value
Evaporation[c] (10^{-7} g·cm^{-2}sec^{-1})	6.6 ± 0.97	4.7 ± 1.5	0.028
BUT[d] (sec)	3.0 ± 2.10	11.0 ± 2.7	0.028
MG[e] score	2.2 ± 0.41	1.0 ± 0.0	0.028

[a]prior to eye warming; [b]following eye warming; [c]tear evaporation rate; [d]fluorescein tear breakup time; [e]meibomian gland orifice obstruction score.

4. DISCUSSION

In this report, we have shown the change of tear stability after 5 min of warm compression administered to MGD patients. The efficacy and change of tear stability were indicated with the improvement of tear evaporation rate, BUT and MG orifice obstruction score. The MG orifice obstruction score indicates the ease of meibum expression. Thus, it suggests secretion of meibum was improved, meibum spread over ocular surface better after treatment and tear evaporation was prevented by it. Tear evaporation measurements may be useful in understanding tear lipid and aqueous dynamics in dry eye syndrome.[13] In this study we did not measure the increase of the tear lipid thickness directly, but the combination of tear interferometry and meibometry may be used to evaluate the change of the tear lipid layer in warm compression therapy.[14,15] Improvements of tear evaporation rate, BUT and MG orifice obstruction indicate warm compression may stabilize the tear, in particular the lipid layer.

REFERENCES

1. P.J. Driver and M.A. Lemp. Meibomian gland dysfunction. *Surv Ophthalmol.* 40:343–67(1996).
2. D.R. Korb and J.V. Greiner. Increase in tear film lipid layer thickness following treatment of meibomian gland dysfunction. *Adv Exp Med Biol.* 350:293–8(1994).
3. J. Shimazaki, M. Sakata and K. Tsubota. Ocular surface changes and discomfort in patients with meibomian gland dysfunction. *Arch Ophthalmol.* 113:1266–70(1995).
4. J. Shimazaki, E. Goto, M. Ono et al. Meibomian gland dysfunction in patients with Sjøgren's syndrome. *Ophthalmology* 105:1485–8(1998).
5. K. Endo, E. Goto, A. Suzuki, et al. An innovative dry eye diagnosis system by utilizing the microbalance technology. *Cornea* supplement to 19:S85 (abstract) (2000).
6. S.H. Lee and S.C. Tseng. Rose bengal staining and cytologic characteristics associated with lipid tear deficiency. *Am J Ophthalmol.* 124:736–50(1997).
7. J.B. Robin, J.V. Jester, J. Nobe et al. *In vivo* transillumination biomicroscopy and photography of meibomian gland dysfunction. A clinical study. *Ophthalmology* 92:1423–6(1985).
8. R.I. Fox, C.A. Robinson, J.G. Curd et al. Sjøgren's syndrome. Proposed criteria for classification. *Arthritis Rheum.* 29:577–85(1986).
9. K.P. Xu, Y. Yagi, I. Toda et al. Tear function index. A new measure of dry eye. *Arch Ophthalmol.* 113:84–8(1995).
10. K. Tsubota, E. Goto, H. Fujita et al. Treatment of dry eye by autologous serum application in Sjøgren's syndrome. *Br J Ophthalmol.* 83:390–5(1999).
11. I. Toda and K. Tsubota. Practical double vital staining for ocular surface evaluation [letter]. *Cornea* 12:366–7(1993).
12. A. Mori. Thermotherapy for meibomian gland dysfunction. *Atarashii Ganka* 18 (3): 317–320 (2001)
13. K. Tsubota and M. Yamada. Tear evaporation from the ocular surface. *Invest Ophthalmol Vis Sci.* 33:2942–50(1992).
14. N. Yokoi, Y. Takehisa and S. Kinoshita. Correlation of tear lipid layer interference patterns with the diagnosis and severity of dry eye. *Am J Ophthalmol.* 122:818–24(1996).
15. N. Yokoi, F. Mossa, J.M. Tiffany, et al. Assessment of meibomian gland function in dry eye using meibometry. *Arch Ophthalmol.* 117:723–9(1999).

AN EVAPORATIVE STRESS TEST FOR BORDERLINE DRY EYE DETECTION

S. Barabino, G. Melica, S. Alongi, G. Calabria, and M. Rolando

Department of Neurologic and Vision Sciences
Ophthalmology R
University of Genoa, Italy

1. INTRODUCTION

Abnormalities in tear film structure depend on alterations in efficacy of tear production, drainage and evaporation systems that may disrupt the three-layer pattern of tears. Any change occurring to tear organization leads to disequilibrium, resulting in a hyperosmolar state. This can occur in both aqueous-deficient and -adequate dry eye syndromes of different pathogenic origins.

In clinical and experimental states of increased tonicity, epithelial cell survival is widely reduced. Interesting information over the last 40 years has been obtained by measuring tear evaporation rates under specific pathologic conditions. For instance, Mishima and Maurice related corneal thickness variations in rabbit eyes with and without lipid layer.[1] We are continuing this research using a modified evaporimeter. It provides indirect measurements and appears to be a reasonably specific test for dry eye. Previous evaluations were performed under static conditions,[3] while our instrument provides an additional and external spurt of air. By inserting an air flow, we effected a stress test.

The present study is the first to use an external stress to determine tear evaporation. We sought better information about ocular surface tear dynamics by evaluating how normal and diseased eyes respond to environmental stress conditions.

Lacrimal Gland, Tear Film, and Dry Eye Syndromes 3
Edited by D. Sullivan *et al.*, Kluwer Academic/Plenum Publishers, 2002

2. METHODS

2.1. Instrumentation

Our system was modified from that described by Rolando and Refojo.[2] It is composed of three main parts The first component was a modified goggle, with a small sensor able to measure humidity and temperature, and two outlets for an air flow conveyed by vinyl pipes. We enhanced the experiment by adding an insulating border (neoprene) around the goggles, contoured to fit snugly over the skin around the globe. The humidity detector (THGM-880) was set for a temperature range of 0°–50° C and a humidity range of 32%–90% RH.The tube outlets, 1.5 cm in length and lying on the same level side by side, are placed into the goggles 2.5 cm from the border. A similar arrangement allows the air flow to move around the closed chamber without hitting the eye directly (such an irritant spur leads to potential measurement error).

The second component is an air pump (Kiorioku −1 Aerator) ensuring a constant rate of air flow at 6 ml/sec. In our system air is then passed through two separate airtight tubes connected to the goggle and sealed up. Global air volume on the eye in the closed chamber is 80 ml.

The third component is the display supplied with a temperature and humidity gauge. In addition to values recorded during the measurement period, % RH, minimum and maximum, are eventually labeled by pressing respective keys (Fig. 1).

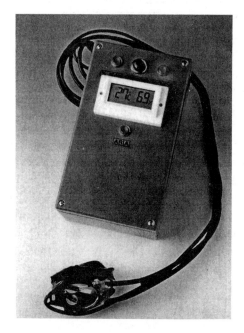

Figure 1. Modified Rolando's evaporimeter.

2.2. Measurements

Patients were instructed to hold the goggles tightly to keep the insulating border sealed. To avoid any air loss, oil cream was applied around the eye. They were asked to look forward and not to blink if at all possible. Lighting was kept dim, and external stimuli were avoided. Evaporation was widely varied by air flow applied into the goggles. During the measurements, the ambient temperature was maintained below 25° C to avoid patient perspiration.

After reading initial values on the gauge, air was pumped (flow rate 6 ml/min.) into the closed chamber and data were recorded at 60 sec after the start of the measurement period. To determine reproducibility, the same eye was measured three times under different conditions of humidity and temperature, and results were compared. During the measurements, it was noticed most of the patients could not resist blinking; however, they were always instructed to try hard to ensure reliability of results during the trial.

Humidity is the ratio percent of water volume (or mass) in a body to volume (or mass) of the body itself. For solid and liquid bodies, humidity is expressed as percentage or gr/kg. In this study the term relative humidity is used and applied as the ratio of water vapor to the amount of saturated water vapor at a certain temperature. It can be expressed as follows:

$$RH = c/c*$$

where c = water vapor concentration in a system and c* = saturated water vapor concentration at the system temperature.

To the extent volume and flow rate of water evaporation from a surface are determined by relative humidity, evaporation itself is proportional to the difference in saturation vapor pressure and actual vapor pressure as follows:

$$E = \frac{(C1 - C2) * V}{A * t}$$

where C1 = initial steam concentration inside the closed chamber at a defined temperature, C2 = changed rate of steam concentration at the final temperature, V = volume of the closed chamber (in ml), t = time in seconds, and A = exposed ocular surface area (in cm^2).

Although the exposed area might be derived by photography, a good correlation exists between corneal surface and eyelid opening.[2] Water evaporation rate derived in this experiment is expressed as $gr/cm^2/sec$.

2.3. Subjects

The 30 subjects with normal eyes included 13 females and 17 males; age range was 15–70 years. Eyes were considered normal if there were no present or previous complaints about the anterior surface of the eye and if there were normal basal secretion. The 30 pathologic eyes included subjects with diagnosis of keratoconjunctivitis sicca, Sjogren syndrome, Sarcoidosis, LES or any other causes of irregularities of the anterior surface.

In response to an established questionnaire, each patient complained of symptoms such as foreign body sensation, itching, burning, redness, heavy eyelids and photofobia. Schirmer test (< 10 mm), BUT test (< 7 sec), Lyssamine Green staining (dry eye syndrome confirmed by Van Bijsterveld classification)[4] and Ferning test (three to four types confirmed by Rolando classification)[5] were previously performed in addition to evaporative measurements.

3. RESULTS

The average evaporation rate in normal eyes was $1.010 \times 10^{-7} \pm 0.357 \times 10^{-7}$ g/cm^2/sec. For a normal range of palpebral aperture (4–9 mm), the amount of moisture collected in the goggles increased with the surface area proportionately. Diseased eyes were considered as a group. The average evaporation rate from the ocular surface was $3.078 \times 10^{-7} \pm 0.778 \times 10^{-7}$ g/cm^2/sec. Difference from normal eyes was statistically significant (P < 0.0001).

4. DISCUSSION

We introduced a simple and reliable method of measuring tear evaporation from the ocular surface under a stress condition. The data were reproducible and highly significant while the measurement itself took just 1 minute. This is shorter than the usual Schirmer test routinely used in eye clinics, and, most important, it is a non-invasive diagnostic tool. It prevents any bias caused by staining or other examinations more frequently performed. e.g., Schirmer test, in which the application of the Schirmer strip disrupts tear-film integrity and increases tear evaporation rate. The 1 min evaporation period is easy for the subject to refrain from blinking during such a short time. In many cases, however, patients could not stop blinking when air was pumped into the goggles. Some blinking has been demonstrated by experimental data not to have appreciably affected measurements, though we always recommend not blinking to reduce this variable.

Our experimental conditions differed from previous reports because former investigations were performed under static conditions where any spur was capable of modifying tear evaporation rate. Our intent was to create a condition of stress similar to physiology in which the ocular surface is constantly exposed to a number of stresses, and the point of the experimental test was to show how normal and diseased eyes deal with such stimuli.

The area of the exposed surface of the eye can be measured photographically, but this method of determining the exposed area of the eye is impractical and time consuming in the clinic. We have, therefore, applied Rolando's correlation between the palpebral aperture and surface area of the exposed eye. Because of the good correlation obtained between these two parameters, it is sufficient to measure only the palpebral aperture to determine the area of the exposed surface of the eye from which evaporation takes place.

By measuring tear evaporation rate under an air flow of 6 ml/min in a closed chamber, we collected some reliable information about eye reaction against environmental conditions. The increased evaporation rate shown by the eye with abnormal tear film is a consequence qualitative or quantitative abnormalities in the tear film structure and, particularly, in the oily layer. As a direct effect of increased evaporation, higher levels of tonicity may occur, and is harmful to lacrimal fluids and epithelial cells.

REFERENCES

1. Mishima S., Maurice D.M: The oily layer of the tear film and evaporation from the corneal surface. *Exp. Eye. Res.*1961;1:39–45.
2. Rolando M, Refojo MF: Tear evaporimeter for measuring water evaporation rate from the tear film under controlled conditions in humans. *Exp. Eye. Res.* 1983;36:25–33.
3. Mathers WD, Daley TE: Tear flow and evaporation in patients with and without dry eye. *Ophthalmology.* 1996;103:664–669.
4. Van Bijsterveld OP: Diagnostic tests in the Sicca Syndrome. *Arch. Ophthal.* 1969;82:10–14.
5. Rolando M, Baldi F, Calabria G: Tear mucus ferning in KCS. Proceedings of the International Tear Film Symposium (Lubbock, Texas, November 1981) in Holly FJ (ED): The preocular tear film in health, disease and contact lens wear. Lubbock: Dry Eye Institute, 1986, 203–210.

TEAR PRODUCTION MEASUREMENT, BASAL OR REFLEX ASSESSMENT?

Jennifer P. Craig,[1,2] Alan Tomlinson,[1] Nicola S. Patterson,[1] Victoria E.H. Reid,[1] and Angus K. McFadyen[3]

[1] Department of Vision Sciences
Glasgow Caledonian University
Glasgow, Scotland
[2] Discipline of Ophthalmology
University of Auckland
Auckland, New Zealand
[3] Department of Mathematics
Glasgow Caledonian University
Glasgow, Scotland

1. INTRODUCTION

Tear production is most commonly measured clinically by the Schirmer test or phenol red thread (PRT) test and in the laboratory setting by fluorophotometry. The Schirmer[1] test, based upon the absorption of tears by a filter paper strip placed in the inferior fornix over a set time period, is most frequently used but is notoriously unrepeatable.[27] This has led to the development of an alternative clinical test, the PRT test that uses a cotton thread impregnated with phenol red dye, giving a color change induced by absorption of tear fluid.[2,7] Indirect comparisons of the results of these tests have been reported.[8] A laboratory test, employing a fluorophotometric technique, utilizes the decay of fluorescence in the tear film as an indicator of tear turnover rate.[9,10] Few reports comparing laboratory and clinical techniques exist, with the exceptions of an evaluation of Schirmer and fluorophotometry,[11] and PRT and fluorophotometry.[12]

Lacrimal Gland, Tear Film, and Dry Eye Syndromes 3
Edited by D. Sullivan *et al.*, Kluwer Academic/Plenum Publishers, 2002

1159

Clearly these methods differ inherently in their mechanisms of measurement and degree of invasiveness; therefore, the relative contribution of reflex and basal tears to the total production measured may vary. These differences bring into question the comparability of measurements. This study directly compared these three techniques in a single population, under conditions in which the degree of reflex tearing could be modified.

2. METHODS

A sample of 40 subjects (20 males, 20 females) with a mean age of 19.8 ± 1.2 years was recruited from students at Glasgow Caledonian University. All subjects were asymptomatic of ocular surface disease or general pathology likely to affect the eye, and none wore contact lenses. The University Ethics Committee granted ethical approval prior to the study and all subjects provided signed, informed consent. Environmental conditions throughout the assessment period were maintained between 21 and 25 °C and 43 and 55% relative humidity. To avoid the effect of any diurnal influence, the time of day of each measurement was controlled for individual subjects to within 1 h.

Topical anaesthesia was used to modify reflex tearing. Prior to tear production measurement, a precise 20-μl volume of either lignocaine 4%, or saline (0.9% sodium chloride) (Minims, Chauvin Pharmaceuticals, UK) as a control, was instilled into the left eye of subjects with an air-displacement pipette. A local hospital pharmacy specially prepared the anaesthetic in a saline base to match the control saline for tonicity and pH. Following instillation, a 5-min-period was allowed for the drop to take effect, for excess fluid to drain from the ocular surface, and to allow any induced reflex tearing to subside.

The order of measurement for each subject was randomized by use of a computer-generated latin-squares matrix, incorporating anaesthetic status and the tear production measurement technique. The subjects attended six sessions, each on a separate day, to undergo the individual measurements. Both the subject and the examiner performing the tear production technique were masked as to the anaesthetic status of the subject.

The Schirmer I test was performed with standardized 5 x 35-mm Whatman No. 41 filter paper strips, (Clement Clarke, UK) in which the notched end was placed in the lateral inferior fornix for 5 min and the wetted length measured.[13] In the PRT test, the tip of commercially available threads (Zone-Quick, Menicon Co Ltd, Japan) was inserted into the lower lateral fornix for the recommended 15 sec. The total wetted length indicated by the color shift induced by the slightly alkaline tear fluid was measured.[7] Automated scanning fluorophotometry (FM-2 Fluorotron TM Master, Ocumetrics Inc., CA) of the tear film was performed with the anterior segment adaptor after instillation of a precise 1 μl volume of 1% fluorescein sodium (Minims, Chauvin Pharmaceuticals, UK).[14] The manufacturer's protocol for measurement, with the modifications recommended by Pearce et al.,[15] was followed.

3. RESULTS

Only the Schirmer and PRT test data, in both anaesthetised and unanaesthetised states, were normally distributed. Table 1 shows mean tear production measurements for each technique with and without topical anaesthesia, including that for the different genders.

Table 1. Means and standard deviations of tear production measurements performed with different techniques, in males and females, in the anaesthetised and unanaesthetised eye

Test		Unanaesthetised			Anaesthetised		
		All	Males	Females	All	Males	Females
Schirmer test (5 min)	mean	19.71	19.50	19.93	17.11	14.94	19.17
(mm)	s.d.	9.78	9.61	10.29	10.26	8.72	11.40
PRT test (15 sec)	mean	20.49	21.74	19.30	22.05	23.25	20.79
(mm)	s.d.	7.49	8.57	6.28	7.46	8.78	5.73
Fluorophotometry	mean	14.21	12.34	16.29	16.04	15.99	16.08
(%)	s.d.	7.43	7.59	6.86	6.90	6.93	7.05

No statistically significant relationships were established between the Schirmer test data and either the PRT test or fluorophotometry data ($P > 0.05$), but a positive correlation was identified between the fluorophotometric values and the PRT test results (Spearman rank correlation, $r = 0.248$, $P = 0.03$), in the unanaesthetised eye. No significant effect of anaesthesia on tear production occurred when measured with the PRT test (Student's paired t-test) or by fluorophotometry (Wilcoxan matched-pairs signed-rank test; $P > 0.05$ in both cases). However, with the Schirmer test, a significant reduction in tear production was found with anaesthesia (Student's paired t-test, $P = 0.02$). In the restricted age group studied, no gender differences in production were observed in any techniques with or without anaesthesia (Student's unpaired t-test or Mann-Whitney U test, $P > 0.05$ in all cases).

4. DISCUSSION

In the unanaesthetised state, only the results of the PRT test and the fluorophotometer were related. This suggested the tests are measuring the same, or closely related, aspects of the tear production. However, this is in conflict with the findings of Tomlinson et al. in which no relationship was found between fluorophotometry and the PRT test.[12] Fluorophotometry, by measuring the rate of decay of fluorescence in the eye of a measured quantity of fluorescein sodium caused by the turnover of new tears, may give a true reflection of tear production.[10] The exact parameter of the tear film assessed by the PRT test has never been fully established. It is most probable that it measures uptake of a small amount of fluid residing in the tear meniscus and may stimulate a low degree of reflex tearing, but it may also reflect the absorption characteristics of the thread, dependent on the biophysics or composition of the tear fluid.[12] The reason why a correlation between the two tests was found in the current but not the previous study is not understood, but may be related to differences in the study samples.

The Schirmer test did not offer results comparable to those of the other tests in the unanaesthetised eye, most likely as a result of the differing composition of the tears in the various tests. The Schirmer test probably produces a more significant degree of reflex tearing initiated by its particularly invasive measurement process. This agrees with the evidence both in comparisons of Schirmer and PRT[8] and fluorophotometry with Schirmer,[11] in which no association existed between the techniques. Indeed, this is further supported by the fact that the only test in the current study to be affected by the instillation of anaesthetic was the Schirmer test, such that tear flow was significantly reduced following topical anaesthesia. This agrees with previous reports.[16,17] The absence of significantly reduced tear production in the other tests following anaesthesia implies the contribution of reflex tears to tear production in the unanaesthetised eyes is relatively small.

Under anaesthesia, the three measurement techniques should give similar results, and this should also occur between the anaesthetised Schirmer and unanaesthetised PRT and fluorophotometry tests. Surprisingly, the latter was not the case and, in the anaesthetised eye, no significant correlations were established between the techniques. Intersubject variations in the depth of anaesthesia achieved following instillation of a single anaesthetic drop could be responsible. Such differences might result in variable degrees of suppression of reflex tearing. Jordan and Baum have proposed from their own and other studies in which the Schirmer test, even with anaesthesia, was unable to measure basal tear secretion[3,18] that an element of reflex tearing is continually present within basic tear flow.[17] It may be a partial reduction in reflex tearing, with intersubject variations in response, is sufficient to mask any relationship between the test results.

The results of the present study suggest the PRT test and fluorophotometry give a closer approximation to basal tear flow than can be measured with the Schirmer test, since the latter induces a significant element of reflex tearing.

REFERENCES

1. O. Schirmer, Studies zur physiologie und pathologie der tränen absonderung und tränen abfuhr, *Graefes Arch Ophthalmol.* 56:127–291 (1903).
2. K. Kurihashi, N. Yanagihara, H. Nishimura, S. Suehiro, and T. Kondo, New tear test – fine thread method. *Pract Otol Kyoto.* 68:533–535 (1975).
3. T.E. Clinch, D.A. Bennedetto, N.T. Felberg, and P.R. Laibson, Schirmer's test, a closer look. *Arch Ophthalmol.* 101:1383–1386 (1983).
4. J.W. Henderson, and W.A. Prough, Influence of age and sex on flow of tears. *Arch Ophthalmol.* 43:224–231 (1950).
5. J.C. Wright, and G.E. Meger, A review of the Schirmer test for tear production. *Arch Ophthalmol.* 67:564–565 (1962).
6. P.A. Asbell, B. Chiang, and K. Li, Phenol red thread test compared to Schirmer test in normal subjects. *Ophthalmology.* 94 (Suppl):128 (1987).

7. H. Hamano, M. Hori, T. Hamano, S. Mitsunaga, J. Maeshima, S. Kojima, H. Kawabe, and T. Hamano, A new method for measuring tears. *CLAO. J.* 9:281–289 (1983).

8. S. Patel, J. Farrell, K.J. Blades, and D.J. Grierson, The value of a phenol red impregnated thread for differentiating between the aqueous and non-aqueous deficient dry eye. *Ophthal Physiol Opt.* 18:471–476 (1998).

9. R.E. Furukawa, and K.A. Polse, Changes in tear flow accompanying aging. *Am J Optom Physiol Opt.* 55:69–74 (1978).

10. M. Göbbels, G. Goebels, R. Breitbach, and M. Spitznas, Tear secretion in dry eyes as assessed by objective fluorophotometry, *German J Ophthalmol.* 1:350–353 (1992).

11. J.R. Occhipinti, M.A. Mosier, J. LaMotte, and G.T. Monji, Fluorophotometric measurement of human tear turnover rate, *Curr Eye Res.* 7:995–999 (1988).

12. A. Tomlinson, K.J. Blades, and E.I. Pearce. What does the PRT test actually measure? *Optom Vis Sci.* 78: 142–146 (2001).

13. P. Cho, and M. Yap, Schirmer test. I. A review, *Optom Vis Sci.* 70:152–156 (1993).

14. FM-2 Fluorotron Master Operator Manual, Ocumetrics Inc., May 1995.

15. E.I. Pearce, B.P. Keenan, and C. McRory, An improved fluorophotometric method for tear turnover assessment, *Optom Vis Sci.* 78: 30–36 (2001).

16. D.W. Lamberts, C.S. Foster, and H.D. Perry, Schirmer test after topical anesthesia and the tear meniscus height in normal eyes. *Arch Ophthalmol.* 97:1082–1085 (1979).

17. A. Jordan, and J. Baum, Basic tear flow. Does it exist? *Ophthalmology.* 87:920–930 (1980).

18. J.L. Baum, Clinical implications of basal tear flow. In: Holly, F.J., Lamberts, D.W., MacKeen, D.L., eds. *The preocular tear film in health, disease and contact lens wear.* Lubbuck, Tx:Dry Eye Institute; 646–652 (1986).

INNOVATIVE DRY EYE DIAGNOSIS SYSTEM USING MICROBALANCE TECHNOLOGY

Koji Endo,[1] Eiki Goto,[2] Atsushi Suzuki,[3] Yoshiaki Fujikura,[1] and Kazuo Tsubota[2]

[1]Analytical Research Center
KAO Corporation
Tochigi, Japan
[2]Department of Ophthalmology
Tokyo Dental College
Chiba, Japan
[3]Health Care Products Research Laboratories No. 2
KAO Corporation
Tokyo, Japan

1. INTRODUCTION

A tear film layer, which covers the ocular surface, is essential for maintaining the integrity of the ocular surface. The tear feature is determined by three factors: (1) tear production by lacrimal glands, (2) drainage through the lacrimal system and (3) evaporation from the ocular surface. Tear dynamics have been evaluated clinically only by tear production using Schirmer's test. Few studies have measured tear evaporation rates[1-3] in spite of the importance in diagnosing dry eye syndromes. The goal of our study is to develop a noninvasive evaluation system for tear dynamics and use it as an alternative to Schirmer's test.

Lacrimal Gland, Tear Film, and Dry Eye Syndromes 3
Edited by D. Sullivan *et al.*, Kluwer Academic/Plenum Publishers, 2002

2. METHODS

2.1. Measurements of Tear Evaporation Rate

Tear evaporation from the ocular surface was evaluated by the following measurement: an eye cap (20-cm^3 volume) was placed tightly covering the eye; air, with a known water content (humidity), was supplied into the cap via a metering pump as a tear evaporation carrier. The tear evaporation rate was measured by calculating the difference between inlet and outlet humidity of the air (ΔH). We adopted a quartz crystal sensor (9 MHz A-T cut quartz crystal, 8 mm in diameter and 0.2 mm in thickness) microbalance with high sensitivity to humidity in this technique. The surface of the quartz crystal sensor was coated with an epoxy resin to enhance the sensitivity to humidity.[4,5] The oscillator circuit inside the probe controlled the quartz crystal humidity sensor. A model 53181A Hewlett Packard frequency counter measured the resonant frequency of the quartz crystal every 0.25 sec, and data were collected and analyzed by a personal computer. The tear evaporation rate, J (g/cm^2·sec), was calculated as follows:

$$J = \Delta H \rho V / A = \Delta F \rho V / kA$$

where ρ is the saturated water content of the air at a given temperature, V is the flow rate of a carrier gas, A is the measuring area and k is sensor constant. In this study, A and V were 13 cm^2 and 150 cm^3/sec, respectively. Fig. 1 illustrates the apparatus used in this study.

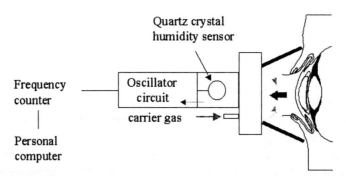

Figure 1. Schematic illustrations of the probe to measure tear evaporation from the ocular surface, including associated frequency counter and personal computer. The probe was made of polytetrafluoroethylene in the cylindrical form, with an inner diameter of 2 cm and a height of 20 cm. A goggle is attached to the part of the device in contact with the skin.

2.2. Subjects

In this study we tested 35 eyes from 21 normal patients and 24 eyes from 24 patients with dry eye symptoms. Either eye or only the right eye was measured.

3. RESULTS AND DISCUSSION

3.1. Measurements of Tear Evaporation Rate

Fig. 2 shows the typical profiles of tear evaporation measurements. In the closed eye state, the humidity sensor responded to the transpiration from skin and reached steady states after 20 sec. In the open eye state, the sensor response showed higher increases in tear evaporation from the ocular surface than the closed eye state and reached steady states after 20 sec. ΔF represents the water evaporation difference between the two steady states, which are open and closed eye. Tear evaporation rates can be calculated from the above equation and then evaluated.

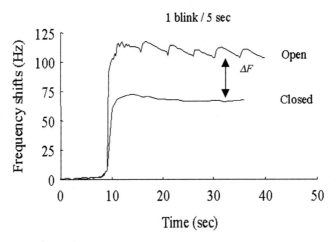

Figure 2. Analysis of tear evaporation profile.

3.2. Relationships between Blink Intervals and Tear Evaporation Dynamics

Fig. 3 shows the relationships between blink intervals and tear evaporation dynamics. In intervals longer than 3 sec, regular oscillations of the output frequency have the same interval as blinks. Therefore, this system is extremely sensitive in evaluating tear evaporation dynamics. Although a ventilated chamber method,[1] closed chamber method[2,3] and modified evaporimeter[6] are currently used for measurement of tear evaporation rates, tear evaporation dynamics have not been discussed. This may be because the humidity sensors used do not have enough sensitivity to detect rapid humidity changes accompanied by blinks. Furthermore, in the case of the closed chamber method, humidity changes in the whole volume of the chamber are measured, so the humidity changes gradually and tear evaporation dynamics to blinks cannot be obtained. In the case of the modified evaporimeter, turbulent air caused by blinking reduces its precision.[6,7]

3.3. Tear Evaporation Rates of Normal and Dry Eye

The tear evaporation rates (10^{-7} g/cm^2·sec) of normal subjects (n = 35) and patients with dry eye symptoms (n = 24) were 4.2 ± 1.3 (mean ± SD) and 4.8 ± 3.4, respectively. The tear evaporation rates of the two groups partially overlapped, and the evaporation rates of the dry eye group varied more significantly. Our results suggest high and low types of tear evaporation rates exist in dry eye symptoms of different etiologies. Studies on the relationships between tear evaporation rates and the causes of dry eye symptoms are now underway.

4. CONCLUSIONS

We have developed a new instrument using quartz crystal microbalance technology to evaluate tear evaporation dynamics from the ocular surface. This system is noninvasive, can easily measure the ocular surface tear evaporation rate and can contribute to the diagnosis of dry eye.

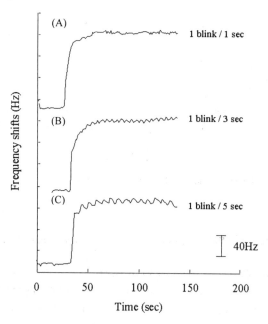

Figure 3. Relationships between blink intervals and tear evaporation dynamics: **(A)** 1 blink/1 sec, **(B)** 1 blink/3 sec, and **(C)** 1 blink/5 sec.

REFERENCES

1.   M. Rolando and M. F. Refojo, Tear evaporimeter for measuring water evaporation rate from the tear film under controlled conditions in humans, *Exp Eye Res*. 36:25 (1983).

2. K. Tsubota and M. Yamada, Tear evaporation from ocular surface, *Invest Ophthalmol Vis Sci.* 33:2942 (1992).

3. W. D. Mathers, G. Binarao and M. Petroll, Ocular water evaporation and the dry eye. A new measuring device, *Cornea* 12: 335 (1993).

4. H. Ito, Humidity sensor by quartz oscillator with epoxy resin as a hygroscopic film, *Proceedings of the 3rd Sensor Symposium* 169 (1983).

5. J. P. Randin and F. Züllig, Relative humidity measurements using a coated piezoelectric quartz crystal sensor, *Sensor and Actuator* 11:319 (1987).

6. G. E. Nilsson, Measurement of water exchange through skin, *Med Biol Eng Comput.* 15:209 (1977).

7. J. C. Seitz and T. S. Spencer, The use of capacitive evaporimetry to measure effects of topical ingredients on transepidermal water loss (TEWL), *J Invest Dermatol.* 78:351 (1982).

THE REPEATABILITY OF DIAGNOSTIC TESTS AND SURVEYS IN DRY EYE

Kelly K. Nichols and Karla Zadnik

The Ohio State University College of Optometry
Columbus, Ohio, USA

1. INTRODUCTION

In the clinical evaluation of dry eye, many diagnostic tests are utilized in making a dry eye diagnosis. Unfortunately, a single "gold standard" diagnostic test does not exist.[1] While the repeatability of many of the clinical tests used to diagnose dry eye have been assessed in normal patients, the repeatability of these tests and of surveys used in dry eye diagnosis has not been evaluated in dry eye patients. In addition to the battery of clinical diagnostic tests assessed in this study, two surveys were evaluated: McMonnies's Dry Eye Questionnaire was developed to assess likelihood of dry eye,[2] and the National Eye Institute Visual Function Questionnaire (NEI-VFQ-25)[3,4] was developed to assess vision-specific quality of life. However, the ocular pain subscale of the latter can be used as an indication of dry eye symptoms. In this study, the repeatability of a battery of dry eye diagnostic tests, the McMonnies's dry eye questionnaire, and the NEI-VFQ-25 were assessed on a sample of dry eye patients.

2. Methods

A battery of dry eye diagnostic procedures was performed on 75 patients on two separate occasions by a single examiner. The tests included the following: symptom assessment, contact lens and medical history, slit-lamp biomicroscopic evaluation of lids, evaluation of meibomian glands, assessment of tear film quality, tear meniscus height, assessment of blink quality, fluorescein tear breakup time, fluorescein and rose bengal staining of the cornea and conjunctiva, phenol red thread test, the Schirmer test, McMonnies's questionnaire, and the NEI-VFQ-25. The entry criterion for the study was a previous dry eye diagnosis (ICD-9 code 375.15), or the presence of moderate-to-severe symptoms (patient reported) on a regular basis.

2.1. Statistical Analyses

Data for the right eye are presented in all analyses. Descriptive data are presented as medians, percentages, ranges, and means ± standard deviations. Cohen's Kappa and weighted Kappa statistics[5,6] were used to determine repeatability of categorical data. A standard weighting system was used for ordered categorical data. A non-weighted Kappa was used to determine the repeatability of staining for each region of the cornea and conjunctiva. The Kappa statistic interpretation proposed by Landis was used in which a Kappa of 0.8 is described as close to perfect agreement, and a Kappa approaching 1.0 regarded as highly unlikely.[7]

The inter-visit mean difference (visit 2 – visit 1) was calculated for continuous data and presented as the mean difference ± standard deviation. Plots of the difference versus mean and the 95% limits of agreement first proposed by Altman and Bland[8,9] are presented for all continuous data. The narrower the 95% limits of agreement, the more repeatable the measurement. This technique can be applied to both the evaluation of repeatability of a measurement and inter-rater agreement. In this study, a single examiner evaluated all patients on both occasions; therefore, this method measures the repeatability of dry eye tests.

3. RESULTS

In this study, the average number of days between the two visits was 7.36 ± 3.10 days and 84% were seen exactly one week apart. The median age of the patients in this sample was 46 years (range 21–81), and 70.7% of the patients were female. Artificial tears were used by 61.3% of the patients, and of those, 79.7% reported that the artificial tears were effective. The results for slit lamp tests and surveys are reported in Table 1, including the number of patients with abnormal results, and the cut-points used.

Table 1. Mean and standard deviation (visit 1) for continuous data. The percent of patients with abnormal test results are also presented

Test	Mean ± SD	Cut-point	Percent Abnormal
Tear meniscus height (mm)	0.3 ± 0.1	≤ 0.3 mm	73.3%
Schirmer I test (mm)	16.9 ± 14.2	≤ 5 mm/5 min.	20.0%
Phenol red thread test (mm)	22.2 ± 7.1	≤ 10 mm/ 10 sec.	5.3%
Tear breakup, mean #1 & #2 (sec.)	5.7 ± 3.4	≤ 10 sec.	87.8%
NEI-VFQ pain subscale (score)	69.5 ± 18.7	< 75 points	69.3%
McMonnies's Questionnaire (score)	21.4 ± 6.8	> 14.5 points	89.3%

Weighted Kappa results for categorical data can be seen in Table 2. The repeatability of the subjective report of dryness ($\kappa_w = 0.62$) and grittiness ($\kappa_w = 0.73$) was moderate to high. In contrast, the repeatability of meibomian gland disease classification ($\kappa_w = 0.20$), presence or absence of inferior corneal fluorescein staining ($\kappa = 0.25$), and inferior conjunctival rose bengal staining ($\kappa = 0.21$) was poor. The inter-visit mean difference and 95% limits of agreement for tear breakup time, the Schirmer test, McMonnies's questionnaire, and the NEI-VFQ pain subscale score were as follows [mean difference (95% limits of agreement)]: tear breakup time [0.06 seconds (–5.71 to 5.83 seconds)], Schirmer test [–4.21 mm (–19.90 to 11.49 mm)], McMonnies's questionnaire score [0.41 (–7.76 to 8.58 points)], and the NEI-VFQ-25 ocular pain subscale score [–1.02 (–36.13 to 33.73 points). Plots of the 95% limits of agreement for the Schirmer test, tear breakup time, and McMonnies's questionnaire can be seen in Figure 1.

Table 2. Weighted Kappa results for categorical data

Test or subjective symptom	κ_w
"Dryness"	0.62
"Grittiness"	0.73
Ocular "soreness"	0.58
Ocular "tiredness"	0.75
"Redness"	0.79
Meibomian gland disease	0.20
Blepharitis	0.21
Lid margin irregularity	0.58
Tear debris	0.25

4. DISCUSSION AND CONCLUSIONS

The patient sample evaluated in this study spans the range of dry eye disease severity, with most individuals in the mild to moderate category. There was good "subjective" repeatability of patient symptoms by interview. In contrast, there was poor repeatability of slit-lamp examination findings. This may be an indication that consistency in clinical evaluation using standard grading schemes is difficult, especially in patients with mild to moderate findings. Further, it is possible that mild changes in slit-lamp findings are common in the short term with dry eye disease. Repeatability of measurements of aqueous production (the Schirmer test and the phenol red thread test) was also poor. However, there appeared to be a mild "trumpet effect" seen in the 95% limits of agreement for the Schirmer test, where the results closer to zero (more abnormal) show a narrower limits of agreement, and results further from zero (more normal) have a wider limits of agreement. This indicates that abnormal results may be more repeatable for the Schirmer test. In this study, the tear breakup time test was the most repeatable clinical test. The NEI-VFQ-25 questions about ocular pain were less repeatable than all other questions. Due to the method of scoring the NEI-VFQ, the pain subscale appears to have poorer repeatability than the overall VFQ score. This is due to the small number of questions (two) in the pain

A. Schirmer I Test

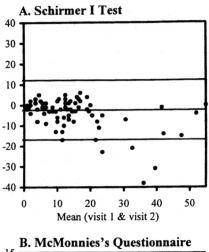

B. McMonnies's Questionnaire

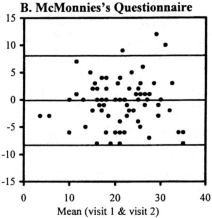

C. Tear Break-up Time

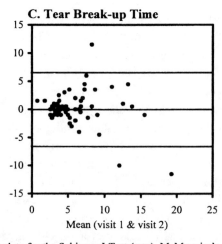

Figure 1. Difference vs. mean plots for the Schirmer I Test (mm), McMonnies's questionnaire (points), and the tear break-up time test (average of two consecutive TBUT measurements in seconds). The y-axis represents the difference between visits (visit 2 – visit 1) and the x-axis denotes the average of the two visits. The horizontal band corresponds with the 95% limits of agreement.

subscale. In the evaluation of patient symptoms, survey instruments with additional pain related questions may yield more repeatable results. In summary, though patient-reported symptoms and surveys are moderately repeatable from visit to visit, many of the procedures used clinically to diagnose and monitor dry eye syndromes are largely unrepeatable.

ACKNOWLEDGMENTS

An American Optometric Foundation Vistakon Ezell Fellowship, the Ohio Lions Eye Research Foundation, and Menicon, Inc. supported this research.

REFERENCES

1. M.A. Lemp. Report of the National Eye Institute/Industry workshop on Clinical Trials in Dry Eyes, *CLAO J.* 21:221–232 (1995).
2. C. McMonnies, A. Ho, D. Wakefield. Optimum dry eye classification using questionnaire responses, *Adv Exp Med Biol.* 438:835–838 (1998).
3. C.M. Mangione, P.P. Lee, J. Pitts, P. Gutierrez, S. Berry, R.D. Hays. Psychometric properties of the National Eye Institute Visual Function Questionnaire (NEI-VFQ). NEI-VFQ Field Test Investigators, *Arch Ophthalmol.* 116:1496–1504 (1998).
4. C.M. Mangione, S. Berry, K. Spritzer, N.K. Janz, R. Klein, C. Owsley, P.P. Lee. Identifying the content area for the 51-item National Eye Institute Visual Function Questionnaire: results from focus groups with visually impaired persons, *Arch Ophthalmol.* 116:227–233 (1996).
5. J. Cohen . A coefficient of agreement for nominal scales, *Educ Psychol Meas.* 20:37–46 (1960).
6. J. Cohen. Weighted kappa: nominal scale agreement with provision for scaled disagreement or partial credit, *Psychol Bulletin.* 70:213–20 (1968).
7. J.R. Landis, G.G. Koch. The measurement of observer agreement for categorical data, *Biometrics.* 33:159–174 (1977).
8. D.G. Altman, J.M. Bland. Measurement in medicine: the analysis of method comparison studies, *Statistician.* 32:307–317 (1983).
9. J.M. Bland, D.G. Altman. Statistical methods for assessing agreement between two methods of measurement, *Lancet.* 1–8476:307-310 (1986).

ASSOCIATION OF CLINICAL DIAGNOSTIC TESTS AND DRY EYE SURVEYS: THE NEI-VFQ-25 AND THE OSDI©

Kelly K. Nichols[1] and Janine A. Smith[2]

[1]The Ohio State University College of Optometry
Columbus, Ohio, USA
[2]The National Eye Institute
National Institutes of Health
Bethesda, Maryland, USA

1. INTRODUCTION

In 1995, the Report of the National Eye Institute/Industry Workshop for Clinical Trials in Dry Eyes defined dry eye as a disorder of the tear film characterized by damage to the interpalpebral ocular surface and symptoms of ocular discomfort.[1] Since then several existing and new surveys have been used to characterize dry eye symptoms: McMonnies' Dry Eye Questionnaire,[2] The National Eye Institute Visual Function Questionnaire (NEI-VFQ),[3] The Dry Eye Questionnaire,[4] Salisbury Eye Evaluation Dry Eye Questionnaire[5] and the Ocular Surface Disease Index (OSDI©).[6]

Evaluation of patient performance on these surveys is important, as these tools may be used in future clinical trials in dry eye. In this study, two surveys were evaluated: OSDI©, a dry eye survey recently developed by Allergan, Inc., and NEI-VFQ-25, developed to evaluate vision-specific quality of life. The results of the survey are converted to subscale scores, including an ocular pain subscale and an overall VFQ score. The pain subscale involves two questions about symptoms of ocular irritation, comparable to symptoms expressed by dry eye patients. In this report, the similarities and differences of patient responses to OSDI© and NEI-VFQ-25 ocular pain subscale score are evaluated and compared to a clinical dry eye diagnosis.

2. Methods

Patients in this analysis were examined to determine eligibility for several ongoing dry eye studies at NEI. The NEI-VFQ (short form) and OSDI© were administered prior to diagnostic testing. Clinical tests were performed in the following order: symptom interview; slit-lamp biomicroscopic evaluation of the meibomian glands, lids and tear film; the Schirmer test (with and without anesthetic); fluorescein tear breakup time; fluorescein staining of the cornea and lissamine green staining of the conjunctiva.

While data were collected on both eyes, a random eye was selected for this analysis. The sensitivity and specificity were calculated using Copenhagen criteria for dry eye (as the "gold standard"): dry eye = van Bijsterveld fluorescein staining ≥ 4 OR Schirmer I ≤ 5 mm/5 min AND symptoms.[7] Receiver operating characteristic (ROC) analysis was performed using sensitivity and specificity and the area under the curve was determined. To compare patient responses to both surveys, the NEI-VFQ scores were inverted; therefore, lower scores indicate fewer dry eye symptoms. The OSDI© scores were categorized to agree with the categories of the NEI-VFQ pain subscale. McNemar's Chi-square test for discordance in patient responses was then used to compare OSDI© and NEI-VFQ pain subscale results.

3. RESULTS

The sample included 70 patients; 97.1% were female. The average age was 44.3 ± 14.9 years (range: 17.4 to 81.0 years). Sixty percent had been previously diagnosed with Sjögren's syndrome and 55.7% reported using artificial tears. The most frequently reported symptoms (by interview) were dryness (41.4%), grittiness (32.9%) and itching (30.0%). Of the sample, 21.4% did not report experiencing any symptoms associated with dry eye.

Table 1. Diagnostic tests for the NEI dry eye data set

Criteria	n (%)
Tear breakup ≤10 sec	64 (92.8%)
Schirmer I ≤ 5 mm/5 min	33 (47.1%)
Symptoms present	54 (77.1%)
Staining ≥ 4	33 (47.1%)
Dry eye diagnosis	33 (47.1%)

Table 2. Frequency of graded results for slit lamp biomicroscopy

Symptom	Grade			
	0	1	2	3
Meibomian blockage	58.0[a]	20.3	13.4	8.7
Lid margin redness	44.1	29.4	19.1	7.4
Tear film debris	49.3	34.8	14.5	1.5
Conj. redness	33.3	37.9	21.2	7.6
Conj. chemosis	81.3	15.6	3.1	0

[a] %

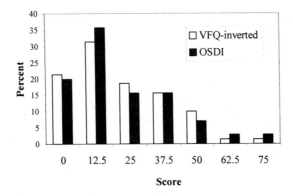

Figure 1. Comparison of the OSDI© and the NEI-VFQ-25 pain subscale scores (100 = worst, 0 = best).

Table 1 displays criteria used to diagnose dry eye. Using the Copenhagen criteria, 47.1% of the patients were diagnosed with dry eye. Symptoms were present in 77.1%. Graded slit lamp examination was unremarkable for many patients (Table 2).

The NEI VFQ-25 pain subscale score (mean ± SD: 78.6 ± 17.8) was lower than the overall VFQ score (89.1 ± 9.7) and all subscale scores, including the general vision subscale score (84.6 ± 13.3). Fifteen patients (21.4%) reported a pain subscale score of 100, which is the highest possible score. The OSDI© is also scaled on a 100-point scale; however, a score of zero is the best (least likely to have dry eye) and 100 is the worst (most likely to have dry eye). Twenty percent of patients (n = 14) had a score of zero on the survey, and the mean score was 17.3 ± 17.5. Fig. 1 is a comparison of the OSDI© and the VFQ pain subscale score. The NEI-VFQ pain subscale score has been inverted so that a score of zero is the best and a score of 100 is the worst. While the distribution appears similar, McNemar's test indicates discordance in patient responses (P = 0.001).

Fig. 2 displays ROC curve analysis of the NEI-VFQ pain subscale score and OSDI©. The Copenhagen criterion was used to classify patients with dry eye. The VFQ pain subscale score that maximizes the sensitivity and specificity (sensitivity + specificity) was 75 points, yielding a sensitivity of 63.6% and a specificity of 67.6%. Using categories similar to the VFQ pain subscale score, the cut point maximizing sensitivity (54.5%) and specificity (64.9%) was 50. The areas under the ROC curves are as follows: NEI-VFQ pain subscale = 0.67, OSDI© = 0.64. The arrows in the figure indicate the cut points.

3. DISCUSSION AND CONCLUSIONS

In this sample, dryness and grittiness were the most frequently reported symptoms. The specificity is high and sensitivity low for the OSDI© and NEI-VFQ pain subscale scores. Patient responses to NEI-VFQ pain subscale questions and OSDI© are similar when scaled in the same direction; however, McNemar's test for discordant pairs (P = 0.001) indicates patients are not answering the two surveys the same way or the surveys are assessing different aspects of symptomatology.

Dry eye is a complex disease in which the symptoms that are the primary factor driving the patient to seek treatment may not correlate with the common clinical assessments currently used to evaluate patients with dry eye. While it is important to use an appropriate survey for diagnosis of dry eye when selecting patients for clinical trials, a different instrument may be more suitable for assessing treatment effects. Further study is required to assess the usefulness of the available questionnaires in clinical trials of dry eye.

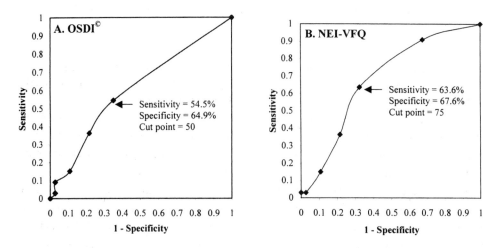

Figure 2. ROC curves for OSDI© (left graph) and NEI-VFQ (right graph). In each plot, the cut point indicated by the arrow yields the maximum sensitivity and specificity (sensitivity + specificity) for each survey.

ACKNOWLEDGMENTS

This research was supported by an American Optometric Foundation Vistakon Ezell Fellowship and NEI.

REFERENCES

1.   M.A. Lemp. Report of the National Eye Institute/Industry Workshop on Clinical Trials in Dry Eyes, *CLAO J.* 21:221–232 (1995).
2.   C.W. McMonnies, A. Ho, D. Wakefield. Optimum dry eye classification using questionnaire responses, *Adv Exp Med Biol.* 438:835–838 (1998).
3.   C.M. Mangione, P.P. Lee, J. Pitts, P. Gutierrez, S. Berry, R.D. Hays. Psychometric properties of the National Eye Institute Visual Function Questionnaire (NEI-VFQ). NEI-VFQ Field Test Investigators, *Arch Ophthalmol.* 116:1496–1504 (1998).

4. K.K. Nichols, C.G. Begley, B. Caffery, L.A. Jones. Symptoms of ocular irritation in patients diagnosed with dry eye, *Optom Vis Sci.* 76:838–844 (1999).

5. K. Bandeen-Roche, B. Munoz, J.M. Tielsch, S.K. West, O.D. Schein. Self-reported assessment of dry eye in a population-based setting, *Invest Ophthalmol Vis Sci.* 38:2469–2475 (1997).

6. R.M. Schiffman, M.D. Christianson, G. Jacobsen, J.D. Hirsch, B.L. Reis. Reliability and validity of the Ocular Surface Disease Index, *Arch Ophthalmol.* 2000;118:615–621 (2000).

7. C. Vitali, S. Bombardieri, H.M. Moutsopoulos, J. Coll, et al. Assessment of the European classification criteria for Sjögren's syndrome in a series of clinically defined cases: results of a prospective multicentre study. The European Study Group on Diagnostic Criteria for Sjögren's Syndrome, *Ann Rheum Dis.* 55:116–121 (1996).

ECONOMIC AND QUALITY OF LIFE IMPACT OF DRY EYE SYMPTOMS IN WOMEN WITH SJÖGREN'S SYNDROME

R. M. Sullivan,[1] J. M. Cermak,[1,2,3] A. S. Papas,[4] M. R. Dana,[1,2,3] and D. A. Sullivan[1,2]

[1]Schepens Eye Research Institute
[2]Harvard Medical School
[3]Brigham and Women's Hospital
[4]Tufts University
School of Dental Medicine
Boston, Massachusetts, USA

1. INTRODUCTION

Sjögren's syndrome is one of the leading causes of dry eye syndromes in the world. The purpose of this investigation was to determine the economic and quality of life impact of dry eye symptoms in women with this autoimmune disorder.

2. MATERIALS AND METHODS

Women diagnosed with primary or secondary Sjögren's syndrome (n = 45; age = 54.3 ± 2.4 years) were recruited from the Sjogren's Syndrome Foundation (Jericho, NY), as well as out-patient clinics at the Brigham and Women's Hospital (Boston, MA) and Tufts University School of Dental Medicine (Boston, MA). After obtaining written informed consent, women were asked to complete without supervision the Dry Eye Disease Impact Questionnaire (Allergan). The vast majority of all questions were answered by subjects and data are based upon the number of responses per question.

Lacrimal Gland, Tear Film, and Dry Eye Syndromes 3
Edited by D. Sullivan *et al.*, Kluwer Academic/Plenum Publishers, 2002

1183

3. RESULTS

To address dry eye symptoms, 90.9% of patients sought help from clinicians at least once during the preceding year. These women averaged 4.9 ± 0.7 (SE) clinical visits during that period, and 69.1 ± 5.5% of these visits were to ophthalmologists (Fig. 1).

Despite clinical assistance, the dry eye symptoms of 67.4% of patients were the same or worse than the previous year. Only 27% of patients were very satisfied with information provided by their health providers about managing and treating their symptoms (Fig. 2). In addition, only 17.9% of patient were very satisfied with their ability to self-manage symptoms (Fig. 3), and only 10.3% were very satisfied with current treatments (Fig. 4).

For symptomatic relief patients often used artificial tears or lubricants "all the time" (Fig. 5). Moreover, 20% of subjects also used other medications because of their dry eye symptoms, including topical or systemic pilocarpine, cyclosporine, corticosteroids, bromhexine, analgesics, antibiotics, non-steroidal anti-inflammatory drugs, decongestants, anti-allergics and anti-depressants.

The average monthly cost of dry eye treatments equaled $28.65 ± 4.65 per patient and these expenses were paid "out of pocket" by most individuals (Fig. 6). As a corollary to dry eye treatments, 53.3% of patients had received punctal occlusion or plugs.

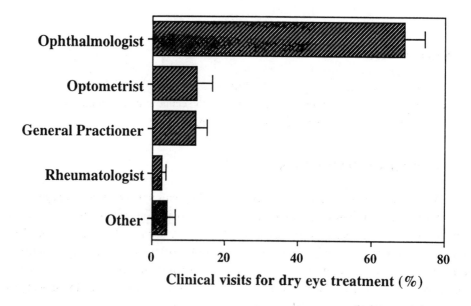

Figure 1. Type of clinician visited most frequently by Sjögren's syndrome patients for the treatment of dry eye symptoms. Columns and bars equal the mean ± SE.

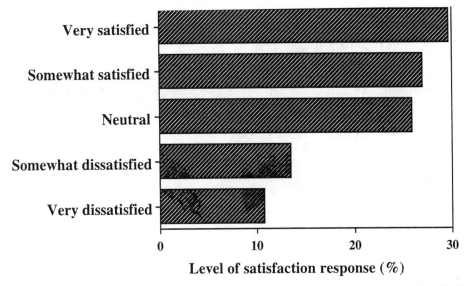

Figure 2. Level of satisfaction of Sjögren's syndrome patients with information received from health provider about managing and treating dry eye symptoms.

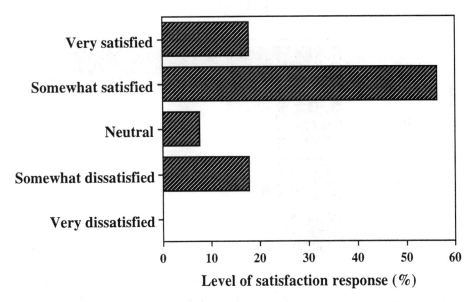

Figure 3. Level of satisfaction of Sjögren's syndrome patients with their ability to self-manage dry eye symptoms.

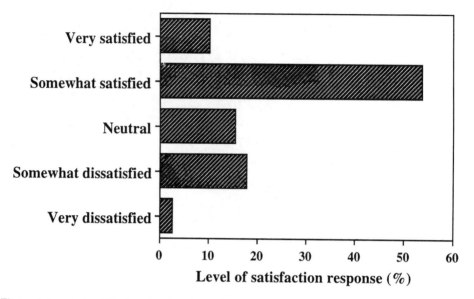

Figure 4. Level of satisfaction of Sjögren's syndrome patients with their current treatment for dry eye symptoms.

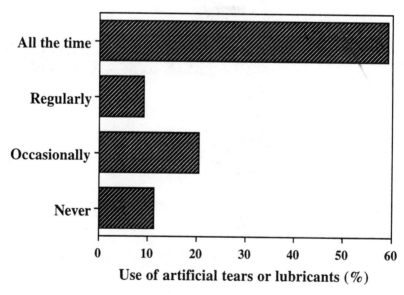

Figure 5. Frequency of use of artificial tears or lubricants by patients with Sjögren's syndrome for symptomatic relief.

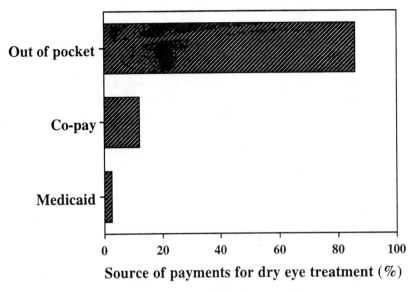

Figure 6. Source of payments for dry eye treatments by patients with Sjögren's syndrome.

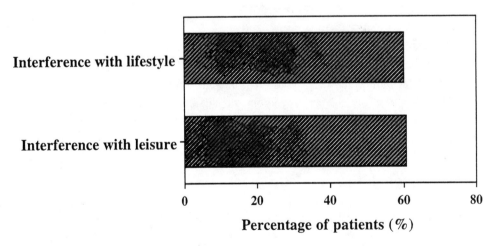

Figure 7. Influence of dry eye symptoms on the lifestyle and leisure activities of Sjögren's syndrome patients.

Table 1. Effect of dry eye symptoms on lifestyle, emotions and leisure activities

Lifestyle	Emotional impact
Changed type of work	Unhappy
Work less	Depressed
Takes more breaks	Uncomfortable
Requires help to perform tasks	Less confident
Takes longer to do things	Frustrated because of pain
Avoids windy, breezy or dry situations	Frustrated with daily activities
Avoids fans, campfires, fireplaces, woodstoves	
Avoids being outside in full sun	
Uses sunglasses more often	**Leisure Activities**
Misses outings with friends and family	Changed type of activities
Spends less time on computers	Reduced frequency of activities
Reads less	Spends less time pursuing activities
Watches less television	Reads less
Can't wear eye makeup	Has to be careful swimming in salt water or
Tries not to drive at night	chlorinated pool

As concerns quality of life, 37.5% of employed patients reported that their dry eye symptoms interfered with their effectiveness at work, and some of these people had to miss work because of symptoms. Furthermore, a majority of patients stated that their dry eye symptoms interfered with leisure activities and affected their lifestyle (Fig. 7, Table 1). Of this latter population, 44.4% felt frustrated and 40.7% felt unhappy or depressed.

4. CONCLUSIONS

Our results show that dry eye symptoms in women with Sjögren's syndrome have a considerable impact on their quality of life. In addition, these symptoms apparently lead to a significant economic burden. Based upon our data, and given the current estimates (>500,000 people) of the Sjögren's syndrome population in the United States, the cost of dry eye treatments for these patients may exceed 150 million dollars of "out of pocket" expenses per year.

ACKNOWLEDGMENTS

Supported by a research grant from Allergan and assistance from the Sjogren's Syndrome Foundation.

PATIENT-REPORTED VERSUS DOCTOR-DIAGNOSED DRY EYE: THE ASSESSMENT OF SYMPTOMS

G. Lynn Mitchell,[1] Kelly K. Nichols,[1] Barbara Caffery,[2]
Robin Chalmers,[3] Carolyn G. Begley,[4] and the DREI Study Group

[1]The Ohio State University College of Optometry
Columbus, Ohio, USA
[2]Toronto, Canada
[3]Atlanta, Georgia, USA
[4]Indiana University School of Optometry
Bloomington, Indiana, USA

1. INTRODUCTION

The definition of dry eye reported by the NEI/Industry workshop includes the presence of ocular symptoms.[1] A clinician often uses patient-reported symptoms in making a dry eye diagnosis.[2,3] The presence or absence of symptoms is also used in epidemiological studies to determine the prevalence of dry eye.[4-6] It is, therefore, very important to understand the symptoms associated with this condition. The assessment of ocular symptom presence or absence is, however, based on patient self-report. This study evaluates discrepancies in reported symptoms between patients diagnosed with dry eye by the doctor and patients who think they have dry eye.

2. METHODS

From July 1998 to July 1999, a newly developed dry eye questionnaire was administered at six clinical sites to 1054 consecutive patients. Participating sites included a large private practice in Toronto, Ontario, and schools and colleges of optometry at Indiana University, The Ohio State University, University of Missouri at St. Louis, the State University of New York and the University of Waterloo, Ontario. No selection process for patients existed, other than being a patient in the center and over 18 years of age. Informed consent was obtained from patients at all sites.

The Dry Eye Questionnaire (DEQ) administered to non-contact lens wearing patients and Contact Lens DEQ (CLDEQ) administered to contact lens wearing patients included questions about medications, allergies and the frequency and diurnal intensity of nine symptoms associated with dry eye. These included discomfort, dryness, visual changes, soreness and irritation, grittiness and scratchiness, foreign body sensation, burning and stinging, light sensitivity and itching. The contact lens wearers were asked to report their symptoms while wearing contact lenses. For each symptom, the patient was asked about the frequency of occurrence of the specific symptom. Patients could respond that the symptom occurred "never", "infrequently", "frequently" or "constantly". The DEQ and CLDEQ also included a self-assessment of whether or not the patient thought he or she had dry eye.

After each patient completed the questionnaire, it was placed in a file in the reception area. A separate doctor diagnosis form, linked by number to the questionnaire, was then placed in the patient's clinic file. The doctor diagnosis form queried the examining doctor as to whether the patient had dry eye or anterior segment pathology. The diagnosis of dry eye was the clinical assessment of the examining doctor and was made without knowledge of the questionnaire responses. Clinicians were free to conduct any clinical tests they felt appropriate to achieve the diagnosis. When the clinic visit was completed, a receptionist or staff person at the site paired each patient's doctor diagnosis form with the corresponding questionnaire.

Patients were divided into the following three groups based on the clinician's and/or patient's self-reported diagnosis of dry eye: 1) both doctor and patient diagnose dry eye, 2) only the doctor diagnoses dry eye, or 3) only the patient diagnoses dry eye.

A kappa statistic was calculated to assess the level of agreement between doctor diagnosis and patient self-diagnosis of dry eye. Contingency table analysis (Chi-square test) was used to compare the distribution of responses across the three defined groups for each symptom frequency. When the assumptions of the Chi-square test could not be met, a Fisher's Exact test was used. Separate logistic regression analyses were used to determine the symptoms significantly related to a doctor's, patient's and doctor and patient's diagnosis of dry eye. In all logistic regression analyses, contact lens wear was used as a controlling variable. Significance was assessed at $\alpha = 0.05$.

3. RESULTS

Of the 193 patients diagnosed with dry eye, only 80 (41%) reported they had dry eye (Table 1). In contrast, 82 (10%) of the 782 patients not diagnosed with dry eye felt they had dry eye. Only one-third of the patients for whom the doctor diagnosed dry eye reported they did not have dry eye(s), while over two-thirds of those not diagnosed by a doctor felt they did not have dry eye(s). The percent of patients who could not determine if they had dry eye was not different between those diagnosed (26%) and those not diagnosed by a

Table 1. Frequency (percent) of patients responding to question about dry eye and doctor's diagnosis

	Do you think you have dry eye(s)?		
Doctor's diagnosis of dry eye	Yes	No	DK
Yes	80 (41%)	63 (33%)	50 (26%)
No	82 (11%)	541 (69%)	159 (20%)

Table 2. Distribution of responses for each ocular symptom, by diagnosis status

	Diagnosis		Response			
Symptom	Status	Never	Infrequently	Frequently	Constantly	P-value
Discomfort	DX+[a] and SDX+[b]	2.8	41.7	50.0	5.6	.003
	DX+ and SDX-[c]	19.4	52.4	24.3	3.9	
	DX-[d] and SDX+	5.2	45.5	45.5	3.9	
Dryness	DX+ and SDX+	4.0	19.7	65.8	10.5	< .001
	DX+ and SDX-	33.6	41.6	20.8	4.0	
	DX- and SDX+	3.8	36.7	53.2	6.3	
Visual	DX+ and SDX+	17.8	50.7	28.8	2.7	.127
Changes	DX+ and SDX-	32.4	35.1	29.7	2.7	
	DX- and SDX+	17.3	48.0	29.3	5.3	
Soreness and	DX+ and SDX+	18.9	51.4	25.7	4.0	.005
Irritation	DX+ and SDX-	35.5	46.7	15.9	1.9	
	DX- and SDX+	23.4	39.0	37.6	0.0	
Grittiness and	DX+ and SDX+	36.4	40.3	22.1	1.3	.139
Scratchiness	DX+ and SDX-	52.4	35.2	12.4	0.0	
	DX- and SDX+	38.2	38.2	22.4	1.3	
Foreign Body	DX+ and SDX+	64.3	23.2	10.7	1.8	< .001
Sensation	DX+ and SDX-	78.5	15.2	5.1	1.3	
	DX- and SDX+	39.0	45.8	11.9	3.4	
Burning and	DX+ and SDX+	40.3	40.3	15.3	4.1	.081
Stinging	DX+ and SDX-	57.0	27.0	16.0	0.0	
	DX- and SDX+	48.0	38.7	13.3	0.0	
Light	DX+ and SDX+	35.6	27.4	24.7	12.3	.065
Sensitivity	DX+ and SDX-	47.6	30.1	19.4	2.9	
	DX- and SDX+	35.4	30.4	30.4	3.8	
Itching	DX+ and SDX+	32.0	44.0	21.3	2.7	.791
	DX+ and SDX-	41.0	38.1	19.0	1.9	
	DX- and SDX+	32.9	48.7	17.1	1.3	

[a]DX+, doctor-diagnosed dry eye; [b]SDX+, self-diagnosed dry eye; [c]SDX- = no self-diagnosed dry eye; [d]DX-, no doctor-diagnosed dry eye

clinician (20%). Excluding those patients who responded don't know (DK), the kappa was 0.407 with a 95% confidence interval from 0.327 to 0.487. This would indicate poor agreement between the doctor and patient with respect to the diagnosis/self-diagnosis of dry eye.

The data displayed in Table 1 was used to create three groups of patients, all of whom could be classified as dry eye suffers: those patients for whom the doctor and patient agree on a diagnosis of dry eye (n = 80); those diagnosed by the doctor, although they do not feel they have dry eye (n = 63), and those who feel they have dry eye(s), although the doctor did not diagnose it (n = 82).

As shown in Table 2, the distributions across frequency of symptom responses were significantly different among the three groups for discomfort, dryness, soreness and foreign body sensation (P < 0.05). Not surprisingly, patients who do not feel they have dry eye(s) were more likely to report "never" or "infrequent" to each of the four symptoms.

Separate logistic regression analyses were performed, controlling for contact lens wear, to determine which symptoms were significantly associated with dry eye diagnosis by the doctor, patient and both doctor and patient. Given the small number of patients who responded that a symptom occurred "constantly", this category was combined with "frequently" in all analyses. Table 3 presents the adjusted odds ratios for the significant

Table 3. Adjusted odds ratios and associated P-values from logistic regression analyses.

Outcome[a] = DX+ and SDX+			
Symptom	Frequency	aOR[b]	P-value
Dryness	Infrequent	8.6	< .001
	Freq/Constant	76.7	< .001

Outcome = DX+ and SDX-			
Symptom	Frequency	aOR	P-value
Dryness	Infrequent	2.7	< .001
	Freq/Constant	10.9	< .001
Soreness and	Infrequent	1.7	.02
Irritation	Freq/Constant	0.7	.31
Burning and	Infrequent	0.8	.36
Stinging	Freq/Constant	2.0	.04

Outcome = DX- and SDX+			
Symptom	Frequency	aOR	P-value
Dryness	Infrequent	6.9	< .001
	Freq/Constant	45.3	< .001
Visual Changes	Infrequent	1.9	.03
	Freq/Constant	1.1	.77
Discomfort	Infrequent	2.5	.05
	Freq/Constant	3.1	.03

[a]DX+, SDX+, DX-, SDX-, as for Table 2; [b]adjusted odds ratio.

symptoms from each model. Only eye dryness was significantly related to both patient and doctor diagnosis of dry eye. In contrast, dryness, soreness and burning are all significantly related to the doctor's dry eye diagnosis. Eye dryness, visual changes and discomfort were all significant in the model with the outcome defined as patient self-diagnosis without doctor diagnosis. In all three models, the most significant predictor of outcome was eye dryness.

4. CONCLUSIONS

Data on symptom frequency from any of the three groups described here could potentially be used to describe the prevalence of dry eye. Unfortunately, for some symptoms, this frequency is not consistent across the three groups. Not surprisingly, patients who think they have dry eye(s) report more frequent eye dryness when compared to patients who do not think they have the condition. However, these patients also report more frequent discomfort, soreness/irritation and foreign body sensation. When determining the prevalence of each symptom and likewise the prevalence of dry eye, it is important to know the patient's self-diagnosis status.

ACKNOWLEDGEMENT

This research was supported by CIBA Vision Corporation.

REFERENCES

1. M.A. Lemp. Report of the National Eye Institute/Industry workshop on clinical trials in dry eyes. *CLAO J.* 1995;21(4):221–32.
2. K.K Nichols and K. Zadnik. Frequency of dry eye diagnostic test procedures used in various modes of ophthalmic practice. *Cornea* 2000;19:477–82.
3. D.R. Korb. Survey of preferred tests for diagnosis of the tear film and dry eye. *Cornea* 2000;19:483–86.
4. O.D. Schein, B. Munoz, J.M. Tielsch, B. Bandeen-Roche, and S. West. Prevalence of dry eye among the elderly. *Am J Ophthalmol.* 1997;124:723–28.
5. M.J. Doughty, D. Fonn, D. Richter, T. Simpson, B. Caffery, and K. Gordon. A patient questionnaire approach to estimating the prevalence of dry eye symptoms in patients presenting to optometric practices across Canada. *Optom Vis Sci.* 1997;74(8):624–31.
6. C.A. McCarty, A.K. Bansal, P.M. Livingston, Y.L. Stanislavsky, and H.R. Taylor. The epidemiology of dry eye in Melbourne, Australia. *Ophthalmology* 1998;105(6):1114–1118.

KERATOCONJUNCTIVITIS SICCA VERSUS DRY MOUTH AND AUTOANTIBODIES IN PRIMARY AND SECONDARY SJÖGREN'S SYNDROME

Kazuko Kitagawa,[1] Takako Nakamura,[1] and Susumu Sugai[2]

[1]Department of Ophthalmology
[2]Department of Internal Medicine
(Hematology and Immunology)
Kanazawa Medical University
Uchinada, Ishikawa, Japan

1. INTRODUCTION

Sjögren's syndrome (SS) is divided into 2 categories: primary SS (1SS) and secondary SS (2SS). 2SS is similar to 1SS but occurs in conjunction with one of several collagen diseases. While SS usually affects tear and saliva production, few reports have compared the degree of keratoconjunctivitis sicca (KCS) in SS to findings from the salivary glands and systemic conditions. The purpose of this study is to compare the degree of KCS with the degree of dry mouth and the level of serum antibody titers in SS patients diagnosed using the Japanese criteria (Table 1). Additionally, because the symptoms of 1SS are not usually differentiated from those of 2SS, we have compared these two forms of SS to each other.

2. SUBJECTS AND METHODS

The subjects were 136 patients diagnosed with SS using the Japanese criteria (Table 1). Eighty-seven of the patients had 1SS, and all were female with ages ranging from 23 to 81 years old. There were 49 patients with 2SS and all were female except for 1 male case, and their ages ranged from 16 to 79 years old. The basic diseases of 2SS were varied:

rheumatoid arthritis, systemic lupus erythematosus, polymyositis, dermatomyositis, progressive systemic sclerosis and mixed connective tissue disease.

The degree of KCS was estimated according to fluorescein and rose bengal scores,[1] and classified into 4 stages: none (grade 0), mild (grade 1), moderate (grade 2), and severe (grade 3). Schirmer test was performed under topical anesthesia. Salivary gland function was assessed by the chewing-gum test, the Saxon test, salivary scintigraphy[2] and lip biopsy (focus score[3]). Serological tests evaluated the levels of IgG, anti-SS-A and anti-SS-B antibody. The correlation of KCS with salivary gland function and serum antibodies was determined by Spearman's correlation coefficient, two sample *t*-test and one-way ANOVA.

Table 1. Japanese Criteria for Sjögren's syndrome

Definite SS[a]	Probable SS[a]
Severe KCS of unknown etiology	Mild KCS of unknown etiology
Specific findings of SS in lacrimal gland or salivary gland biopsy	Decrease of salivary secretion (chewing-gum test < 10ml/10min)
Specific finding of SS in sialography	Recurrent or chronic salivary gland swelling of unknown etiology

[a]All show sensation of dryness with unknown etiology and at least one of the listed traits

3. RESULTS

The mean age of the patients with 1SS (53.8 ± 13.2 years, mean ± SD) was significantly higher than that of the 2SS patients (47.6 ± 14.8 years, P < 0.05). The proportion of patients with each grade of KCS in both SS groups is shown in Table 2.

Table 2. The proportion of patients with each degree of KCS in 1SS and 2SS

Category	KCS Grade			
	0	1	2	3
1SS	19 (22)[a]	25 (29)	23 (26)	20 (23)
2SS	14 (29)	9 (18)	13 (27)	13 (27)

[a]Number of cases (%)

There was a significant correlation (P < 0.05) for 1SS patients between KCS grade and Schirmer test, Saxon test, focus scores, salivary scintigraphy, and anti-SS-A antibody titlers. No significant correlation existed for 1SS KCS grade and chewing gum test, IgG or Anti-SS-B antibody titers. For 2SS patients, no significant correlations existed between KCS grade and any of the examinations.

Direct comparisons between 1SS and 2SS KCS grades revealed significant differences among the grades and between the two categories of SS (Fig.1). Schirmer values in 1SS were reduced significantly according to the progression of KCS. In contrast, for 2SS there was no clear correlation between Schirmer test scores and KCS grade; however, the lowest value was obtained in grade 3. In the chewing-gum test, the volume in grade 2 for 1SS was significantly lower than that of grade 1. In grade 3, the volume in 2SS was higher than 1SS. For lip biopsy and salivary scintigraphy, only the grade 3 KCS score for 2SS was significantly less (better) than for 1SS. In serological tests, the IgG level was higher in grades 1 and 2 than in grade 0 KCS. Anti-SS-A antibody increased significantly from grade 1 to grade 3 KCS for 1SS. Such an increase was not observed in 2SS. In contrast, anti-SS-B antibody showed a tendency to increase according to grade of KCS, but there were no statistical differences because of the large standard deviations.

4. DISCUSSION

Half of the patients with SS had no or only mild KCS. For these individuals the ocular findings alone might have failed to diagnose SS in patients who only showed abnormalities in their auto-antibodies and salivary glands. Since KCS might appear after salivary or serum abnormalities, a more sensitive method to detect the ocular abnormalities of SS is desirable. Further investigations like impression cytology, brush cytology, the concentration of tear fluid components such as lactoferrin and lysozyme, and a nasal stimulated Schirmer test should be required.[4] Other diagnostic methods should also be developed.

The age of the subjects with 2SS was significantly less than those with 1SS. This may be because the patients with 2SS have more life-threatening disorders, and their life span might be shorter. The correlation of KCS with other findings was more significant in 1SS than in 2SS. In particular, anti-SS-A antibody is strongly correlated with KCS in 1SS but not 2SS. Although the character of this antibody has not been completely clarified, it correlates with sicca syndrome as previously reported.[4]

In 1SS, although lymphoplastic disorders may progress, the main lesions are initially restricted to the exocrine glands, namely salivary and lacrimal glands, and these lesions may correlate well with each other. For 2SS, in contrast, more fatal systemic disorders progress and require intensive therapy such as corticosteroids or imunosuppressant agents. Under such conditions, the lesions of exocrine organs are only very partial symptoms and may be modified by those treatments. Discussions concerning KCS of SS usually do not differentiate between 1SS and 2SS. However, it would be better to recognize the differences exist in sicca syndromes, and so that ocular pathologies, especially KCS, more evaluated more sensitively as required for the diagnosis of SS.

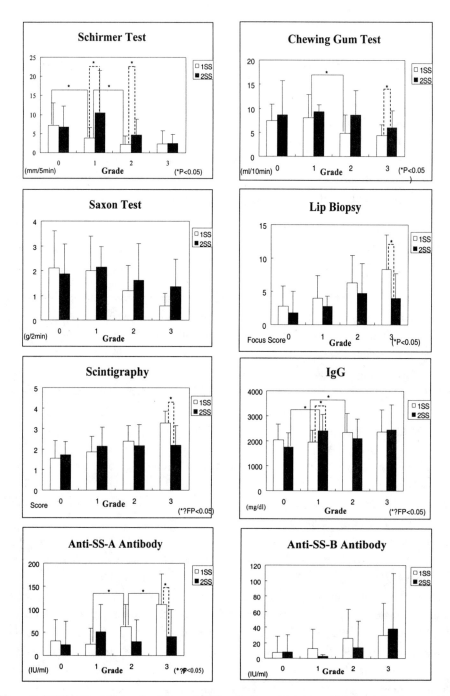

Figure 1. Relationship between the degree of KCS and other parameters which are defined as exocrine function and auto antibodies: Schirmer test, Chewing-Gum test, Saxon test, lip biopsy (focus score), salivary scintigraphy, and the levels of serum IgG, anti-SS-A antibody and anti-SS-B antibody. Significant differences between the columns in the same graph are shown with the connection of the lines; *P < 0.05.

REFERENCES

1. O.P. van Bijsterveld: Diagnostic tests in the sicca syndrome. *Arch. Ophthalmol.* 82:10 (1969)
2. Y. Ogawa, S. Sugai: The image diagnosis of salivary glands in patients with Sjogren's syndrome-salivary scintigraphy and sialography-. *Jap. J. Clinical Med.* 53:2363(1995)
3. J.S. Greenspan, T.E. Daniel, N. Talal, et al.: The histopathology of Sjogren's syndrome in labial salivary gland biopsies. *Oral Surg. Oral Med. Oral Pathol.* 37:217(1974)
4. K. Kitagawa, N. Kohda, S. Sugai, T. Ogawa: Correlation of corneoconjunctival lesions to salivary disorders and aoutoantibodies in primary Sjogren's syndrome. *Rinsho Ganka (Jpn. J. Clin. Ophthalomol.)* 51:1913(1997) 104:110(2000)

TEAR SPREADING RATES: POST-BLINK

Helen Owens and John R. Phillips

Department of Optometry and Vision Science
University of Auckland
Auckland, New Zealand

1. INTRODUCTION

The tear film plays a crucial role in the areas of contact lens wear and dry eye. However, despite its accessibility, there is no "gold-standard" test for assessing its functional quality. Current clinical tests are often poorly correlated with each other[1] and may be subjective and/or invasive in nature.[2] While much effort has been expended in assessing tear volume and stability, significantly less attention has been paid to the mechanism of tear film formation. Early work using motion-photography of naturally occurring lipid particles within the tear film noted that, immediately following a blink, the tears spread upwards over the cornea with a velocity of about 10 mm/sec.[3] More recent work analysing tear spreading in relation to the thickness of the tear film[4] proposed that the key factors influencing spreading were eyelid velocity, tear meniscus, tear viscosity, and surface tension.

In this work we report a video-photographic mehod that gives a quantitative description of tear spreading, and we explore its potential for assessing tear quality. We demonstrate that manipulations of the tear constituents result in quantifiable changes in the spreading nature of tears.

2. METHODS

Ten subjects (3 male, 7 female; age range: 20–50 years) participated in the first part of the study. None of the subjects wore contact lenses, and none reported symptoms of dry eye. Informed consent was obtained from all subjects.

Lacrimal Gland, Tear Film, and Dry Eye Syndromes 3
Edited by D. Sullivan *et al.*, Kluwer Academic/Plenum Publishers, 2002

1201

Standard video recordings of the tear film were made using a Sony Handycam (model CCD-TR3-3E), which was accurately centred in front of one ocular of a Nikon slit lamp biomicroscope (Model FS-3V) and securely attached to the body of the slit lamp. An overall magnification of 130x was used for all recordings. Subjects were required to fixate a small crosswire target placed anterior to the eye not under examination. The illumination system of the slit lamp was set at 40 degrees to the microscope system and an illuminated aperture of 1 mm by 2 mm (H x V) was directed at an area on the subject's cornea, just below the margin of the pupil. Levels of illumination were chosen to avoid reflex tearing from dazzle. The temperature and humidity of the room did not vary significantly (Temp: 22°C ± 3°C; humidity 55% ± 10%). The experimenter recorded the movement of naturally occurring particles in the lipid film over a period of about one minute while the subject blinked naturally. Velocity profiles were determined for a minimum of three blinks for each subject and averaged. This determination involved replaying the video and following particles frame-by-frame to determine the distance moved every 0.04 sec, from which velocity was calculated. In addition to normative data, preliminary results are reported from subjects in which manipulation of each of the tear film components was attempted. Table 1 summarises the methods used to achieve this essentially artificial change in the tear film components.

Table 1. Methods used to manipulate the individual components of the tear film. A silicone punctum plug (Oasis, Glendora, CA. USA) was fitted to a patient with Sjögren's Syndrome to mimic an increased aqueous component.

Tear Layer	Increase	Decrease
Lipid	Lower eyelid squeeze	Induced ectropion (lower eyelid held away)
Aqueous	Punctum plug treatment	Sjögren's Syndrome
Mucin	0.1% sodium hyaluronate (1 drop)	20% Acetylcysteine rinse (4 drops, 5 mins apart)

3. RESULTS

The tear velocity profiles for the ten normal subjects are graphically presented in Fig. 1A. The average initial velocity of particles, 40 msec after a blink, was found to be 6.46 ± 2.2 mm/sec, (mean ± SD, n = 10). The decay of particle velocity is best described by a logarithmic function, as is shown in Fig. 1B for the mean data from the ten subjects. The time to particle stabilisation (zero velocity) may be extrapolated from the equation of the fitted line. Mean stabilisation time was 1.01 ± 0.3 sec (mean ± SD).

Table 2 presents a summary of the results for the various manipulations of the tear film described in Table 1. Data are summarised in terms of the initial particle velocity and stabilisation time.

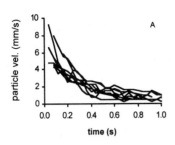
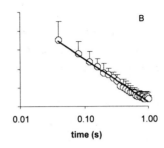

Figure 1. (A) Tear particle velocity versus time following a blink for 10 normal subjects (linear time scale). (B) Tear particle velocity (Mean +1 SD) for 10 subjects plotted with a logarithmic time scale.

Table 2. Summary of results from manipulation of the tear layers in terms of initial particle velocity and time to stabilisation (zero velocity)

Tear layer	Method used	Particle velocity change relative to baseline	Change in stabilisation time relative to baseline
↑ Lipid	Lid squeeze	↓ 55% (6.5 to 3mm/s)	unchanged
↓ Lipid	Ectropion, detergent	↓ 61% (6 to 2.3mm/s)	↓ 15%
↑ aqueous	Punctum plug	↑ (0.4 to 12mm/s)	↑ to 0.95s
↓ aqueous	Sjögren's	↓ 75% (rel to norm. data)	↓ 90%
↑ mucin	Sodium hyaluronate	equivocal	unchanged
↑ mucin?	Irritant inhalation	unchanged	↓ 71% (from 1.4 to 0.4s)
↓ mucin	Acetylcysteine	↑15% (7.1 to 8.2mm/s)	unchanged

4. DISCUSSION

We have demonstrated that lipid particle motion immediately following a blink may be summarised by two measures: the initial velocity and time to stabilisation. These descriptors are evidently consistent for a group of normal individuals and may alter in specific ways in response to manipulations of the tear constituents, as shown in Table 2. For reduced aqueous volume in Sjögren's syndrome, initial particle velocity and stabilisation time were severely reduced relative to normal, whereas following punctum plug installation, both descriptors returned to near-normal values.

Increased lipid concentration from lid-squeeze reduced initial velocity while leaving stabilisation time unchanged. Decreased lipid concentration created by induced ectropion had the unexpected effect of reducing stabilisation time. However, induced ectropion results in a loss of the protective lipid layer, the presence of which is known to retard evaporation rates.[5] Without the protective lipid layer, it is likely that the aqueous component was rapidly depleted, leaving a concentrated and viscous mucus component. We might expect an increased mucus concentration to result in a shorter stabilisation time.

Data for changes in mucus content were equivocal. Acetylcysteine (used as a mucolytic agent) resulted in an increase in initial particle velocity but did not change the stabilisation time. It is possible that excessive stinging and consequent reflex tearing induced by the agent may have been responsible for its dilution and consequent lack of effect. The mucomimetic agent sodium hyaluronate also failed to produce consistent results in a group of nine normal subjects. The reason for this is unclear, although it may be that a higher concentration of the agent is necessary to produce a repeatable effect.

Although attempts to change mucin concentration were equivocal, we have also investigated the effect of inhaling an irritant (onion vapour), which may be construed, at least in part, as imitating an increased mucus situation. The irritant created excessive reflex tearing and blinking, resulting in increased aqueous and lipid in the initial stages. Recordings of particle motion were not possible until subjects' reflex blinking had reduced, allowing the excessive tearing to subside. We propose that a dramatic reduction in stabilisation time could be attributed to an irritant-induced mucus increase[6] and blink-induced lipid excess.[7]

We conclude that our analysis of lipid particle velocity is a technique which offers a repeatable measure of tear film function. Its major advantages are that it is non-invasive and it provides a quantifiable measure of the tear film in vivo. Further work in assessing its viability in detecting aetiologically different dry eye states is indicated, and we are currently developing the software to expedite data collection and analysis.

REFERENCES

1. R.K. Madden, J.R. Paugh and C.Wang. Comparative study of two non-invasive tear film stability techniques. Curr Eye Res. 13: 263 (1994).
2. C.A. Paschides, G. Kitsios, K.X. Karakostas, C. Psillas and H.M. Moutsopoulos. Evaluation of tear breakup time, Schirmer's-I test and rose bengal staining as confirmatory tests for keratoconjunctivitis sicca. Clin Exp Rheum. 7: 155 (1989).
3. R.E. Berger, and S. Corrsin. A surface tension gradient mechanism for driving the pre-corneal tear film after a blink. J Biomech. 7: 225 (1974).
4. H. Wong, I. Fatt & C. Radke. Deposition and thinning of the human tear film. J Coll Int Sci. 184: 44 (1996).
5. J.M. Tiffany and A.J. Bron. Role of tears in maintaining corneal integrity. Trans Ophthalmol Soc U K. 98: 335 (1978).
6. J.P. McCulley, M.B. Moore and A.Y. Matoba. Mucus fishing syndrome. Ophthalmology. 92: 1262 (1985).
7. D.R. Korb, et al. Tear film lipid layer thickness as a function of blinking. Cornea. 13: 354 (1994).

NEAR VISION ACCOMMODATION IN HORIZONTALITY WITH VDT: WHY LOW BLINKING AND DRY EYE?

Juan Murube and Eduardo Murube

Rizal Foundation for Research in Ophthalmology
University of Alcala
Madrid, Spain

1. INTRODUCTION

Users of current computer Visual Display Terminals (VDT) frequently complain of eye dryness, ocular surface redness, eye itching, burning, irritation, sandiness, headache, glare, neck pain, fatigue and tiredness[1-6]. These same users do not feel these symptoms when reading a book. In this study we examine different positional changes of the eyes and head during both plesiovision, i.e., downward gazing near vision, and horizontal near vision. As VDT occupy a large volume of space, they are generally placed on office tables in a vertical position, obliging the users to look at them horizontally. This study presents the positional changes of the eyes when focusing from a distance (teleopsis) with the eyes in a horizontal position, when focusing near (plesiopsis) with the eyes gazing downward, and when focusing near with the eyes in a horizontal position.

2. MATERIAL AND METHODS

Ten healthy medical students (5 males and 5 females), 23 to 30 years of age, were videotaped while (1) looking at a television display positioned at eye level approximately 5 metres in front of them; (2) reading a book positioned on a table at about chest level, approximately 30 cm in front of their eyes (infravergent near vision); (3) reading text on a desktop VDT positioned at eye level, about 30 cm in front of their eyes (horizontal near vision), and (4) manipulating and observing a shell, a knotted thread, and a book. During the study, the room temperature was 23°C, the relative humidity 55%, and the air draft less than 0.03 m/sec.

Lacrimal Gland, Tear Film, and Dry Eye Syndromes 3
Edited by D. Sullivan *et al.*, Kluwer Academic/Plenum Publishers, 2002

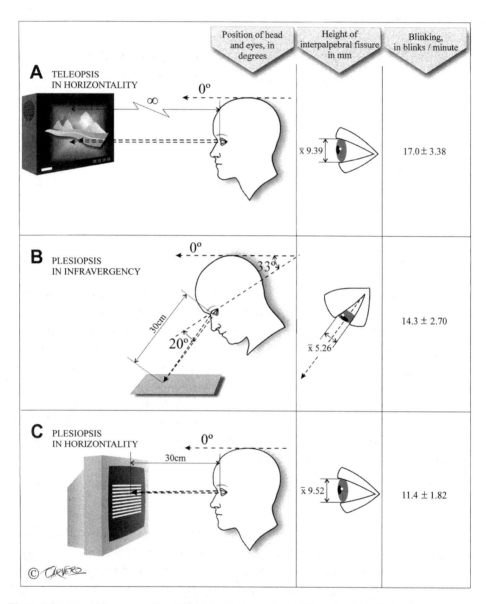

Figure 1. Position of the eyes and head, height of the interpalpebral fissure, and blinking rate in 3 different situations. (**A**) Vision at distance (teleopsis) in primary position of the gaze, (**B**) Close vision (plesiopsis) in downgaze, (**C**) Close vision (plesiopsis) in horizontality.

The subjects were unaware that our interest was centred on their eyes during the videotaping of their left sides, including left eye movements. Each subject was videotaped for 5 to 10 min for each of the first 3 tests, and for about 1 min for the fourth test. The tapes were then displayed on a TV monitor, and the height of the palpebral fissure in the vertical meridian between blinks, as well as the number of blinks, were counted only when the subjects maintained their gazes for 1 min or more during the first 3 tests. The positions of the head and eyes in relation to horizontality were also measured when subjects focused near and downward while reading a book. Heads were maintained in the vertical position, and the eyes gazed horizontally when subjects focused on the TV screen or the desktop VDT.

3. RESULTS

When looking horizontally at a distance (TV display), the upper lids covered the vertical corneal meridians about 1 mm, and the lower lids were tangent to the limbus. The resulting lid apertures were 9.39 ± 0.47 mm·(mean ± sd). The number of blinks/min was 17.00 ± 3.38 (Fig. 1A).

When looking near with infraconvergency (book on table), the inclination of the heads in relation to the theoretical horizontal axes was 33.4 ± 7.91 degrees, and the inclination of the eyes in relation to the axial plane of the heads was 20.16 ± 3.94 degrees, resulting in a total eye inclination of 53.5 ± 8.67 degrees. The upper lids covered the vertical corneal meridians about 4 mm, and the lower lids, about 1 mm. The resulting lid apertures were 5.26 ± 0.46 mm. The number of blinks/min was 14.3 ± 2.70 (Fig. 1B).

When looking near with the eyes in a horizontal position (at the desktop computer VDT), the eyelid apertures were 9.52 ± 0.36 mm, and the number of blinks/min was 11.4 ± 1.82 (Fig. 1C).

When the subjects untied a knotted thread, or observed a shell or a book, they placed the items 25–40 cm in front of their eyes at chest level and at about their mid-sagittal plane.

4. DISCUSSION

The height of the ocular interpalpebral fissures averaged 9.39 mm when gazing from a distance, 5.26 mm during downward near focusing, and 9.52 mm during horizontal near focusing. Therefore, the height of the interpalpebral fissure, and consequently the ocular surface exposure, is about the same in horizontal convergence and horizontal distant vision. However it is is much narrower during downward near vision. This translates to ocular surface exposures (including cornea, medial and lateral conjunctival trigoni and lacrimal lake) of about 2 cm^2 during distance gazing and to about 1 cm^2 during downward near vision,[7] demonstrating that horizontally positioned eyes have greater ocular surface exposure, and consequently, greater fluid evaporation.

Maximum blinking frequencies (average 17 per min) were observed when subjects focused from a distance with their eyes in the primary position. Medium blinking frequencies (average 14.3 per min) were observed when the subjects focused near at a downward angle, and minimum blinking frequencies (average 11.4/min) were observed when subjects focused near horizontally. Other researchers have also reported decreases in blink rates in subjects focusing on VDTs, both at a distance and while focusing near at a downward angle.[1,8-10] Lower blinking frequencies in horizontal plesiopsis during desktop VDT use induces greater evaporation and results in dry eye.[11]

We assume that the slow blink rates observed when subjects focus near at a downward angle results from small palpebral fissures, and consequently, slower fluid evaporation rates. But because the horizontal near and distant fissures are about the same when gazing horizontally, there is no apparent explanation for the difference in blink rate between them of approximately 5.6 blinks/min.

In addition to the well known processes of synkinesis that occur during near vision, e.g., accommodation, convergence and miosis, the present study demonstrates that there are other posturological changes as well. For instance, reading a book on a table is most frequently associated with a 30–35 degree flexion of the neck, a 15–25 degree infravergence of eyes in the orbits, and significant descent of both lids, specially the upper one. The descent of the lids is important because a 10° rotation of the eye displaces the ocular surface about 2 mm, translating to a 4 mm descent of the cornea with an infraversion of 20°. Therefore, in plesiopsis while gazing downward, the upper lid descends about 7 mm and the lower lid about 3 mm, covering the cornea about 4 mm by the upper lid and 1 mm by the lower lid. These postural changes can vary widely between reading a paper held in the hands or placed on the table, during writing, while performing manual tasks, or in relation to the height of the table, the chair or the height of the individual. However, in all these situations, individuals essentially maintain a cervical flexion, an ocular infraversion, and some degree of lid descent.

Gazing horizontally is a natural position for human eyes during distance vision, but it is not natural during near vision. The most natural position for near vision appears to be at a downward angle. For instance, when a stone is given to an ape for observation, or when a book is given to a person to read, the objects are reflexively positioned at chest level in the sagittal plane of their head, resulting in infraconvergence, cervicoflexion, and lid descent. Convergence of the eyes while gazing in a horizontal plane does not appear to be a natural position when maintained for more than a few seconds. Thus, horizontal near vision is unnatural in humans, and is normally maintained only in forced circumstances, and for short intervals. The use of a desktop VDT seems to be the first wide-spread situation in the history of the species where humans maintain horizontal convergence for long periods of time.

To explain positional changes observed during plesiopsis in the horizontal position, we hypothesize that during near vision, not only is there the in a synkineses of accommodation, miosis and convergence that is generally understood, there is also down

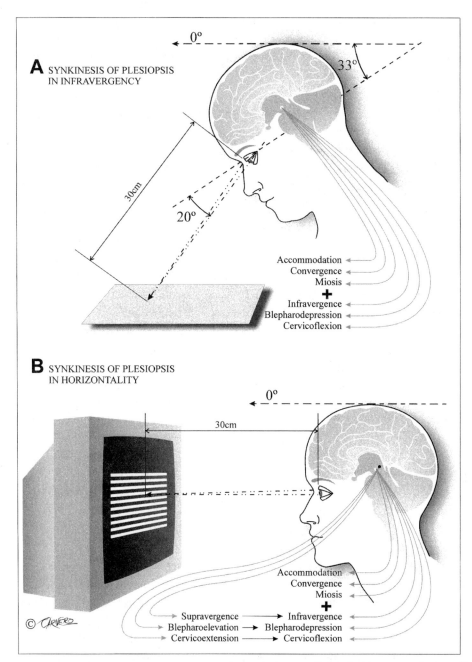

Figure 2. Hypothesis to explain the smaller blinking rate when a subject is looking at near horizontality versus looking at near down gaze. (**A**) The physiologic near vision is in downgaze with corresponding synkinesis: accommodation, convergence and miosis, plus oculodepression, lid descent, and cervicoflexion, (**B**) When looking in near horizontality, the 3 last synkinesis can not be suppressed, and they must be counteracted by activating the center for eye elevation, producing diskinesies and contractures.

gaze positioning, eyelid descent, and cervical flexion (Fig. 2A). Other synkineses of lesser degree are also present. It is possible that all these synkineses are interdependent, and triggered simultaneously by unrestrainable primary or secondary synaptic associations in the hypothetic nucleus of Perlia.

Therefore, it is our supposition that there are 6 synkineses (accommodation, oculoconvergency, miosis, oculodepression, lid descent, and cervical flexion) involved in horizontal near focusing that are triggered simultaneously by the activation of the center for plesiovision. However, to maintain the horizontal position of the eyes, the contraction of the muscles controlling the latter three synkineses, whose simultaneous activation can not be suppressed, must be counteracted by the contraction of their opponent muscles. At that point, the center for conjugate eye elevation, with its corresponding associations, is activated to return the visual axis to a horizontal plane, triggering synkinesis of elevation (eye elevation, lid elevation, inhibition of blinking, neck extension, etc.). Infraconvergence, due mainly to the contraction of the inferior recti (but also of the superior obliques and medial recti), is counteracted mainly by the contraction of the superior recti; the descent of the upper lid, associated with near vision, is counteracted by the contraction of the lid elevators; and cervical flexion is counteracted by cervical extension, etc. These responses provoke fatigue, muscular contractions, wide interpalpebral fissures, diminished blink rates, and increased tear evaporation when horizontal near vision is maintained.

From the results of this study we conclude that horizontal near focusing for a long intervals results in an unnatural position, and that the use of current desktop VDT forces this position. We submit, that focusing near at a downward angle at liquid crystal displays positioned on the table, or at a laptop VDTs, are more physiologic.

We also believe that examination of near binocular vision in persons with binocular fusion, and the study of near vision eye positioning in strabismic children must not be done with tests that force horizontal focusing of eyes. These tests should be performed in the natural position of infravergence, with the tests placed at the level of the chest, permitting natural cervical flexion, downward convergence, and corresponding equilibrium of the extrinsic oculomotor muscles.

Furthermore, the eyelashes of the upper lid occlude 1 – 2 mm of the interpalpebral fissure. When in horizontal vision, the eyelashes do not interfere with the vision of the central cornea, but when in down gaze, they occlude and difract the light entering the upper half of the central cornea. This make us suppose that in physiologic near vision only the image formed with the beam coming from the lower part of the cornea (and reaching the photoreceptors from down to up) is well focused. The plesiopia in horizontality allows to the light entering through the upper part of the cornea to focuse, which, may be, is not physiologic.

REFERENCES

1. Yaginuma Y, Yamada H, Nagai H. Study of the relationship between lacrimation and blink in VDT work. Ergonomics 1990; 33:799–809.
2. Yamano T, Yamada H, Nagai H, Marumoto T, Sekiryu T. Present status and proposals for asthenopia in workers with visual display terminals (VDT). Rinsho Ganka 1986; 40:735–738.
3. Yamano T, Yamada H, Nagai H, Sekiryu T. Study of asthenopia. Jap Rev Clin Ophthalmol 1986; 80:727–731.
4. Rolando M, Martinoli C, Salano R, Copello F, Ravazzoni L. Ocular surface stress bi environmental conditions on VDT workers. A climate controlled chamber study. In: Miglior M et al (eds.) The Lacrimal System 1992. Ghedini. Milano 1993. pp 69–72.
5. Tsubota K, Nakamori K. Effects of ocular surface area and blink rate on tear dynamics. Arch Ophthalmol 1995;113:155–158.
6. Hagan S, Lory B. Prevalence of dry eye among computer users. Optom Vis Sci 1998; 75:712–713.
7. Murube J. Dacryologia Basica. Royper. Madrid. 1981, p 467.
8. Tsubota K, Nakamori K. Dry eyes and video display terminals. N Eng J Med 1983; 328:584 .
9. Foreman J. San Francisco passes ordinance regulating VDT use. Arch Ophthalmol 1991; 109:477.
10. Takano H, Takamura E. Yoshino K, Tsubota K. The increase of the blink interval in ophthalmic procedures. In: Sullivan DA, Dartt DA, Meneray MA (eds.). Lacrimal Gland, Tear Film, and Dry Eye Syndromes. Adv Exp Med Biol 1994, Vol. 350. pp. 525–527.
11. Korb DR, Baron DF, Herman JP, et al. Tear film lipid layer thickness as a function of blinking. Cornea 1994;13:354–359.

ANALYSIS OF TEAR PROTEIN PATTERNS FOR THE DIAGNOSIS OF DRY EYE

Franz H. Grus, Perihan Sabuncuo, Simone Herber, and Albert J. Augustin

Department of Ophthalmology
University of Mainz, Germany

1. INTRODUCTION

Dry eye syndrome occurs very frequently in the industrial world. In the United States, one of five people, 59 millions patients, suffer from symptoms of this disease (Eagle Vision, Yankelovich and Partners, 1997), and the number of patients has doubled in the last 10 years.[1-4] Diagnosis and treatment of dry eye syndrome are challenging. In fact, a curative therapy of this disease does not exist. Dry eye is a disease with various symptoms resulting from aqueous, mucin or lipid deficiency. Different diseases, among them Sjögren's syndrome, can cause this deficiency, but most patients do not have accompanying disorders.[5,6]

Diagnosis of dry eye syndrome is commonly based on clinical tests such as the basal secretory test (BST, Schirmer's test). However, diagnosis is complicated by the fact that this test has no good correlation with the course of the disease and other clinical tests available. This study analyzed the electrophoretic patterns of tears from patients with dry eye disease (DRY) and healthy subjects (CTRL) to diagnose dry eye and compare them to the BST results and subjective symptoms of the patients.

2. METHODS

In this study 408 eyes were examined (153 DRY, 255 CTRL). Patients were grouped primarily according to BST results. For each patient, a clinical score based on subjective symptoms such as discomfort, foreign body sensation, burning, irritation and photophobia

Lacrimal Gland, Tear Film, and Dry Eye Syndromes 3
Edited by D. Sullivan *et al.*, Kluwer Academic/Plenum Publishers, 2002

1213

was calculated. The tear proteins from non-stimulated tears were separated by sodium-dodecyl-sulfate poly-acrylamide gel electrophoresis (SDS-PAGE). Digital image analysis was performed by ScanPacK (Biometra, Goettingen, Germany) and a densitograph was created for each electrophoretic lane (Fig. 1). From each densitograph a data vector was created. The data were analyzed using a multivariate analysis of discriminance and an artificial neural network (ANN, artificial intelligence). These procedures led to a classification of tears based solely on their protein patterns and were described elsewhere in detail.[7-10]

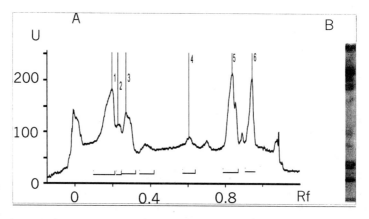

Figure 1. Densitograph (**A**) and photograph (**B**) of an electrophoretic lane of the DRY group. Scanner units (U) are plotted vs. the Rf-values (relative mobility). 1: lactoferrin and sIgA, 2: albumin, 3: heavy chain sIgA, 4: alpha sIgA, 5: lipocalin and 6: lysozyme.

3. RESULTS

In both groups a complex staining pattern was obtained. The number of peaks was significantly increased in DRY compared to controls ($P < 0.05$). The discriminant analysis showed a significant difference between the tear protein patterns of the DRY and CTRL groups ($P < 0.01$). The quality of discrimination was shown by the canonical roots. Fig. 2 demonstrates the canonical roots for the DRY and CTRL groups. The difference between the main protein peaks of both groups was statistically significant ($P < 0.05$), but it was insufficient to be used in a classification procedure. However, the tear patterns could be classified for diagnostic purposes using the densitographic raw data. When classifying unknown samples not included in the training procedure, ANN exceeded the conventional statistical analysis (89% ANN, 73% analysis of discriminance) in its accuracy.

The mean BST values in the CTRL group were significantly higher than the DRY group classifying the patients using their protein patterns. Using the electrophoretical patterns only, the analysis yielded to a significant difference in the mean clinical scores (P

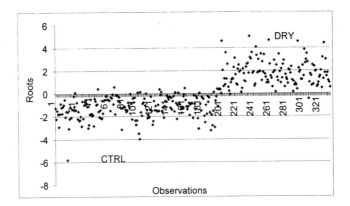

Figure 2. First canonical roots of the DRY and CTRL groups were plotted for each patient. The canonical roots show a good separation between the electrophoretic lanes of the DRY and CTRL groups.

< 0.02) of the groups. The clinical scores of the groups were not significantly different from each other (P < 0.236) if the diagnosis of dry eye were performed by the BST test only.

4. DISCUSSION

The results show significant differences between the DRY and CTRL groups. The tear samples could be classified based on their protein patterns. Diagnosis of dry eye based solely on the electrophoretic patterns had a better correlation with the subjective symptoms of patients compared to the BST results. Thus, the electrophoretic tear evaluation is a fast and useful approach for diagnosing dry eye. Furthermore, the analysis of discriminance can quantify the difference of individual tear samples from the control group. As a result, the procedure could provide an objective analysis of the efficacy of therapeutic treatments on the tear film composition.

REFERENCES

1. O.D.Schein, B. Munoz, J.M. Tielsch, K. Bandeen-Roche, S. West. Prevalence of dry eye among the elderly. *Am J Ophthalmol.*1997; 124: 723–728.
2. K.B. Bjerrum. Keratoconjunctivitis sicca and primary Sjoegren's syndrome in a Danish population aged 30–60 years. *Acta Ophtahlmol Scand.* 1997; 75:281–286.
3. B.E. Damato, D. Allan, S.B. Murray, W.R. Lee. Senile atrophy of the human lacrimal gland: the contribution of chronic inflammatory disease. *Br J Ophthalmol.* 1984; 68: 674–680.
4. T. Hikichi, A. Yoshida, Y. Fukui, et al.. Prevalence of dry eye in Japanese eye centers. *Graefes Arch Clin Exp Ophthalmol.* 1995; 233: 555– 558.

5.   A.F.M. de Roeth. Low flow of tears: the dry eye. *Am J Ophthalmol.* 1952; 35:782–787.

6.   N.J. Van Haeringen. Clinical biochemistry of tears. *Cur Eye Res.* 1981; 26:84–96.

7.   F.H. Grus, A.J. Augustin, N.G. Evangelou, K. Toth-Sagi. Analysis of tear-protein patterns as a diagnostic tool for the detection of dry eyes in diabetic and non-diabetic dry eye patients. *Eur J Ophthalmol.* 1998; 8,2: 90–97.

8.   F.H. Grus, A.J. Augustin. Analysis of tear protein patterns by a neural network as a diagnostical tool for the detection of dry eyes. *Electrophoresis* 1999; 20: 875–880.

9.   F.H. Grus, A.J. Augustin. Proteinanalytische Methoden bei der Diagnostik des Sikka-Syndroms. *Der Ophthalmologe* 2000; 97:54–61.

10.   F.H. Grus, A.J. Augustin, F. Koch, C.W. Zimmermann, J. Lutz. Autoantibody repertoires in animals with lens-induced uveitis after different therapies using megablot technique: a comparison. *Adv Ther.* 1996; 13,4:203.

THE SIGNIFICANCE OF SALIVARY GLAND ULTRASONOGRAPHY IN THE DIAGNOSIS OF DRY EYE SYNDROME

Iwona Switka - Więcławska,[1] Lidia Portacha,[1] Tadeusz Kęcik,[1] and Hanna Markiewicz[2]

[1]Department of Ophthalmology
[2]Department of Dento- Maxillo- Facial Radiology
Medical University
Warsaw, Poland

1. INTRODUCTION

The differential diagnosis between Sjögren's (SDE) and non-Sjögren's dry eye (NSDE) is critical to the prevention of severe complications in Sjögren's syndrome, and to favorable prognoses in both forms of the disease. The purpose of this study is to determine the value of salivary gland ultrasonography (US) to ophthalmologists as a tool in the diagnosis of both SDE and NSDE.

2. MATERIALS AND METHODS

Three hundred and eight patients, 271 women and 37 men, aged 39–99 years (average 74 years), were examined for signs and symptoms of dry eye disease. A careful history was taken from each patient regarding ocular and other physical symptoms. The diagnosis of dry eye was based on following tests: complete ocular examination, fluorescein staining, rose bengal staining, tear breakup time (BUT), and unanesthetized Schirmer test (Schirmer 1 test). All patients had symptoms of dry eye in various stages from mild to severe. Subsequently, examination of the parotid and submandibular salivary glands by US with a scanner operating at a frequency of 7.5 MHz. Finally, all patients were examined by a rheumatologist to confirm or exclude Sjögren's syndrome.

Lacrimal Gland, Tear Film, and Dry Eye Syndromes 3
Edited by D. Sullivan *et al.*, Kluwer Academic/Plenum Publishers, 2002

3. RESULTS

Symptoms typical for Sjögren's syndrome were found by salivary gland US in 93 women and 7 men. All of these patients, aside from those with keratoconjunctivitis sicca or xerostomia, had rheumatoid arthritis. Bilateral changes were found in salivary glands of 92 patients, and unilateral changes were found in 8 patients. Fig. 1 shows the localization of these changes.

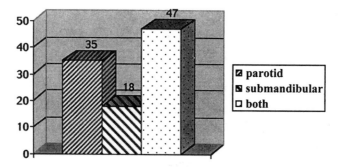

Figure 1. Number of patients with echogenic findings by US of parotid, submandibular or both salivary glands.

In Sjögren's syndrome, the sonographic image is highly dependent on the duration and the degree of inflammation of the glandular parenchyma. Patients suffering from SDE have enlarged, non-homogeneous, hypoechogenic salivary glands (Fig. 2).[2,7,8]

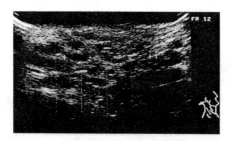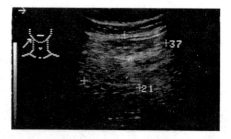

Figure 2. Parotid **(left)** and submandibular **(right)** salivary glands with hypoechogenic foci of patient with Sjögren's syndrome. Numerous hypoechogenic circumscript masses are detected within the parenchyma and may correspond to cystic duct extensions or enlarged intraglandular lymph nodes. Salivary gland structure appears "cloudy" in the sonograph.[8]

In all patients, a positive Sjögren diagnosis by US was independently confirmed on examination by a rheumatologist. Primary SDE was diagnosed in 44 patients, and secondary SDE was diagnosed in 56 patients. In 205 patients, salivary gland US did not show characteristic changes for SDE (Fig. 3).Within this group, despite normal US findings, 2 patients (0.9%) were diagnosed with Sjögren's syndrome by a rheumatologist. Sjögren's syndrome was excluded in the remaining 203 patients with negative US findings.

4. DISCUSSION

Dry eye syndrome consists of a variety of disorders that are differentiated by etiology, pathogenesis, clinical findings, and prognosis. Among dry eye disorders, there are those with a severe ocular course, like pemphigoid and SS, but there are also those with quite mild courses, for example NSDE.[3,4] SS is an autoimmune disorder characterized by a chronic inflammation of the lacrimal and salivary glands that results in keratoconjunctivitis sicca and xerostomia (sicca complex).[1,6] Sicca complex, in the absence of associated connective tissue disease, is known as primary Sjögren's syndrome. Secondary Sjögren's syndrome is defined as a sicca complex in association with other autoimmune disorders.[5,6] In most cases, the ocular outset of this disorder is quite mild, as in NSDE, but it can be exacerbated later, even by corneal perforation.[3,4]

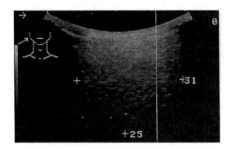

Figure 3. Normal sonographic image- similar homogeneous echogenicity with sharp borders.

The diagnosis of dry eye syndrome is difficult, and is typically based on both clinical signs and symptoms, as well as on specific diagnostic tests. Ocular findings sometimes correlate with oral symptoms of xerostomia, as well as rheumatologic evaluation and corresponding abnormal serological test results. Many of these tests are very expensive with poor accessibility. US of salivary glands is a quick, non-invasive diagnostic test that is complimentary to traditional diagnostic tools.

Sonography has established itself as the primary imaging technique in the diagnosis of salivary gland diseases.[2,8] The salivary glands are arranged in pairs, and due to their superficial location, are easily accessible to sonographic diagnosis. Under normal conditions, they exhibit similar homogeneous, intermediate echogenicity with sharp borders.[2,7,8] In Sjögren's syndrome, hypoechogenic foci can be detected. Although normal

US findings cannot exclude the disease, a diagnosis of the disease is less likely in the absence of positive US findings.

In conclusion, ultrasonography of salivary glands is an easy, quick, non-invasive test that may objectively show alterations in salivary gland parenchyma. Examination of patients indicates that this test is a valuable preliminary test in diagnosing dry eye syndrome. Hypoechogenic foci found in US are indicative of Sjögren's syndrome. Normal US findings cannot exclude the disease, but in the absence of positive findings, Sjögren's syndrome is less likely.

REFERENCES

1.   S.Carson, J.Scinka , and W.Total : The Sjögren Syndrome handbook foundation. Inc Port Washington, New York (1989)
2.   D. E. March , V.M. Reo, D. Zwillenberg : Computed tomography of salivary glands in Sjögrens Syndrome , Arch Otolaryngol. Head & Neck Surg ; 115: 105 (1989)
3.   L .Portacha : Do_wiadczenia w_asne w leczeniu zmian ocznych w zespole Sjögrena. Klin Oczna 86: 485 (1984)
4.   M..E. Prost : Leczenie zaburze_ stabilno_ci filmu _zowego. Nowa Medyc 2: 29 (1995)
5.   M. Semp, E. Morgnard :The dry eye. It comprehensive guide. Springer Verlag Berlin, Heidelberg New York (1992)
6.   N. G. Snebold : Noninfections orbital inflammations and vasculitis; in Albert & Jakobiec: Principl and Practice of Ophthal 3: 1923(1994)
7.   G. E. Valvassori, F. M. Mafee, and L.B.Carter: Imaging of the head and neck: 475 (1995)
8.   T. J. Vogl, J.Balzer, M.Mack, W.Steger: Differential diagnosis in head and neck imaging :245 (1999)

THE EFFECT OF NASAL MUCOSAL STIMULATION ON SCHIRMER TESTS IN SJÖGREN'S SYNDROME AND DRY EYE PATIENTS

Aya Fujisawa,[1] Kazuko Kitagawa,[1] and Susumu Sugai[2]

[1]Department of Ophthalmology
[2]Department of Internal Medicine (Hematology and Immunology)
Kanazawa Medical University
Uchinada, Ishikawa, Japan

1. INTRODUCTION

The Schirmer test evaluates tear quantity and reflects the aqueous component of tear fluid. This test distinguishes dry eye patients, including patients with Sjögren's syndrome (SS), from normal subjects. It is difficult, however, to differentiate patients without SS (non-SS) from patients with SS using this test only. Tsubota reported the importance of the Schirmer test with nasal stimulation and noted the test was useful to determine the reserve function for aqueous tear production.[1]

Further tests are necessary to determine the state of the lacrimal glands, because few methods can determine lacrimal dysfunction. In addition, reliable examinations, such as lip biopsy for evaluating damage to the salivary glands, are usually difficult to perform.[2] The Schirmer test with nasal stimulation reflects the degree of lacrimal lesions and seems to be useful for evaluating the severity of dry eye and for differentiating non-SS dry eye from SS. This study evaluated the usefulness of the Schirmer test with nasal stimulation in the diagnosis of patients with SS.

2. METHODS

The subjects were 40 patients with dry eye who visited our hospital from March 1999 to August 1999. Of the 40, 22 were diagnosed with SS using the Japanese criteria,[3] and all were female with an average age of 55.7±12.3. Among the SS patients, 16 cases

Lacrimal Gland, Tear Film, and Dry Eye Syndromes 3
Edited by D. Sullivan *et al.*, Kluwer Academic/Plenum Publishers, 2002

1221

were female with an average age of 55.7±12.3. Among the SS patients, 16 cases were primary (1SS) and 6 were secondary (2SS). Of the 2SS patients, 4 had a basal disease of systemic lupus erythematosus, 1 rheumatoid arthritis and 1 mixed connective tissue disease. Of the 40, 18 (13 female, 5 male) suffered from dry eye with negative autoantibodies (non-SS). The average age of the non-SS group was 53.3±10.0.

The degree of keratoconjunctivitis sicca (KCS) was estimated according to meniscus height, fluorescein and rose bengal scores[4] and tear breakup time. The Schirmer test was performed under topical anesthesia, and the Schirmer test with nasal stimulation was performed under topical anesthesia according to Tsubota.[1] A cotton-tipped applicator was inserted into the nose until the tip reached the nasal membrane of the ethmoid sinus on the side with the lower Schirmer values (Fig. 1). Salivary gland function was estimated using the following examinations: chewing gum test, Saxon test, salivary scintigraphy and lip biopsy (focus score[5]). Serological tests evaluated the levels of IgG, IgA, IgM, antinuclear antibody, anti-SS-A antibody and anti-SS-B antibody. Correlations of the Schirmer values with nasal stimulation and other disorders of the salivary glands and of the serological tests were investigated. A statistical analysis was performed using Spearman's correlation coefficient and *t* test.

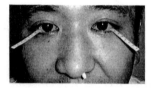

Figure 1. Schirmer test with nasal stimulation. A cotton-tipped applicator was inserted into the nose.

3. RESULTS

Although no difference between fluorescein scores of SS and non-SS patients occurred, the rose bengal scores were significantly higher in SS (Fig. 2). Although the Schirmer values without nasal stimulation could not distinguish SS from non-SS, the values with nasal stimulation were significantly lower in SS than non-SS (Fig. 3). While a significant correlation was observed among the Schirmer values with nasal stimulation compared to the rose bengal scores for both eyes and Saxon test scores (Figs. 4 and 5), none was seen with the chewing gum test and other tests. The anti-SS-A antibody titers showed no correlation with the Schirmer values, both with and without nasal stimulation. However, when comparing the amount of increment after stimulation, the values in patients with positive antibody titers were significantly lower than those of SS patients with negative antibody titers (Fig. 6). The sensitivity and specificity of the Schirmer values of nasal stimulation were investigated to determine cut-off values at several points that help differentiate SS from Non-SS. Increments rather than direct values also showed

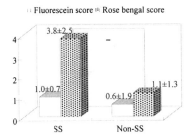

Figure 2. Comparison of fluorescein and rose bengal scores of SS and Non-SS. The rose bengal scores were significantly higher in SS(*P < 0.05).

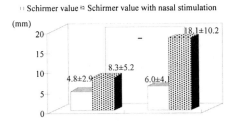

Figure 3. Comparison of Schirmer values with and without nasal stimulation. The values with nasal stimulation were significantly lower in SS, although the values without stimulation did not show significant difference (*P < 0.05).

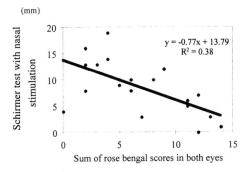

Figure 4. Correlation of Schirmer values with nasal stimulation and rose bengal scores. A significant correlation was observed between them (P < 0.05).

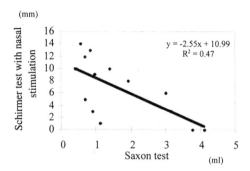

Figure 5. Comparison of Schirmer values with nasal stimulation and Saxon test. A significant correlation was observed between them (P < 0.05).

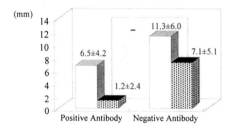

Figure 6. Comparison of Schirmer values and anti-SS-A antibody titers. The amount of increment in patients with positive antibody titers was significantly lower than that in patients with negative antibody titers (P < 0.05).

Table 1. Cut-off value of increment of Schirmer values with nasal stimulation. When the cut-off value was set at 4 mm, the sensitivity and specificity scores were optimal

Cut-off value	Sensitivity (%)	Specificity (%)
2 mm	45.5	88.9
4 mm	68.2	88.9
6 mm	77.3	61.1
8 mm	90.9	38.3

higher percentages with nasal stimulation. When the cut-off value was set at 4 mm, the sensitivity and specificity scores were optimal: 68.2% and 88.9%, respectively (Table 1).

4. DISCUSSION

When comparing the ocular findings of SS to those of non-SS, rose bengal scores were significantly higher in SS, although no difference in fluorescein scores occurred. The intensive staining of the bulbar conjunctiva with rose bengal suggests SS. However, even with mild or no staining, a diagnosis of SS cannot be ruled out, because almost half the patients diagnosed with salivary and serum abnormalities showed minimal or no rose bengal staining.[2] While the Schirmer values showed no difference between SS and non-SS, the Schirmer values with nasal stimulation were significantly lower in SS. Thus, the reserve function of the lacrimal glands for producing aqueous tear is lower in SS, and a decreased reserve function might produce ocular surface damage that is indicated by rose bengal staining. This may explain the good correlation between rose bengal scores and Schirmer values with nasal stimulation. The Saxon test measures the weight of the salivery fluid when sterile gauze is chewed for 2 min and relates more significantly with the Schirmer values with nasal stimulation than the chewing gum test. The two tests for measuring salivary fluid are different in nature.

Comparing Schirmer values of subjects with and without positive titers of anti-SS-A antibody, no significant correlation was seen in Schirmer tests with or without nasal stimulation. However, the increment after stimulation was significantly lower in SS patients. The lower increment correlates with higher titers of anti-SS-A antibody. Keratoconjunctivitis sicca becomes more severe in SS patients with positive anti-SS-A antibody,[2] and the results obtained here showed a lower increment after nasal stimulation strongly related to the presence of a positive antibody and deterioration of ocular surface lesions. No correlation existed between the values of nasal stimulation and other examinations of SS. It is necessary to investigate 1SS and 2SS differently, because better correlation can be expected in 1SS than in 2SS.[3]

The sensitivity and specificity of Schirmer values with nasal stimulation for differentiating SS from non-SS showed the best values (almost 70%, 90%, respectively) when the cut-off value was set at 4 mm of the increment. The Schirmer test with nasal stimulation is a good test for evaluating damage to the lacrimal glands, since the reserve function of the glands is induced, and differentiating SS from non-SS.

REFERENCES

1. K. Tsubota: The importance of the Schirmer test with nasal stimulation. *Am. J. Ophthalmol.* 111:106(1991)
2. K. Kitagawa, N. Kohda, S. Sugai, and T. Ogawa: Correlation of corneoconjunctival lesions to salivary disorders
and aoutoantibodies in primary Sjogren's syndrome. *Rinsho Ganka (Jpn. J. Clin. Ophthalomol.)* 51:1913(1997)

3. K. Kitagawa, T. Nakamura, and S. Sugai: Keratoconjunctivitis sicca versus dry mouth and autoantibodies in primary and secondary Sjogren's syndrome (in this proceeding)
4. O.P. van Bijsterveld: Diagnostic tests in the sicca syndrome. *Arch. Ophthalmol.* 82:10(1969)
5. J.S. Greenspan, T.E. Daniel, N. Talal, and Sylyester P.A.: The histopathology of Sjogren's syndrome in labial salivary gland biopsies. *Oral Surg. Oral Med. Oral Pathol.* 37:217(1974)

A SIMPLE METHOD OF DETECTING THE ANTI-INFLAMMATORY ACTION OF TOPICAL DRY EYE TREATMENTS

D. L. MacKeen[1,2] and H.-W. Roth [2,3]

[1]MacKeen Consultants Ltd.
Bethesda, Maryland, USA
[2]Georgetown University Medical Center
Washington, DC, USA
[3]Institut für wissenschaftliche Kontaktoptik
Ulm, Germany

1. INTRODUCTION

Clinical criteria for the diagnosis of dry eyes are limited; most patients are diagnosed on the basis of subjective responses. This makes assessment of dry eye treatments difficult in clinical practice and research, and additional clinical tests are needed. Morgan et al.[1] reported an increase in corneal temperatures in dry eye patients; measurements were made with (infrared) IR thermography. Presumably the increased temperature is due to the increased blood flow associated with the accompanying inflammation.

Using a thermistor unit designed for the measurement of skin temperatures, we found that lid temperatures of the closed eye also increased in dry eye patients. Using this sign, we then assessed the effects of various treatments in patients with moderate dry eye. Measurements with the thermistor were inexpensive, simple to perform and took less than 1 sec. Initial measurements of patients were similar and repeatable. Although the temperature changes of patients with severe dry eye seem to be greater and of longer duration, we excluded them from this study, because of lid temperature variability, the probability of other confounding medical problems, and the difficulty of recruiting sufficient numbers needed for a clinical study.

Lacrimal Gland, Tear Film, and Dry Eye Syndromes 3
Edited by D. Sullivan *et al.*, Kluwer Academic/Plenum Publishers, 2002

1227

2. METHODS

A total of 272 male and female patients (544 eyes) from 35 to 75 years of age participated in this study. Each met the following criteria: mild dry eye with chronic conjunctivitis, central corneal thickness between 475 and 525 microns, and closed lid surface temperature of 37.0 ± 0.2°C. Corneal thickness measurements were by micropachometry using a modified Ophtho-Sonic® pachometer.[2] Lid surface temperature was taken with a Thermoprobe® that was calibrated before each set of measurements. Patients were asked to close their lids; the thermistor probe was then touched gently to the approximate midpoint of each upper lid. Triplicate measurements were made and averaged for each eye. This was done before instillation of saline, commercially available artificial tears (AT, 30 patients each) or applications to the lower lid of DEO™, a patented[3] formulation of calcium carbonate in a petrolatum-based ointment (22 patients) or petrolatum (Vaseline®, 10 patients).

AT formulations were sterile solutions adjusted to tonicity with electrolytes. Each contained the following polymers[4]: I, polyvinyl pyrrolidone (PVP) and polyvinyl alcohol (PVA)*; II, PVP and hydroxymethylcellulose (HPMC); III, PVP and PVA*; IV, Carbapol 940; V, PVP; VI, Heparin; VII, Poloxamer 407 (Tetronic 1304). Subsequent measurements of each eye were taken at 10-min intervals until the lid temperature returned to initial values. For statistical analysis, the initial values (37.0 ± 0.2°C) were adjusted to 37.0°C. (*Different concentrations-different commercial solutions).

3. RESULTS

Fig. 1 shows the time-dependent temperature curves for all test items recorded with the thermistor at 10-min intervals in 271 patients. The maximum time of return to baseline temperatures (RTB) was 110 min following lower lid application of DEO™. The RTB for aqueous preparations ranged from 50 to 90 min, and 50 min for saline. Vaseline reached a maximum decrease in lid temperature (MTD) at 30 min, then rose nearly linearly to baseline at 90 min; the temperature was greater than 36.5°C after 50 min.

MTD was exhibited following DEO™; the temperature decrease exhibited an inverted plateau, i.e., less than 35.5°C between 10 and 80 min (Fig. 1). MTD for aqueous preparations was seen at 20 min and was identical for saline and AT II. The values for the remainder of the commercial aqueous AT were grouped. All reached 36.5°C or higher by 60 min; the MTD of one commercial product was 36.7°C.

Data from subjects not included in this report were obtained from patients with more severe dry eye conditions. These data echoed those reported from the IR thermometer findings, e.g., the temperature-lowering effect of DEO™ lasted for approximately 4 h, but the effects of aqueous instillations were not increased. We also tested similar patients using the thermistor. These patients could not provide the homogeneity needed for statistically meaningful study, largely because of their variability. IR thermography was performed on a male with severe dry eye using an Everest Model 310 IR thermometer before and after application of DEO™ to the skin of the lower lid. Triplicate measurements were taken at intervals over a 5–6 h period following application of DEO™ and the values averaged. The initial temperature of this subject varied, but the temperature decrease persisted for more than 4 h (Table 1).

Table 1. IR thermometer measurements of upper lid temperatures of a subject with severe dry eye before and after application of DEO™ to the lower lid

	0 h	1 h	4 h	5 h	6 h
Session 1	39.1°C	37.1°C	36.8 °C	36.9 °C	N.D.
Session 2	37.3 °C	36.8 °C	36.6 °C	N.D.	37.0 °C

4. DISCUSSION

Lid temperature measurements with a thermistor provide another needed diagnostic value in the assessment and treatment of dry eye patients. Its cost is low, and measurements are simple and quick to perform. The investigator or assistants could readily make the measurements while the subject was seated in the examination chair; each of the triplicate measurements took about 1/10 sec. This enabled rapid measurements of numerous patients at frequent intervals. All these factors make such a method appropriate for any clinician.

All of the commercially available AT exhibited similar curves; the temperature-lowering effect lasted 10 or 20 min longer than saline, presumably due to the composition of aqueous polymers. Only one, AT II, had an MDT equal to saline. Two solutions (AT III and IV) oscillated in the first 40 min; the reasons for this are not readily apparent from the polymeric components.

As previously reported[5] the average central corneal thickness (CCT) is 545 ± 20 μ. In patients with mild to moderate dry eye, this value is decreased as the result of excessive cell water loss. Instillations of aqueous preparations can rapidly increase the CCT presumably by osmosis. Depending on the formulation, temporary rehydration of all epithelial cells occurs. Because of its larger surface area, the tarsal and bulbar conjunctivas may act as reservoirs to prolong the rehydration of the exposed corneal epithelial cells.

4. CONCLUSIONS

This additional method of quantifying a symptom of dry eye treatment is simple, non-invasive, and requires relatively inexpensive equipment. Expanded available dry eye tests,

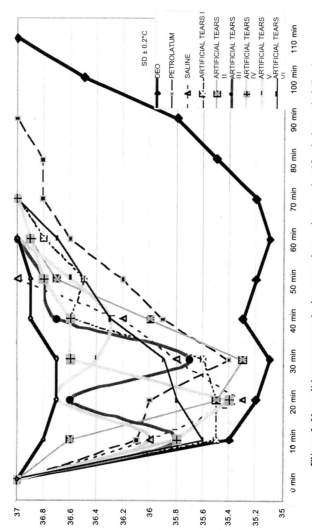

Figure 1. Upper lid temperatures in dry eye patients prior to and at 10-min intervals following a single instillation or application of various dry eye treatments.

including lid temperature measurements to assess inflammation, could enhance research and multicenter clinical investigations.

Saline or commercial AT instillations cause transient decreases of lid temperature. This appears to be a local cooling resulting from the admixture of a relatively large volume of room temperature water and overheated tears. Certain water-soluble mixtures prolonged the temperature-lowering effect.

DEO™ applied to the lateral lower lid skin resulted in a significant decrease of lid temperature lasting over an hour. The greater the initial temperature, the greater the temperature decrement and duration of action. These findings suggest that the initial response from the application of DEO™ results from two actions: (1) chemical and pharmacological actions of calcium carbonate shown to increase with the duration of supracutanaeous movement into the eye[5-8], and (2) a concomitant minor effect of the slow improvement of the lipid layer by the input of petrolatum, which presumably reduces friction during the blink. The lubricating action of petrolatum without any anti-inflammatory activity is seen from the 20-min delay to reach an MTD of DEO™ and the nearly linear rise from that point to 37°C. This rise paralleled the aqueous tear preparations containing polymers. These test results are limited to measurements made in a single day and single applications, and do not reflect the cumulative action reported by Tsubota et al.[6] and Roth[7] for calcium carbonate ointment applied to the lid skin.

We previously tested the effect of DEO™ against preserved and unpreserved saline and Lens Fresh on dry eye patients using micropachometry.[8] We concluded:

"Different dry eye formulations show different effect on the corneal thickness. There are two sets of actions based on normalization of corneal thickness: rapid onset and short duration of effect observed with an instilled aqueous preparation vs. slow onset and prolonged effect from a supracutaneously applied ointment formulation."

A long-term test of DEO™ in comparison with AT is planned using lid temperature as a criterion for success. Further long-term research is planned to contrast the lid temperature-lowering effects of DEO™ and aqueous AT. Finally, plans for *ex vivo* tests of a current hypothesis regarding the anti-inflammatory action of calcium carbonate are planned for the near future. The rationale for this investigation involves phagocytized calcium carbonate interference with the intracellular catheptic processing of antigens, an assumption based on the work of Miskuochi et al.[9]

ACKNOWLEDGMENTS

The authors wish to thank Austin Mircheff for helpful discussions regarding aspects of calcium carbonate's pharmacology.

REFERENCES

1. P.B. Morgan, A.B. Tullo, N. Efron. Infrared thermography of the tear film in dry eye. *Eye* 9:615–8 (1995).

2. H.W. Roth, D.L. MacKeen. Micropachometric differentiation of dry eye syndromes, in *Lacrimal Gland and Dry Eye Syndromes*. D.A. Sullivan, ed. Plenum Press, NY (1994).

3. D.L. MacKeen, US Patents 5,290,572 , 5,366,739, 5,595,764, 5,830,508 and patents pending. Australian, Canadian, Mexican and European patents and patents pending.

4. J. Murube, A. Paterson, E. Murube. Classification of artificial tears I and II, in *Lacrimal Gland and Dry Eye Syndromes 2*. D.A. Sullivan, ed. Plenum Press, NY (1998).

5. H.W. Roth. Hornhautpachometrie beim Gesunden, Erkrankten und beim Kontaktlinsentrager. Enke Verlag, Stuttgart (1994).

6. K.Tsubota, Y. Monden, Y. Yagi, M. Goto, S. Shimmura. New treatment of dry eye: the effect of calcium ointment through eyelid skin delivery. *Brit J Ophthalmol.* 83(7):767–770 (1999).

7. H.W. Roth. Use of corneal thickness changes to compare the efficacy of conventional eye drops with supracutaneous treatment of dry eye, in *Lacrimal Gland and Dry Eye Syndromes 2*. D.A. Sullivan, ed., Plenum Press, NY (1998).

8. D.L. MacKeen, H.W. Roth, and P.D. MacKeen , Supracutaneous treatment of dry eye patients with calcium carbonate, in *Lacrimal Gland, Tear Film and Dry Eye Syndromes 2*. D.A. Sullivan, ed. Plenum Press, NY(1998).

9. T. Miskuochi., S.T. Yee, M. Ksai, T. Kakiuchi, D. Muno, E. Kominami. Both cathepsin B and cathepsin D are necessary for processing ovalbumin as well as for degradation of class II HMC invariant chain. *Immunol Lett.* 43:189–193 (1994).

EFFICACY OF SODIUM HYALURONATE EYE DROPS OF DIFFERENT OSMOLARITIES IN THE SYMPTOMATIC TREATMENT OF DRY EYE PATIENTS

Giovanni Milazzo[1], Vincenzo Papa[1], Pasquale Aragona[2,] Simona Russo[1], Pietro Russo[1], and Alessandro Di Bella[1]

[1]Clinical Research Department SIFI S.p.A.,
Catania, Italy
[2]Institute of Ophthalmology
University of Messina, Italy

1. INTRODUCTION

High tear film osmolarity is often associated to keratoconjunctivitis sicca (KCS).[1] This increased osmolarity (up to 30–40 mosm/l) may be, at least in part, responsible for both symptoms and signs of KCS. In addition in vitro studies have demonstrated that such high osmolarity may be toxic to corneal epithelium.[2] Since the treatment of KCS is mainly based on the use of artificial tears solutions and/or ocular lubricants, an attempt has been made to reduce such osmolarity by applying hypotonic solutions.[2] However, the clinical results obtained by using hypo-osmolar or normally osmolar substances have been contradictory.[1,3,4] Sodium hyaluronate (NaHa) solutions have beneficial effects in the relief of symptoms and conjunctival epithelial conditions in dry eye patients.[5,6] The present study was designed to assess the effects of two NaHa solutions, with different molecular weight and tonicity, on symptom relief in KCS patients.

2. PATIENTS AND METHODS

This was a multi-centered, double masked, crossover clinical trial conducted in 158 patients with clinically well-defined dry eye syndrome. Only patients with history of dry eye for at least two months, tear film abnormalities (tear breakup time [BUT] < 10 sec and/or Schirmer's test < 5.0 mm in one or both eyes) and ocular surface damage (rose bengal

Lacrimal Gland, Tear Film, and Dry Eye Syndromes 3
Edited by D. Sullivan *et al.*, Kluwer Academic/Plenum Publishers, 2002

1233

and/or fluorescein test score of at least 3 in at least one eye) were included in the study. Patients received two ophthalmic solutions, one hypotonic (215 mOsm/l) and the other isotonic (305 mOsm/l), both containing NaHa. Treatments were given over a period of 28 days for each study medication. A 7 day wash out period separated the two treatments. The viscosities of the two solutions were similar. Clinical symptoms and signs of ocular surface damage (stinging, burning, foreign body sensation, heaviness of the lids, photophobia, global ocular symptoms, tear BUT, fluorescein stain, rose bengal stain and anesthetized Schirmer's test) were performed at the end of each treatment period.

3. RESULTS

Data from 139 patients who completed the study were used for the statistical analysis. Efficacy analysis was based on pooled results from both treatment arms for each of the two compounds used. Both the hypotonic and isotonic solutions of NaHa were found to be effective for the resolution of symptoms in dry eye patients. Most of the maximally obtainable relief occurred within one month of treatment. For most symptoms and signs, there was no statistical difference between the two compounds (Fig. 1). However, tear BUT was significantly improved by the hypotonic solution (P = 0.021). The local tolerance was good and no adverse events were reported.

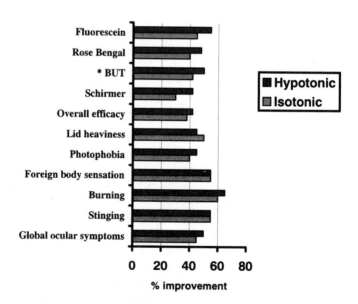

Figure 1. Efficacy of isotonic and hypotonic solutions on symptoms and signs of dry eye. Data are expressed as % improvement over baseline. *P = 0.021.

4. DISCUSSION

The treatment of dry eye is aimed to stabilize the tear film. To date no ideal tear substitute is available for relief of symptoms in such disorders. Damage to the ocular epithelium surface seems to be associated to KCS, and it was suggested that increased tear osmolarity could be responsible for such damage.[1,2] However, clinical results obtained by using hypo-osmolar or normally osmolar substances have been contradictory. Some studies have demonstrated subjective improvements in eyes receiving hypotonic solutions[1,3] but these findings were not confirmed in other reports.[4] In the present study we tested two NaHa containing solutions with different osmolarity and found that both were equally effective in ameliorating both signs and symptoms of KCS. In particular our study showed that ocular surface damage, evaluated by either rose bengal or fluorescein stain, may be reduced with NaHa containing artificial tears despite their osmolarity. Interestingly, during the one week saline wash-out phase, symptoms worsened markedly in both treatment arms, suggesting that the improvement could indeed be related to the viscosity enhancing agent NaHa. In conclusion, although an increased tear osmolarity has been found in dry eye patients and it has been considered toxic for the corneal epithelium,[1,2] the results of the present study indicate that greater importance should be ascribed to the substances which constitute the tear substitute solution rather than their physical properties, such as osmolarity, since a reduction of tear osmolarity could be obtained also by the administration of isotonic solutions. From this perspective, NaHa, for its well known properties such as non-Newtonian characteristics, efficacy on wound healing and anti-inflammatory action,[8,9] can be considered particularly useful for the treatment of dry eye.

REFERENCES

1.   M. Rolando, M.F. Refojo, and K.R. Kenyon. Increased tear evaporation in eyes with keratoconjunctivitis sicca, *Arch. Ophthalmol.*101: 557 (1983).

2.   J.P. Gilbard, J.B. Carter, D.N. Sang, M.F. Refojo, L.A. Hanninen, and K.R. Kenyon. Morphologic effect of hyperosmolarity on rabbit corneal epithelium, *Ophthalmology* 91: 1205 (1984).

3.   J.P. Gilbard, and R.L. Farris. Tear osmolarity and ocular surface disease in keratoconjunctivitis sicca, *Arch. Ophthalmol.* 97:1642 (1979).

4.   P. Wright, M. Cooper, and A.M. Gilvarry. Effect of osmolarity of artificial tear drops on relief of dry eye symptoms: BJ6 and beyond, *Br. J. Ophthalmol.* 71:161 (1987).

5.   J.D. Nelson, and L.R. Farris. Sodium hyaluronate and polyvinyl-alchol artificial tears preparations. A comparison in patients with keratoconjunctivitis sicca, *Arch. Ophthalmol.* 106:484 (1988).

6.   B.B. Sand, K. Marner, and M.S. Norn. Sodium hyaluronate in the treatment of keratoconjunctivitis sicca. A double masked clinical trial, *Acta Ophthalmol.* 67:181 (1989).

7.   S. Shimmura, M. Ono, K. Shinozaki, I. Toda, E. Takamura, Y. Mashima, and K. Tsubota. Sodium hyaluronate eyedrops in the treatment of dry eyes, *Br. J. Ophthalmol.*79:1007 (1995).

8.   P.H. Wiegel, S.J. Frost, C.T. Mc Gary, and R.D.LeBouef. The role of hyaluronic acid in inflammation and wound healing, *Int. J. Tissue React.* 10:355 (1988).

9.   T.Nishida, M.Nakamura, H. Mishima, and T. Otori. Hyaluronan stimulates corneal epithelial migration, *Exp. Eye Res.* 53:753 (1991).

TOPICAL NON-PRESERVED DICLOFENAC THERAPY FOR KERATOCONJUNCTIVITIS SICCA

M. Rolando, S. Barabino, S. Alongi, and G. Calabria

Department of Neurological and Vision Sciences
Ophthalmology R
University of Genoa, Italy

1. INTRODUCTION

Keratoconjunctivitis sicca, regardless of origin, is associated with variable levels of inflammation of the ocular surface.[1] Chronic inflammation induces and maintains the metaplastic changes that typically develop in the conjunctival epithelium of patients with this disease.[2,3] Although inflammatory phenomena associated with KCS can be immunomediated and/or induced by local prostaglandin secretion, it is unknown which of the two mechanisms has the prevailing role in the genesis of injury. Nevertheless, it is reasonable to assume that inflammatory prostaglandins are liberated as a result of epithelial damage. In support of this hypothesis, a recent study demonstrated the efficacy of a therapy consisting of cycles of topically administered steroids in the treatment of KCS(4). However, the study also reported notable risks to other aspects of ocular structural health from the well known side effects of corticosteroids, e.g., intraocular hypertension and cataracts. A therapy with non-steroidal and non-preserved anti-inflammatory agents in association with tear substitutes could be an effective therapy for the reduction of inflammation of the ocular surface and its associated symptoms.

2. MATERIALS AND METHODS

Twelve patients (1 male, average age 49.6 yrs) suffering from KCS in both eyes were included in this pilot study. The diagnosis of KCS was based on the presence of typical symptoms and objective test results. Each patient exhibited at least two of the following subjective symptoms: burning, foreign body sensation, pain, blurred vision improved by

Lacrimal Gland, Tear Film, and Dry Eye Syndromes 3
Edited by D. Sullivan *et al.*, Kluwer Academic/Plenum Publishers, 2002

1237

substitute tear instillation or photophobia. Objective tests include tear production (Schirmer I test <5.5 mm/5 min), break up time (BUT < 7 sec), positive staining of the inferior exposed ocular surface with 1% Lissamine green (Van Bjsterveld score scale).[5] In addition, the distinction between patients with Sjogren and non-Sjogren dry eyes was determined by serologic[6] and clinical[7] evaluation for the presence of collagen or inflammatory disease. The subjects who suffered from non-dry eye related inflammation of the ocular surface, such as allergical, viral or bacterial conjunctivitis and ocular surgery, were excluded from the study. Specifically, patients with Rosacea, staphylococcal blepharitis and meibomian glands diseases were excluded.

Impression cytology was performed in both eyes of each patient following the Tseng[9] modification of the technique originally described by Nelson.[8] The results were classified according to the criteria outline by Tseng.[9] In lieu of their previous substitute tear therapy, patients were treated with unpreserved Diclofenac 0.1% (Voltaren Ofta monodose, Ciba Vision, Italy) in one eye and unpreserved 0.3% hydroxypropylmethylcellulose with 0.1% dextran substitute tears (Dacriosol monodose, Alcon, Italy) in the controlateral eye, 4 times a day for 15 consecutive days. A visual analogue scale (VAS) made up from a 10 cm line on which the patient could mark the level of his symptoms (Fig. 1). It was used to evaluate the changes in the symptoms in each eye from the beginning to the end of the study in each patient. Lissamine green staining and impression cytology tests were repeated at the end of the study. The Mann-Whitney statistical test was used to analyze the results.

Worse   Better

Figure 1. Visual Analogue Scale (100 mm).

3. RESULTS

The results of VAS, Lissamine green staining and impression cytology are summarized in Fig. 2. Analysis of the subjective symptom results indicates that Diclofenac 0.1% significantly improved the symptoms of KCS (P = 0.03). The ocular surface condition, evaluated by Lissamine green staining, was also significantly improved after treatment (P = 0.05). Although not statistically significant, visual inspection of the impression cytology results also suggested an improvement after treatment. There were no significant differences between Sjogren and non-Sjogren patients.

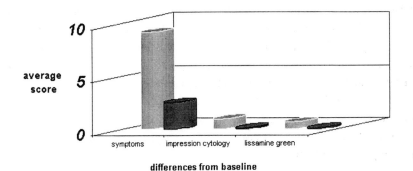

Figure 2. Differences from baseline scores for symptoms, impression cytology and Lissamine green staining of the ocular surface. Unpreserved 0.3% hydroxypropylmethylcellulose with 0.1% dextran (**gray bars**); Unpreserved 0.1% Diclofenac (**black bars**).

4. DISCUSSION

In our study the short-term topical use of unpreserved 0.1% Diclofenac was effective in improving symptoms and signs of KCS. This therapy was used in a group of patients where the therapy with substitute tears did not effectively improve the chronic ocular discomfort, and consequently, the quality of life of the patient. After 2 weeks of treatment we noticed an appreciable improvement ($P = 0.03$) in symptoms and therefore, in quality of life of the patients.

Lissamine green staining is an indicator of conjunctival epithelium damage.[10] After 2 weeks of therapy, Lissamine green staining indicated that treatment with Diclofenac improved the condition of the conjunctival epithelium and therefore the ocular surface. Pflugfelder[3] previously reported that there is decreased goblet cell density in the temporal bulbar conjunctiva in KCS. At the end of our study we noticed an increase of goblet cell density in this area, even if not statistically significant.

The rationale for the use of Diclofenac in KCS is related to recently published information on the pathogenesis of the disease. Inflammation of the ocular surface is either the origin of injury or can exacerbate an existing injury in eyes with KCS.[11] The primary insult that gives rise to inflammation is unclear. However, the inflammatory cascade results in the expression of high levels of pro-inflammatory cytokine mRNA in conjunctival cells of patients with Sjogren syndrome[12] and in general with dry eye.[1,13] Corticosteroid treatment improves symptoms of these diseases appreciably, probably by interrupting the inflammatory cascade. Chronic treatment with these drugs, however, resulted in important side effects as pointed out in this and other studies.[4]

Diclofenac, like other non-steroidal anti-inflammatory drugs, acts by inhibiting the activity of cyclooxygenase on arachidonic acid[14] that in turn inhibits the production of pro-inflammatory cytokines on the ocular surface. The use of a single dose unpreserved treatment regimen avoids the well known and harmful effects of benzalkonium chloride on the ocular surface.[15] Although rare, pathologic ocular surface changes as result of topically administered non-steroidal anti-inflammatory drugs after ocular surgery have been recently described.[16] The number of patients in our study is too small to assess the complete safety of such a treatment in dry eyes. However, dry eye patients treated with topically administered non-steroidal anti-inflammatory drugs should be monitored for the appearance of corneal changes. Although further studies are necessary to establish the safety and the efficacy of long term therapy, the results of this study are encouraging and indicate that Diclofenac in association with tear substitutes can be effective for the treatment of patients with KCS.

REFERENCES

1. Brignole F, Pisella PJ, Goldschild M, De Saint Jean M, Goguel A, Baudoin C. Flow cytometric analysis of inflammatory markers in conjunctival epithelial cells of patients with dry eye. Invest Ophthalmol Vis Sci;41,6:1356 (2000).
2. Bron AJ, Mengher LS. The ocular surface in keratoconjunctivitis sicca. Eye; 3:428 (1989).
3. Rolando M, Terragna F, Giordano G, Calabria G. Conjunctival surface damage distribution in keratoconjunctivitis sicca. An impression cytology study. *Ophthalmologica* ;200:170 (1990).
4. Marsh P, Pflugfelder SC. Topical nonpreserved methylprednisolone therapy for Keratoconjunctivitis sicca in Sjogren syndrome Ophthalmology; 106:811 (1999).
5. Van Bijsterveld OP. Diagnostic tests in the sicca syndrome Arch Ophthalmol;82:10 (1969).
6. Ferreri G, Santamaria S. Autoimmunità in oftalmologia. Apparato lacrimale. Boll Ocul; 74 suppl. 1:333 (1995).
7. Prause JU, Manthorpe R, Oxholm P, Schiodt M. Definition and criteria for Sjogren Syndrome used by contributors to the first international seminar on Sjogren Syndrome. Scand J Rheumatol,suppl. 61:17 (1986).
8. Nelson JD, Havener VR, Cameron JD. Cellular acetate impressions of the ocular surface. Arch Ophthalmol;101:1869 (1983).
9. Tseng SC. Staging of conjunctival squamous metaplasia by impression cytology. Ophthalmology;92:728 (1985).
10. Pflugfelder SC. Advances in the diagnosis and management of keratoconjunctivitis sicca. Current opinion in Ophthalmology,9;IV 50 (1998).
11. Stern ME, Beuerman RW, Fox RI, et al. The pathology of dry eye: the interaction between the ocular surface and lacrimal glands. Cornea;17:584 (1998).
12. Jones DT, Monroy D, Ji Z, Atherton SS, Pflugfelder SC. Sjogren's syndrome: cytokine and Epstein-Barr virus gene expression within the conjunctival epithelium. Invest Ophthalmol Vis Sci, 35:3493 (1994).
13. Tsubota K, Fujihara T, Saito K, Takeuchi T. Conjunctival epithelium expression of HLA-DR in dry eye patients. Ophthalmologica;213 (1):16 (1999).
14. Ku EC, Lee W, Kothari HV,Scholer DW. Effect of diclofenac Sodium on the arachidonic acid cascade. Am J Med;80 (4B):18 (1985).
15. Ichijima H, Petrol WM, Jester JV, Cavanagh HD. Confocal microscopic studies of living rabbit cornea treated with benzalkonium chloride. Cornea;11:221 (1992).
16. Lin J, Rapuano C, Laibson P et al. Corneal melting associated with use of topical nonsteroidal anti-inflammatory drugs after ocular surgery (letter). Arch Ophthalmol;118:1129 (2000).

THE EFFECT OF BOTULINUM TOXIN A TREATMENT ON TEAR FUNCTION PARAMETERS AND ON THE OCULAR SURFACE

Jutta Horwath-Winter, Jutta Berglöff, Ingrid Flögel, Eva-Maria Haller-Schober, Klaus Müllner, and Otto Schmut

University Eye Clinic
Graz, Austria

1. INTRODUCTION

Botulinum toxin A, a dichain protein, is one of seven neurotoxins produced by *Clostridium botulinum*. The light-chain endopeptidase component of botulinum toxin acts in the end-plate of motor neurons where it cleaves SNAP-25, a protein involved in the fusion of acetylcholine-containing vesicles with the presynaptic membrane.[1] When injected locally, it causes muscular paralysis by interfering with the release of acetylcholine at neuromuscular junctions. Neuromuscular function is restored after collateral sprouting of axons and regeneration of the end plate, thus limiting the action of botulinum toxin A to a 3-month period.

Botulinum toxin injections are a well established treatment for blepharospasm and hemifacial spasm. An increase in tear secretion and a decrease in dry eye symptoms have also been reported after periorbital botulinum toxin injections for blepharospasm.[2] On the other hand, Price et al.[3] did not find any changes in Schirmer test results 1 week after injection compared with measurements taken prior to injection. In the present study, the effects of periorbital botulinum toxin A injections on tear function parameters and on the ocular surface are investigated in patients suffering from essential blepharospasm and hemifacial spasm.

2. PATIENTS AND METHODS

Botulinum toxin A (Dysport, Porton Products, GB) was administered to 10 patients, 6 males and 4 females with an age range of 50 to 93 years (mean 72.2 years), 9 with essential

Lacrimal Gland, Tear Film, and Dry Eye Syndromes 3
Edited by D. Sullivan *et al.*, Kluwer Academic/Plenum Publishers, 2002

1241

blepharospasm and 1 with hemifacial spasm. The duration of the spasm was 3 months to 20 years. A standard dose of 120 units per eye[4] was administered by subcutaneous injection into 4 periorbital locations, 20 units medially and 40 units laterally into the junction between the palpebral and orbital parts of both the upper and lower orbicularis oculi muscle of each eye.

Patients were given a subjective questionnaire. Ocular examinations including tear breakup time, Schirmer test without local anesthesia, and rose bengal staining were performed prior to treatment, 1 week, and 1 and 3 months after injection with botulinum toxin A. Breakup time (FBUT) was measured by adding one drop of fluorescein, obtained by wetting a fluorescein strip with non-preserved sterile saline, to the inferior fornix. The mean value of a total of three measurements was recorded. Schirmer test was performed without anesthesia using a commercially available 5 x 35 mm paper strip. Rose bengal staining was performed by adding one drop of 1% rose bengal solution to the inferior fornix, and classified according to the van Bijsterveld scoring system.[5] Impression cytology was performed in all patients before treatment and 3 months after botulinum toxin A treatment. Sheets of cellulose acetate filter paper (type VC, 0.10 μm, VCWP 04700 Millipore Corp. Bedford US) were used to collect cells from the superior and inferior bulbar conjunctiva. The specimens were stained using the procedure previously described by Tseng[6] and examined under a light microscope.

Data were analyzed using the Wilcoxon rank test ($p < 0.01$).

3. RESULTS

Nine of the 10 patients complained of dry eye symptoms or were using artificial tears. Dry eye was determined objectively in 4 patients by clinical evaluation (Schirmer test < 5 mm, rose bengal staining score ≥ 3.5 or FBUT < 5 sec). Impression cytology demonstrated a moderate to severe desquamation of the epithelial cells in 6 patients and a decrease in the number of goblet cells in 7 patients. Inflammatory cells were found in 9 patients prior to treatment.

Although botulinum toxin treatment relieved the blepharospasm (mean duration 2.5 months) in each patient, only 2 patients reported a subjective improvement of dry eye symptoms. Five patients noticed no difference in dry eye symptoms, and 3 reported an increase in symptoms, including 2 with sagging of the lower lid, leading to exposure of the ocular surface. The other experienced lower eyelid laxity leading to an entropium.

One week and 1 month after injection, FBUT tended to increase but the difference was not statistically significant (Fig.1). Schirmer test results were significantly reduced ($p < 0.01$) after 1 and 3 months (Fig.2). Rose bengal staining appeared to increase 1 week and 1 month after the injection; however, these measurements were not significantly different from measurements taken prior to botulinum injections (Fig.3).

The cellular morphology of the ocular surface was examined by impression cytology. After 3 months no changes were present in the ocular surfaces of 6 patients. There was an improvement in the cellular picture in two patients, while in another two patients the pathologic changes increased.

4. DISCUSSION

Benign essential blepharospasm is characterized by repeated forceful spasmodic contractions of the orbicularis oculi muscles, frequently resulting in prolonged eyelid closure and severe visual disability. It has been hypothesized that the essential blepharospasm is prolonged due to a vicious cycle[7] within an as yet unidentified central control area located in the region of either the basal ganglia, midbrain or brainstem. Multifactorial stimuli such as light, emotion, stress, and other psychologic factors contribute to the cycle.

Blepharospasm is often associated with reduced tear secretion,[3,8] and our results confirm these findings. Many of the patients with blepharospasm in this study also tested positive for dry eye using the Schirmer test, rose bengal staining and impression cytology. It is difficult to determine whether dry eye causes blepharospasm because the rate of blinking increases to compensate for tear film instability or deficiency.[9] On the other hand,

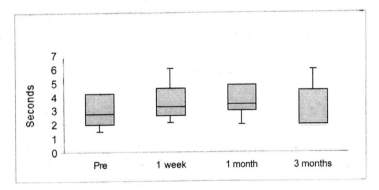

Figure 1. FBUT values before, 1 week, 1 month and 3 months after treatment.

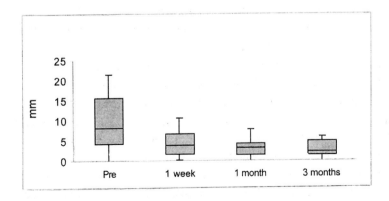

Figure 2. Schirmer test values before, 1 week, 1 month and 3 months after treatment.

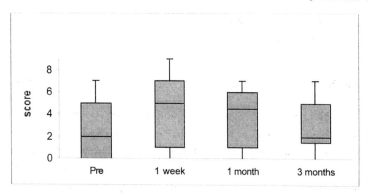

Figure 3. Rose bengal staining score before, 1 week, 1 month and 3 months after treatment.

blepharospasm could cause dry eye since the periodic blinking action of the eyelids is crucial for the maintenance and renewal of the precorneal tear film.[10] Electrophysiologic studies[11] have confirmed abnormalities in the electrically-induced blink reflex in patients with blepharospasm, even in the absence of lid spasm. Manning et al[11] observed that blink rate was slow in both opening and closing the eyelids and did not improve following treatment with botulinum toxin. In contrast, botulinum toxin treatment significantly altered blink lid-lowering kinematics, and a significant decrease in the peak velocity of the blink down-phase was observed.

Botulinum toxin A has been suggested for dry eye therapy because of the resultant reduction in lacrimal drainage after treatment.[12] Paralysis of the orbicularis oculi muscle acts on the canaliculi and induces a decreased pump function during blinking. Therefore, botulinum toxin injection significantly lowered the blink output 3 weeks after treatment.[12] The slight eyelid retraction, the decrease in vertical movement of the upper eyelid and the decreased horizontal sliding of the lower eyelid affect the apposition of the medial parts of the eyelids during blinking and could contribute to a reduction in lacrimal drainage. This would result in an increased tear volume around the eye, explaining the observed improvements in Schirmer test measurements.

One explanation for the differences in our Schirmer test measurement results after treatment with botulinum toxin A compared to those of previous studies is that the patients in this study had a mean age of 72.2 years, and lacrimal drainage capacity decreases with increasing age.[12] Another reason could be a direct pharmacological effect of botulinum toxin on lacrimal gland tear production. Botulinum toxin not only affects neuromuscular junctions but also autonomic cholinergic transmission during the systemic manifestation of botulism. Clinical presentation includes mydriasis or delayed pupillary reaction, accommodation paresis, reduced salivary secretion and lacrimation, constipation, disturbed micturition and cardiovascular regulation. Therefore, potential indications for botulinum toxin A treatment in disorders of the autonomic nervous system include diseases associated with sudomotor or secretomotor hyperactivity such as pathologic tearing (crocodile tears), hypersalivation, and hyperhidrosis.[13]

Our results demonstrate significantly reduced Schirmer test measurements after botulinum toxin A injection. This could be due to inhibition of the autonomic cholinergic transmission. However, the clinical relevance of this objective finding is unknown because it did not correlate with dry eye symptoms. The lack of improvement in dry eye symptoms after treatment for blepharospasm suggests that it is an underlying condition. Although botulinum toxin improves blepharospasm, it does not improve the physiology of the blink reflex or restore normal tear function.

In our patients, treatment with botulinum toxin A did not improve the tear function and ocular surface morphology. Dry eye symptoms may have even worsened in some cases due to increased corneal exposure resulting from poor blinking and lagophthalmos. This is consistent with the findings of Dutton et al.[14] that dry eye is the most common side effect of botulinum toxin therapy. Increased lacrimation and photophobia are also caused by increased corneal exposure resulting from lower eyelid laxity. The absence of contractive forces around the walls of the lacrimal system may produce both, epiphora and an associated keratoconjunctivitis, probably as a result of a stagnant tear meniscus that provides microorganisms with a good environment for replication and allows inflammatory products to accumulate.[15]

REFERENCES

1. J. Blasi, E.R. Chapman, E. Link, Botulinum neurotoxin A selectively cleaves the synaptic protein SNAP-25, *Nature*. 365: 160–163 (1993).

2. H. Spiera, P.A. Asbell, D.M. Simpson, Botulinum toxin increases tearing in patients with Sjögren`s syndrome: A preliminary report, *J. Rheumatol*. 24: 1842–3 (1997).

3. J. Price, J. O`Day, A comparative study of tear secretion in blepharospasm and hemifacial spasm patients treated with botulinum toxin, *J. Clin. Neuroophthalmol*. 13: 67–71 (1993).

4. F. Grandas, J. Elston, N. Quinn, C.D. Marsden, Blepharospasm: a review of 264 patients, *J. Neurol. Neurosurg. Psychiatry*. 51: 767–772 (1988).

5. O.P. van Bijsterveld, Diagnostic tests in the sicca syndrome, *Arch. Ophthalmol*. 82:10–14 (1969).

6. S.C.G. Tseng, Staging of conjunctival squamous metaplasia by impression cytology, *Ophthalmology*. 92: 728–733 (1985).

7. R.L. Anderson, B.C.K. Patel, J.B. Holds, D.R. Jordan, Blepharospasm: past, present, and future, *Ophthal. Plast. Reconstr. Surg*. 14: 305–317 (1998).

8. N. Shorr, S.R. Seiff, J. Kopelman, The use of botulinum toxin in blepharospasm, *Am. J. Ophthalmol*. 99: 542–546 (1985).

9. K. Tsubota, S. Hata, Y. Okusawa, F. Egami, T. Ohtsuki, K. Nakamori, Quantitative videographic analysis of blinking in normal subjects and patients with dry eye, *Arch. Ophthalmol*. 114: 715–720 (1996).

10. M.G. Doane, Blinking and the mechanics of the lacrimal drainage system, *Ophthalmology*. 88: 844–851 (1981).

11. K.A. Manning, C. Evinger, P.A. Sibony, Eyelid movements before and after botulinum therapy in patients with lid spasm, *Ann. Neurol*. 28: 653–660 (1990).

12. S. Sahlin, E. Chen, Evaluation of the lacrimal drainage function by the drop test, *Am. J. Ophthalmol*. 122: 701–708 (1996).

13.   M. Naumann, W.H. Jost, K.V. Toyka, Botulinum toxin in the treatment of neurological disorders of the autonomic nervous system, *Arch. Neurol.* 56: 914–916 (1999).

14.   J.J. Dutton, E.G. Buckley, Long-term results and complications of botulinum A toxin in the treatment of blepharospasm, *Ophthalmology*. 95: 1529–1534 (1988).

15.   A.A. Afonso, L. Sobrin, D.C. Monroy, M. Selzer, B. Lokeshwar, S.C. Pflugfelder, Tear fluid gelatinase B activity correlates with IL-1alpha concentration and fluorescein clearance in ocular rosacea, *Invest. Ophthalmol. Vis. Sci.* 40: 2506–2512 (1999).

EFFICACY OF AUTOLOGOUS SERUM TREATMENT IN PATIENTS WITH SEVERE DRY EYE

Etsuko Takamura, Kazumi Shinozaki, Hiromi Hata, Junko Yukari, and Sadao Hori

Dept of Ophthalmology
Tokyo Women's Medical University
Tokyo, Japan

1. INTRODUCTION

Many tear components regulate ocular surface epithelial cell proliferation, differentiation and maturation, and are therefore critical in maintaining the ocular surface. Lack of either basic or reflex tearing results in severe dry eye conditions. Sjøgren's syndrome (SS) dry eye involves decreased reflex tearing.[1] When treating severe dry eye, the primary goals are to reverse ocular surface desiccation and effect and preserve the stability of the tear component. Since many ocular surface regulatory components, including vitamin A and growth factors such as epidermal growth factor[2,3] and TGF-β, are also present in serum, it is not surprising autologous serum is an effective treatment for severe SS and Stevens-Johnson syndrome dry eye.[4-6] In this study, we confirm the efficacy of autologous serum treatment for severe dry eye.

2. MATERIALS AND METHODS

Twenty-six female patients, mean age 57.2 ± 13.2 years (mean ± SD), with moderate to severe dry eye symptoms that did not respond to conventional treatment were recruited for this study. Twenty-one of these were diagnosed with SS and 5 with non-SS dry eye. Reflex tearing was measured by the Schirmer test with nasal stimulation,[1] results of which were less than 10 mm in all cases. Patients were also prescreened for the infectious diseases HIV, HCV, and HBV.

Lacrimal Gland, Tear Film, and Dry Eye Syndromes 3
Edited by D. Sullivan *et al.*, Kluwer Academic/Plenum Publishers, 2002

1247

2.1. AUTOLOGOUS SERUM

A total of 40 ml of blood, procured by venepuncture into a non-heparinized tube, was centrifuged at 3000 rpm for 10 min. Under sterile conditions, the serum was carefully separated and diluted to 20% with saline. Patients were instructed to store their autologous serum solutions in a dark and cool place (refrigerator) during use and store the rest in a freezer until required. Solutions were stored in the refrigerator (4°C) for up to 1 month, and in the freezer (-20°C) for up to 3 mo. Patients instilled autologous serum drops directly into the eyes 4 or 8 times per day, in addition to following a more conventional treatment regimen including frequent preservative-free artificial tears, highly viscous hyaluronic acid 5 times a day and ocular ointments at bedtime.

2.2. OCULAR SURFACE EVALUATION

The efficacy of autologous serum treatment was evaluated by patient-reported symptoms of ocular discomfort, rose bengal staining, fluorescein staining and the Schirmer test before and 2 mo after treatment. Symptoms of ocular discomfort were determined by a questionnaire in which patients graded their symptoms as worse, stable, improved or remarkably improved. The intensity of rose bengal staining in the cornea and temporal and nasal conjunctiva were graded on a scale of 0 to 3 points. Therefore, the maximum score for each eye was 9. The intensity of fluorescein staining in the cornea was graded on a scale of 0 to 3 points. Squamous metaplasia was evaluated by impression cytology of the nasal bulbar conjunctiva. The specimens were stained with PAS and evaluated by Nelson's classification.

2.3. CRITERIA OF DRY EYE AND SS

The diagnosis of dry eye was based on the following three criteria: (1) symptoms of dry eye, (2) abnormalities of tear dynamics (Schirmer test was less than 5 mm) and (3) abnormalities of the ocular surface (rose bengal score of more than 3, and/or fluorescein score of more than 1). The diagnosis of SS was made based on Fox and Saito's 1994 criteria.

3. RESULTS

The mean autologous serum treatment duration was 8.3 mo ± 5.2 (SD). Improvements in symptoms of ocular discomfort were reported by 76.9% of the patients. Among the 4 times/day instillation group, 57.5% reported symptomatic improvements, and among the 8 times/day instillation group, 94.1% reported symptomatic improvements. Rose bengal and fluorescein staining were significantly decreased in both groups. Two months after treatment, improvements in squamous metaplasia were observed by impression cytology.

Only one patient reported itching of the lid margin after treatment, but other than that, no severe adverse effects of the treatment occurred.

3.1. CASE REPORT

A 59-year-old female had suffered from SS for 4 years. She had severe dry eye and dry mouth (a lip biopsy confirmed the presence of sialadenitis), and was positive for serum auto-antibodies including rheumatoid factor, antinucleus factor and anti-SS-A and anti-SS-B antibodies. Although she used frequent preservative-free artificial tears, highly viscous hyaluronic acid eye drops and special glasses for dry eye patients that had side panels and moist inserts for maintaining humidity around the eyes, the treatments did not sufficiently improve symptoms or the condition of her ocular surfaces. Prior to autologous serum treatment, her Schirmer value was 1 mm, Schirmer test with nasal stimulation was less than 5 mm in both eyes, rose bengal score was 7 and fluorescein score was 3. After treatment with autologous serum solution 4 times per day, her ocular discomfort decreased, and rose bengal and fluorescein scores decreased to 5 and 2, respectively. The patient then switched to a treatment regimen of 8 times per day, and her dry eye symptoms and signs improved remarkably. Prior to autologous serum treatment, severe grade 3 squamous metaplasia was observed by impression cytology. Two months after instillation of autologous serum 8 times per day, squamous metaplasia was improved.

4. DISCUSSION

A 20% solution of autologous serum was sufficient to supply a beneficial concentration of vitamin A and growth factors to the ocular surface. Instillation of the solution 8 times per day was more effective than 4 times per day for decreasing ocular discomfort and improving ocular surface conditions. Punctal plug insertion for the treatment of severe dry eye results in a remarkable, rapid and sustained increase in the tear volume and tear components. Instillation of serum provides similar benefits to punctal plug insertion without undergoing an invasive procedure. One drawback of the treatment is that the patients need to provide a continuous supply of blood for a long period. Despite this drawback, after considering the benefits of symptomatic relief with the negative aspects of conventional treatments, most patients continued treatment. We look forward to future development of artificial tears containing essential tear components that will eliminate the need for serum collection, while providing symptomatic relief comparable to that of autologous serum.

REFERENCES

1.   K. Tsubota, The importance of the Schirmer test with nasal stimulation, *Am J Ophthalmol*. 111: 106–108 (1991).

2.   Y. Ohashi, M. Motokura, Y. Kinoshita et al., Presence of epidermal growth factor in human tears, *Invest Ophthalmol Vis Sci.* 30: 1879–1887 (1989).

3.   G. Van Setten, T. Tervo, K. Tervo et al., Epidermal growth factor (EGF) in ocular fluids: presence, origin and therapeutical concentration, *Acta Ophthalmol.* 202: 54–59 (1992).

4.   R. Fox, R. Chan, J. Michelson et al., Beneficial effects of artificial tears made with autologous serum in patients with keratoconjunctivitis sicca, *Arthritis Rheum.* 27: 459–461 (1984).

5.   K. Tsubota, E. Goto, H. Fujita et al., Treatment of dry eye by autologous serum application in Sjøgren's syndrome, *Br J Ophthalmol.* 83: 390–395 (1999).

6.   K. Tsubota, E. Goto, S. Shimmura et al., Treatment of persistent corneal epithelial defect by autologous serum application, *Ophthalmology.* 106: 1984–1989 (1999).

OCULAR SAFETY OF INS365 OPHTHALMIC SOLUTION, A P2Y$_2$ AGONIST, IN PATIENTS WITH MILD TO MODERATE DRY EYE DISEASE

Benjamin R. Yerxa,[1] Mohan Mundasad,[2] Robin N. Sylvester,[1] JoAnn C. Gorden,[3] Mark Cooper,[2] and Donald J. Kellerman[1]

[1]Inspire Pharmaceuticals
Durham, North Carolina, USA
[2]Simbec Research Ltd.,
Merthyr Tydfil, United Kingdom
[3]Regulatory Consultant
Overland Park, Kansas, USA

1. INTRODUCTION

Activation of P2Y$_2$ receptors in various tissues results in secretion of ions, fluid and mucin.[1] *In situ* hybridization experiments on fresh primate tissue confirm the presence of mRNA coding for the gene expression of the P2Y$_2$ receptor in the conjunctiva and the meibomian gland.[2] Topical ocular administration of nucleotides causes increased chloride and mucin secretion, as well as fluid transport.[3-5] These pharmacologic findings lead to the hypothesis that P2Y$_2$ receptor agonists may have therapeutic efficacy in diseases of impaired ocular hydration.

INS365, a chemically stable, selective P2Y$_2$ receptor agonist, has been evaluated in numerous preclinical studies for safety and efficacy and in 50 healthy volunteers for safety and tolerability. We now report the results of the first ocular safety and tolerability study of INS365 in patients with mild to moderate dry eye disease.

The first study of INS365 in humans was a double-masked, placebo-controlled, randomized, within subject paired-comparison, dose-escalation study in five cohorts of 10 healthy subjects.[6] The concentrations of INS365 Ophthalmic Solution were 0.5, 1.0, 2.0, and 5.0% and given three times over 6 h. Safety was assessed by general and ophthalmic examinations and symptomatology. Unanesthetized Schirmer testing was performed in the

Lacrimal Gland, Tear Film, and Dry Eye Syndromes 3
Edited by D. Sullivan *et al.*, Kluwer Academic/Plenum Publishers, 2002

1251

last cohort of 10 subjects to evaluate the acute effects of INS365 on tear secretion. The study demonstrated no significant differences in the number of subjects with ocular events or in any safety parameter in placebo-treated eyes compared to INS365-treated eyes. Unanesthetized Schirmer testing showed no significant acute effects of INS365 on tear secretion vs. placebo in healthy subjects in which reflex tearing often produced maximal Schirmer values. These results show that stimulation of ocular surface $P2Y_2$ receptors is not associated with ocular tolerability issues in healthy subjects.

2. METHODS

2.1. Dry Eye Study Objectives

This study had two objectives. The first was to evaluate the safety and tolerability of INS365 when applied topically as eyedrops in patients with mild to moderate dry eye disease. The second was to evaluate the pharmacologic action of INS365 in the eye in the final dosing cohort only

2.2. Entry Criteria

Subjects for this study were of either sex, at least 18 years old, and had a best corrected visual acuity of at least 20/100 in each eye. Each had mild to moderate dry eye disease in the judgment of the investigator as defined by the presence of two out of five specified symptoms (foreign body sensation, ocular dryness, burning or pain, photophobia, vision fluctuating with blink) in at least one eye. Each subject had a corneal fluorescein staining score ≥ 3 out of a possible 15 (5 areas each with scale of 0–3). Subjects in cohort 5 (see below) had an unanesthetized Schirmer score < 8 mm in at least one eye.

Subjects with the following criteria were excluded from the study: (1) unable to stop any concomitantly prescribed or over-the-counter ocular drug, including dry eye products, (2) had intraocular surgery or ocular laser surgery in the past 90 days, (3) had Sjögren's syndrome, uveitis, or other ocular pathology leading to dry eye as a secondary feature of other primary pathology, (4) had worn contact lenses and did not agree to remove them 24 h prior to dosing and through the day after dosing, (5) had an intraocular pressure ≥ 22 mmHg at screening, and (6) had taken part in any other clinical trial within 90 days prior to screening.

2.3. Study Design

This was a single-center, in-patient study, with double-masked, randomized, placebo-controlled, parallel-group, rising-dose protocol. The dose escalation determined by opinion of investigator and medical monitor from masked data. The study was comprised of five cohorts that were treated as described in Table 1.

Table 1. Cohort treatment protocols

	Treatment
Cohorts 1–4 8 subjects each	Five ocular instillations over an 8-h period
	0.5%, 1.0%, 2.0% or 5.0% INS365 solution or placebo instilled in both eyes
Cohort 5[a] 30 subjects	Five ocular instillations over an 8-h period
	5.0% INS365 solution or placebo instilled in both eyes

[a]maximum tolerated dose from Cohorts 1–4

2.4. Assessments

Safety criteria were assessed by (1) physical examination, (2) clinical laboratories, (3) vital signs, (4) ocular symptomatology, (5) visual acuity, (6) biomicroscopy, (7) intraocular pressure, (8) ophthalmoscopy, and (9) adverse events. Pharmacologic activity was assessed by unanesthetized Schirmer test and anesthetized Schirmer test.

3. RESULTS

Cohort demographics are given in Tables 2 and 3.

3.1. OCULAR SAFETY PARAMETERS

3.1.1. Ocular Symptomatology (10 min to 24 h post-dose). Patients scored the severity of prespecified ocular symptoms (foreign body sensation, ocular dryness, burning or pain, photophobia, vision fluctuating with blink) and any other reported symptom on a scale of 0 = none, +1 = mild, +2 = moderate, +3 = severe. As part of the entry criteria, patients were required to have two of the five predefined symptoms in at least one eye.

Table 2. Demographics of Cohorts 1–4

	Placebo n = 8	0.5% INS365 n = 6	1.0% INS365 n = 6	2.0% INS365 n = 6	5.0% INS365 n = 6	Total n = 32
Gender n (%)						
Male	1 (13%)	3 (50%)	3 (50%)	3 (50%)	1 (17%)	11 (34%)
Female	7 (88%)	3 (50%)	3 (50%)	3 (50%)	5 (83%)	21 (66%)
Age (yrs)						
Mean	59.3	60.0	57.7	54.8	61.8	58.8
SD	7.0	15.6	3.9	12.4	4.7	9.3

Table 3. Demographics of Cohort 5.

	Placebo n = 11	5.0% INS365 n = 17	Total n = 28
Gender n (%)			
Male	5 (45%)	8 (47%)	13 (46%)
Female	6 (55%)	9 (53%)	15 (54%)
Age (yrs)			
Mean	65.7	65.0	65.3
SD	12.0	9.1	10.1

None of the symptoms reported were considered clinically significant in the opinion of the investigating physician. A reduction in the severity of all ocular symptoms was noted for each treatment group, which persisted 5 to 8 days. No notable differences were detected between treatment groups for ocular symptomatology.

3.1.2. Biomicroscopy. No subjects had a treatment-emergent biomicroscopy finding or worsening of any of the following structures at any post-dose time point up to 24 h: eyelid edema; conjunctival edema; corneal edema, staining and endothelial changes; anterior chamber cells or flare or lens pathology. No clinically meaningful shifts were noted for any remaining biomicroscopy parameters for which only the following treatment-emergent findings were reported:

Eyelid erythema:

- 1 subject (5.0% INS365) had a shift from "none" predosing to "mild" in the left eye 10 min after the first dose.
- Conjunctival erythema: 1 subject (1.0% INS365) had a shift from "none" predosing to "mild" in the left and right eyes at 4 h, 10 min and 24 h.
- Conjunctival discharge: 4 subjects (1 in the 2.0% INS365 group and 3 in the 5.0% INS365 group) had transient mild conjunctival discharge at one or more post-treatment assessment time points that was not present at predosing.

3.1.3. Other Clinical Assessments. No clinically significant changes in visual acuity or intraocular pressure results were noted for any subject during the study period. No study-emergent abnormal ophthalmoscopic results were found. Administration of INS365 Ophthalmic Solution five times in an 8-h period at doses up to 5.0% did not produce clinically significant changes in physical examination, vital signs, 12-lead ECG or clinical laboratories.

3.2. Adverse Events

Table 4 summarizes the adverse events associated with this study.

Table 4. Number of subjects reporting adverse events

Body System/ Adverse Event	Placebo n = 19	INS365 0.5% n = 6	1.0% n = 6	2.0% n = 6	5.0% n = 23	All Treatments n = 60
Subjects with Any Adverse Event	8[a] (42%)	1 (17%)	2 (33%)	1 (17%)	4 (17%)	16 (27%)
Central and Peripheral Nervous System Disorders						
Headache	5 (26%)	0	2 (33%)	1 (17%)	1 (4%)	9 (15%)
Vision Disorders						
Conjunctivitis	0	1 (17%)	0	0	1 (4%)	2 (3%)
Eye Pain	1 (5%)	0	0	0	1 (4%)	2 (3%)
Ocular Hemorrhage	0	0	0	0	1 (4%)	1 (2%)
Respiratory Disorders						
Upper Respiratory Tract Infection	1 (5%)	0	1 (17%)	0	1 (4%)	3 (5%)
Urinary System Disorders						
Urinary Tract Infection	1 (5%)	0	0	0	0	1 (2%)
Body as a Whole - General Disorders						
Back Pain	1 (5%)	0	0	0	0	1 (2%)
Fatigue	1 (5%)	0	0	0	0	1 (2%)

[a]n (%)

3.3. Safety Summary

All doses (0.5, 1.0, 2.0 and 5.0% administered five times in an 8-h period) were well-tolerated in healthy subjects. No notable difference occurred in the incidence or severity of adverse events following administration of INS365 compared to placebo (normal saline).

3.4. Effect Of INS365 On Unanesthetized And Anesthetized Schirmer

All subjects in Cohort 5 entered the study with an unanesthetized Schirmer score < 8 mm in at least one eye. The unanesthetized Schirmer test was performed 10 min after the first dose (time point = 10 min) and 30 min after the third dose (time point = 4.5 h). The anesthetized Schirmer was performed 10 min after the fifth dose (time point = 8.5 h).

The unanesthetized Schirmer results were higher (indicating more tear fluid secretion) than the anesthetized results. This agrees with the expectation that the unanesthetized eye is more susceptible to reflex tearing due to the irritating effects of the Schirmer test strip. Mean unanesthetized Schirmer test scores improved compared to prestudy in both treatment groups at the 10-min and 30-min post-dose time points, although the difference was not statistically significant. In the dry eye population with unanesthetized Schirmer test scores < 8 mm, the INS365 group improved over the placebo group at 10 min post-dose (5.9 mm vs. 4.0 mm) and at 30 min post-dose (5.7 mm vs. 4.1 mm) as shown in Fig. 1.

No differences were seen in the anesthetized Schirmer test scores.

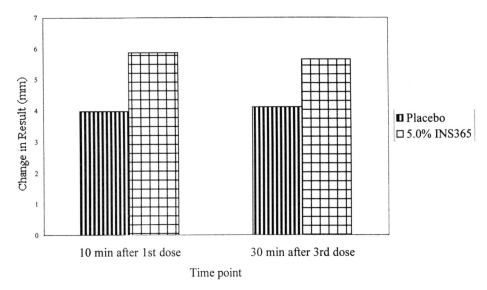

Figure 1. Mean change in unanesthetized Schirmer test results from pre-study to post-study dry eye population.

4. CONCLUSION

INS365 in concentrations ranging from 0.5% to 5.0% was as well-tolerated when topically administered as normal saline solution five times daily to the eyes of patients with mild to moderate dry eye disease. No cumulative effect existed; the second through fifth instillations were not associated with any greater tolerability problems than the first. Further investigation of these concentrations in patients with dry eye is supported by these findings.

Based on the safety findings in this study, another study in patients with dry eye was conducted with the same doses. The study was a double-masked, randomized, placebo-controlled, parallel-group, dose-ranging study in patients with moderate to severe dry eye disease. The trial has been completed and a final report is pending.

REFERENCES

1. B.R. Yerxa and F.L. Johnson. P2Y$_2$ receptor agonists: structure, activity and therapeutic utility. *Drugs of the Future.* 24:759–769 (1999).
2. B.R. Yerxa, V.Z. Zhang, L. Sheridan et al. Cellular localization of P2Y$_2$ receptor gene expression in primate ocular tissues by nonisotopic *in situ* hybridization. *Invest Ophthal Vis Sci.* 41:S540 Abstract #2630 (2000).
3. U. Li, K. Kuang, Q. Wen et al. Rabit conjunctival epithelium transports fluid. *Invest Ophthal Vis Sci.* 40:S90 Abstract #482 (1999).

4.   J.E. Jumblatt and M.M. Jumblatt. Regulation of ocular mucin secretion by P2Y$_2$ nucleotide receptors in rabbit and human conjunctiva. *Exp Eye Res.* 67:341–346 (1998).

5.   B.R. Yerxa, P.P. Elena, T. Caillaud et al. INS365, a P2Y$_2$ receptor agonist, increases Schirmer scores in albino rabbits. *Invest Ophthal Vis Sci.* 40:S540 Abstract #2848 (1999).

6.   M.V. Mundasad, G.D. Novack, V.E. Allgood, R.M. Evans, J.C. Gorden and B.R. Yerxa. Ocular safety of INS365 ophthalmic solution: a P2Y$_2$ agonist, in healthy subjects. *J Ocul Pharmacol Ther.* (in press).

OCULAR SURFACE RECONSTRUCTION WITH AMNIOTIC MEMBRANE TRANSPLANTATION IN CHEMICAL BURN

Myrna S. Dos Santos, Daniela Fairbanks, Erik A. Pedro,
Marcelo C. Cunha, Denise de Freitas, and Jose A. P.Gomes

Federal University of Sao Paulo-Brazil (UNIFESP)
Paulista School of Medicine
Department of Ophthalmology

1. INTRODUCTION

Chemical burn of the ocular surface is considered an important cause of limbal stem cell deficiency,[1] which determines different degrees of conjuctival re-growth, neovascularization, chronic inflammation and epithelial defects on the cornea. It can also damage the conjuctival epithelium and goblet cell, which may induce severe dry eye with keratinization, symblepharon formation and scarring of the eyelids.

Five important factors must be considered in the approach of ocular surface cicatricial diseases: limbal stem cell deficiency, conjuctival deficiency, dry eye, inflammation and basement membrane destruction.[2] The ideal treatment of ocular surface diseases should recover anatomic and physiologic structure, with reconstruction of the corneal and conjuctival epithelium, and keep an appropriate biological micro-environment.[3] A recent approach to the restoration of the basement membrane and damaged stromal matrix in ocular surface disorders is the transplantation of preserved amniotic membrane (AM).[2]

In 1995, Kim and Tseng[1] reported the use of AM in an experimental model of chemical injury. Since then, AM has been used as an effective adjunct for ocular surface reconstruction of several disorders.[4-9] Its properties include induction of re-epithelialization through adhesion and migration of epithelial cells, reduction of inflammation, cicatricial and angiogenic processes, and being immunologically inert.[9] Nevertheless, some controversies about its mechanism of action and reproducibility of its results remain. The purpose of this study was to report our outcomes using AM transplantation with or without

Lacrimal Gland, Tear Film, and Dry Eye Syndromes 3
Edited by D. Sullivan *et al.*, Kluwer Academic/Plenum Publishers, 2002

1259

associated limbal transplantation for ocular surface reconstruction in patients with chemical burn.

2. PATIENTS AND METHODS

2.1. Patients

Fifteen patients with total (12 eyes) and partial (3 eyes) limbal deficiency caused by ocular chemical injury underwent AM transplantation with or without limbal transplantation (12 and 3 eyes respectively). Patient profiles are summarized in Table 1. Fourteen (93.3%) patients were males and 1 (6.7%) female, with mean age of 38.5 years old. The mean time between injury and surgery was 6.8 years.

Table 1. Pre-operative characteristics of patients with chemical burn

N	Sex/Age (yo)	Duration (years)	Previous surgeries	Limbal deficiency	Stroma	Schirmer I	Fluoresceine
1	M/63	20	-	T	Th/V	-	ED
2	M/36	1.5	-	T	Th/V	Nl	P
3	M/37	10	PK /ECCE	T	Th/V	Nl	P
4	F/24	2	-	T	Th/V	Nl	P
5	M/65	3	-	T	Th/V	Nl	ED
6	M/34	3	PK /ECCE	T	Th/V	Nl	ED
7	M/22	2	-	T	Th/V	Nl	ED
8	M/22	10	PK / LTx	P	Th/V	Nl	P
9	M/47	3	-	T	Th/V	<5mm	ED
10	M/49	8	-	T	Th/V	<5mm	P
11	M/77	26	ECCE	P	Th/V	<5mm	P
12	M/12	0,2	-	P	Th/V	Nl	P
13	M/40	3	PK/Teno	T	Th/V	<5mm	P
14	M/33	4	PK/ECCE	T	Th/V	Nl	ED
15	M/26	7	Conj flap	T	Th/V	-	P

M = male; F = female; yo = years-old; LTx = limbal transplantation; PK = penetrating keratoplasty; Conj flap = conjunctival flap; Teno = tenoplasty; ECCE = extracapsular cataract extraction; Th. = stromal thinning; V = neovascularization; Nl = normal; ED = epithelial defect; P = punctate keratopathy; T = total; P = partial.

2.2. Surgical Methods

The amniotic membrane was obtained at the time of Cesarean section and preserved at -80°C in glycerol and cornea culture medium at a ratio of 1:1, using the method proposed by Tseng et al.[1], with some modifications. Except for one patient who received general anesthesia, all surgeries were performed with the patients receiving peribulbar anesthesia. Superficial keratectomy associated with total or sectorial AM transplantation was performed in all cases. For total limbal deficiency, limbal transplantation (autologous or heterologous) was performed simultaneously. When necessary, penetrating keratoplasty (PK) and cataract extraction were also performed.

2.3. Post-operative

After surgery, topical 0.1% dexametasona and 0.3% ofloxacin were used. In the cases of heterologous limbal transplantation or PK, systemic immunosuppression with cyclosporine A (5mg/kg/day) was done .

Patients were instructed to use preservative-free artificial tears every hour and to wear special dry eye glasses. Lacrimal punctum occlusion was performed when necessary and the most severe cases were also instructed to use autologous serum drops.

3. RESULTS

The mean follow-up time was 11.2 months (range 2–19 mo). Satisfactory ocular surface reconstruction (clinical/cytological) was obtained in 13 (87%) eyes, with reduced inflammation and vascularization and a mean epithelialization time of 3 weeks. Surgical failure was observed in 2 (13%) severe cases. Except for 2 eyes that maintained visual acuity, a significant visual improvement was observed in all cases. Complication was observed in 1 patient, who developed total necrosis of the graft 8 days after surgery. The post-operative results are summarized in the Table 2.

Table 2. Surgical results of the ocular surface reconstruction with AM in chemical burn

N	VA pre-op	VA post-op	Simultaneous Procedures	Follow-up (months)	Epithelialization time (weeks)	Fluoresceine	Results
1	HM	20/200	LTx	6	4	-	S
2	HM	20/40	LTx	17	4	P	S
3	CF0,5m	20/50	LTx/PK	15	3	-	S
4	HM	20/80	LTx	12	2	-	S
5	CF1m	CF1m	LTx	17	3	P	S
6	HM	20/400	LTx	18	3	P	S
7	CF1m	CF2m	LTx	18	3	P	S
8	CF2m	CF3m	-	12	3	P	S
9	HM	20/60	LTx/PK	12	3	P	S
10	LP	LP	LTx	2	-	ED	F/C
11	CF1m	CF4m	-	7	3	P	S
12	20/200	20/30	-	4	2	-	S
13	PL	HM	LTx/PK/ECCE	11	4	P	F
14	HM	20/80	LTx/PK	4	3	P	S
15	HM	CF1m	LTx	6	4	-	S

VA = visual acuity; HM = hand motion; CF = count finger; m = meter; LP = light perception; LTx = limbal transplantation; PK = penetrating keratoplasty; ECCE = extracapsular cataract extraction ; ED = epithelial defect; P = punctate keratopathy; S = success (surface reconstruction with corneal epithelium phenotype + stromal transparency); F = failure (recurrence of epithelial defect, vascularization and inflammation); C = complication.

4. CONCLUSIONS

This study confirms that AM transplantation alone is effective for the reconstruction of the ocular surface in eyes with chemical burn and partial limbal stem cell deficiency. In cases of total limbal stem cell deficiency, a combination of AM and limbal stem cell transplantantion is necessary. Comparative and controlled studies with longer follow-up are necessary to establish the efficacy of AM in these cases.

REFERENCES

1. J.C.Kim, S.C.G.Tseng, Transplantation of preserved human amniotic membrane for surface reconstruction in severely damaged rabbit corneas. *Cornea* 14:473–484 (1995).
2. S.C.G. Tseng and K. Tsubota, Important Concepts for Treating Ocular Surface and Tear Disorders. *Am J Ophthalmol* 124:825–835 (1997).
3. K. Tsubota, Y. Satake, M. Ohyama, I.Toda, Y. Takano, M. Ono, N. Shinozaki, J. Shimazaki, Surgical reconstruction of the ocular surface in advanced ocular cicatricial pemphigoid and Stevens-Johnson syndrome. *Am J Ophthalmol* 122:38–52 (1996).
4. S.C.G. Tseng, P. Prabhasawat, K. Barton, T. Gray, D. Meller, Amniotic membrane transplantation with or without limbal allografts for corneal surface reconstruction in patients with limbal stem cell deficiency. *Arch Ophthalmol* 116:431–441 (1998).
5. J. Shimazaki, H-Y. Yang, K. Tsubota, Amniotic membrane transplantation for ocular surface reconstruction in patients with chemical and thermal burns. *Ophthalmology* 104:2068–2076 (1997).
6. S. Lee and S.C.G. Tseng, Amniotic membrane transplantation for persistent epithelial defects with ulceration. *Am J Ophthalmol* 123:303–312 (1997).
7. P. Prabhasawat, K. Barton, G. Burkett, S.C.G. Tseng, Comparison of conjunctival autografts, amniotic membrane grafts and primary closure for pterygium excision. *Ophthalmology* 104:974–985 (1997).
8. S.C.G. Tseng, P. Prabhasawat and S-L. Lee, Amniotic membrane transplantation for conjunctival surface reconstruction. *Am J Ophthalmol* 124:765–774 (1997).
9. J.A.P Gomes, C.M. Komagome, J.D. Pena, N.C. Santos, A.A. Lima-Filho, D. Freitas, Clinical and ultrastructural aspects of amniotic membrane for ocular surface reconstruction. *Invest Ophthalmol Vis Sci* 40:S328 (1999).

SURGICAL RECONSTRUCTION OF THE TEAR MENISCUS AT THE LOWER LID MARGIN FOR TREATMENT OF CONJUNCTIVOCHALASIS

Norihiko Yokoi, Aoi Komuro, Jiro Sugita, Yo Nakamura, and Shigeru Kinoshita

Department of Ophthalmology
Kyoto Prefectural University of Medicine
Kyoto, Japan

1. INTRODUCTION

Conjunctivochalasis is an important cause of several types of ocular surface discomforts, including irritation, intermittent epiphora, dryness, and blurred vision,[1-6] and is often associated with with dry eye.[5] Since conjunctivochalasis most frequently protudes into the tear meniscus at the lower lid margin, the disease often interferes with the tear meniscus lacrimal pathway,[9,10] and tear fluid reservoir functions.[11] In addition, conjunctivochalasis-related redundant conjunctiva can irritate the ocular surface like a foreign body. Accordingly, conjuctivochalasis may explain patient complaints related to tear meniscus dysfunction or foreign body sensation. In this report, we propose that conjunctivochalasis causes tear meniscus dysfunction, and although several surgical procedures have been previously developed for treatment of this disease,[1,3,4,6-8] we introduce a new surgical treatment for removal of redundant conjuctiva while simultaneously restoring the original tear meniscus at the lower lid margin.[8]

Lacrimal Gland, Tear Film, and Dry Eye Syndromes 3
Edited by D. Sullivan *et al.*, Kluwer Academic/Plenum Publishers, 2002

2. METHODS

2.1. Patients

Forty-six patients (46 eyes) with symptoms attributable to conjuctivochalasis-related foreign body sensation or tear meniscus dysfunction, enrolled for our new conjunctivochalasis operation. Of these, there were 12 males and 34 females, with ages ranging from 50–87 years, and a mean age of 67.6 yrs ±7.2 SD. The disease was detected by fluorescein-assisted slit lamp examination. Of the 46 subjects, 16 had conjunctivochalasis alone, 18 had conjunctivochalasis with aqueous-deficient dry eye (AD), 2 had conjunctivochalasis with AD and pinguecula, 5 had conjunctivochalasis with meibomian gland dysfunction (MGD), 1 had conjunctivochalasis with MGD and AD, 1 had conjunctivochalasis with lymphangiectasia, 1 had conjunctivochalasis with lymphangiectasia and pinguecula, 1 had conjunctivochalasis with unoperated nasolacrimal obstruction, and 1 had conjunctivochalasis with operated nasolacrimal obstruction. The main symptoms were intermittent epiphora (21 eyes), especially in those patients with conjunctivochalasis alone; foreign body sensation (23 eyes), especially in those patients with dry eye; and prominent mucus discharge (2 eyes) in patients with dry eye. Patients were observed postsurgically for 6–21 months with an average postsurgical observation duration of 15.2 months.

2.2. A New Operation for Conjunctivochalasis

The aim of our new operation was to remove redundant conjunctiva that causes tear meniscus dysfunction, while restoring the original tear meniscus. The operation consisted of the following steps (Fig. 1): [1] Topical anesthetic eyedrops were instilled. [2] Redundant conjunctiva was squeezed from the limbus down to the conjunctival fornix, while the extent of conjunctival redundancy was confirmed by dividing the conjunctivochalasis into three portions along the lower lid margin: the lower cornea, the inferotemporal lower conjunctiva, and the inferonasal conjunctiva. [3] For resection, the bulbar conjunctiva was marked circumferentially 2 mm from the limbus, with endpoints that extended radially toward the lower punctum and lateral canthus. Radial markings were also placed downward from the circumferencial marking at the 5 and 7 o'clock positions. [4] A subconjunctival local anesthesia was administered. [5] The circumferential incision was made using the markings as guides, and then radial incisions were made toward the lower punctum and lateral canthus at the 5 and 7 o'clock positions. After the circumferential incision, and at the same time that the radial incisions were made, a tenonectomy was performed under the redundant conjunctiva. [6] The redundant conjunctiva at the lower corneal position was first drawn upward, and then resected along the marking. After resection, interrupted sutures of 9-0 silk were inserted. Care was taken not to raise the fornicial conjunctiva which could cause shortening of the bulbar conjunctiva. After suturing, the inferotemporal and inferonasal conjunctiva were drawn in centrally, and resected in accordance with their redundancy by the same method described for resection of the lower cornea. [7] Finally, interrupted sutures were inserted to close the inferonasal and temporal portions. In complicated cases, other ordinal surgical procedures for pingecula and pterygium were also performed.

2.3. Postsurgical Treatment

Patients were treated postoperatively with antibiotics (ofloxacin eyedrops 4 times per day) in combination with steroid eyedrops (fluorometholone four times per day, or in cases receiving expanded tenonectomy, betamethasone four times per day). Stitches were removed within 3 weeks after the operation at the latest, although stitches often fell out spontaneously.

2.4. Evaluation of Operation Outcome

The efficacy of the surgery was determined by evaluating the status of the main symptoms, complications both during and after surgery, and tear meniscus reconstruction by fluorescein staining along the lower lid margin from the lateral canthus to the lower punctum. Symptoms were evaluated by a patient grading system as follows: remarkable improvement, moderate improvement, unchanged or worsened.

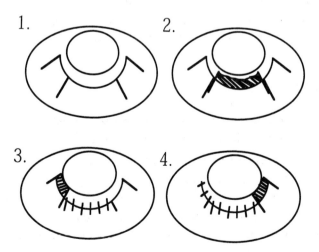

Figure 1. Operation steps: (1) Marking, conjunctival incision, and resection of subconjunctival tissue. (2–3). Resection of redundant conjunctiva at the lower portion and suturing with 9-0 silk. (3–4). Resection of redundant conjunctiva at the inferonasal and temporal portions, and suturing.

3. RESULTS

3.1. Symptom Evaluation

Patients reported a remarkable improvement in 22 eyes (47.8%), a moderate improvement in 17 eyes (37.0%), and no change in 7 eyes (15.2%). There were no reports of worsened symptoms. Approximately 84.8% of the patients obtained some symptomatic improvement. A greater than moderate improvement was observed in 19 of the 23 eyes that experienced foreign body sensation prior to surgery, in 18 of the 21 eyes that exhibited intermittent epiphora prior to surgery, and in 2 of the 2 eyes that exhibited mucus discharge prior to surgery.

3.2. Postoperative Complications

In all 46 cases (100%), some degree of subconjunctival hemorrhage was observed either during or after surgery. In 3 of the earlier cases (6.5%), wound cleavage occurred either after unexpected excessive subconjunctival hemorrhage during surgery, or as a result of excess conjunctival resectioning. In one case (2.2%), excess removal of subconjunctival tissue caused secondary lymphangiectasia that resolved after needle puncture.

3.3. Tear Meniscus Reconstruction

Fluorescein staining confirmed complete tear meniscus reconstruction in 30 eyes (65.2%). In 16 eyes (34.8%), although there was complete reconstruction of the lower corneal areas, incomplete reconstruction of the inferonasal or temporal areas resulted in residual conjunctivochalasis with minor interrupted menisci. A representative eye, before and after surgery, is shown in Fig 2. Minor incomplete reconstruction occurred in a few of the earlier cases; however, these cases decreased as technical skill improved. No recurrence of conjunctivochalasis was observed within the follow-up period.

4. DISCUSSION

Conjunctivochalasis is frequently encountered clinically, however, it is often overlooked or treated with eye drops that do not improve symptoms. Although there have been a number of reports on treating this disease surgically with good outcomes,[3,4,7,8] the philosophy behind treatment in present study differs from that of previous surgical treatments. We believe that conjunctivochalasis not only causes ocular "foreign body" discomfort but tear meniscus dysfunction as well; therefore, we sought to remove redundant conjunctiva and reconstruct the tear meniscus simultaneously. This study demonstrates that the reconstruction of the tear meniscus at the lower lid margin results in a remarkable improvement in symptoms, confirming that this surgery is much more

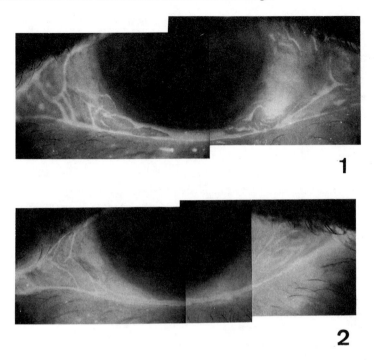

Figure 2. Representative case: **(1)** Before operation, **(2)** After operation. Complete reconstruction of tear meniscus at lower lid margin.

effective than current medical treatments for conjunctivochalasis. Further, although there is considerable variation in the specific manifestations of conjunctivochalasis, our surgical treatment appears to be effective for all forms.

After surgery, patients experienced particular improvement in intermittent epiphora, the most common symptom of conjunctivochalasis. It has been proposed that intermittent epiphora results when meniscus tear flow is blocked by redundant conjunctiva.[3,6] However, in a previous study from our lab using fluorophotometry, no change in tear flow was observed, even after meniscus reconstruction surgery.[12] An alternative explanation for intermittent epiphora is that the interruption of tear meniscus caused by conjuctivochalasis attenuates tear-retentive function of the meniscus.[13–14]

Interestingly, in conjunctivochalasis with aqueous-deficient dry eye, the sensation of the presence of a foreign body was relieved by tear meniscus reconstruction. It is possible that in dry eye conditions, conjunctivochalasis may behave like a foreign body that attenuates the efficiency of artificial tears by interrupting the tear-retentive function of the meniscus. Alternatively, an ectopic tear meniscus formed over the redundant conjunctiva may cause meniscus-induced tear instability[13–14] and worsen the ocular surface damage in dry eye.

In view of the low complication rate and the remarkable symptomatic improvements, this new surgical treatment is recommended when eyes are compromised by conjunctivochalasis. However, further investigation is needed to confirm our theory on the deformation of menisci causing interrupted tear flow, and whether the results of our new surgical method[8] are sufficient evidence to treat conjunctivochalasis as one of the ocular surface diseases.

REFERENCES

1. W.L. Hughes. Conjunctivochalasis. *Am J Ophthalmol.* 25: 48–51 (1942).
2. S.L. Bosniak, and B.C. Smith. Conjuctivochalasis. *Adv Ophthalmic Plast Reconstr Surg.* 3: 153–155 (1984).
3. D. Liu. Conjunctivochalasis: a cause for tearing and its management. *Ophthalmic Plast Reconstr Surg.* 2: 25–28 (1986).
4. F. Serrano, and M.L. Mora. Conjunctivochalasis: a surgical technique. *Ophthalmic Surg.* 20: 883–884 (1989).
5. R.B. Grene. Conjunctival pleating and keratoconjuctivitis sicca. *Cornea* 10: 367–368 (1991).
6. D. Meller, and S.C.G. Tseng. Conjunctivochalesis: Literature review and possible pathophysiology. *Surv Ophthalmol.* 43:225–232 (1998).
7. I. Otaka, and N. Kyu. A new surgical technique for management of conjunctivochalasis. *Am J Ophthalmol.* 129: 385–387 (1999).
8. N. Yokoi, and S. Kinoshita. New surgical treatment for conjunctivochalasis to restore normal tear meniscus at the lower lid margin. *Atarashii Ganka (J. Eye)* 17: 573–576 (2000).
9. M.G. Doane. Blinking and the mechanics of the lacrimal drainage system. *Ophthalmology* 88: 844–851 (1981).
10. M.A. Lemp, and H.H. Weiler. How do tears exit? *Invest Ophthalmol Vis Sci.* 24: 619–622 (1983)
11. F.J. Holly. Physical chemistry of the normal and disordered tear film. *Trans Ophthalmol Soc UK.* 104: 374–380 (1985).
12. A. Komuro, N. Yokoi, and S. Kinoshita. Effect of conjunctivochalasis on tear turnover. *Atarashii Ganka (J. Eye)* 17: 581–583 (2000).
13. J. McDonald, and S. Brubaker. Meniscus-induced thinning of tear films. *Am J Ophthalmol.* 72: 139–146 (1971).
14. T.R. Golding, A.S. Bruce, and J.C. Mainstone JC: Relationship between tear-meniscus parameters and tear-film breakup. *Cornea* 16: 649–661 (1997).

THE EFFECT OF PUNCTAL OCCLUSION ON TEAR LACTOFERRIN IN AQUEOUS DEFICIENT DRY EYE PATIENTS

Vernon Reese[1] and Pierce R. Youngbar[2]

[1]Belair Road
Baltimore, Maryland, USA
[2]Touch Scientific
Raleigh, North Carolina, USA

1. INTRODUCTION

Many dry eye patients undergo punctal occlusion to relieve the signs and symptoms of keratoconjunctivitis sicca (KCS). Punctal occlusion inhibits lacrimal fluid loss from the tear film by blocking the naso-lacrimal duct. This improves the symptoms in patients with tear-deficient dry eye, but it is unknown as to whether there is a concomitant improvement in the biochemistry of the tear film. This study analyzes tear lactoferrin levels in response to punctal occlusion in patients with tear-deficient dry eye.

2. METHODS

The lactoferrin concentration in tears of patients with chronic signs and symptoms of KCS was analyzed using an enzyme-linked immuno-sorbent assay (ELISA; Touch Scientific, Raleigh, NC) that is FDA approved for in-office diagnostics.[1] Tear samples were collected from the inferior marginal tear strip using calibrated glass capillary tubes. The prepared samples were placed on a solid-phase matrix that immobilized the lactoferrin for subsequent probing with an enzyme-labeled conjugate. The complexed enzyme was then quantified using a chromogenic substrate with a reflectance photometer. As reported in the 1995 NEI report on dry eye,[2] those patients with low lactoferrin levels were determined to have tear-deficient dry eye due to acinar cell dysfunction. Recruited patients underwent punctal occlusion, and their tear lactoferrin concentrations were retested at two

Lacrimal Gland, Tear Film, and Dry Eye Syndromes 3
Edited by D. Sullivan *et al.*, Kluwer Academic/Plenum Publishers, 2002

1269

subsequent follow-up office visits. The first follow-up visits were at 2–6 weeks, with an average of 27 days, following the occlusion procedure. The second follow-up visits were at 9 – 16 weeks, with an average of 94 days, following the occlusion procedure. Signs and symptoms of KCS were also measured at all follow-up visits.

3. RESULTS

Eight patients complaining of dry eye symptoms with initial lactoferrin levels below 1 mg/ml (Fig. 1) were pre-qualified for the study. Punctal occlusion was performed on each patient, and at two subsequent office visits, tear lactoferrin levels were reevaluated. Elevations in tear lactoferrin concentrations and improvements in symptoms were observed at the first follow-up visit. Lactoferrin levels were above 1 mg/ml by the second visit, with continued amelioration of dry eye symptoms.

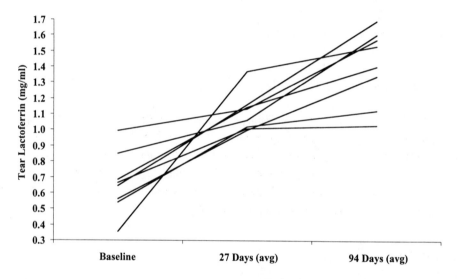

Figure 1. Tear lactoferrin levels in aqueous deficient dry eye patients before and after punctual occlusion.

4. CONCLUSIONS

Although an improvement in signs and symptoms following punctal occlusion was expected, the consistent increase in lactoferrin concentrations was unanticipated. The mechanism responsible for normalizing lactoferrin levels in these patients can only be theorized. It is possible that, by preserving the tear proteins *in situ*, a dynamic equilibrium between acinar cell function, ocular surface stress, and tear film chemistry is achieved, resulting in greater lacrimal gland functional efficiency, and eventually normal stasis is reached. These data demonstrate that tear biochemistry can be utilized to identify the etiology of tear-deficient dry eye diseases that respond favorably to punctal occlusion treatment.

A follow-up study is planned that will compare and contrast the effects of punctal occlusion in aqueous deficient and evaporative dry eye patients. It is also of interest whether the positive results of punctal occlusion are maintained long-term.

REFERENCES

1. C. McCollum, G. Foulks, B. Bodner, J. Shepard, K. Daniels, V. Gross, L. Kelly and H.D. Cavanagh, Rapid assay of lactoferrin in keratoconjunctivitis sicca. *Cornea,* 13(6):505–508 (1994).
2. M.A. Lemp, Report of the National Eye Institute/Industry Workshop on clinical trials in dry eyes. *CLAO J,* 21(4):221–232 (1995).

SILICON PUNCTAL PLUG INSERTION IN THE TREATMENT OF SEVERE DRY EYE

Kazumi Shinozaki, Etsuko Takamura, Jyunko Yukari,
and Sadao Hori

Department of Ophthalmology
Tokyo Women's Medical University
Tokyo, Japan

1. INTRODUCTION

Silicone punctal plug insertion is often faster acting than eyedrops in treating severe dry eye.[1] Punctal plug dropout may occur, and lacrimal punctal swelling or granulation may develop afterlong term identical plug use. We evaluated the long-term efficacy, safety, and stability of silicone punctal plugs in patients that were followed for 2 or more years after punctal plug insertions

2. METHODS

Eighty-three lacrimal puncta in 43 eyes of 22 patients with severe dry eye were studied. One male and 2 females, aged 56.0 years ± 13.7 (SD) were observed for a mean of 42.5 months ±7.2. The criteria used to categorize severe dry eye were rose bengal staining (6–9), fluorescein staining (2–3), Schirmer I test (< 5 mm), and ineffectiveness of topical treatments. Tear dynamics were evaluated by the Schirmer I test (without anesthesia) before and after punctal plug insertion. The ocular surface was examined using the double vital staining method, before and after punctal plug insertion. The degree of rose bengal staining in the temporal and nasal conjunctiva and cornea was graded on a scale of 0 to 3, so that the maximum score for each eye was 9. Fluorescein staining of the cornea was rated from 0 to 3.

Conjunctival sac bacterial cultures were performed to evaluate the bacteriological safety before and after punctal plug insertion. Cultures were obtained by rubbing a swab over the conjunctival sac, and streaking the specimens over chocolate agar, 5% sheep blood agar and Sabouraud's agar plates. Complications including bacterial conjunctivitis, corneal ulcer, and swelling or granulation of lacrimal punctum were observed and recorded. Eagle™ lacrimal plugs were used, and natural punctal plug dropout was monitored to evaluate their stability.

3. RESULTS

The Schirmer I test increased (Fig. 1), and rose bengal and fluorescein staining decreased from 1 day after punctal plug insertion (Fig. 2). Cultures showed increased bacterial levels, but mainly in normal bacterial flora (Figs. 3 and 4). The incidences of several complications that were observed in the eyes over the 2 years after punctal plug insertion are shown in Table 1. The rate of punctal plug dropout was 50% by the end of the first year (Fig. 5).

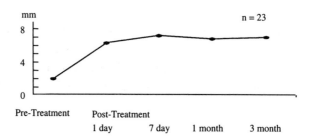

Figure 1. Schirmer I test before and after punctal plug insertion.

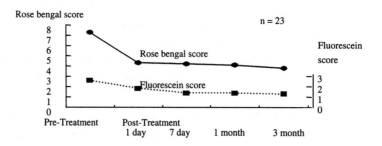

Figure 2. Rose bengal staining and fluorescein staining before and after punctal plug insertion.

Bacteria levels

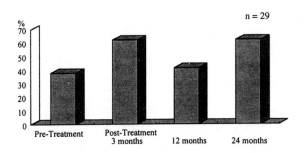

Figure 3. Bacteria levels before and after punctal plug insertion.

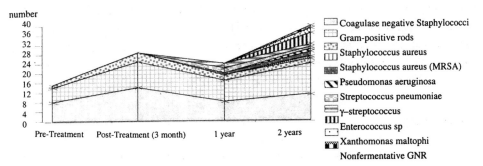

Figure 4. Bacteria flora before and after punctal plug insertion.

Table 1. Complications associated with silicon plug treatment (43eyes)

Complication	n	%
Bacterial corneal ulcer	0	0.0
Bacterial conjunctivitis	19	44.2
Swelling of lacrimal punctum	3	7.0
Granulation of lacrimal punctum	0	0.0

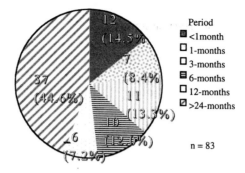

Figure 5. Natural dropout frequency.

4. DISCUSSION

Results of the Schirmer I test increased dramatically after punctal plug insertion. In addition, the ocular surface conditions of severe dry eye patients showed remarkable improvement starting one day following the insertion of the punctal plugs, demonstrating the effectiveness of the treatment. There were higher levels of bacteria in lacrimal sac cultures following treatment, but mainly normal bacterial flora were observed.[2] The culture results indicated that silicone lacrimal plug insertion was comparatively bacteriologically safe, and serious complications did not occur. Because bacterial conjunctivitis and swelling of lacrimal puncta were observed in eyes that had plugs for over 2 years, we conclude that plugs should be exchanged within 2 years after insertion. The natural dropout frequency of silicone punctal plugs was also acceptible.[3] Therefore, silicone punctal plugs are effective in the treatment of dry eye, safe under follow-up care, and moderately stable.

REFERENCES

1. H.Y. Yang, S. Igarashi, E. Goto et.al, Treatment of dry eye with silicone punctum plug, J. Eye. 14:1825–1830 (1997)
2. K. Shinozaki, E. Takamura, Bacterial flora in the conjunctival sac of dry eye, gankarinnsyouihou. 92:1095–1097(1998)
3. J.P. Gilbard, S.R. Roiss, D.T. Azar et al, Effect of punctal occlusion by freeman silicone plug insertion on tear osmolarity in dry eye disorders. CLAO J. 15:216–218 (1989)

APPLICATION OF ATELOCOLLAGEN SOLUTION FOR LACRIMAL DUCT OCCLUSION

Jun Onodera[1], Akihiko Saito[2], Joseph George[1], Toru Iwasaki[1], Hiroshi Ito[1], Yu Aso[1], Takashi Hamano[3], Atushi Kanai[4], Teruo Miyata[1], and Yutaka Nagai[1]

[1]Koken Bioscience Institute
Tokyo, Japan
[2]Triangle Animal Hospital
Tokyo, Japan
[3]Department of Ophthalmology
Osaka University Medical School
Osaka, Japan
[4]Department of Ophthalmology
Juntendo University Medical School
Tokyo, Japan

1. INTRODUCTION

Dry eye is a condition of dry, irritated, burning or gritty feeling in the eyes. It is mainly caused by ocular surface diseases, immunomodulation or injuries that affect tear secretion or composition.[1,2] The diagnosis and treatment of dry eyes have improved dramatically during recent years. Application of artificial tears is usually carried out as a method of treatment, but it requires frequent application and provides only temporary effect. In addition, it leads to epithelial cell toxicity, changes in epithelial membrane permeability and increased chances of eye infections.[3,4] Punctal occlusion prevents the discharge of natural tears from the lacrimal punctum.[5,6] Punctal occlusion prolongs the duration of tears on the ocular surface of the eye and improves the symptoms of dye eye significantly. The conventional method of punctal occlusion is the application of solid-type punctal plugs such as collagen-rod, silicone or plastic plugs. Such plugs often cause an unpleasant foreign-body sensation, corneal epithelial cell damage, granulation and accidental dropout.[7,8] To overcome such problems, we developed 3% atelocollagen solution, which forms fibrils at 37°C and neutral pH, for lacrimal duct occlusion as a new method for the treatment of dry eye. Atelocollagen is capable of forming fibrils (gel) at

Lacrimal Gland, Tear Film, and Dry Eye Syndromes 3
Edited by D. Sullivan *et al.*, Kluwer Academic/Plenum Publishers, 2002

1277

body temperature.[9] The gel formed from 3% atelocollagen solution is smooth and solid, so that when it is injected into the lacrimal duct, it forms fibrils and occludes the lacrimal canal. In the present investigation, we have studied the effect of atelocollagen solution for lacrimal duct occlusion using beagle dogs as an animal model.

2. METHODS

Atelocollagen was extracted from calf skin with pepsin digestion.[10] The extracted collagen was purified by isoelectric precipitation and filtration. The purified collagen was sterilized by micro-filtration. The atelocollagen solution was adjusted to 3% in 0.1 M phosphate buffer, pH 7.4.

Six beagle dogs (12 eyes) were used to evaluate the effect of atelocollagen on lacrimal duct occlusion. The dogs basal tear secretion, residual tear volume and tear turnover rate were studied on days 0–7 as control values. The third eyelid, which secretes 30–60% of basal lacrimal secretion,[11] was removed on day 8 under general anesthesia to reduce the basal tear volume. On day 56, after anesthetizing the animals with sodium pentobarbital, atelocollagen solution was injected into the lower lacrimal punctum until it overflowed through the upper punctum. The atelocollagen solution was applied in both eyes of all six dogs.

Schirmer tear test (STT)[12] was used to assess the basal tear secretion level before and after removal of the third eyelid. STT paper was obtained from Showa Pharmaceuticals (Tokyo, Japan). To avoid over-sensitization, 1 drop of Benoxil (Santen Pharmaceuticals, Osaka, Japan) was added to eye, and after 5 min the tear-testing paper was held to the lower eyelid for about 1 min and extent of the wet portion was measured.

Residual tear volume was determined before and after application of atelocollagen by phenol red thread tear (PRT, Showa Pharmaceuticals, Tokyo, Japan) test.[13] The cotton thread was held for 15 sec in the lower eyelid and length of the red color portion, which is directly proportional to the residual tear volume, was measured.

The tear turnover rate (TTR) was measured as described by Shimizu et al.[14]. To measure TTR, 5 µl FITC-labeled dextran (MW = 4400, Sigma, St. Louis, MO, USA) was applied in the eye and fluorescence intensity on the corneal surface was measured at 4, 6, 8, 10 and 12 min using an anterior fluorometer (Kowa, Tokyo, Japan). The values were converted to fluorescein concentration and the natural logarithm of the fluorescein concentration was plotted versus time. TTR was calculated from the regression curve.

The mean and standard error were calculated from the data. The results were evaluated using one-way analysis of variance (ANOVA). The mean values in each group were compared using contrast analysis method. The value of $P < 0.05$ was considered statistically significant.

3. RESULTS

The value of STT decreased significantly ($P < 0.05$) from day 19 onward after removal of the third eyelid, and remained at a reduced level after injection of atelocollagen (Fig. 1). The residual tear volume decreased significantly ($P < 0.001$) after removal of the third eyelid (Fig. 2). However, after injection of atelocollagen, the residual tear volume increased considerably ($P < 0.01$).

TTR decreased significantly ($P < 0.05$) after removal of the third eyelid (Fig. 3). Punctal occlusion with atelocollagen resulted in a further decrease of TTR values ($P < 0.001$).

4. DISCUSSION

The purpose of this investigation was to evaluate the effect of atelocollagen solution as a novel method for occlusion of the lacrimal duct. Immunochemical studies of collagen revealed that telopeptides are the major immunologic sites in native tropocollagen.[15] Atelocollagen obtained by pepsin digestion is a poor antigen[16] and is best suitable for occlusion of the lacrimal duct because of its poor antigenicity and excellent tissue compatibility.

In the present study we selected beagle dogs as an animal model because dogs are a good mammalian model for ocular studies and easy to maintain under laboratory conditions. It was necessary to remove the third eyelid of the dogs, which secretes 30–60% of basal tear secretion,[11] to create the environment of dry eye. STT proved the basal tear secretion decreased considerably after removal of the third eyelid. In this study we confirmed occlusion of the lacrimal duct by PRT and TTR test.

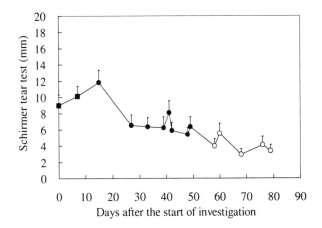

Figure 1. Basal tear secretion level before (■) and after (●) removal of the third eyelid and after application of 3% atelocollagen solution into the lacrimal duct (O) (Mean ± Standard error).

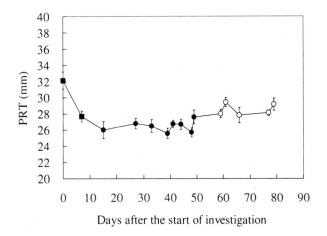

Figure 2. Residual tear volume before (■) and after (●) removal of the third eyelid and after application of 3% atelocollagen solution into the lacrimal duct (O) (Mean ± Standard error).

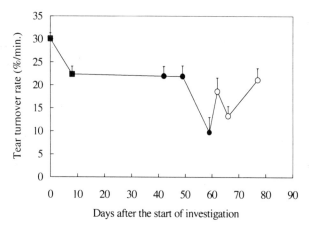

Figure. 3. TTR before (■) and after (●) removal of the third eyelid and after application of 3% atelocollagen solution into the lacrimal duct (O) (Mean ± Standard error).

Occlusion of the lacrimal duct by 3% atelocollagen solution is effective up to 3 weeks. When 3% atelocollagen solution is administered into the lacrimal duct of rabbits, occlusion is effective up to 7 weeks.[17] Importantly, the rabbit has a considerably bigger lacrimal sac compared to that of the dogs.

Our results suggest occlusion of the lacrimal duct by 3% atelocollagen solution is an effective method for treatment of dry eye. The atelocollagen solution can be injected into the lacrimal duct, regardless of its size, to form the soft fibrils without foreign-body sensation or corneal cell injury.

REFERENCES

1. J.D. Nelson, H. Helms, R. Fiscella, Y. Southwell, and J.D. Hirsch, A new look at dry eye disease and its treatment. *Adv. Ther.* 17:84 (2000).

2. J.D. Nelson, Dry eye syndromes. In. *Current Practice in Ophthalmology*, A.P. Schachat, ed., Mosby-Year Book, Missouri, 49 (1992).

3. F.A. Kaszli, and G.K. Krieglstein, Tear film deficiencies, pharmacology of eye drops and toxicity. *Curr. Opin. Ophthalmol.* 7:12 (1996).

4. J.P. Gilbard, S.R. Rossi, and K.G. Heyda, Ophthalmic solutions, the ocular surface, and a unique therapeutic artificial tear formation. *Am. J. Ophthalmol.* 107:348 (1989).

5. J. Murube and E. Murube, Treatment of dry eye by blocking the lacrimal canaliculi. *Surv. Ophthalmol.* 40:463 (1996).

6. E.J. Cohen, Punctal occlusion. *Arch. Ophthalmol.* 117:389 (1999).

7. S. Rumelt, H. Remulla, and P.A.D. Rubin, Silicone punctal plug migration resulting in dacryocystitis and canaliculitis. *Cornea.* 16:377 (1997).

8. C.N.S. Soparkar, J.R. Patrinely, J. Hunts, J.V. Linberg, R.C. Kersten, and R. Anderson, The perils of permanent punctal plugs. *Am. J. Ophthalmol.* 123:120 (1997).

9. D.L. Helseth, Jr. and A. Veis, Collagen self-assembly *in vitro*. *J. Biol. Chem.* 256:7118 (1981).

10 F. DeLustro, R.A. Condell, M.A. Nguyen, J.M. McPherson, A comparative study of the biologic and immunologic response to medical devices derived from dermal collagen. *J. Biomed. Mater. Res.* 20:109 (1986).

11.   L.C. Helper, W.G. Magrane, J. Koehm, and R. Johnson, Surgical induction of keratoconjunctivitis sicca in the dog. *J. Am. Vet. Med. Assoc.* 165:172 (1974).

12.   D.B. Harker, A modified Schirmer tear test technique. *Vet. Rec.* 86:196 (1970).

13.   M.H. Brown, J.C. Galland, H.J. Davidson, and A.H. Brightman, The phenol red thread tear test in dogs. *Vet. Comp. Ophthalmol.* 6:274 (1996).

14.   A. Shimizu, N. Yokoi, K. Nishida, S. Kinoshita, and K. Akiyama, Fluorophotometric measurement of tear volume and tear turnover rate in human eyes. *J. Jpn. Ophthalmol. Soc.* 97: 1047 (1993).

15.   R. Timpl, Immunological studies on collagen. In: *Biochemistry of Collagen,* G.N. Ramachandran, and A.H. Reddi, ed., Plenum Press, New York, 319 (1976).

16.   H. Furthmayr, and R. Timpl, Immunochemistry of collagens and procollagens. *Int. Rev. Connect. Tissue Res.* 7: 61 (1976).17.J. Onodera, T. Iwasaki, A. Saito, H. Ito, T. Hamano, A. Kanai, and Y. Nagai, The application of atelocollagen solution to a lacrimal duct occlusion. *Jpn. J. Artif. Organs.* 28:225 (1999).

ATELOCOLLAGEN PUNCTAL OCCLUSION FOR THE TREATMENT OF THE DRY EYE

Takashi Hamano

Department of Ophthalmology
Osaka University
Osaka, Japan

1. INTRODUCTION

Punctal occlusion is the most effective treatment for dry eye that is currently available clinically. The most popular method of punctal occlusion is the use of punctal plugs. The purpose of this study is to evaluate the efficacy of a new method, atelocollagen puctal occlusion, for the treatment of dry eye. Atelocollagen, a pepsin digested collagen from calf skin, is a viscous solution that gels at body temperature, and has been used clinically for implantation in dermatology and dentistry.[1,2] The main difference between the two methods is that atelocollagen forms a temporary occlusion while the punctal plug is more permanent.

2. METHODS

A total of 31 subjects raging in age from 30 to 72 years, average 50 years, were tested. Sixteen of the subjects were dry eye patients, of which 11 females and 1 male were diagnosed with Sjögren's syndrome, and 3 females and 1 male were diagnosed with non-Sjögren's syndrome.

The atelocollagen for this study was provided by Koken Co. Ltd., Tokyo, Japan. Atelocollagen is a translucent, viscous solution under 10°C, and it gels at body temperture by fiber renaturation. A 3.0% solution of atelocollagen in phosphate buffer was used in this study. The test materials were provided by the manufacturer as ready–to–use steriled cartridges.The amount of the atelocollagen was measured (34.8 ± 12.5 mg per eye) in each case, and solution was injected into the lower and upper puncta of each eye. Treated eyes

Lacrimal Gland, Tear Film, and Dry Eye Syndromes 3
Edited by D. Sullivan *et al.*, Kluwer Academic/Plenum Publishers, 2002

1283

were kept closed for 15 minutes to prevent the solution from leaking out before the collagen solidified, 10 – 15 min at body temperature. Rose bengal staining and symptoms were scored and recorded just before, and 2 to 3 weeks after, injection of atelocollagen.

3. RESULTS

In 13 eyes, scores for rose bengal staining decreased 2 to 3 weeks after injection of atelocollagen, in 14 eyes the scores did not change, and in 4 eyes the scores increased. Significant improvement in rose bengal scoring was observed after the treatment (Wilcoxon test, $P = 0.0125$). Fifteen of the 16 dry eye patients reported symptomatic relief, though the duration of the symptomatic relief varied from a few days to a few months.

4. DISCUSSION

Punctal occlusion is the most effective method for the treatment of the dry eye that is currently available clinically. There are several advantages of atelocollagen punctal occlusion. First, it is easy to administer. Scaling and dilation of the punctum are not necessary. Second the procedure does not produce granuloma. Punctal plugs are reported to be the cause of the granuloma formation. Since atelocollagen is a soft gel in the punctum and canaliculus, it does not cause granuloma formation. Third, occlusion is temporary. In many cases of dry eye, permanent punctum occlusion is not necessarily desirable. Atelocollagen does not stay in the lacrimal duct permanently. Finally, atelocollagen gel is removed easily by irrigation. According to our preliminary clinical trial, the atelocollagen punctal occlusion is a promising candidate for the treatment of the dry eye.

REFERENCES

1.   Ishikawa T,Ohura T,Ogura T,Hoshi M,Honda K,Iida K: Clinical study of injectable atelocollagen from bovine calf skin. Nishinihon Hifuka 47, 1985
2.   Onodera J,Iwasaki T,Saito A,Ito H,Hamano T,Kanai A,Nagai Y: The application of atelocollagen solution to a lacrimal duct occlusion. Jpn J Aritf Organs 28,225–229, 1999

FIBERSCOPIC OBSERVATION OF CANALICULI AFTER PUNCTAL PLUG EXTRUSION

Kazuhiro Shimizu,[1] Norihiko Yokoi,[2] and Shigeru Kinoshita[2]

[1]Department of Ophthalmology
Osaka Medical College
Osaka, Japan
[2]Department of Ophthalmology
Kyoto Prefectural University of Medicine
Kyoto, Japan

1. INTRODUCTION

Punctal plug therapy is effective in treating severe aqueous-deficient dry eye. However, the formation of granulation tissue within the canaliculus sometimes causes plug extrusion.[1-5] When this occurs, it is essential to evaluate the extent of the canalicular obstruction by granulation tissue before considering subsequent therapies, including insertion of a replacement plug or surgical occlusion of the punctum. Currently, there is no appropriate method for evaluating the degree of canalicular blockage after punctal plug extrusion. In the present study, we evaluated the effectiveness of a fiberscope system in the examination of canaliculi interiors in patients with extruded punctal plugs.

2. METHODS

We examined thirty puncta in 20 eyes of 12 female and 3 male patients with severe dry eye, mean age 60.5 years, whose canaliculi had been previously occluded with punctal plugs (FCI Ophthalmics, France). Ten of the 15 patients had Sjögrens syndrome, 4 had a form qf dry eye other than Sjögrens, and 1 patient had Stevens-Johnson syndrome. The mean duration of punctal plug placement was 4.6 months (1–17 months).

In most cases, a fiberscope system (FVS-3000M, M&M Co. Ltd., Japan) fitted with a 0.72 mm semi-rigid fiberscope (SRC-72S, M&M Co. Ltd., Japan) was used. However,

Lacrimal Gland, Tear Film, and Dry Eye Syndromes 3
Edited by D. Sullivan *et al.*, Kluwer Academic/Plenum Publishers, 2002

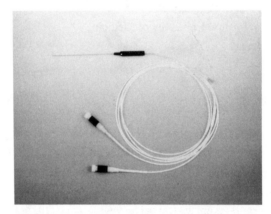

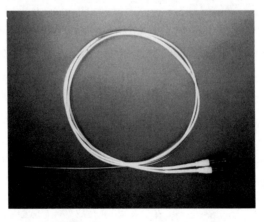

Figure 1. Fiberscope system (FVS-3000M, M&M Co. Ltd., Japan): **(1)** Main unit of a fiberscope system; **(2)** 0.72 mm semi-rigid fiberscope (SRC-72S, M&M Co. Ltd., Japan); **(3)** 0.45-mm flexible fiberscope (FM-45, M&M Co. Ltd., Japan).

when puncta were small, the system was fitted with a 0.45 mm flexible fiberscope (FM-45, M&M Co. Ltd., Japan; Figs. 1-1, 1-2, 1-3). Before the examination, a local anesthesia of 0.4% oxybuprocaine hydrochloride was instilled into each eye. During the examination, the punctum was first illuminated with the fiberscope, and then the fiber tip was advanced into the canaliculus. If further insertion was possible, the interior of the canaliculus was observed. Images were recorded sequentially on video.

3. RESULTS

Among the 30 puncta (20 eyes) examined using the fiberscope, the interior of the canaliculus of 8 puncta (5 eyes) was clearly visualized (Fig. 2-1). In these canaliculi, there was no granulation tissue formation, and normal intracanalicular structure was preserved. Clear images were obtained with both fibersope types. In 4 puncta (2 eyes), images were obtained at approximately 2 – 3 mm from the punctum, but no deeper images were obtained. In these cases, the canaliculus seemed to be narrowed in spite of the opening at the punctum. There were 18 puncta (13 eyes) in which fiberscopic images could not be completely obtained at a depth of 2 mm, indicating that the canaliculus was completely occluded (Fig. 2-2).

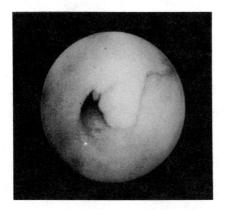 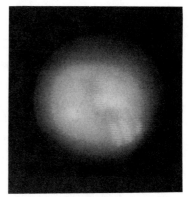

Figure 2. Fiberscope images of puncta from dry eye patients: **(1)** Open canaliculus; **(2)** Occluded canaliculus.

4. DISCUSSION

Superficial punctate keratitis can be improved by inserting plugs into the puncta that effectuate the retention of tears. However, due to narrowing of the puncta, repeated extrusion of reinserted punctal plugs is common. In these cases, aside from suturing the punctum closed, there are no other treatment options available. The ability to observe the interior of the punctum and canaliculus and to identify the presence of obstructions, could potentially clarify subsequent treatment options in cases where punctal plugs had been extruded. Using a fiberscope, we sought to investigate this hypothesis.

The fiberscope was originally developed to visualize the interior of the lactiferous ducts, and is effective in the identification of lactiferous duct papillomas and cancer. When visualizing the interior of lactiferous ducts for purposes of imaging, a semi-rigid fiber with a diameter slightly larger than that of a flexible fiber is preferred. Since dedicated fiberscopes for examining puncta are yet to be developed, we used a fiberscope system that was intended for observation during dacryocystorhinostomy.[6] In dacryocystorhinostomy, a rigid fiberscope of 4 mm diameter is usually used. This is too large for the canaliculus, so we used either a semi-rigid or a flexible fiber instead. The difference between these two fibers for purposes of canaliculus visualization is insignificant since adequate images can be obtained even with flexible fibers.

In theory, superior images should be obtained when canaliculi are dilated. Therefore, during the imaging procedure, canaliculi were irrigated via an irrigation tube fitted to the rear end of the semi-rigid fiberscope. In our experience, no substantial improvement was observed using this procedure, possibly due to the rigidity of the canaliculus itself which rendered irrigation less effective.

Visualization of the position of the tip of the fiberscope as it is advanced within the canaliculus is critical. Fortunately, the light from the fiberscope tip passes through the skin and can be observed from outside. Under these conditions, it is relatively simple to follow the progress of the fiberscope into the canaliculus and accurately judge the site of occlusion.

We had hoped that fiberscopic examination would allow the morphology of granulation tissue to be ascertained; however, while visualization of healthy canaliculi was easy, we found that good images could not be obtained when the lumen was narrowed as a result of factors such as the formation of granulation tissue. In those cases, it was even difficult to recognize the presence of a wall immediately in front of the fiberscope tip. However, such problems might be solved if the fiberscope were improved.

This study demonstrates the effectiveness of the fiberscope system for assessing the degree of canaliclular occlusion after punctal extrusion. This fiberscope system may also be useful for determining future therapeutic strategies for treatment of dry eye.

REFERENCES

1. H. Sato, K. Aoi, N. Yokoi and S. Kinoshita. Evaluation of punctal plug treatment for severe dry eye. Atarashii Ganka (Journal of the Eye, Japan) 16(6): 843–846 (1999).
2. JM. Freeman. The punctum plug: Evaluation of a new treatment for the dry eye. Trans Am Acad Ophthalmol Otolaryngol 70: 874–879 (1975).
3. MW. Rebecca, F. Robert and HK. Jay. The tretment of aqueous deficient dry eye with removable punctal plugs. Ophthalmology 94: 514–518 (1987).
4. JP. Gilbard, SR. Roiss and DT. Azar. Effect of punctal occlusion by freeman silicone plug insertion on tear osmolarity in dry eye disorders. CLAO J 15: 216–218 (1989).
5. M. Juan and M. Eduardo. Treatment of dry eye by blocking the lacrimal canaliculi. Surv Ophthalmol 40: 463–480 (1996).
6. K. Kurihashi. Application of the endoscope in dacryocystorhinostomy. Japanese Journal of Ophthalmic Surgery 9: 25–32 (1996).

Gene, Stem Cell and Surgical Therapy of the Ocular Surface

NEURAL STEM CELLS FOR CNS REPAIR: STATE OF THE ART AND FUTURE DIRECTIONS

Kook In Park,[1,2] Mahesh Lachyankar,[1] Sahar Nissim,[1] and Evan Y. Snyder[1]

[1] Departments of Neurosurgery, Pediatrics, and Neurology
Children's Hospital
Harvard Medical School
Boston, Massachusetts, USA
[2] Department of Pediatrics
Yongdong Severance Hospital
Yonsei University College of Medicine
Seoul, South Korea

1. INTRODUCTION

Approximately 10–15 years ago, we and a small group of other investigators recognized cells with an unanticipated degree of plasticity and multipotency existed in the CNS. These cells are now called neural stem cells (NSCs). They probably represented the cellular basis for a great deal of unrecognized plasticity programmed into the developing and post-developmental CNS. It is likely that therapeutic advantage might be realized by harnessing this plasticity.[1] Two complementary goals drive the motivation to study NSCs both in situ and following their isolation and propagation ex vivo. The first is to understand the processes of commitment, differentiation, migration and plasticity in the mammalian nervous system during development and degeneration. The second is to use such cells for therapeutic purposes (at all ages). The pursuit of one goal typically nurtures the other, and our lab pursues these two overlapping areas of investigation. In fact, our hypothesis has been that, by first understanding the fundamental biology of pluripotent NSCs (whose role is to mediate organogenesis, homeostasis and reconstitution), and then exploiting those

Lacrimal Gland, Tear Film, and Dry Eye Syndromes 3
Edited by D. Sullivan *et al.*, Kluwer Academic/Plenum Publishers, 2002

1291

properties for therapeutic ends, novel strategies may emerge for redressing CNS dysfunction. Our experiments are performed in vitro and in vivo following transplantation of clonally related, genetically tagged NSCs into normal and abnormal developing (fetal, neonatal) and adult rodent and primate brain, spinal cord and peripheral nervous system.[2,3]

2. EXTRINSIC CUES MOLD INTRINSIC DEVELOPMENTAL PROGRAMS

In studying the mechanisms by which NSCs differentiate in vitro or become integrated into the developing and degenerating, immature and mature, normal and abnormal nervous system (following transplantation across a range of regions and ages), we have learned NSCs derive many instructive cues from their environment. However, while such extrinsic cues (e.g., such diffusible factors as neurotrophic agents) clearly play a role, NSCs possess autonomous differentiation programs that are then inhibited, dysinhibited and/or modified by specific environmental cues to which the system is programmed to respond at particular points of susceptibility, e.g., exit from the cell cycle. One such internal program is for an NSC to become a neuron (glutamatergic or GABA'ergic) by default. Intriguingly, the fate choice by one NSC (either by default, instruction or stochastic processes) then actively influences that of juxtaposed, clonally related, ostensibly identical sister NSCs by contravening or modifying their intrinsic programs.[3] Through cell-cell contact by membrane-mediated/associated factors, an NSC that has already committed to a neuronal lineage will actively inhibit that differentiation option for an uncommitted sister NSC, even in situations in which that NSC would ordinarily assume a neuronal phenotype. The NSC is instructed to become a glial cell or join a pool of reserve progenitor cells. Juxtaposed sister cells assuming this support role for the neuronal progeny of the clone do produce trophic factors, e.g., glial-derived growth factor, that the neuron does not make for itself. Conversely, the neuron bears receptors to such factors that the producer cells do not. To put this in a broader context, the fabric of the brain may be constructed by a series of such ever-more-refining interactions in which symbiotic interactions constitutively emerge.

An example of a further refinement of this type may be observed when nurr-1, a ventral mesencephalic-determining transcription factor, is overexpressed in pluripotent NSCs and predisposes them to become mesencephalic-like neurons.[4] Yet the final molding of its dopaminergic phenotype is consummated only when such cells are then exposed to type 1 astrocytes specifically derived from the ventral mesencephelon and not any other region.

Therefore, based on such observations, to the list of external instructive cues determining the differentiation of an NSC, the phenotype of contiguous, neighboring cells must be added. Furthermore, the idea that intrinsic, genetically based internal programs require tailoring by external environmental variables is applicable to fate determination by stem cells of other organ systems.

3. NEURODEGENERATIVE ENVIRONMENTS CONSTITUTE AN EXTRINSIC CUE

While various external factors may play a role in determining the differentiation of NSCs, our studies suggest a category not previously anticipated: a neurodegenerative environment and/or an environment lacking in a particular neural cell type itself provides cues instructive to the fate of an NSC. Through studying a series of rodent mutations and injury models (and, more recently, primate equivalents of some of these), we learned that a pluripotent NSC will, during key temporal windows, shift its differentiation fate to compensate for a missing cell type. The process by which this occurs remains an active area of investigation for this may constitute a fundamental developmental mechanism with therapeutic implications. In situations in which particular cell types fail to develop properly, a progeny counting system or a sensitivity on the part of the NSC to a lack of a critical number of given cell types or the signals they elaborate might exist.[5]

That neurodegenerative, often apoptotic,[6] environments may spontaneously direct the proliferation, migration and differentiation of multipotent NSCs towards repopulation of the injured region suggests cytokines, chemokines, extracellular matrix, adhesion and other traditionally non-neural molecules may also play an important role. Being able to use homogenous clones of well-characterized, easily traced NSCs as reporter cells that can mimic the behavior of endogenous NSCs and progenitors with which they are intermixed has allowed us to conclude acute degeneration exerts tropic and trophic effects on stem cells, both host and donor. Because these reporter cells sense and respond to subtle changes in neural environments (via altered migratory and differentiation fates), some neurodegenerative conditions may change non-neurogenetic into neurogenetic environments, at least transiently. This phenomenon occurs in a range of animal models, including apoptotic degeneration in adult neocortex,[6] asphyxial damage to the cerebrum,[7] motor neuron degeneration in the spinal cord,[7] Purkinje cell degeneration in cerebellum[1] and dopaminergic degeneration in striatum. Again these observations suggest a significant degree of plasticity may be programmed into the nervous system at the level of the stem cell as a fundamental developmental mechanism, one that may clearly have therapeutic utility. Indeed, we have determined endogenous neural progenitors do attempt to respond in an identical manner. Seemingly stereotypical immutable developmental patterns and migratory routes will be diverted in response to injury to deliver such cells to areas of need.[8] Although these intrinsic self-repair mechanisms may often be inadequate to reverse functional impairment in the most severe clinical conditions, the response may be augmented by supplying exogenous cells and factors.

4. NEURAL STEM CELLS ARE IDEALLY SUITED FOR CELL AND GENE REPLACEMENT IN NEURODEGENERATIVE ENVIRONMENTS

Transplantation of exogenous NSCs is an effective tool for gene therapy and repair of the CNS. From work pursued in various animal models of cerebral and spinal dysfunction and neurodegeneration, we accumulated substantial evidence NSCs may be used for gene transfer,[9-11] to replace dysfunctional and maldeveloped neural cell types (e.g., neurons,[5-7] oligodendrocytes[12,13]), structures (e.g., myelin[12,13]) and other molecules (e.g., extracellular matrix) and to rescue abnormal cytoarchitecture. Our lab studies models more for the stem cell biology they may help elucidate than diseases they represent. With this guiding principle, NSC-mediated approaches have been directed at a set of prototypical models of neural dysfunction, including spinal motorneuron degeneration, hypoxic-ischemic injury, traumatic injury to the spine and head, abnormalities of migration and lamination, neurogenetic degenerative conditions of childhood (e.g., lysosomal storage diseases,[9,10] inborn metabolic errors, leukodystrophies[13]) and adulthood (e.g., basal ganglia dysfunction, Alzheimer's disease[14]) and brain tumors.[15,16] In this context, we have also explored some dynamics by which NSCs may serve as unique and better-controlled vehicles for gene therapy. For example, we have quantified and devised simple techniques for combating retrovirally transduced transgene nonexpression in cellular vectors by insuring multiple copies of a gene have been integrated into the genome of a given NSC. We have transformed NSCs into unique engraftable, migratory packaging/producer cells for retroviral vectors.[16-18] We have demonstrated NSCs may form the interface between cell biology and tissue engineering/material science. For instance, we have used them to populate biosynthetic scaffolds[19] that can bridge gaps in cerebrum and spinal cord and hence promote neurite outgrowth and possible connectivity. NSCs may serve as the glue that brings and holds together therapeutic possibilities derived from many disciplines.

In addition to describing shifts in differentiation and developmental patterns, we have directed much attention to elucidating the mechanisms by which NSCs are drawn to areas of degeneration and their fates redirected. Our working hypothesis has been that injury promotes a re-expression of developmental molecular signals to which NSCs respond appropriately similar to their normal embryonic expression.[20] Using this hypothesis to guide and direct our search, we are using gene chip microarrays and proteomics to aid in the identification and isolation of candidate migratory, attractant and repulsive molecules. Among the genes that may be altered by injury and degeneration are those encoding cell cycle regulatory proteins, cytokines and cytokine receptors. Observations such as these may link NSC biology to organogenesis, immunobiology, hematology, endocrinology, angiogenesis, tumor biology, etc. Furthermore, because NSCs are efficient vehicles for transferring exogenous gene products to the CNS, they are also being used for gene therapy and to assess the impact of the ectopic, excessive and/or inappropriate expression of developmentally important or plasticity-related genes in vivo.

Finally, we have succeeded in isolating, propagating and transplanting human NSCs (hNSCs)[21-23] that follow developmentally appropriate programs in vivo. They, too, can express foreign genes and replace degenerating or underdeveloped neurons. As the first published clones of engraftable hNSCs (and the first reported solid organ stem cells of human origin[24]), they are being assessed for their effectiveness as gene therapy and cell replacement vehicles in preclinical situations in rodents and monkeys. These studies in animal models of neurodegeneration and injury might promote rapid advancement toward novel molecular and/or cellular replacement clinical therapies for some developmental, degenerative and acquired human neurological dysfunctions. In fact, these actual hNSCs may have the potential for direct human application.

REFERENCES

1. E.Y. Snyder, D.L. Deitcher, C. Walsh, S. Arnold-Aldea, E.A. Hartweig, C.L. Cepko, Multipotent neural cell lines can engraft and participate in development of mouse cerebellum, *Cell* 68:33 (1992).
2. V. Ourednik, J. Ourednik, K.I. Park, E.Y. Snyder, Neural stem cells: a versatile tool for cell replacement and gene therapy in the CNS, *Clin. Genet.* 46: 267 (1999).
3. E.Y. Snyder, Neural stem-like cells: developmental lessons with therapeutic potential, *The Neuroscientist* 4: 408 (1998).
4. J. Wagner, P. Akerud, D. Castro, P.C. Holm, E.Y. Snyder, T. Perlmann, E. Arenas, Type 1 astrocytes induce a midbrain dopaminergic phenotype in *Nurr*1-overexpressing neural stem cells, *Nature Biotech.* 17:653 (1999).
5. C.M. Rosario, B.D. Yandava, B. Kosaras, D. Zurakowski, R.L. Sidman, E.Y. Snyder, Differentiation of engrafted multipotent neural progenitors towards replacement of missing granule neurons in meander tail cerebellum may help determine the locus of mutant gene action, *Development* 124:4213 (1997).
6. E.Y. Snyder, C.H. Yoon, J.D. Flax, J.D. Macklis, Multipotent neural precursors can differentiate toward replacement of neurons undergoing targeted apoptotic degeneration in adult mouse neocortex, *Proc. Natl. Acad. Sci. USA* 94:11663 (1997).
7. K.I. Park, S. Liu, J.D. Flax, S. Nissim, P.E. Stieg, E.Y. Snyder, Transplantation of neural progenitor and stem-like cells: developmental insights may suggest new therapies for spinal cord and other CNS dysfunction, *J. Neurotrauma* 16:675 (1999).
8. K.I. Park, E.Y. Snyder, Injury shifts developmental patterns to promote establishment of lost neural cells in "non-neurogenic" CNS regions, *Soc. Neurosci. Abstr.* (in press).
9. E.Y. Snyder, R.M. Taylor, J.H. Wolfe, Neural progenitor cell engraftment corrects lysosomal storage throughout the MPS VII mouse brain, *Nature* 374:367 (1995).
10. H.D. Lacorazza, J.D. Flax, E.Y. Snyder, M. Jendoubi, Expression of human β-hexosaminidase α-subunit gene (the gene defect of Tay-Sachs disease) in mouse brains upon engraftment of transduced progenitor cells, *Nature Medicine* 4:424 (1996).
11. F.J. Rubio, Z. Kokaia, A. del Arco, M.I. Garcia-Simon, E.Y. Snyder, O. Lindvall, J. Satrustegui, A. Martinez-Serrano, BDNF gene transfer to the mammalian brain using CNS-derived neural precursors, *Gene Therapy* 6: 1851 (1999).
12. B.D. Yandava, L.L. Billinghurst, E.Y. Snyder, Global cell replacement is feasible via neural stem cell transplantation: evidence from the *shiverer* dysmyelinated mouse brain, *Proc. Natl. Acad. Sci. USA* 96:7029 (1999).
13. L.L. Billinghurst, R.M. Taylor, E.Y. Snyder, Remyelination: cellular and gene therapy, *Sem. in Ped. Neurol.* 5: 211 (1998).
14. L. Doering, E.Y. Snyder, Cholinergic expression by a neural stem cell line grafted to the adult medial septum/diagonal band complex, *J. Neurosci. Res.* 61:597 (2000).
15. K.S. Aboody, A. Brown, N.G. Rainov, K.A. Bower, S. Liu, W. Yang, J.E. Small, U. Herrlinger, V. Ourednik, P.M. Black, X.O. Breakefield, E.Y. Snyder, Neural stem cells display extensive tropism for pathology in adult brain: evidence from intracranial tumors, *Proc. Natl. Acad. Sci. USA* 97:12846 (2000).
16. U. Herrlinger, C. Woiciechowski, K.S. Aboody, A.H. Jacobs, N.G. Rainov, E.Y. Snyder, X.O. Breakefield, Neural stem cells for delivery of replication-conditional HSV-1 vectors to intracerebral gliomas, *Molecular Therapy* 1:347 (2000).

17. W.P. Lynch, A.H. Sharpe, E.Y. Snyder, Neural stem cells as engraftable packaging lines can optimize viral vector-mediated gene delivery to the CNS: evidence from studying retroviral *env*-related neurodegeneration, *J. Virol.* 73:6841 (1999).

18. D.J. Poulsen, C. Favara, E.Y. Snyder, J. Portis, B. Chesebro, Increased neurovirulence of polytropic mouse retroviruses delivered by inoculation of brain with infected neural progenitor cells, *Virology* 263:23 (1999).

19. K.I. Park, E. Lavik, Y.D. Teng, R. Langer, E.Y. Snyder, Transplantation of neural stem cells (NSCs) seeded in biodegradable polyglycolic acid (PGA) scaffolds into hypoxic-ischemic (HI) brain injury, *Soc. Neurosci. Abstr.* (in press).

20. V. Ourednik, J. Ourednik, K.I. Park, Y.D. Teng, K.A. Aboody, K.I. Auguste, R.M. Taylor, B.A. Tate, E.Y. Snyder, Neural stem cells are uniquely suited for cell replacement and gene therapy in the CNS, in *Novartis Foundation Symposium: Neural transplantation in neurodegenerative disease: current status and new directions*, D.J. Chadwick, J.A. Goode, eds., Wiley & Sons, NY (2000).

21. J.D. Flax, S. Aurora, C. Yang, C. Simonin, A.M. Wills, L. Billinghurst, M. Jendoubi, R.L. Sidman, J.H. Wolfe, S.U. Kim, E.Y. Snyder, Engraftable human neural stem cells respond to developmental cues, replace neurons, and express foreign genes, *Nature Biotech.* 16:1033 (1998).

22. A. Villa, E.Y. Snyder, A. Vescovi, A. Martinez-Serrano, Establishment and properties of a growth factor-dependent, perpetual neural stem cell line from the human CNS, *Exp. Neurol.* 161:67 (2000).

23. A.L. Vescovi, E.Y. Snyder, Establishment and properties of neural stem cell clones: plasticity *in vitro* and *in vivo*, *Brain Pathology* 9:569 (1999).

24. E.Y. Snyder, A.L. Vescovi, Stem cells: possibilities or perplexities, *Nature Biotech.* 18:827 (2000).

PARTICLE-MEDIATED GENE TRANSFER TO OCULAR SURFACE EPITHELIUM

Winston W.-Y. Kao

Department of Ophthalmology
University of Cincinnati
Cincinnati, Ohio, USA

1. INTRODUCTION

When molecular cloning was introduced in the early 1970s, gene therapy became a real possibility in curing congenital diseases caused by gene mutations, e.g., thalassemia.[1] Since then, advances in technology have extended in attempts to cure acquired diseases such as cancer, wound healing, metabolic diseases and immune disorders.[2-9] Nevertheless, gene therapy is still in its infancy. Although some promising results have been obtained, in general no procedure can yet be used to successfully cure congenital or acquired diseases. Despite technological advances sustaining the field of gene therapy, few individual patients have benefited from it so far. The reasons for its limited success are due to difficulties in efficient gene transfer, and sustained and controlled expression of the transgenes leading to production of therapeutic quantity of proteins, as well as complications of the introduction of foreign genes to humans.[10,11] For example, the widely used viral vectors can produce considerable side effects of host immune responses to the viral proteins, which prohibit its repetitive uses for effective treatment.[12-17] The other techniques–non-viral vehicles, i.e., liposomes, and physical means, i.e., particle-mediated gene transfers[14] and electric pulse–also have not yielded satisfactory results.[15, 18-20]

The objective of gene therapy by introducing a foreign gene to the somatic cells is to modulate cellular functions by allowing cells to gain or lose specific functions related to the disease process, and ablation of cells in cancer therapy. The strategy to use one or more of these properties depends on the nature of the diseases. For example, gain and loss of functions are common strategies to treat cancers, such as expression of suicidal toxin, re-expression of tumor suppressor genes and ablation of anti-apoptotic genes.[3,4,8,21-25]

Lacrimal Gland, Tear Film, and Dry Eye Syndromes 3
Edited by D. Sullivan *et al.*, Kluwer Academic/Plenum Publishers, 2002

1297

Ocular surface tissue is an ideal model system to examine the efficacy of gene therapy strategy due to its easy access. However, little success has been reported for application of gene therapy to cure eye diseases. This may be primarily because treating eye diseases has the same problems as using gene therapy for other diseases. These include: 1. effective gene delivery to the target tissue, 2. cell type-specific gene expression, 3. sustained expression of the transgene and 4. minimal adverse effects of the procedure.

2. METHODS OF GENE DELIVERY

Methods of gene delivery to the somatic cells of a living organism can be classified into three major categories: (1) viral vectors, (2) liposomes and (3) physical methods. The most commonly used viral vectors include DNA virus and retrovirus. The transfection of the foreign gene into a somatic cell by viral vectors is mediated via cell surface receptors. Expression efficiency using the viral vectors is usually high; however, many cell types may express the virus receptors, which results in low tissue specificity.[16,17,26,27] The gene transfer mediated by the DNA virus can only be transiently expressed because integration of transgenes into host genome is very rare. However, it does provide an advantage that cell division is usually not required for the use of such foreign genes as most somatic cells are quiescent. Retrovirus has a high frequency of integration of the foreign DNA into the host genome on cell division; therefore, sustained expression of the transgene can be achieved.[27-29] However, gene transfer using the DNA virus and retrovirus causes the complications of host immune responses, which significantly diminish their effectiveness for gene therapy.

The non-viral particles of liposomes/ligand-mediated transfer have also been tried in animal models. The technique of liposomes usually does not require a specific receptor; therefore, it is very low in tissues specificity, and transfection efficiency is also quite low.[14,15,18,19,27] This technique only allows transient expression of the transgene. The efficiency of liposome transfection can be greatly enhanced by conjugation with ligands for cell surface receptors, e.g., integrins.[14] The techniques usually do not trigger host immune responses. Thus, it is possible to use liposomes as gene delivery vehicles repeatedly. Nevertheless, ligand-mediated gene delivery still can cause considerable inflammatory reactions.[30]

Use of physical methods of gene delivery to living organisms is a relatively new approach as a gene therapy strategy. Physical methods include particle mediator gene transfer and electric pluses.[20,31-37] Both techniques do not require cell receptors for transfer of the gene into the cells. Transfection efficiency is usually low, but gene expression does not require cell growth. The integration of transgenes to genome is very infrequent, thus it does not lead to sustained expression. The electric pulses and bombardment with gold particles also cause a certain degree of tissue damage and cell death, but these two procedures do not induce host immune responses.

2.1. Gene Gun

Our laboratory has adapted the Gene Gun technique to deliver genes to the ocular surface epithelium such as the cornea and conjunctiva.[36] The principle of Gene Gun that employs bombardment of gold particles coated with plasmid DNA by pressure helium is straightforward.[38–40] The DNA coated on the gold particle enters the nucleus via an unknown mechanism where the transgene is transcribed, and eventually leads to protein synthesis. We have used Gene Gun to identify the *cis*-regulatory DNA elements of the cornea-specific keratin 12 gene (K12).

2.1.1. In Vivo Gene Transfer

Gold particles of 0.6 µm, 1.0 µm or 1.6 µm (5 µg DNA per mg of gold particles) were coated with plasmid DNA purified by Qiagen® columns (Qiagen, Chatsworth, CA) according to the procedures recommended by the manufacturer of the Helios™ Gene Gun System (BIO-RAD, Hercules, CA). One ml of gold suspension was then loaded to a Tezel Tubing® attached on the tubing preparation station. The tubing coated with gold was dried by nitrogen air flow and cut into 0.5-inch segments. Thus, each segment contained 0.5 mg gold and 2.5 µg reporter plasmid DNA.[36]

All animal experiments were performed according to the ARVO resolution on the use of animals in vision research. New Zealand white rabbits (about 2 kg) were anesthetized with ketamine (30 mg/kg) and xylazine (3 mg/kg). One drop of 0.5% proparacaine/HCl was applied to cornea.[36] Back hair was clipped and residual hair removed by treatment with NAIR® (Carter-Wallace, NY). The Gene Gun was held against corneas, conjunctivas and skin to bombard gold particles into tissues: 1 delivery per individual cornea, 2 deliveries onto opposite sites of individual conjunctiva, and up to 12 sites on the skin of individual rabbits. Samples were collected 48 h after delivery and subjected to further experiments for determination of β-galactosidase (β-gal) activities.

2.1.2. Determination of β-gal Activities

Excised tissue specimens were minced with a razor blade, and 0.5 ml extraction buffer (0.25 M Tris-HCl, pH 7.4, 0.1% Tween 20) was added. The samples were subjected to three freeze-thaw cycles, 5 min on dry ice and 5 min at 37°C. The supernatants were collected by centrifugation at 13,000 x g, 4°C for 10 min. Aliquots of supernatant were incubated in a 0.3-ml mixture containing 50 mM 2-mercaptoethanol, 1 mM $MgCl_2$, 1.33 mg/ml o-nitrophenyl β-galactopyranoside and 0.1 M phosphate buffer, pH 7.0, at 37°C for 1 to 5 h. To terminate the reaction, 0.7 ml of 1 M Na_2CO_3 was added.[36] The enzyme activity was determined by comparing the optical density at 460 nm to that of purified β-gal (Boehringer and Mannheim, Indianapolis, IN).

To determine β-gal activity in situ, the excised tissues were fixed in 4% paraformaldehyde in 0.1 M phosphate buffer, pH 7.4, at 4°C for 4 h, and then rinsed with phosphate buffer saline three times. The specimens were incubated in a solution containing

0.5 mg/ml X-gal, 150 mM NaCl, 1.3 mM MgCl$_2$, 3 mM potassium ferrocyanide, 3 mM potassium ferricyanide and 44 mM HEPES, pH 7.4, at 37°C for 16 h. The enzyme reaction in each sample was examined with a stereo-microscope (Olympus, Melville, NY).

As with many other techniques, one needs to optimize the conditions for the gene delivery to ocular surface tissues, i.e., cornea and conjunctiva. Therefore, a series of experiments was performed to find a condition that efficiently delivers plasmid DNA while keeping the tissue damage to a minimum. We found the optimal conditions for expression of the reporter gene activity in the conjunctival, corneal and epidermal epithelium (Figs. 1, 2 and 3). The optimal conditions for gene transfer are as follows: cornea, 0.6 μm gold particles, 0.5 mg gold at 150 psi; conjunctiva, 1.6 μm gold particles, 0.5 mg gold at 150 psi; skin, 1.6 μm gold particles, 0.5 mg gold at 400 psi, respectively (Table 1). At these conditions most of the gold particles stay in the epithelium of each tissue type as judged by histological examination. Fig. 4 shows an example of histochemistry of β-gal in which the reporter gene activities were detected in corneal epithelial cells.

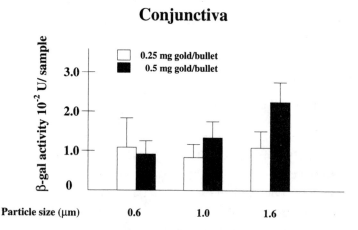

Figure 1. Effects of gold particle sizes on transfection efficiency. Gold particles, 0.6, 1.0 and 1.6 μm in diameter, were coated with pCMV-β-gal plasmid DNA (5 μg/mg gold). 0.25 mg and 0.5 mg of gold particles were coated to 0.5 inch of Tefzel® tubing (Bio-Rad, Hercules, CA). The gold particle were delivered to conjunctiva at 200 psi. The conjunctiva was dissected 48 h after transfection and subjected to enzyme assays for β-gal activities.

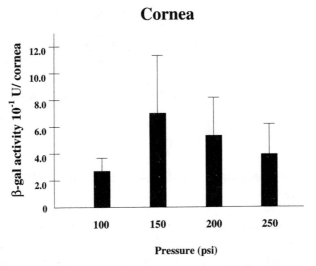

Figure 2. Effects of pressure on transfection efficiency in cornea. Gold particles, 0.6 μm, were coated with pCMV-β-gal plasmid DNA (5 μg/mg gold) and delivered to cornea (0.5 mg gold particle/shot) at different pressures: 100, 150, 200 and 250 psi. The enzyme activities in corneas were determined as described in Fig.1.

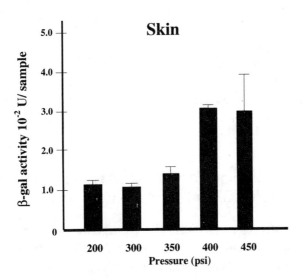

Figure 3. Effects of pressure on transfection efficiency in skin. Gold particles, 1.6 μm, were coated with pCMV-β-gal plasmid DNA (5 μg/mg gold) and delivered to skin (0.5 mg gold particle/shot) at different pressures: 200, 300, 350, 400 and 450 psi. The enzyme activities in corneas were determined as described in Fig. 1.

Table 1. Optimal conditions for gene transfer by Gene Gun

Tissue	Particle size	Amount of gold	Pressure
Cornea	0.6 μm	0.5 mg/bullet	150 psi
Conjunctiva	1.6 μm	0.5 mg/bullet	150 psi
Skin	1.6 μm	0.5 mg/bullet	400 psi

3. IDENTIFICATION OF CORNEA-SPECIFIC *CIS*-REGULATORY ELEMENTS OF K12 PROMOTER

Fig. 5 demonstrates the expression pattern of K12 in corneal epithelial cells.[41] Expression of the K3/K12 keratin pair signifies the cornea-type epithelial differentiation.[41-46] However, the mechanism(s) regulating cornea-specific K12 expression are unknown. To better understand cornea-specific expression, it is vital to identify the *cis*-regulatory elements essential for corneal epithelial cell-specific K12 expression. To accomplish this, we cloned the murine *Krt1.12* gene, which spans 6,567 base pairs of the murine genome and contains eight exons. Chromosome mapping reveals that like other members of the type I keratin family, murine *Krt1.12* is within the *Krt1* complex of mouse chromosome 11.[47] We have cloned 8 kb mouse genomic DNA 5' of the *Krt1.12* gene. Interestingly, the very 5' end of this 8-kb genomic DNA fragment contains exons 6 and 7 of mouse keratin 21. The 5-kb sequence upstream of the *Krt1.12* gene contains four Pax-6 homeobox binding sites between -910 and -2,000 bp.[36] We have tested several K12 promoter-reporter gene constructs by in vitro transfection of large T-antigen transformed corneal epithelial cells. In addition, we have generated transgenic mouse lines in our attempts to characterize the cornea-specific K12 promoter. Unfortunately, neither approach has lead to identification of *cis*-regulatory elements governing cornea-specific K12 expression.[36]

3.1. IN VIVO EXPRESSION OF K12PR-β-GAL REPORTER GENES IN RABBIT CORNEAL, CONJUNCTIVAL AND EPIDERMAL EPITHELIA

Due to the lack of appropriate cell lines for characterization of the K12 promoter, we used the newly developed Gene Gun technique to deliver K12pr-β-gal reporter gene constructs (Figs. 6 and 7) to rabbit corneas, conjunctivas and skin. Figs. 6 and 7 show the level of β-gal activity exhibited by selected K12 promoter-driven β-gal reporter gene constructs in corneal tissue. The 0.2 KZ shows lower β-gal activity in comparison with other reporter genes containing longer K12 promoter regions. The 1.0 KZ, 2.0 KZ, 2.5 KZ and 5.0 KZ constructs (all contain Pax-6 homeobox binding sites) exhibit higher activity than 0.2 and 0.6 KZ reporter genes. Several Pax-6 binding sites are within the 5' flanking

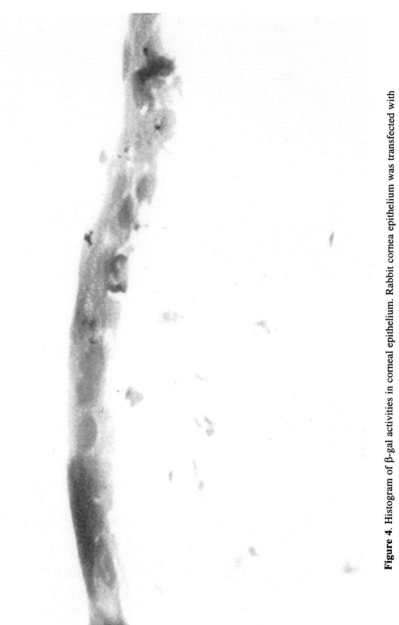

Figure 4. Histogram of β-gal activities in corneal epithelium. Rabbit cornea epithelium was transfected with pCMV-β-gal reporter gene as described in Fig. 2. The enucleated eyes were fixed in 4% paraformaldehyde in 0.1 M phosphate buffer, pH 7.4, for 4 h and stained with X-gal at 37°C overnight. The cornea was subjected to histological examination in paraffin sections.

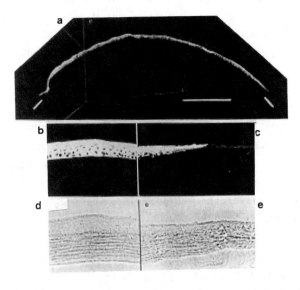

Figure 5. Immunofluorescent staining of corneal epithelium with anti-K12 antibodies. Frozen sections of mouse cornea containing epithelium and stroma were subjected to immunofluorescent staining with anti-K12N antibodies as previously described.[47] Photographs in Panels D and E are phase-contrast pictures of Panels B and C, respectively. Anti-K12N antibodies were used in Panels A–C.

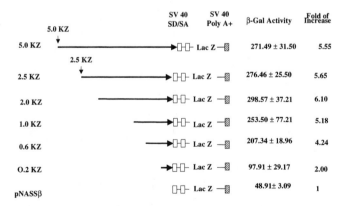

Figure 6. In vivo β-gal expression by K12 promoter-β-gal constructs in rabbit cornea. Rabbit corneas were bombarded with 0.6-μm gold particles coated with KZ constructs at 150 psi. Tissues were collected 48 h after delivery and analyzed for β-gal activity.

region from −1.0 to -2.0 kb of the *Krt1.12* gene, which may serve as an enhancer for K12 expression.[36] The observations indicate silencer elements exist between −2.0 and −2.5 kb upstream of the *Krt1.12* gene. Interestingly, only 0.2 KZ expresses significant β-gal activity in conjunctiva and skin (Fig. 7). To confirm cornea-specific expression of β-gal activity from each construct, rabbit corneas and conjunctivas were subjected to histochemical staining with X-gal. Fig. 8 shows pCMVβ generates a positive reaction in the cornea and conjunctiva, while pNASSβ shows no enzyme activities in either tissue. Significantly, 0.2 KZ shows a positive reaction in the cornea and conjunctiva. In contrast, β-gal activity from 0.6 KZ, 1.0 KZ and 2.5 KZ constructs was only detected in the cornea. In addition, only 0.2 KZ showed activity in the skin. These observations indicate the -0.2-kb 5' to *Krt.1.12* gene may contain basic regulatory elements for epithelial cell expression of the β-gal reporter gene, and a 0.6-kb or longer promoter sequence from the *Krt1.12* gene contains the necessary *cis*-regulatory elements for cornea-specific K12 expression. Enzyme assays and histochemistry from our biolistic experiments show a K12 promoter fragment 600 bp or longer is sufficient to drive β-gal expression specifically and exclusively in corneal epithelium. This clearly supports the assertion that promoter elements responsible for cornea specificity lie within the region 5' to the first 200 bp of this promoter. Our results indicate 1.5 KZ has a lower promoter activity than the 1.0-KZ and 2.5-KZ constructs, which may indicate the existence of a silencer element only manifesting itself in vivo (Figs. 6 and 7). Further studies are necessary to establish the validity of this hypothesis.

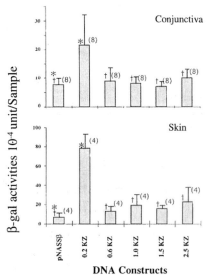

Figure 7. In vivo β-gal expression by K12 promoter-β-gal constructs in rabbit cornea, conjunctiva and skin. Top panel: Conjunctiva, 1.6-μm gold particles coated with KZ constructs were delivered to bulbar conjunctivas at 150 psi. Bottom panel: Skin, 1.6-μm gold particles were delivered to rabbit skin at 400 psi. Tissues were collected 48 h after delivery and analyzed for β-gal activity. Bars indicate standard deviations. The numbers in parentheses are number of specimens examined. *Significant difference between pNASSβ and individual K12 promoter constructs. †No significant difference.

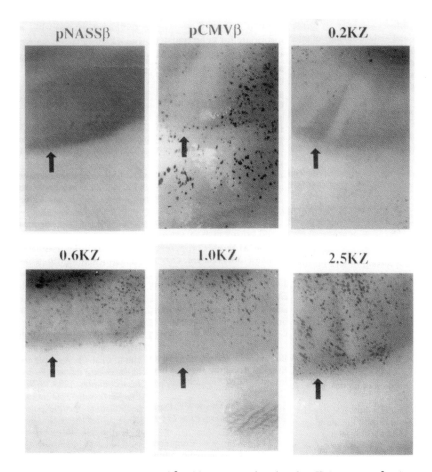

Figure 8. In situ histochemical staining of β-gal in cornea and conjunctiva. K12 promoter-β-gal constructs were coated onto 1.6-μm gold particles and delivered to corneal/conjunctival junctions at 150 psi. The tissues were collected at 48 h and subjected to staining with X-gal as described previously.[36] pNASSβ generates no positive reaction in cornea and conjunctiva. pCMVβ and 0.2 KZ generate positive reaction in cornea and conjunctiva. Expression of β-gal by constructs (0.6KZ, 1.0KZ and 2.5KZ) that contain 600 bp or longer 5' flanking sequence of *Krt1.12* gene is restricted in cornea. Arrows indicate the boundary between cornea and conjunctiva

The identification of the corneal epithelial cell-specific promoter has been a difficult task, because of the lack of established cell lines that maintain corneal epithelial cell phenotypes, e.g., expression of K12. To overcome this difficulty, we have used in vivo particle-mediated gene transfer in rabbit corneas, conjunctivas and skin with β-gal reporter gene constructs to identify the mouse K12 promoter.

Our results suggest the Helios™ Gene Gun is the method of choice for in vivo identification of regulatory *cis*-DNA elements of the K12 gene and/or other corneal epithelial cell-specific genes in lieu of the expensive and laborious transgenic mouse model. In addition to easy access, corneal epithelial cells are metabolically active and can express high levels of reporter genes compared to other tissue. Thus, the K12 promoter can be used to prepare other reporter genes, e.g., growth factors, receptors and cytokines in constructs, which can be expressed in a cornea-specific manner to examine their roles in wound healing and maintenance of corneal physiology.

4. CONCLUSION

Many strategies have been tested for gene therapy. However, few patients have been cured by this treatment. In general, gene therapy is still facing major obstacles, i.e., efficient delivery of genes to target tissue, sustained cell type-specific transgene expression and host immune response to viral vectors. Ocular surface tissue is an ideal model for development of gene therapy strategy, because of its easy access for gene delivery and evaluation of efficacy. The particle-mediated gene transfer is a suitable technique for in vivo delivery of transgenes to corneal and conjunctival epithelium. The K12-promoter reporter gene can be specifically, but transiently, expressed by corneal epithelium. Therefore, it is necessary to transduce limbal stem cells with a reporter gene that becomes integrated into the host genome to achieve the goal of gene therapy. It is also necessary to design reporter genes that can be readily translocated into the nucleus and integrated to the somatic cell genome. Recently, Dr. David Dean has reported a SV-40 enhancer sequence that can facilitate the translocation of plasmid DNA into the nuclei of transfected cells in culture.[48–50] Inclusion of such enhancer elements may improve gene delivery efficiency. Further studies are needed to provide better understanding of mechanisms of DNA recombination governing integration of transgenes to genome. Thus, an appropriate vector can be designed to enhance integration of transgenes to the host genome for sustained transgene expression, while it reduces the risks of adverse effects disrupting normal cell functions.

ACKNOWLEDGEMENTS

This study was supported in part by NIH EY10556, the Ohio Lions Eye Research Foundation, Columbus, OH, and Cincinnati Eye Bank, Cincinnati, OH.

REFERENCES

1. W.F.Anderson, Gene therapy, JAMA 246:2737 (1981).
2. P.Anklesaria, Gene therapy: a molecular approach to cancer treatment, Curr.Opin.Mol.Ther. 2:426 (2000).
3. T.Bettinger and M.L.Read, Recent developments in RNA-based strategies for cancer gene therapy, Curr.Opin.Mol.Ther. 3:116 (2001).
4. M.S.Dilber and G.Gahrton, Suicide gene therapy: possible applications in haematopoietic disorders, Journal of Internal Medicine 249:359 (2001).
5. S.Gojo, D.K.Cooper, J.Iacomini, and C.LeGuern, Gene therapy and transplantation, Transplantation 69:1995 (2000).
6. D.B.Kohn, Gene therapy for genetic haematological disorders and immunodeficiencies, Journal of Internal Medicine 249:379 (2001).
7. R.J.Levy, S.A.Goldstein, and J.Bonadio, Gene therapy for tissue repair and regeneration, Adv.Drug Deliv.Rev. 33:53 (1998).
8. R.Morishita, M.Aoki, S.Nakamura, J.Higaki, Y.Kaneda, and T.Ogihara, Gene therapy for cardiovascular disease using hepatocyte growth factor, Annals of the New York Academy of Sciences 902:369 (2000).
9. G.S.Stein, J.B.Lian, J.L.Stein, and A.J.van Wijnen, Bone tissue specific transcriptional control: options for targeting gene therapy to the skeleton, Cancer 88:2899 (2000).
10. K.Anwer, A.Bailey, and S.M.Sullivan, Targeted gene delivery: a two-pronged approach [In Process Citation], Crit Rev.Ther.Drug Carrier Syst. 17:377 (2000).
11. T.Pap, R.E.Gay, and S.Gay, Gene transfer: from concept to therapy, Current Opinion In Rheumatology 12:205 (2000).
12. S.A.Lukashok and M.S.Horwitz, New perspectives in adenoviruses, Current Clinical Topics In Infectious Diseases 18:286 (1998).
13. H.S.Pandha, L.A.Martin, A.S.Rigg, P.Ross, and A.G.Dalgleish, Gene therapy: recent progress in the clinical oncology arena, Curr.Opin.Mol.Ther. 2:362 (2000).
14. S.L.Hart, C.V.Arancibia-Carcamo, M.A.Wolfert, C.Mailhos, N.J.O'Reilly, R.R.Ali, C.Coutelle, A.J.George, R.P.Harbottle, A.M.Knight, D.F.Larkin, R.J.Levinsky, L.W.Seymour, A.J.Thrasher, and C.Kinnon, Lipid-mediated enhancement of transfection by a nonviral integrin- targeting vector, Human Gene Therapy 9:575 (1998).
15. A.G.Schatzlein, Non-viral vectors in cancer gene therapy: principles and progress, Anticancer Drugs 12:275 (2001).
16. D.F.Larkin, H.B.Oral, C.J.Ring, N.R.Lemoine, and A.J.George, Adenovirus-mediated gene delivery to the corneal endothelium, Transplantation 61:363 (1996).
17. F.C.Marini, III, Q.Yu, T.Wickham, I.Kovesdi, and M.Andreeff, Adenovirus as a gene therapy vector for hematopoietic cells [In Process Citation], Cancer Gene Ther. 7:816 (2000).
18. P.L.Katsel and R.J.Greenstein, Eukaryotic gene transfer with liposomes: effect of differences in lipid structure [In Process Citation], Biotechnol.Annu.Rev. 5:197 (2000).
19. P.H.Tan, W.J.King, D.Chen, H.M.Awad, M.Mackett, R.I.Lechler, D.F.Larkin, and A.J.George, Transferrin receptor-mediated gene transfer to the corneal endothelium, Transplantation 71:552 (2001).
20. Y.Oshima, T.Sakamoto, I.Yamanaka, T.Nishi, T.Ishibashi, and H.Inomata, Targeted gene transfer to corneal endothelium in vivo by electric pulse, Gene Ther. 5:1347 (1998).
21. M.Aghi, F.Hochberg, and X.O.Breakefield, Prodrug activation enzymes in cancer gene therapy [In Process Citation], J.Gene Med. 2:148 (2000).
22. M.Caraglia, A.Budillon, G.Vitale, G.Lupoli, P.Tagliaferri, and A.Abbruzzese, Modulation of molecular mechanisms involved in protein synthesis machinery as a new tool for the control of cell proliferation, European Journal of Biochemistry 267:3919 (2000).
23. E.S.Hickman and K.Helin, The p53 tumour suppressor protein, Biotechnol.Genet.Eng Rev. 17:179 (2000).

24. M.M.Hitt and J.Gauldie, Gene vectors for cytokine expression in vivo, Curr.Pharm.Des 6:613 (2000).

25. J.H.Wolfe, M.S.Sands, J.E.Barker, B.Gwynn, L.B.Rowe, C.A.Vogler, and E.H.Birkenmeier, Reversal of pathology in murine mucopolysaccharidosis type VII by somatic cell gene transfer, Nature 360:749 (1992).

26. X.Wang, B.Appukuttan, S.Ott, R.Patel, J.Irvine, J.Song, J.H.Park, R.Smith, and J.T.Stout, Efficient and sustained transgene expression in human corneal cells mediated by a lentiviral vector, Gene Ther. 7:196 (2000).

27. S.N.Yeung and F.Tufaro, Replicating herpes simplex virus vectors for cancer gene therapy, Expert.Opin.Pharmacother. 1:623 (2000).

28. J.Kampmeier, A.Behrens, Y.Wang, A.Yee, W.F.Anderson, F.L.Hall, E.M.Gordon, and P.J.McDonnell, Inhibition of rabbit keratocyte and human fetal lens epithelial cell proliferation by retrovirus-mediated transfer of antisense cyclin G1 and antisense MAT1 constructs, Human Gene Therapy 11:1 (2000).

29. M.R.Mautino, N.Keiser, and R.A.Morgan, Inhibition of human immunodeficiency virus type 1 (HIV-1) replication by HIV-1-based lentivirus vectors expressing transdominant Rev, Journal of Virology 75:3590 (2001).

30. J.Norman, W.Denham, D.Denham, J.Yang, G.Carter, A.Abouhamze, C.L.Tannahill, S.L.MacKay, and L.L.Moldawer, Liposome-mediated, nonviral gene transfer induces a systemic inflammatory response which can exacerbate pre-existing inflammation, Gene Ther. 7:1425 (2000).

31. T.Sakamoto, Y.Oshima, K.Nakagawa, T.Ishibashi, H.Inomata, and K.Sueishi, Target gene transfer of tissue plasminogen activator to cornea by electric pulse inhibits intracameral fibrin formation and corneal cloudiness, Human Gene Therapy 10:2551 (1999).

32. J.B.Martin, J.L.Young, J.N.Benoit, and D.A.Dean, Gene transfer to intact mesenteric arteries by electroporation, Journal of Vascular Research 37:372 (2000).

33. C.Andree, W.F.Swain, C.P.Page, M.D.Macklin, J.Slama, D.Hatzis, and E.Eriksson, In vivo transfer and expression of a human epidermal growth factor gene accelerates wound repair, Proc.Natl.Acad.Sci.USA 91:12188 (1994).

34. L.Cheng, P.R.Ziegelhoffer, and N.S.Yang, In vivo promoter activity and transgene expression in mammalian somatic tissues evaluated by using particle bombardment, Proc.Natl.Acad.Sci.USA 90:4455 (1993).

35. A.L.Rakhmilevich, J.Turner, M.J.Ford, D.McCabe, W.H.Sun, P.M.Sondel, K.Grota, and N.S.Yang, Gene gun-mediated skin transfection with interleukin 12 gene results in regression of established primary and metastatic murine tumors, Proc.Natl.Acad.Sci.USA 93:6291 (1996).

36. A.Shiraishi, R.L.Converse, C.Y.Liu, F.Zhou, C.W.Kao, and W.W.Kao, Identification of the cornea-specific keratin 12 promoter by in vivo particle-mediated gene transfer, Invest Ophthalmol.Vis.Sci. 39:2554 (1998).

37. W.Xiao and J.L.Brandsma, High efficiency, long-term clinical expression of cottontail rabbit papillomavirus (CRPV) DNA in rabbit skin following particle-mediated DNA transfer, Nucleic Acid Research 24:2620 (1996).

38. E.F.Fynan, R.G.Webster, D.H.Fuller, J.R.Haynes, J.C.Santoro, and H.L.Robinson, DNA vaccines: protective immunizations by parenteral, mucosal, and gene-gun inoculations, Proc.Natl.Acad.Sci.USA 90:11478 (1993).

39. D.C.Tang, M.DeVit, and S.A.Johnston, Genetic immunization is a simple method for eliciting an immune response, Nature 356:152 (1992).

40. N.S.Yang, J.Burkholder, B.Roberts, B.Martinell, and D.McCabe, In vivo and in vitro gene transfer to mammalian somatic cells by particle bombardment, Proc.Natl.Acad.Sci.USA 87:9568 (1990).

41. C.Y.Liu, G.Zhu, A.Westerhausen-Larson, R.L.Converse, C.W.C.Kao, T.T.Sun, and W.W.Y.Kao, Cornea-specific expression of K12 keratin during mouse development, Current Eye Research 12:963 (1993).

42. C.Chaloin-Dufau, T.T.Sun, and D.Dhouailly, Appearance of the keratin pair K3/K12 during embryonic and adult corneal epithelial differentiation in the chick and in the rabbit, Cell Differentiation & Development 32:97 (1990).

43. W.Y.Chen, M.M.Mui, W.W.Y.Kao, C.Y.Liu, and S.C.G.Tseng, Conjunctival epithelial cells do not transdifferentiate in organotypic cultures: expression of K12 keratin is restricted to corneal epithelium, Current Eye Research 13:765 (1994).

44. P.D.Moyer, A.H.Kaufman, Z.Zhang, C.W.Kao, A.G.Spaulding, and W.W.Kao, Conjunctival epithelial cells can resurface denuded cornea, but do not transdifferentiate to express cornea-specific keratin 12 following removal of limbal epithelium in mouse, Differentiation 60:31 (1996).

45. R.L.Wu, G.Zhu, S.Galvin, C.Xu, T.Haseba, C.Chaloin-Dufau, D.Dhouailly, Z.G.Wei, Kao WY, and et al, Lineage-specific and differentiation-dependent expression of K12 keratin in rabbit corneal/limbal epithelial cells: cDNA cloning and northern blot analysis, Differentiation 55:137 (1994).

46. G.Zhu, M.Ishizaki, T.Haseba, R.L.Wu, T.T.Sun, and W.W.Y.Kao, Expression of K12 keratin in alkali-burned rabbit corneas, Current Eye Research 11:875 (1992).

47. C.Y.Liu, G.Zhu, R.L.Converse, C.W.C.Kao, H.Nakamura, S.C.G.Tseng, M.M.Mui, J.Seyer, M.J.Justice, M.E.Stech, G.M.Hansen, and W.W.Y.Kao, Characterization and chromosomal localization of the cornea- specific murine keratin gene Krt1.12, Journal of Biological Chemistry 269:24627 (1994).

48. D.A.Dean, J.N.Byrd, Jr., and B.S.Dean, Nuclear targeting of plasmid DNA in human corneal cells, Current Eye Research 19:66 (1999).

49. S.Li, F.C.MacLaughlin, J.G.Fewell, M.Gondo, J.Wang, F.Nicol, D.A.Dean, and L.C.Smith, Muscle-specific enhancement of gene expression by incorporation of SV40 enhancer in the expression plasmid, Gene Ther. 8:494 (2001).

50. J.Vacik, B.S.Dean, W.E.Zimmer, and D.A.Dean, Cell-specific nuclear import of plasmid DNA, Gene Ther. 6:1006 (1999).

IN VIVO GENE TRANSFER INTO CORNEAL EPITHELIAL PROGENITOR CELLS BY VIRAL VECTORS

Tsutomu Igarashi,[1,3] Koichi Miyake,[1,2] Noriko Suzuki,[1] Hiroshi Takahashi,[3] and Takashi Shimada[1,2]

[1]Department of Biochemistry and Molecular Biology
[2]Division of Gene Therapy Research
Center for Advanced Medical Technology
[3]Department of Ophthalmology
Nippon Medical School
Tokyo, Japan

1. INTRODUCTION

Gene therapy is defined as the treatment of diseases by the transfer of new genes into patient's cells. This therapeutic strategy was originally developed to treat genetic disorders. However, acquired disorders such as cancer and other non-genetic diseases are also important potential targets for gene therapy. The first gene therapy clinical trial was started in 1990, and since then more than 3,000 patients have been treated by gene transfer techniques. Many ocular diseases have been proposed as potential targets for gene therapy. Table 1 shows the current status of ocular gene therapy research.[1] Various types of vectors containing therapeutic genes or marker genes have been used to transfer genes into ocular target cells. However, it has generally been difficult to achieve efficient and stable gene expression in these target cells. Clinical trials of ocular gene therapy have yet to be conducted.

Establishment of a safe and efficient gene delivery system is a key factor in the success or failure of gene therapy treatments.[2] Table 2 lists the characteristics of viral vectors. Mouse retroviruses and adenoviruses have been already used in the treatment of patients in organ systems other than the eye. Retroviral vectors integrate stably into chromosomes, but replication-competent viruses are pathogenic, and random integration may result in insertional mutagenesis. Another problem is that the retroviral vector is only

Lacrimal Gland, Tear Film, and Dry Eye Syndromes 3
Edited by D. Sullivan *et al.*, Kluwer Academic/Plenum Publishers, 2002

1309

able to infect actively dividing cells. Thus, it is difficult to transduce most ocular target cells in vivo. On the other hand, adenovirus transduction efficiency is very high, and they can infect non-dividing cells. However, gene expression is transient, and the only adenovirus vector currently in use is cytotoxic and immunogenic.

Table 1. Ocular gene therapy

Disease	Target Cell	Vector	Gene
Retinitis Pigmentosa	RPE	Adeno, AAV, HIV	CNTF, bcl-2, c-fos, PDE, GDNF, PEDF
Glaucoma	Retinal Ganglion Cell, Trabecular Meshwork	Liposome, Adeno	Bax Antisense, BDNF
Retinoblastoma	Retinal blastoma Cell	Retro, Herpes	HSV-tk, Ribonucleotide Reductase
Retinal and Choroidal Vasculature	Vascular Endothelial cell	HVJ-Liposome, Retro	Antisense Oligonucleotide against VEGF, Fibronectin
Retinal Detachment	Fibroblast	Liposome	HSV-tk, E2F Decoy Oligonucleotide
Bullous Keratopathy	Corneal Endothelium	Liposome(in vivo) Adeno, HSV, AAV, (ex vivo)	Marker Gene
Corneal Epithelial Stem Cell Deficiency	Corneal Epithelium	Adeno, Retro, HIV(ex vivo)	Marker Gene

To overcome the problems associated with the viral vectors currently in use, we are working to develop two new viral vectors, one derived from adeno-associated virus (AAV) and the other derived from human immunodeficiency virus (HIV). These two vectors are still under development, but initial studies indicate that they have potentially important biological features. AAV is non-pathogenic and has a single-stranded DNA genome, and recombinant AAV vectors are able to transduce non-dividing cells. A disadvantage of AAV is its extremely low efficiency of vector packaging. HIV vectors were originally developed for use in gene therapy treatment of AIDS.[3] HIV vectors pseudotyped with VSVG protein can be used to transfer genes into various non-dividing cells, including neuronal and hematopoietic stem cells.[4] A serious disadvantage of HIV vectors is the pathogenicity of their parent virus. However, if this safety problem can be overcome, HIV vectors should prove to be very useful in gene therapy of ocular diseases.

2. METHODS

In this study, we examined the utility of viral vectors for gene transfer into corneal epithelial stem cells that are thought to reside in the basal cell layer of the limbal

Table 2. Characteristics of viral vectors

	MLV	Adeno	AAV	HIV (Lenti)
Pathogenicity	+	+	-	+
Toxicity	-	+	-	-
Efficiency	Med	Very high	Med	High
Tranduction to non-dividing cell	-	+	+	+
Integration	+	-	±	+
Expression	Low	High	Med	Med
Duration	Long	Short	Med	Long
Packaging Efficiency	High	High	Low	Low

epithelium.[5] An adenoviral vector containing the eGFP marker gene was generated by homologous recombination in 293 cells, and was concentrated by cesium chloride banding;[6,7] the final titer was 1×10^9 pfu/ml. An AAV vector was synthesized by plasmid transfection into 293 adenovirus-infected cells, and was concentrated by a combination of column chromatography and ultrafiltration;[8] the final titer was 5×10^{11} particles/ml. An HIV vector was generated by plasmid transfection into 293T cells, and was concentrated by ultrafiltration and ultracentrifugation;[4] the final titer was 1×10^8 TU/ml.

Fig. 1 shows HIV vector-mediated gene transfer into various ocular target cells. HIV vectors containing the eGFP marker gene were directly injected into various sites in the eye, and expression of eGFP was assayed three days after injection. Corneal endothelia and trabecular meshwork were successfully transduced by injection into the anterior chamber (Fig. 1A, B). The HIV vectors were also highly effective at transducing retinal pigmental epithelia (Fig. 1D).

In a preliminary trial of in vivo gene transfer into the cornea, topical application of the viral vectors resulted in no significant gene transfer into epithelial cells, indicating that they cannot pass through the mucin layer and the superficial cells of the intact cornea. To increase the likelihood of direct contact between vector particles and basal layer cells, the lower limbal epithelium was linearly scraped with a blade. Small plastic rings were prepared by slicing yellow tips. Rings plastered with silicone grease were placed on the open eyes of anesthetized rats, and were then filled with 20 μl of vector solution. The ring was left in place for 30 minutes. The cornea was excised, and GFP expression in the corneal epithelia was assayed by fluorescence microscopy.

3. RESULTS AND DISCUSSIONS

In the cases of all three vectors, three days after transduction treatment, eGFP-positive cells were detected among the epithelial cells. The adenoviral vector transduced epithelial

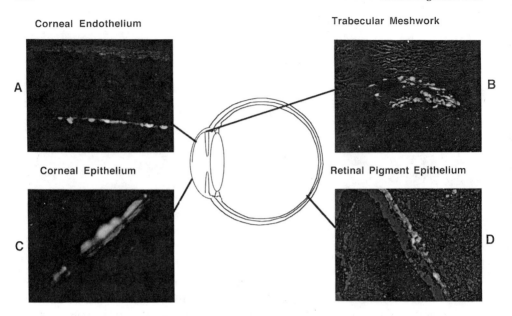

Figure 1. HIV vector mediated gene transfer into ocular cells.

cells with greater efficiency than the AAV vector or the HIV vector. However, in cells transduced by adenovirus vectors, eGFP expression could not be detected 2 weeks after treatment. These findings are in good agreement with data from a previous study, in which adenovirus-mediated transgene expression within human corneas in situ remained at high levels for only 7 days and diminished to undetectable levels by 28 days after treatment[9].

AAV vector-mediated eGFP expression in the epithelial cells was detectable 2 weeks after treatment, but transduction efficiency was very low. EGFP-positive cells could not be detected 6 weeks after treatment. The low transduction efficiency of the AAV vector appears to have been due to the relatively low titer used in this study. Another factor may be low efficiency of chromosomal integration of the AAV genome. It was believed that AAV vectors were able to integrate into the chromosomes of non-dividing cells;[10] however, recent reports have shown that most of the AAV vector genome present in target cells is in episomal (extrachromosomal) form.[11,12] If the AAV genome is converted to the double-stranded form, these DNA molecules can serve as templates for transcription, even though they are in episomal form. Gene expression of episomal AAV DNA appears to be relatively stable in transduced non-dividing cells. However, in dividing cells, the genome quickly becomes diluted, leading to reduced gene expression. Thus, AAV vectors may not be an appropriate choice for gene transfer into dividing epithelial cells.

HIV-mediated eGFP expression was detected 3 days after treatment (Fig. 1-C), and eGFP expression continued in the basal layer cells of corneal epithelia for at least 6 weeks after treatment. These results strongly suggest that the HIV vectors integrated stably into the chromosomes of long-lived epithelial cells. A previous study using a radiolabeling technique demonstrated that corneal basal epithelial cells migrate for 4 to 6 days to suprabasal postmitotic cells, and are then lost from superficial terminally differentiated cells by desquamation.[13,14] Since surface markers for epithelial stem cells are not available, direct confirmation of stem cell transduction is difficult at present. However, our data strongly suggest that HIV-based vectors are able to transduce corneal epithelial stem cells.

4. CONCLUSIONS

In conclusion, the results of this study indicate that lentiviral vectors can efficiently transfer transgenes to corneal epithelial cells, and that the resultant transgene expression in vivo can continue for a relatively long term. These vectors may be useful in the treatment of various corneal diseases, including recurrent corneal erosion, dry eyes, and stem cell deficiency due to Stevens-Johnson syndrome. Furthermore, stable genetic modification of corneal epithelial cells is likely to be useful in studies of their stem cell biology.

REFERENCES

1.   Hauswirth WW, Beaufrere L. Ocular gene therapy: quo vadis? *Invest Ophthalmol Vis Sci.* 41:2821(2000).
2.   Vile R, Russell SJ. Gene transfer technologies for the gene therapy of cancer. *Gene Ther.* 1:88(1994).
3.   Shimada T, Fujii H, Mitsuya H, Nienhuis AW. Targeted and highly efficient gene transfer into CD4+cells by a recombinant human immunodeficiency virus retroviral vector. *J Clin Invest.* 88:1043(1991).
4.   Naldini L, Blomer U, Gallay P, et al. In vivo gene delivery and stable transduction of nondividing cells by a lentiviral vector.. *Science.* 272:263(1996).
5.   Zieske JD. Perpetuation of stem cells in the eye. *Eye.* 8:163(1994).
6.   Kanegae Y, Makimura M, Saito I. A simple and efficient method for purification of infectious recombinant adenovirus. *Jpn J Med Sci Biol.* 47:157(1994).
7.   Miyake S, Makimura M, Kanegae Y, et al. Efficient generation of recombinant adenoviruses using adenovirus DNA-terminal protein complex and a cosmid bearing the full-length virus genome. *Proc Nalt Acad Sci U S A .* 93:1320(1996).
8.   Tamayose K, Hirai Y, Shimada T. A new strategy for large-scale preparation of high-titer recombinant adeno-associated virus vectors by using packaging cell lines and sulfonated cellulose column chromatography. *Hum Gene Ther.* 7:507(1996).
9.   Oral HB, Larkin DF, Fehervari Z, et al. Ex vivo adenovirus-mediated gene transfer and immunomodulatory protein production in human cornea. *Gene Ther.*4:639(1997).
10. Fisher-Adams G, Wong KK, Jr., Podsakoff G, Forman SJ, Chatterjee S. Integration of adeno-associated virus vectors in CD34+human hematopoietic progenitor cells after transdution. *Blood.* 88:492(1996)..
11. Giraud C, Winocour E, Berns KI. Recombinant junctions formed by site-specific integration of adeno-associated virus into an episome. *J Virol.* . 69:6917(1995).

12. Malik AK, Monahan PE, Allen DL, Chen BG, Samulski RJ, Kurauchi K. Kinetics of recombinant adeno-associated virus-mediated gene transfer *J Virol.* 74:3555(2000).

13. Hanna C. Cell production and migration in the epithelial layer of the cornea. *Archives of ophthalmology.* 64:536(1960).

14. Haskjold E, Refsum SB, Bjerknes R. Cell renewal of the rat corneal epithelium. A method to compare corresponding corneal areas from individual animals. *Acta Ophthalmol(Copenh).* 66:533(1988).

GENE THERAPY FOR THE PREVENTION OF CORNEAL HAZE AFTER PHOTOREFRACTIVE/PHOTOTHERAPEUTIC KERATECTOMY EXCIMER LASER SURGERY

Ashley Behrens and Peter J. McDonnell

Department of Ophthalmology
University of California Irvine
Irvine, California, USA
Department of Ophthalmology
Doheny Eye Institute
University of Southern California School of Medicine
Los Angeles, California, USA

1. INTRODUCTION

A decade ago the first patient was successfully treated using gene therapy to correct a severe combined immunodeficiency.[1] Since then hundreds of patients have been treated with this therapy for many pathologic conditions. However, progress in developing clinical protocols has been slow, in part due to a lack of safe and efficient delivery systems for the therapeutic genes.[2]

The eye is an ideal target for gene therapy because of its accessibility and relative immunologic privilege. Advances in molecular biology have led to precise isolation of specific genes related to ocular disease.[3,4] It is therefore logical to focus attempts of gene modulation on this organ.[5–10]

2. GENE TRANSFER METHODS

Two approaches are used to apply ocular gene therapy: the in vivo or ex vivo strategy.[11,12] The first involves transfer of genetic information directly to the patient, and the second removal of cells from the patient with the subsequent transfection of those cells

Lacrimal Gland, Tear Film, and Dry Eye Syndromes 3
Edited by D. Sullivan *et al.*, Kluwer Academic/Plenum Publishers, 2002

1315

in vitro. In addition, specific methods to introduce the therapeutic gene into the cell are required to achieve its integration in the host. These methods may be grouped in two generic approaches: (1) nonviral mediated transfection and (2) viral vector transfection. In the nonviral group, microinjection-,[13] liposome-,[14,15] peptide-,[16] electric pulse-(electroporation)[17] or cell surface receptor-mediated gene transfer have been described.[18] In the second group, different viral vectors are used to achieve gene transfer across the cell membrane, such as retrovirus,[19] adenovirus,[20,21] adeno-associated virus,[22,23] lentivirus[24] and poxvirus.[25,26]

An additional factor in vector engineering is cell targeting. Proposed methods are physical and molecular biological targeting.[27] Physical targeting consists of the attachment to the delivery vehicle of ligands that bind to the target cells by specific surface receptors. In molecular biological targeting the target cell selectively expresses the therapeutic gene through selective promoters. Another method of targeting involves the extracellular matrix binding properties of the vector that may ensure the exposure to specific sites of injury.[28,29]

3. OCULAR GENE THERAPY

Ocular disease occurs frequently as a result of uncontrolled cellular proliferation. Gene therapy therefore has been directed against cell replication in certain degenerative and post-surgical conditions. Special attention has focused on retroviral vectors, since their infectivity depends on events associated with the host cell cycle. These vectors are not capable of infecting cells in the non-replicating phase of the cell cycle (stationary cells).[30] Thus, the only cells susceptible to gene transfer may be those actively replicating.

3.1. Gene Therapy for Corneal Haze Prevention after Excimer Laser Surgery

Several experimental models have studied the efficacy of retroviral vectors bearing varied genes to avoid cell proliferation, especially in cancer research.[12,19,31,32] In corneal laser surgery, a target of these vectors is keratocytes, which have been directly involved in corneal haze development after photorefractive keratectomy (PRK)/phototherapeutic keratectomy (PTK).[33–38] The corneal opacity after excimer laser superficial ablation proportionally increases according to the depth of ablation attempted.[39] Recently, the presence of crystalline cytoplasmic deposits with diffractive properties has been reported after corneal laser surgery in proliferating keratocytes.[40] A theoretical approach may include the post-surgical transfection of these cells by topical application of gene therapy to selectively avoid cell proliferation.

3.2. Corneal Haze Prevention: Background Studies

Seitz et al. evaluated the potential of a retroviral vector (murine Moloney leukemia virus) bearing a ß-galactosidase gene as a marker to test the in vitro transduction efficiency into human keratocytes.[41] They found a transduction efficiency of 12% to 18% in their

simplex thymidine kinase (HStk) gene.[42] An excimer laser PTK was performed in New Zealand white rabbits, with a subsequent topical application of the vector. A higher in vivo transduction efficiency of 25% to 40% was found. Likewise, the cytostatic/cytotoxic effects after topical administration of ganciclovir resulted in a significant inhibition of keratocyte proliferation and lower corneal haze, compared to the control vector group.

Kampmeier et al. exposed rabbit keratocytes and human fetal epithelial lens cells in vitro to the same vector bearing cell cycle control genes antisense cyclin G1 (aG1) and antisense "ménage a trois" (aMAT1).[43] A decrease in the epithelial lens cell proliferation may also be beneficial to reduce the incidence of secondary cataract formation after primary cataract surgery. The potential advantage of these genes over HStk is their cytocidal effect once they are integrated in the host's DNA, without requiring ganciclovir. A transduction efficiency of 34% in keratocytes and 20% in human fetal epithelial cells was found, with successful downregulation of cyclin G1 and MAT1 protein expression after 24 h. The cytostatic effects of the aG1 and aMAT1 gene transfer were more evident after day 5 of vector application.

3.3. Corneal Haze Prevention: Present Studies

Based on previous experience, a collagen-binding retroviral vector bearing a mutant cyclin G1 gene (dnG1) was selected to evaluate its in vivo (rabbit model) efficacy to prevent corneal haze after PTK.[29] The results were compared to a collagen-binding "empty" vector group. The vector was administered after surgery in topical eyedrops of 40 μl each, every 10 min for 2 h in 2 consecutive days. A modified objective corneal haze assessment method with a digital camera and an imaging software (Scion Image for Windows, Scion Corp., Frederick, MD) was performed at days 0, 14 and 21 after surgery.[42,44] The digital images were transformed to grayscale pictures, and the different tonalities quantitatively analyzed and compared using a statistical software (SPSS 10.0 for Windows, SPSS Inc., Chicago, IL).

Transduction efficiency was approximately 35% using this vector (unpublished data). Less corneal haze was observed at all study time points in the dnG1-treated group compared to controls (Fig. 1). The re-epithelialization rate was similar in both groups, and no permanent epithelial defects were observed at the end of the observation period (Fig. 2). At histology, central keratocyte counts were significantly lower in the dnG1 group, with less subepithelial stromal scarring (Fig. 3).

Ocular toxicity following a modified Draize score,[45] and vector biodistribution studies are underway to assess the safety of this therapy for clinical environment use. In our preliminary data, the topical application did not show any secondary effects compromising the ocular surface. The present vector allows keratocyte proliferation inhibition without the need of an additional drug (ganciclovir) to achieve cytocidal/cytostatic effects. The total administration time may be accomplished in 4 h during the first 2 days after surgery. The anatomic location of the cornea allows vector application in a straightforward approach.

Preoperative   **Postoperative (Day 21)**

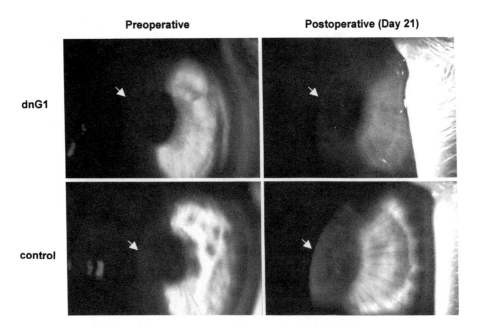

Figure 1. Corneal haze development after excimer laser PTK. The arrows show the slit lamp illumination on the anterior surface of the cornea to visualize the opacity. Note the clear corneal image at postoperative day 21 in the corneas treated with the collagen/binding retroviral vector bearing the dnG1 gene compared to preoperative (first row). The control group showed marked haze at postoperative day 21 compared to preoperative clear corneas (second row).

24 h postoperative   **72 h postoperative**

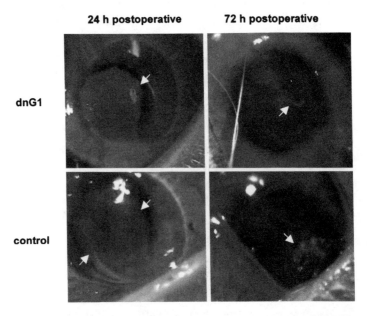

Figure 2. Re-epithelialization after PTK. Fluorescein staining reveals the extent of the epithelial defect between arrows. The retroviral vector-dnG1 treated cornea (first row) showed a similar pattern and rate of epithelial defect closure compared to controls (second row). The epithelial defect closed completely after 96 h in both groups.

DnG1   **Control**

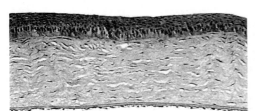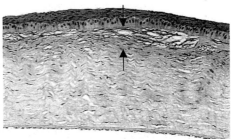

Figure 3. Histology sections of the central cornea. The retroviral vector-dnG1 treated cornea shows a regular epithelial layer, with normal cellular phenotype from the superficial to the basal layers. A more regular superficial stromal pattern, with less keratocytes and fibrotic reaction is observed, compared to controls. A disarranged collagen distribution pattern was observed in control corneas (between arrows).

4. CONCLUSIONS

In summary, gene therapy is a promising treatment for several ocular diseases. Refinement in administration methods and targeting of compromised cells in certain genetic diseases need further research before a clinical application is established.[46] However, the corneal tissue is ideal for a localized and controlled exposure to vectors, without the complications of a systemic administration of genetic material. This may enable further clinical work to assess the safety of this therapy for other corneal pathologic conditions.

REFERENCES

1.   W.F. Anderson, Human gene therapy, *Science* 256:808 (1992).
2.   W.F. Anderson, Human gene therapy, *Nature* 392:25(1998).
3.   N.G. Della, Molecular biology in ophthalmology. A review of principles and recent advances, *Arch Ophthalmol.* 114:457(1996).
4.   M.A. Musarella, Gene mapping of ocular diseases, *Surv Ophthalmol.* 36:285(1992).
5.   D.J. Zack, Ocular gene therapy. From fantasy to foreseeable reality, *Arch Ophthalmol.* 111:1477(1993).
6.   J. Bennett, and A.M. Maguire, Gene therapy for ocular disease, *Mol Ther.* 1:501(2000).
7.   A.A. Okada, and J.V. Forrester, Ocular inflammatory disease in the new millennium, *Arch Ophthalmol.* 118:116(2000).
8.   M.F. Cordeiro, G.S. Schultz, R.R. Ali, S.S. Bhattacharya, and P.T. Khaw, Molecular therapy in ocular wound healing, *Br J Ophthalmol.* 83:1219(1999).
9.   M.B. Reichel, T. Hudde, R.R. Ali, and P. Wiedemann, Gene transfer in ophthalmology, *Ophthalmologe* 96:570(1999).
10.   K. Csaky, and R. Nussenblatt, Gene therapy in the treatment of ocular inflammation, *Springer Semin Immunopathol.* 21:191(1999).
11.   T.J. Stout, Gene therapy in ocular disease, *Ophthalmology* 102:1415(1995).

12.   T. Murata, H. Kimura, T. Sakamoto, R. Osusky, C. Spee, T.J. Stout, D.R. Hinton, and S.J. Ryan,
      Ocular gene therapy: experimental studies and clinical possibilities, *Ophthalmic Res.* 29:242(1997).

13.   D. Lechardeur, K.J. Sohn, M. Haardt, P.B. Joshi, M. Monck, R.W. Graham, B. Beatty, J. Squire, H.
      O'Brodovich, and G.L. Lukacs, Metabolic instability of plasmid DNA in the cytosol: a potential barrier
      to gene transfer, *Gene Ther.* 6:482(1999).

14.   K. Anwer, C. Meaney, G. Kao, N. Hussain, R. Shelvin, R.M. Earls, P. Leonard, A. Quezada, A.P.
      Rolland, and S.M. Sullivan, Cationic lipid-based delivery system for systemic cancer gene therapy,
      Cancer Gene Ther. 7:1156(2000).

15.   Y. Horiguchi, W.A. Larchian, R. Kaplinsky, W.R. Fair, and W.D. Heston, Intravesical liposome-
      mediated interleukin-2 gene therapy in orthotopic murine bladder cancer model. *Gene Ther.*
      7:844(2000).

16.   G.U. Dachs, C. Coralli, S.L. Hart, and G.M. Tozer, Gene delivery to hypoxic cells in vitro, *Br J Cancer*
      83:662(2000).

17.   Y. Oshima, T. Sakamoto, I. Yamanaka, T. Nishi, T. Ishibashi, and H. Inomata, Targeted gene transfer
      to corneal endothelium in vivo by electric pulse, *Gene Ther.* 5:1347(1998).

18.   L. Vayssea, I. Burgelina, J.P. Merliob, and B. Arveilera, Improved transfection using epithelial cell
      line-selected ligands and fusogenic peptides, *Biochim Biophys Acta* 1475:369(2000).

19.   E.M. Gordon, and W.F. Anderson, Gene therapy using retroviral vectors, *Curr Opin Biotechnol.*
      5:611(1994).

20.   F. Malecaze, B. Couderc, S. de Neuville, B. Serres, J. Mallet, V. Douin-Echinard, S. Manenti, F.
      Revah, and J.M. Darbon, Adenovirus-mediated suicide gene transduction: feasibility in lens epithelium
      and in prevention of posterior capsule opacification in rabbits, *Hum Gene Ther.* 10:2365(1999).

21.   B. Mashhour, Gene therapy in ophthalmology, *Bull Acad Natl Med.* 180:645(1996).

22.   T. Hudde, S.A. Rayner, M. De Alwis, A.J. Thrasher, J. Smith, R.S. Coffin, A.J. George, and D.F.
      Larkin, Adeno-associated and herpes simplex viruses as vectors for gene transfer to the corneal
      endothelium, *Cornea* 19:369(2000).

23.   J. Bennett, A.M. Maguire, A.V. Cideciyan, M. Schnell, E. Glover, V. Anand, T.S. Aleman, N.
      Chirmule, A.R. Gupta, Y. Huang, G.P. Gao, W.C. Nyberg, J. Tazelaar, J. Hughes, J.M. Wilson, and
      S.G. Jacobson, Stable transgene expression in rod photoreceptors after recombinant adeno-associated
      virus-mediated gene transfer to monkey retina, *Proc Natl Acad Sci U S A* 96:9920(1999).

24.   X. Wang, B. Appukuttan, S. Ott, R. Patel, J. Irvine, J. Song, J.H. Park, R. Smith, and J.T. Stout,
      Efficient and sustained transgene expression in human corneal cells mediated by a lentiviral vector,
      Gene Ther. 7:196(2000).

25.   M. Brown, D.H. Davies, M.A. Skinner, G. Bowen, S.J. Hollingsworth, G.J. Mufti, J.R. Arrand, and
      S.N. Stacey, Antigen gene transfer to cultured human dendritic cells using recombinant avipoxvirus
      vectors, *Cancer Gene Ther.* 6:238(1999).

26.   M.M. Hitt, and J. Gauldie. Gene vectors for cytokine expression in vivo, *Curr Pharm Des.*
      6:613(2000).

27.   K. Anwer, A. Bailey, and S.M. Sullivan, Targeted gene delivery: a two-pronged approach, *Crit Rev
      Ther Drug Carrier Syst.* 17:377(2000).

28.   F.L. Hall, E.M. Gordon, L. Wu, N.L. Zhu, M.J. Skotzko, V.A. Starnes, and W.F. Anderson, Targeting
      retroviral vectors to vascular lesions by genetic engineering of the MoMLV gp70 envelope protein,
      Hum Gene Ther. 8:2183(1997).

29.   F.L. Hall, L. Liu, N.L. Zhu, M. Stapfer, W.F. Anderson, R.W. Beart, and E.M. Gordon, Molecular
      engineering of matrix-targeted retroviral vectors incorporating a surveillance function inherent in von
      Willebrand factor, *Human Gene Ther.* 11:983(2000).

30.   D.G. Miller, M.A. Adam, and A.D. Miller, Gene transfer by retrovirus vectors occurs only in cells that
      are actively replicating at the time of infection, *Mol Cell Biol.* 10:4239(1990).

31. M.J. Skotzko, L.T. Wu, W.F. Anderson, E.M. Gordon, and F.L. Hall, Retroviral vector-mediated gene transfer of antisense cyclin G1 (CYCG1) inhibits proliferation of human osteogenic sarcoma cells, *Cancer Res.* 55:5493(1995).

32. D.S. Chen, N.L. Zhu, G. Hung, M.J. Skotzko, D.R. Hinton, V. Tolo, F.L. Hall, W.F. Anderson, and E.M. Gordon, Retroviral vector-mediated transfer of an antisense cyclin G1 construct inhibits osteosarcoma tumor growth in nude mice, *Hum Gen Ther.* 8:1667(1997).

33. F.E. Fantes, K.D. Hanna, G.O. Waring 3d, Y. Pouliquen, K.P. Thompson, and M. Savoldelli, Wound healing after excimer laser keratomileusis (photorefractive keratectomy) in monkeys, *Arch Ophthalmol.* 108:665(1990).

34. W.C.S. Wu, W.J. Stark, and W.J. Green, Corneal wound healing after 193 nm excimer laser keratectomy, *Arch Ophthalmol.* 109:1426(1991).

35. K.D. Szerenyi, X.W. Wang, K. Gabrielian, and P.J. McDonnell, Keratocyte loss and repopulation of anterior corneal stroma after de-epithelialization, *Arch Ophthalmol.* 112:973(1994).

36. M. Campos, S. Raman, M. Lee, and P.J. McDonnell, Keratocyte loss after different methods of de-epithelialization, *Ophthalmology* 101:890(1994).

37. M.C. Corbett, J.I. Prydal, S. Verma, K.M. Oliver, M. Pande, and J. Marshall, An in vivo investigation of the structures responsible for corneal haze after photorefractive keratectomy and their effect on visual function, *Ophthalmology* 103:1366(1996).

38   T. Møller-Pedersen, H.F. Li, W.M. Petroll, H.D. Cavanagh, and J.V. Jester, Confocal microscopic characterization of wound repair after photorefractive keratectomy, *Invest Ophthalmol Vis Sci.* 39:487(1998).

39. T. Seiler, and P.J. McDonnell, Excimer laser photorefractive keratectomy, *Surv Ophthalmol.* 40:89(1995).

40. T. Møller-Pedersen, H.D. Cavanagh, W.M. Petroll, and J.V. Jester, Stromal wound healing explains refractive instability and haze development after photorefractive keratectomy, *Ophthalmology* 107:1235(2000).

41. B. Seitz, E. Baktanian, E.M. Gordon, W.F. Anderson, L. LaBree, and P.J. McDonnell, Retroviral vector-mediated gene transfer into keratocytes: in vitro effects of polybrene and protamine sulfate, *Graefes Arch Clin Exp Ophthalmol.* 236:602(1998).

42. B. Seitz, L. Moreira, E. Baktanian, D. Sánchez, B. Gray, E.M. Gordon, W.F. Anderson, and P.J. McDonnell, Retroviral vector-mediated gene transfer into keratocytes in vitro and in vivo, *Am J Ophthalmol.* 126:630(1998).

43. J. Kampmeier, A. Behrens, Y. Wang, A. Yee, W.F. Anderson, F.L. Hall, E.M. Gordon, and P.J. McDonnell, Inhibition of rabbit keratocyte and human fetal lens epithelial cell proliferation by retrovirus-mediated transfer of antisense cyclin G1 and antisense MAT1 constructs, *Hum Gene Ther.* 11:1(2000).

44. M.J. Maldonado, V. Arnau, R. Martínez-Costa, A. Navea, F.M. Mico, A.L. Cisneros, and J.L. Menezo, Reproducibility of digital image analysis for measuring corneal haze after myopic photorefractive keratectomy, *Am J Ophthalmol.* 123:31(1997).

45. W.W. Hauswirth, and L. Beaufrere, Ocular gene therapy: quo vadis? *Invest Ophthalmol Vis Sci.* 41:2821(2000).

46. M. Imayasu, T. Moriyama, J. Ohashi, H. Ichijima, and H.D. Cavanagh, A quantitative method for LDH, MDH and albumin levels in tears with ocular surface toxicity scored by Draize criteria in rabbit eyes, *CLAO J.* 18:260(1992).

EX VIVO PRESERVATION AND EXPANSION OF HUMAN LIMBAL EPITHELIAL STEM CELLS ON AMNIOTIC MEMBRANE FOR TREATING CORNEAL DISEASES WITH TOTAL LIMBAL STEM CELL DEFICIENCY

Scheffer C. G. Tseng,[1,2] Daniel Meller,[1,3] David F. Anderson,[1]
Amel Touhami,[1] Renato T. F. Pires,[1] Martin Grüterich,[1]
Abraham Solomon,[1] Edgar Espana,[1] Helga Sandoval,[1]
Seng-Ei Ti,[1] and Eiki Goto[1]

[1]Department of Ophthalmology
Bascom Palmer Eye Institute
[2]Department of Cell Biology and Anatomy
University of Miami School of Medicine
Miami, Florida, USA
[3]Department of Ophthalmology
University of Essen
Essen, Germany

1. INTRODUCTION

Stem cells (SC) for the corneal epithelium are located exclusively at the limbus, i.e., the anatomic junction between the cornea and conjunctiva.[1] Limbal epithelial SC are the ultimate source of regeneration of the entire corneal epithelium under normal and injured states.[2,3] The epithelial phenotype of the limbal basal epithelium does not express corneal epithelial-specific keratin 3,[1,4] keratin 12[5,6] and connexin 43.[7] As shown in cell-cycle kinetic studies, some portions of the limbal basal epithelial cells are slow-cycling[8,9] and label-retaining.[8,10] Other studies further confirmed limbal epithelial SC have greater growth potential in explant cultures[11,12] and higher clonogenicity when cocultured on 3T3 fibroblast-feeder layers[13-17] than corneal epithelial cells, and their proliferative potential is resistant to tumor-promoting phorbol esters.[8,10,18]

Lacrimal Gland, Tear Film, and Dry Eye Syndromes 3
Edited by D. Sullivan *et al.*, Kluwer Academic/Plenum Publishers, 2002

1323

When the SC-containing limbal epithelium is partially[19,20] or totally[21,22] damaged, the corneal surface is invariably covered by ingrowing conjunctival epithelial cells with goblet cells, and the corneal stroma becomes vascularized with chronic inflammation. These pathologic signs signify a process of conjunctivalization, and can be found in many corneal diseases with limbal SC deficiency.[23] Clinically, transplantation of an autologous or allogeneic source of limbal epithelial SC is necessary to restore vision and the normal corneal surface in these diseases.[24,25] Previously, we noted transplantation of preserved human amniotic membrane (AM) can restore vision and corneal surface for patients with partial limbal SC deficiency,[26,27] suggesting AM may help expand the residual limbal epithelial SC in vivo. Recently, clinical[28,29] and experimental[30] studies have further shown ex vivo expanded limbal epithelial cells on AM can be used as a graft to treat limbal SC deficiency. We provide herein additional evidence AM might be an ideal matrix for ex vivo preservation and expansion of limbal epithelial SC. This new evidence supports the clinical efficacy of our preliminary trial in rabbits and human patients. The significance of this finding is further discussed.

2. METHODS

2.1. Explant Cultures on Amniotic Membrane

Human corneas were obtained from the Florida Lions Eye Bank (Miami, FL). After careful removal of excessive sclera, iris, corneal endothelium, conjunctiva and Tenon's capsule, the remaining tissue was placed in a culture dish and exposed for 5 – 10 min to Dispase II (1.2 U/ml in Mg^{2+}- and Ca^{2+}-free Hank's balanced salt solution) at 37°C under humidified 5% CO_2. Following one rinse with DMEM containing 10% FBS, the tissue was subdivided by a trephine into three portions, i.e., the central cornea (7.5 mm in diameter), the peripheral cornea (7.5 mm to 1 mm within the limbus) and the limbus (the remainder). Each of these three portions was then cut into cubes of approximately 1 x 1.5 x 2.5 mm by a scalpel. Preserved human AM was kindly provided by Bio-Tissue (Miami, FL) and fastened onto a culture insert as recently reported.[31] A tissue cube was placed on the center of AM (Fig. 1A) and cultured in a medium described by Jumblatt and Neufeld[32] made of an equal volume of HEPES-buffered DMEM containing bicarbonate and Ham's F12, and supplemented with 5% FBS, 0.5% dimethyl sulfoxide, 2 ng/ml mouse EGF, 5 μg/ml insulin, 5 μg/ml transferrin, 5 ng/ml selenium, 0.5 μg/ml hydrocortisone, 30 ng/ml cholera toxin, 5% FBS, 50 μg/ml gentamicin and 1.25 μg/ml amphotericin B. Cultures were incubated at 37°C under 5% CO_2 and 95% air and the medium was changed every 2 to 3 days while the extent of each outgrowth was monitored with a phase contrast microscope.

2.2. BrdU Labeling

When the outgrowth reached 5 to 8 mm in diameter, explant cultures were incubated with a fresh medium containing 10 μM BrdU for 1 or 7 days. Some limbal cultures after 7-

day labeling were chased for 14 days by switching to the BrdU-free medium. All samples were fixed in cold methanol and processed for immunostaining. In a subset of experiments human limbal epithelial cells (HLEC) on AM were treated for 24 h with 1.0 μg/ml PMA, then continuously labeled with BrdU for 7 days and subcutaneously transplanted in nude mice and chased for 9 days in vivo.

2.3. Subcutaneous Implantation in Athymic Balb/c Mice

After the skin covering the rectus abdominis was undermined to expose an area measuring approximately 1.5 cm x 1.5 cm, HLEC cultured on AM for 2–3 weeks were implanted subcutaneously onto the fascial surface of the muscle, and the skin flap was then closed with a running, coated Vicryl® 7–0 suture. Postoperatively, gentamicin ointment was applied once daily to the wound. A firm subcutaneous nodule formed during an 8-day period was excised with the surrounding skin and muscle for histology and immunostaining.

2.4. Immunostaining

For BrdU labeling, the flat-mount preparation of the epithelial outgrowth on AM was used directly, and the tissue from athymic mice subjected to frozen sections of 6-μm thickness in OCT®. Immunostaining was carried out according to a protocol provided by the manufacturer with a mouse anti-BrdU antibody followed by a VECTASTAIN Elite Kit. The slides were counterstained with eosin. Under 400x magnification, positive nuclei were counted among the total nuclei within the entire field, and a total of 2 to 4 fields were counted per specimen. The labeling index was expressed as the number of positively labeled nuclei to the number of all nuclei x 100%. To study the corneal epithelial differentiation, we used immunostaining with AE5 (1:100) monoclonal antibody to keratin 3.[1]

2.5. TRANSPLANTATION OF EX VIVO EXPANDED LIMBAL EPITHELIAL CELLS WITH AMNIOTIC MEMBRANE

We created limbal SC deficiency in one eye of 15 rabbits using a model previously described.[33] After 2 to 3 months, when clinical signs of limbal SC deficiency were fully manifested and proven by impression cytology,[23] a limbal biopsy was performed on the normal fellow eye. The limbal specimen measured in 1 x 2 mm was prepared for explant culture on amniotic membrane as described above. After 2 to 3 weeks of culturing, the outgrowth and membrane were then transplanted to the limbal-deficient cornea where the superficial pannus was removed by suturing with 9–0 Vicryl sutures. A similar procedure was also performed in five patients. Three suffered from unilateral chemical burns and received autologous limbal epithelial SC transplantation, while two suffered from bilateral limbal SC deficiency caused by multiple endocrine deficiency[23] and pseudopemphigoid, respectively, and received allogeneic transplantation from the living related donor.

3. RESULTS

3.1. Explant Outgrowth

No outgrowth occurred in 22 central corneal explants tested, while some existed in 2 of 24 (8.3%) peripheral corneal explants. In contrast, abundant outgrowth was noted in 77 of 80 (96.2%) limbal explants (P < 0.0001, Fisher's Exact Test). These results suggest AM preferentially supports the outgrowth of epithelial progenitor cells derived from the limbus. For the limbal explant grown on AM (Fig. 1A), the rate of epithelial outgrowth was slow the first week, but became rapid from then on and frequently reached a size of 2 to 3 cm in 2 to 3 weeks. The basal epithelial cells of the limbal outgrowth were uniformly small, cuboidal and with a scanty cytoplasm (Fig. 1B), similar to what we have previously noted in conjunctival epithelial cells grown in the same culture system.[31] The outgrowth reached confluence, i.e., the limit of the membrane fastened to a 35-mm insert, in 4 to 5 weeks.

3.2. Epithelial Differentiation

As reported,[1] AE5 antibody, which recognizes K3 keratin, stained the suprabasal limbal epithelium (Fig. 1C) and the full thickness of the central corneal epithelium (Fig. 1D), but not the conjunctival epithelium (not shown). AE5 antibody stained suprabasal cell layers of human limbal epithelial cells cultured on AM for 13 or 21 days (not shown). This result indicates the resultant phenotype retained a limbal origin, was predominantly basal epithelial cells and remained undifferentiated. After subcutaneous transplantation in athymic Balb/c mice for 8 days, they formed a stratified epithelium consisting of one layer of small round-to-cuboidal basal epithelial cells, three to four layers of wing cells and three to four layers of flattened superficial layers. The staining by AE5 antibody was negative in the basal cell layer, but positive in suprabasal and superficial epithelial cells (Fig. 1E), a pattern identical to that of the normal human limbal epithelium.

3.3. Cell-Cycle Analysis

To determine the cell cycle, we labeled the S phase with BrdU, an analogue of thymidine, for 24 h in 2- to 3-week-old cultures. In all limbal explants tested, the majority (65%) had a uniformly low labeling index (3.3 ± 3.3%, n = 13) (Fig. 2A), while 35% had a mixed pattern with areas of a moderately high labeling index (44.3 ± 15.3%, mean ± SE, n = 7) (Fig. 2B), (P < 0.0001, unpaired t test). In contrast, the outgrowth from all peripheral corneal explants had a uniformly high labeling index (61.7 ± 2.2%, n = 2), indistinguishable from that shown in Fig. 2B. These results indicated 24-h BrdU labeling

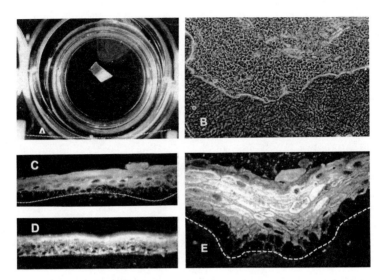

Figure 1. Limbal explant outgrowth on amniotic membrane (**A**), showing uniformly small basal epithelial cells (**B**). The expression of K3 keratin was negative in the limbal basal epithelium in vivo (**C**). The dotted line indicates the basement membrane. K3 keratin expression was positive throughout the full thickness of the corneal epithelium (**D**). The stratified epithelium after subcutaneous transplantation in nude mice showed negative expression of K3 keratin in the basal epithelium (**E**). The dotted line indicates the basal epithelium.

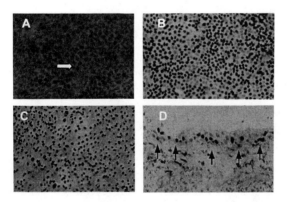

Figure 2. Representative photographs of BrDU labeling for 24 h (**A, B**) showing the high labeling index (**A**) and low labeling index (**B**), indicated by an arrow), and for 7 days (**C**). Basally located labeling-retaining cells (indicated by arrows) in the stratified epithelium after nude mice transplantation (**D**).

predominantly labeled rapid-cycling progenitor cells in the peripheral cornea. The inclusion of peripheral corneal transient amplifying cells might cause the patchy pattern of high labeling index in some limbal explants. To confirm the low labeling index of the limbal outgrowth was indeed a result of slow-cycling progenitor cells but not due to post-mitotic differentiated cells, we labeled 13 limbal explants continuously for 7 days. Of these, 8 (61.5%) showed a high labeling index of 75.5 ± 10.9% (Fig. 2B), while the remaining 5 (38.5%) still showed a low labeling index of 10.01 ± 10.8% ($P < 0.0001$, unpaired t test). This result indicated the majority of limbal epithelial explants had a slower cell cycle, and some had a cell-cycle length longer than 7 days. In a separate experiment, human limbal epithelial cells grown on AM were treated for 24 h with 1.0 μg/ml PMA, which resulted in increased cellular desquamation. The remaining adherent cells proliferated to form a cohesive sheet after 7 days, when they were continuously labeled with BrdU, resulting in a high labeling index (Fig. 2C) and supporting the notion they were PMA-resistant progenitor cells. These cells and AM were transplanted to the subcutaneous space in athymic Balb/c mice and chased for 9 days. The resultant stratified epithelium showed the majority of label-retaining cells were at the basal layer and some were scattered in the suprabasal cell layers (Fig. 2D). Together, these data support the notion AM cultures preferentially permit expansion of limbal epithelial progenitor cells that retain the slow-cycling property, and PMA treatment does not inhibit their proliferative capacity. After continuous labeling by BrdU, these progenitor cells still retain labels following a prolonged period of chase in vivo.

3.4. TRANSPLANTATION IN RABBITS

The model of limbal SC deficiency was successfully created in all rabbits, which showed diffuse corneal vascularization and conjunctivalization (Fig. 3A). All corneas of 10 rabbits receiving transplantation of rabbit AM alone remained limbal deficient (similar to that shown in Fig. 3A) as evidenced by conjunctivalization with goblet cells (Fig. 3B). In contrast, 3 of 15 rabbits (20%) receiving ex vivo expanded limbal epithelial cells on AM restored a normal avascular clear cornea (Fig. 3C) and a corneal epithelial phenotype (Fig. 3D), while 10 corneas showed a mixed corneal and conjunctival phenotype (not shown), and the remaining 1 cornea showed total conjunctivalization, i.e., failure (not shown). This difference is statistically significant.

3.5. TRANSPLANTATION IN HUMAN PATIENTS

Table 1 summarizes the results of five patients who received limbal epithelial cells expanded by amniotic membrane. The three receiving autologous limbal epithelial cells from the fellow eyes showed improved vision and their corneas restored intact, smooth and quiet surfaces. Fig. 4 shows one such example in a 26-year-old female suffering from a

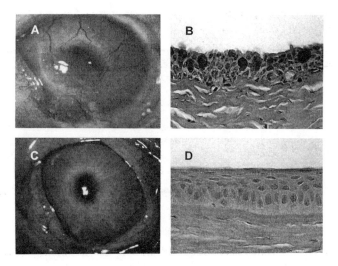

Figure 3. (A) Preoperative appearance of limbal SC deficiency showing vascularization. (B) Conjunctivalization of the cornea following AMT alone in the control. (C) One year postoperative appearance of the same rabbit shown in (A) receiving ex vivo expanded limbal SC on AM. (D) Restoration of corneal epithelial phenotype after surgery.

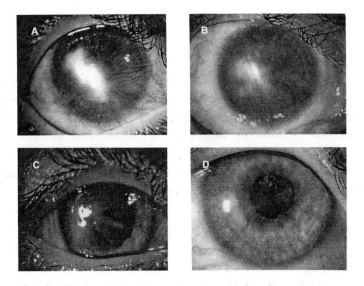

Figure 4. (A) Preoperative appearance of Case #1. (B) Fourteen months after surgery of Case #1. (C) Preoperative appearance of Case #5. (D) Nine months after surgery of Case #5.

Table 1. Summary of clinical results of transplantation of ex vivo expanded limbal epithelial SC on AM using autologous and allogeneic sources of limbal biopsy

No	Age/Sex	Dx	Pre-op	Post-op	F/U	Outcome	Note
			Autologous				
1	26/F	Chemical	3/200	20/80	14	success	
2	33/M	Chemical	CF	20/200	3.8	success	
3		Chemical	CF	20/70	9.7	success	residual scar
			Allogenic				
4	51/M	Trachoma/	CF	20/100	1		from son
		OCP?		20/200	9.3	failure	
5	20/M	Multiple ED	20/200	20/40	9	success	from mother

chemical burn to her right eye with total limbal SC deficiency (Fig. 4A). The transplantation of ex vivo expanded limbal epithelial cells from her left eye biopsy using AM resulted in a stable surface and improved vision (Fig. 4B). One of the two patients receiving allogeneic limbal epithelial cells was suffering from multiple endocrine deficiency (Fig. 4C). His cornea was markedly improved after receiving cells expanded from the limbal biopsy of his mother (Fig. 4D). The other patient with pemphigoid failed.

4. DISCUSSION

Here we present experimental evidence to support the hypotheses that AM cultures preferentially preserve and expand limbal epithelial SC. The outgrowth from limbal explants was preferentially promoted when compared to those derived from the peripheral cornea and central cornea. The majority of human limbal epithelial cells grown on AM formed a monolayer of uniformly small and compact cuboidal cells that had scanty cytoplasm. They weakly expressed cornea-specific keratin 3 on the suprabasal layers. Following subcutaneous transplantation in nude mice, they became markedly stratified and exhibited a phenotype resembling that of the normal limbal epithelium in vivo, i.e., the lack of expression of keratin 3 in the basal epithelium.[1,5] These data indicate the progenitor cells of human limbus are indeed preserved during the expansion by AM cultures.

To prove these ex vivo expanded epithelial progenitor cells possess SC characteristics, we performed cell-cycle analysis by BrdU labeling. In vivo, under the normal state the corneal epithelium incorporates pulse-administered [³H] thymidine, suggesting it contains more rapid-cycling cells.[8,14,34] This notion was also illustrated by a uniformly high labeling index of 61% in the outgrowth of all peripheral corneal explants following 24-h BrdU labeling. A similar high labeling index of 35% was noted in some areas of the limbal outgrowth tested, suggesting these limbal explants might have included transient amplifying cells from the peripheral cornea. Alternatively, they might have permitted some differentiation of limbal SC into rapid cycling cells. Notable was the finding that the majority of the limbal explants (65%) showed a low (less than 5%) labeling index, which was increased to 61.5% after a 7-day continuous labeling period. This finding confirmed the low labeling index noted after 24-h labeling was indeed due to slow cycling of the progenitor cells and not a result of post-mitotic differentiation. Collectively, these findings indicate the outgrowth of limbal explants predominantly contains slow-cycling progenitor cells, and 24-h BrdU labeling is a technique useful to distinguish them from rapid-cycling progenitor cells.

After continuous labeling for 7 days, many labeled progenitor cells still retained their labels despite 9 days of chase when transplanted in nude mice. The detection of basally located label-retaining cells offers another piece of evidence confirming these progenitor cells were slow cycling in generating their offspring. It should be noted these progenitor cells resisted a brief treatment with PMA, a phorbol ester tumor promoter that causes a divergent response to SC and transient amplifying cells (TAC) in epidermal keratinocytes[35,36] and ocular surface epithelia.[8,10,18] We noted exposure to PMA for 24 h increased cell desquamation in some adherent cell layers, indicating the existence of more differentiated transient amplifying cells, which cease mitosis and undergo terminal differentiation on exposure to this tumor promoter.[10,18,35,36] However, the remaining adherent cells proliferated as evidenced by their continuous growth into a larger epithelial sheet and incorporation of BrdU for 7 days. These findings provide more evidence to support the notion that a subpopulation of expanded human limbal epithelial cells are indeed SC, which proliferate and retain labels after treatment with such a tumor promoter.[8,10,18]

Previously, epithelial SC have been expanded by cocultured 3T3 fibroblasts.[13,15,17,37-39] Such ex vivo expanded epithelial SC have been used to reconstruct skin following burns[40] and in eyes with total limbal SC deficiency.[41] The fact that limbal epithelial SC can be preserved and expanded by AM cultures without the inclusion of 3T3 fibroblasts will reduce the potential risk of using a mouse-derived cell line. Such ex vivo expanded limbal SC on AM also make it easier to transfer to the recipient eye during surgery. The clinical efficacy of this new approach of transplanting limbal epithelial SC for treating corneal diseases has been reported in a short-term study in rabbits[30] and clinical patients[28,29] with unilateral partial or total limbal SC deficiency. Here we confirmed this efficacy in a long-term rabbit study and a small series of human patients with unilateral and bilateral total limbal SC deficiency. This new approach provides an advantage over the conventional limbal conjunctival autograft[42] by reducing risks to the donor eye inasmuch as a small biopsy of approximately less than 1.5 mm², not two large strips of the limbal tissue, will be removed.

In conclusion, strong experimental evidence supports the hypothesis that AM cultures preferentially preserve and expand limbal epithelial SC that retain their in vivo characteristics of nondifferentiation, slow cycling, label retaining and resistance to a phorbol ester tumor promoter. Studies to elucidate how such a culture system achieves this novel action may help unravel the secret of how the "stemness" of epithelial SC is maintained. This culture system may serve as a first step toward engineering various epithelial tissues and developing new therapeutics targeted at epithelial SC in the future.

ACKNOWLEDGEMENTS

The nonclinical part of this report was supported by a grant (EY 06819) from National Eye Institute, National Institute of Health.

REFERENCES

1. Schermer A, Galvin S, Sun T-T. Differentiation-related expression of a major 64K corneal keratin *in vivo* and in culture suggests limbal location of corneal epithelial stem cells. *J Cell Biol.* 1986;103:49–62.
2. Tseng SCG. Regulation and clinical implications of corneal epithelial stem cells. *Mol Biol Rep.* 1996;23:47–58.
3. Tseng SCG, Sun T-T. Stem cells: ocular surface maintenance. In: Brightbill FS, ed. *Corneal Surgery: Theory, Technique, and Tissue.* St. Louis: Mosby; 1999:9–18.
4. Kiritoshi A, SundarRaj N, Thoft RA. Differentiation in cultured limbal epithelium as defined by keratin expression. *Invest Ophthalmol Vis Sci.* 1991;32:3073–3077.
5. Chen WYW, Mui M-M, Kao WW-Y, Liu C-Y, Tseng SCG. Conjunctival epithelial cells do not transdifferentiate in organotypic cultures: expression of K12 keratin is restricted to corneal epithelium. *Curr Eye Res.* 1994;13:765–778.
6. Liu C-Y, Zhu G, Converse R, Kao CW-C, Nakamura H, Tseng SCG, Mui M-M, Seyer J, Justice MJ, Stech ME, Hansen GM, Kao WW-Y. Characterization and chromosomal localization of the cornea-specific murine keratin gene *Krt1.12. J Biol Chem.* 1994;260:24627–24636.
7. Matic M, Petrov IN, Chen S, Wang C, Dimitrijevich SD, Wolosin JM. Stem cells of the corneal epithelium lack connexins and metabolite transfer capacity. *Differentiation.* 1997;61:251–260.
8. Cotsarelis G, Cheng SZ, Dong G, Sun T-T, Lavker RM. Existence of slow-cycling limbal epithelial basal cells that can be preferentially stimulated to proliferate: implications on epithelial stem cells. *Cell.* 1989;57:201–209.
9. Tseng SCG, Zhang S-H. Limbal epithelium is more resistant to 5-fluorouracil toxicity than corneal epithelium. *Cornea.* 1995;14:394–401.
10. Lavker RM, Wei ZG, Sun TT. Phorbol ester preferentially stimulates mouse fornical conjunctival and limbal epithelial cells to proliferate *in vivo. Invest Ophthalmol Vis Sci.* 1998;39:301–307.
11. Ebato B, Friend J, Thoft RA. Comparison of central and peripheral human corneal epithelium in tissue culture. *Invest Ophthalmol Vis Sci.* 1987;28:1450–1456.
12. Matic M, Petrov IN, Wang C, Dimitrijevich SD, Wolosin JM. Stem cells of the corneal epithelium lack connexins and metabolite transfer capacity. *Differentiation.* 1997;61:251–260.
13. Lindberg K, Brown ME, Chaves HV, Kenyon KR, Rheinwald JG. *In vitro* preparation of human ocular surface epithelial cells for transplantation. *Invest Ophthalmol Vis Sci.* 1993;34:2672–2679.
14. Lavker RM, Dong G, Cheng SZ, Kudoh K, Cotsarelis G, Sun TT. Relative proliferative rates of limbal and corneal epithelia. Implications of corneal epithelial migration, circadian rhythm, and suprabasally located DNA-synthesizing keratinocytes. *Invest Ophthalmol Vis Sci.* 1991;32:1864–1875.

15. Wei Z-G, Lin T, Sun T-T, Lavker RM. Clonal analysis of the *in vivo* differentiation potential of keratinocytes. *Invest Ophthalmol Vis Sci.* 1997;38:753–761.

16. Tseng SCG, Kruse FE, Merritt J, Li D-Q. Comparison between serum-free and fibroblast-cocultured single-cell clonal culture systems: evidence showing that epithelial anti-apoptotic activity is present in 3T3 fibroblast conditioned media. *Curr Eye Res.* 1996;15:973–984.

17. Pellegrini G, Golisano O, Paterna P, Lambiase A, Bonini S, Rama P, De Luca M. Location and clonal analysis of stem cells and their differentiated progeny in the human ocular surface. *J Cell Biol.* 1999;145:769–782.

18. Kruse FE, Tseng SCG. A tumor promoter-resistant subpopulation of progenitor cells is present in limbal epithelium more than corneal epithelium. *Invest Ophthalmol Vis Sci.* 1993;34:2501–2511.

19. Chen JJY, Tseng SCG. Corneal epithelial wound healing in partial limbal deficiency. *Invest Ophthalmol Vis Sci.* 1990;31:1301–1314.

20. Chen JJY, Tseng SCG. Abnormal corneal epithelial wound healing in partial thickness removal of limbal epithelium. *Invest Ophthalmol Vis Sci.* 1991;32:2219–2233.

21. Huang AJW, Tseng SCG. Corneal epithelial wound healing in the absence of limbal epithelium. *Invest Ophthalmol Vis Sci.* 1991;32:96–105.

22. Kruse FE, Chen JJY, Tsai RJF, Tseng SCG. Conjunctival transdifferentiation is due to the incomplete removal of limbal basal epithelium. *Invest Ophthalmol Vis Sci.* 1990;31:1903–1913.

23. Puangsricharern V, Tseng SCG. Cytologic evidence of corneal diseases with limbal stem cell deficiency. *Ophthalmology.* 1995;102:1476–1485.

24. Tseng SCG. Conjunctival grafting for corneal diseases. In: Tasman W, Jaeger EA, eds. *Duane's Clinical Ophthalmology.* Philadelphia: J. B. Loppincott Co.; 1994:1–11.

25. Holland EJ, Schwartz GS. The evolution of epithelial transplantation for severe ocular surface disease and a proposed classification system. *Cornea.* 1996;15:549–556.

26. Tseng SCG, Prabhasawat P, Barton K, Gray T, Meller D. Amniotic membrane transplantation with or without limbal allografts for corneal surface reconstruction in patients with limbal stem cell deficiency. *Arch Ophthalmol.* 1998;116:431–441.

27. Anderson DF, Ellies P, Pires RTF, Tseng SCG. Amniotic membrane transplantation for partial limbal stem cell deficiency: long term outcomes. *Br J Ophthalmol.* 2001;in press.

28. Tsai RJF, Li L-M, Chen J-K. Reconstruction of damaged corneas by transplantation of autologous limbal epithelial cells. *N Eng J Med.* 2000;343:86–93.

29. Schwab IR, Reyes M, Isseroff RR. Successful transplantation of bioengineered tissue replacements in patients with ocular surface disease. *Cornea.* 2000;19:421–426.

30. Koizumi N, Inatomi T, Quantock AJ, Fullwood NJ, Dota A, Kinoshita S. Amniotic membrane as a substrate for cultivating limbal corneal epithelial cells for autologous transplantation in rabbits. *Cornea.* 2000;19:65–71.

31. Meller D, Tseng SCG. Conjunctival epithelial cell differentiation on amniotic membrane. *Invest Ophthalmol Vis Sci.* 1999;40:878–886.

32. Jumblatt JE, Neufeld AH. Beta-adrenergic and serotonergic responsiveness of rabbit corneal epithelial cells in culture. *Invest Ophthalmol Vis Sci.* 1983;24:1139–1143.

33. Tsai RJF, Sun T-T, Tseng SCG. Comparison of limbal and conjunctival autograft transplantation for corneal surface reconstruction in rabbits. *Ophthalmology.* 1990;97:446–455.

34. Haddad A. Renewal of the rabbit corneal epithelium as investigated by autoradiography after intravitreal injection of ^{3}H-thymidine. *Cornea.* 2000;19:378–383.

35. Yuspa SH, Ben T, Hennings H, Lichti U. Divergent responses in epidermal basal cells exposed to the tumor promoter 12-O-tetradecacanoylphorboAB-13-acetate. *Cancer Res.* 1982;42:2344–2349.

36. Furstenberger G, Gross M, Schweizer J, Vogt I, Marks F. Isolation, characterization and *in vitro* cultivation of subfractions of neonatal mouse keratinocytes: effects of phorbolesters. *Carcinogenesis.* 1986;7:1745–1753.

37. Rheinwald JG, Green H. Serial cultivation of strains of human epidermal keratinocytes: the formation of keratinizing colonies from single cells. *Cell.* 1975;6:331–337.

38. Barrandon Y, Green H. Three clonal types of keratinocytes with different capacities for multiplication. *Proc Natl Acad Sci USA.* 1987;84:2302–2306.

39. Kobayashi K, Rochat A, Barrandon Y. Segregation of keratinocyte colony-forming cells in the bulge of the vibrissa. *Proc Natl Acad Sci USA.* 1993;90:7391–7195.

40. Green H, Kehinde O, Thomas J. Growth of cultured human epidermal cells into multiple epithelia suitable for grafting. *Proc Natl Acad Sci USA.* 1979;76:5665–5668.

41. Pellegrini G, Traverso CE, Franzi AT, Zingirian M, Cancedda R, De Luca M . Long-term restoration of damaged corneal surface with autologous cultivated corneal epithelium. *Lancet.* 1997;349:990–993.

42. Kenyon KR, Tseng SCG. Limbal autograft transplantation for ocular surface disorders. *Ophthalmology.* 1989;96:709–723.

CORNEAL BLINDNESS FROM END-STAGE SJÖGREN'S SYNDROME AND GRAFT-VERSUS-HOST DISEASE

Claes H. Dohlman[1], Eric J. Dudenhoefer,[1] Bilal F. Khan,[1] and Jan G. Dohlman[2]

[1]Massachusetts Eye and Ear Infirmary
Boston, Massachusetts, USA
[2]East Boston Neighborhood Health Center
East Boston, Massachusetts, USA

1. INTRODUCTION

In rheumatoid arthritis (RA), sterile thinning, ulceration and perforation of the cornea are well-known clinical entities.[1-4] Characteristically, the patient presents almost totally lacking ocular pain, inflammation or corneal infiltration. These events can occur even in eyes with normal tear secretion but are most common in elderly patients with dry eye, or Sjögren's syndrome (SS). Surgery and treatment with corticosteroids can be triggering factors, thus keratoplasty in advanced RA is often complicated by sterile melt.[5] This phenomenon of often unprovoked tissue breakdown is probably related to the immune status and age of the patient, but few details are known.

Graft–versus–host disease (GVHD) may also involve a concomitant dry eye condition, epitheliopathy and melt of the stroma.[6] The severity of the conjunctival disease corresponds to the severity of the systemic disease. With corneal involvement, the visual outcome can be catastrophic.[7]

In end-stage severe SS and dry and inflamed GVHD, standard corneal transplantation is virtually hopeless. The outcome of keratoplasty in these situations is similar to that in end-stage Stevens-Johnson syndrome (SJS) or ocular cicatricial pemphigoid (OCP) in which this procedure almost invariably fails.[8] This leaves only a keratoprosthesis (KPro) as a possibility for restoring vision (Fig. 1). This modality can be quite successful in non-cicatrizing and low-inflammation categories of diseases such as graft failures from corneal degenerations, dystrophies and infections.[9] In presumed autoimmune conditions such as

Lacrimal Gland, Tear Film, and Dry Eye Syndromes 3
Edited by D. Sullivan *et al.*, Kluwer Academic/Plenum Publishers, 2002

1335

SJS and OCP, however, prognosis is much more guarded,[10] and SS and GVHD might fall in the same category. Here we present three cases that illustrate the difficulties with KPro surgery in these syndromes.

2. PATIENT SUMMARIES

Patient 1 developed RA at age 24 and is now "burned out," with absolute dryness in both eyes. The left eye (OS) was lost to glaucoma and vision in the right eye (OD) was light perception with inaccurate projection. A KPro Type I was placed in OD in 1994. In the beginning severe fibrin reaction in the anterior chamber ensued but it cleared on methotrexate. The interior remained clear for 6 years and allowed a visual acuity between 20/50 to 20/25 (Fig. 2) . Melt around the stem of the KPro began after 3 years, however, eventually requiring eight repairs with corneal tissue. Finally the leak became irreparable and the eye remained soft for 2 months. Remicade® (Centocor, Inc., Malvern, PA), a monoclonal antibody against tumor necrosis factor-α (TNF-α), was then started intravenously (3.8 mg/kg). TNF-α is a crucial cytokine that in turn increases the level of other pro-inflammatory cytokines. Within weeks the leak healed and healthy conjunctiva started to build up. However, a retinal detachment was found, which had probably developed during the leak, and proved irreparable.

Patient 2 had long-standing RA with visual loss in both eyes (OU) from severe keratoconjunctivitis sicca, corneal ulceration, scarring, melts and failed keratoplasties OU. Vision in OD was eventually lost and in OS it decreased to light perception with projection. She underwent placement of a Type I KPro plus Ahmed shunt and blepharoplasty in April 1999. Within 8 months visual acuity improved in that eye to 20/20. However, in April 2000, she developed a large, quiet melt with leak, requiring repair around the KPro with a new corneal graft. By August 2000 a recurrent leak occurred from new necrosis and melt, requiring surgical removal of the Type I device and replacement with a Type II through the lid Kpro (Fig. 3). Remicade® infusion was started pre-operatively. Her vision improved immediately to 20/70 and subsequently to 20/20. She has had no further complications since that time.

Patient 3 developed GVHD after bone marrow transplantation for chronic myelogenic leukemia 10 years earlier at age 27, resulting in absolute ocular dryness, recurrent corneal melting and recurrent failed keratoplasties bilaterally. Visual acuity measured as light perception with projection OU, OS greater than OD. He underwent a Type I KPro plus extracapsular cataract extraction plus blepharoplasty OS in August 1999 (Fig. 4). Ahmed shunt placement was done subsequently for intractable glaucoma OS. Within 1 month after KPro placement, tissue melt around the stem required repair with corneal tissue. Quiet (non-inflamed), concurrent melt OD also required repair, Gunderson flap and complete tarsorrhaphy. For the past year, the eyes have remained quiet, however, without melt. Visual acuity in OS has remained between 20/50 and 20/20.

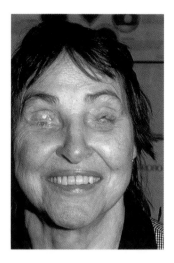

Figure 3. (a) Sjögren's syndrome. Vision in only eye (left) is hand movements. (b) Immediate postoperative appearance (Type II). Later vision became 20/20.

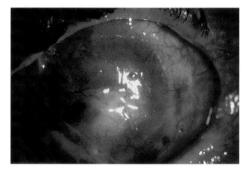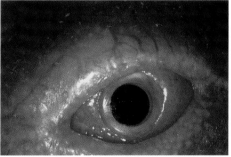

Figure 4. Graft-versus-host disease. (a) Vision hand motion after repeated perforations. (b) Keratoprosthesis Type I, glaucoma shunt and lid repair. Vision after 1 1/2 years is 20/25.

Figure 1. Keratoprostheses: Dohlman-Doane Type I left, Type II right.

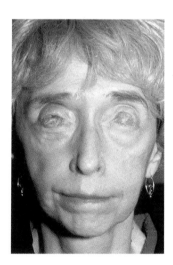

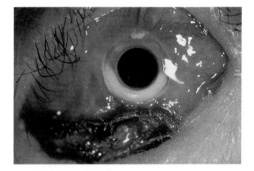

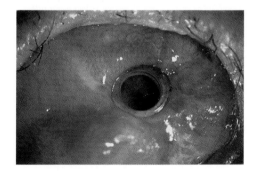

Figure 2. Sjögren's syndrome with completely dry eyes. Vision lost in left eye. **(a)** 6 years following keratoprosthesis (Type I). Vision 20/40. **(b)** Necrosis between plates below, leak. **(c)** Healing of perforation, build-up of conjunctiva after Remicade.®

3. DISCUSSION

These three patients' corneas had a very marked tendency to necrotize and melt around the KPro. In the two SS patients, the stroma next to the stem gradually loosened its firm grip and started to melt towards the periphery, eventually resulting in a leak of aqueous fluid. This situation poses a great threat in terms of possible further complications such as endophthalmitis or retinal detachment. In the first case, eight surgical repairs were necessary to keep the eye intact. In spite of these frequent calamities, vision was kept at the 20/50–20/25 level for 6 years. Case number 2 has been followed only for 18 months and has already required two major repairs. Present vision is 20/30. The GVHD patient had a history of repeated melts in both eyes and perforations before the KPro surgery. Vision is presently 20/20 after 15 months.

The course of KPros in these three cases differed substantially from KPro implantation in standard graft failure cases. The latter, in relatively non-inflamed, non-autoimmune, non-cicatrizing eyes, survive much longer without severe complications. Thus, a recent series of 18 patients in this category showed a two-thirds survival with a vision of 20/200–20/20 after 5 years.[10] KPro surgery in SS, and probably in GVHD, has a quite guarded prognosis, most likely similar to that in SJS. To improve that outcome, new ways must be sought to block the inflammatory and enzymatic cascades that cause so much tissue damage around the device.

In the prevention and treatment of sterile immune-mediated corneal ulcers in general, the use of systemic immunosuppressive agents has resulted in improved outcome of keratoplasty in RA.[5] Other investigators have found such regimens without effect.[11] Topical therapy with cyclosporine, on the other hand, has been useful adjunctive therapy for RA-associated corneal ulceration.[11,12,13] Such ulcers were arrested by the topical application of a 2% solution of cyclosporine given four times daily for 1 month, followed by a twice daily regimen for at least 6 months.[11] It was recommended the patients be monitored for renal toxicity.

We routinely use corticosteroids, given topically or by sub-Tenon route or, rarely, systemically, following KPro surgery. Inflammation is effectively suppressed but the well-known inhibition of stromal repair can cause loosening of the tissue. In animals, corticosteroids enhance extrusion of Kpro.[14] We have routinely used mexdoxyprogesterone as a 1% topical suspension in our KPro patients since 1992.[15] It suppresses collagenase synthesis and can retard corneal ulceration in animals.[16] Its efficacy in the clinical setting has never been proved, however. We have also sporadically used tetracycline as a 1% suspension for topical use.[15] It is a direct collagenase inhibitor that, if given systemically, can also retard corneal ulceration in animals.[17]

Recently drops have been introduced that are directed against inflammatory cytokines of the type elevated in RA, particularly TNF-α. Remicade® periodically given intravenously controls RA. After starting Remicade, our first patient's chronically leaking eye healed almost immediately, and within 2 months the conjunctival tissue around the KPro thickened and seemed healthy (the retinal detachment had probably occurred much earlier). Also, the second RA patient did well on the drug, but it had to be stopped due to diagnosis of metastatic breast cancer. All three cases illustrate the complexities in visual rehabilitation of end-stage dry eyes, but the anecdotal observations are encouraging and Remicade® certainly deserves further testing.

REFERENCES

1. Gudas, P. P., Altman, B., Nicholson, D. H., Corneal perforations in Sjögren's syndrome, *Arch Ophthalmol.* 90:470 (1973).
2. Krachmer, J. H., Laibson, P.R., Corneal thinning and perforation in Sjögren's syndrome, *Am J Ophthalmol.* 78:917 (1974).
3. Pfister, R. R., Murphy, G. E., Corneal ulceration and perforation associated with Sjögren's syndrome, *Arch Ophthalmol.* 98:89 (1980).
4. Jabs, D. A., Ocular manifestations of the rheumatic diseases, in: *Duane's Clinical Ophthalmology*, rev ed Vol 4., W. Tasman, E. A. Jaeger, eds. JB Lippincott, Philadelphia (1992).
5. Bernauer, W., Ficker, L. A., Watson, P. G., Dart, J. K. G., Management of corneal perforations associated with rheumatoid arthritis, *Ophthalmology* 102:1325 (1995).
6. Franklin, R., Kenyon, K., Tutschka, P., Ocular manifestations of graft-versus-host disease, *Ophthalmology* 90:4 (1983).
7. Arcoker-Mettinger, E., Skorpik, F., Grabner, G., et al: Manifestations of graft-versus-host disease following allogeneic bone marrow transplantation, *Eur J Ophthalmol.* 1:28 (1991).
8. Tugel-Tutkun, I., Akova, Y. A., Foster, C. S., Penetrating keratoplasty in cicatrizing conjunctival diseases, *Ophthalmology* 102:576 (1995).
9. Dohlman, C. H., Power, W. J. Keratoprosthesis, in: *Principles and Practice of Ophthalmology* 2nd ed., D. Albert and F. Jakobiec, eds., Philadelphia, WB Saunders (1999).
10. Yaghouti, F., Nouri, M., Abad, J. C., Power, W. J., Doane, M. G., Dohlman, C. H., Keratoprosthesis: Preoperative prognostic categories. *Cornea*, in press.
11. Kervick, G. N., Pflugfelder, S. C., Haimovici, R., Brown, H. Tozman, E., Yee, R., Paracentral rheumatoid corneal ulceration. Clinical features and cyclosporine therapy. *Ophthalmology* 99:80 (1992).
12. Kruit, P. J., VanBalen, A. T., Stilma, J. S., Cyclosporine A treatment in two cases of corneal peripheral melting syndrome. *Doc Ophthalmol.* 59:33 (1985).
13. Zierhut, M., Thiel, T. H., Weidle, E. G., et al., Topical treatment of severe corneal ulcers with Cyclosporine A. Graefes Arch. *Clin Exp Ophthalmol.* 227:30 (1989).
14. Nouri, M., Al-Merjan, J., Abad, J. C., Yaghouti, F., Khan, B. F., Doane, M. G., Refojo, M. F., Gipson, I., Dohlman, C. H., An animal model for a keratoprosthesis to test parameters for survival. In preparation.
15. Dohlman, C. H., Doane, M. G., Some factors influencing outcome after keratoprosthesis surgery. *Cornea* 13:214 (1994).
16. Newsome, D. A., Gross, J., Prevention by medroxyprogesterone of perforation in the alkali-burned rabbit cornea: inhibition of collagenocytic activity. *Invest Ophthalmol Vis Sci.* 16:21 (1977).
17. Seedor, J. A., Perry, H. D., McNamara, T. F., Golub, L. M., Buxton, D. F., Guthrie, D. S., Systemic tetracycline treatment of alkali-induced corneal ulceration in rabbits. *Arch Ophthalmol.* 105:268 (1987).

Closing Session

CLOSING ADDRESS: THE ACADEMIC PERSPECTIVE

John L. Ubels

Department of Biology
Calvin College
Grand Rapids, Michigan, USA

Sixteen years ago last week, the First International Tear Film Symposium was held in Lubbock, Texas, and we are now at the conclusion of the Third International Conference on the Lacrimal Gland, Tear Film and Dry Eye Syndromes. If we look back to 20 years ago when I first became involved in research on the ocular surface, we can see the amount of research on dry eye has grown impressively. In 1980 only about a dozen abstracts related to this subject were presented at ARVO, while the 2000 ARVO meeting included about 120. In Lubbock in 1984, 100 presentations were made, and at this meeting you have heard and seen 240! It is clear our knowledge about the lacrimal gland, tears and dry eye disease is substantial; the amount of information in print on the subject is voluminous.

What have academic basic scientists contributed to this database? I do not have time to review the literature extensively; however, I will mention a few highlights: At Lubbock there were reports on effects of neurotransmitters on lacrimal secretion, and early work on subcellular fractionation of lacrimal gland was presented. Now we know much more about mechanisms of lacrimal secretion, signaling pathways, membrane trafficking and the role of androgens in the lacrimal gland.

At Lubbock there were reports on the glycoprotein composition of ocular mucins. Now we know ocular surface epithelial cells, as well as goblet cells, produce mucins. Most important, the genes for mucins have been cloned and we know which mucins are produced by each cell type. At Lubbock early reports concerned analysis of meibomian lipid composition, and interest in the biophysics of the lipid layer existed. At this meeting evidence for the role of androgens in meibomian gland function and disease has been presented.

Two decades ago emphasis centered on the composition of tear fluid. Now we understand tears do not simply wet the ocular surface, but electrolytes, vitamins and

Lacrimal Gland, Tear Film, and Dry Eye Syndromes 3
Edited by D. Sullivan *et al.*, Kluwer Academic/Plenum Publishers, 2002

1341

growth factors in tears are essential for the health of this surface. It is now also evident inflammatory mediators in tears can cause ocular surface inflammation, and many of us are convinced ocular surface inflammation is perhaps the most important component of dry eye disease.

As a result of the work of academic investigators, we now know much about the biology of the lacrimal system and ocular surface. It has been a stimulating and productive two decades of research, but now I must raise a difficult question. What effect has all this research had on the ways in which dry eye disease is treated? Has all our work on tear composition, neurotransmitters, receptors, transporters, signaling molecules, signaling pathways, genes and gene products made a difference for the dry eye patient?

Twenty years ago dry eye was treated with artificial tears and punctal plugs. Several improved artificial tear products are now on the market, but over the past few decades the methods of treatment of dry eye have not changed substantially. At one time it was suggested retinoids might be used to treat ocular surface disease, but this has proven ineffective. Pending further trials, we don't yet know the fate of cyclosporin, and having no effective drugs, the clinician must still rely on artificial tears and punctal plugs. I am afraid the dry eye patient may begin to question the usefulness and applicability of the past 20 years of research.

This question of utility may make us uncomfortable, and some may think it should not be raised at all. As basic scientists we believe our logical hypotheses, well-designed experiments and elegant data are legitimate ends in themselves. We value knowledge for it's own sake. We see basic research as a noble vocation, especially in the academic sphere where we pass this knowledge to our students and prepare the next generation of scientists. However, in today's climate we **will** face the question of utility. My NIH grant on interactions of retinoids and androgens in lacrimal gland was recently renewed because the reviewers liked the science. They considered the hypothesis reasonable and design of the experiments good; however, to quote the summary statement, "Enthusiasm [was] somewhat diminished by the lack of strong justification for the utility of this effort." I believe this is a comment that we as academicians will increasingly see in reviews of our proposals. We have chosen to study dry eye disease, and as such, we are to a certain extent performing applied science. Part of our obligation is to do work that is useful to clinicians, the pharmaceutical industry and ultimately, most importantly, to patients.

I do not mean to end this meeting on a negative note. This is an exciting time. We have done a very good job in our labs working on a difficult and complicated disease, and this will continue. I am not here to make specific recommendation for research priorities. Rather, my challenge to academic investigators is this: Between now and when this society meets again, let us move forward, using our database and our future studies, to do all we can to move our laboratory research toward an effective treatment for dry eye disease, and ultimately to prevention and cure. Patients with Sjogren's syndrome and other dry eye conditions are counting on us.

INDUSTRY PERSPECTIVE

Benjamin R. Yerxa

Inspire Pharmaceuticals
Durham, North Carolina, USA

First I would like to thank the organizers for a great meeting, especially the Sullivans for their tremendous effort and attention to details. It is nice to go to a meeting where every session is excellent and relevant to our work. I would also like to commend the organizers for inviting speakers from neighboring disciplines to point out parallels in other fields.

When I was asked to give a 5-minute talk on the industry perspective, I was honored by such an invitation, but later realized that 5 minutes without slides or data to talk about is like an eternity. But now that we are at the close of the meeting, it's clear that 5 minutes is not long enough to sum up all the terrific science presented here at the meeting. So I thought I would try to describe my impressions in broad strokes. It would be a cop out to just say dry eye is a complex disease–we all know that. From a drug development point of view, and certainly for the patients' sake, I wish it were as simple as saying "humuhumunukunukuapua..."

But seriously, I could walk away from this meeting thinking that it would be completely crazy to even think about developing drugs to treat dry eye, because the signs and symptoms don't correlate, and the objective measures are highly specialized and extremely variable. I wish we could approach dry eye with a single objective target, like lowering IOP in glaucoma patients. In reality though, we have to deal with the fact that this is a multi-faceted disease in which there is an inflammatory component, a tear quantity component and a somewhat mystifying tear quality component.

We could argue all day about pie charts showing the different types of dry eye: aqueous deficient versus evaporative versus inflammatory dry eye, etc... but it would still boil down to the fact that clinicians have very little in their arsenal to treat patients with.

At the end of this meeting though, *I am not discouraged* by the complexity; in fact, I am delighted to see lots of innovative techniques being used in current research, because

Lacrimal Gland, Tear Film, and Dry Eye Syndromes 3
Edited by D. Sullivan *et al.*, Kluwer Academic/Plenum Publishers, 2002

1343

these techniques provide the data that generates the future pipeline for drug discovery in dry eye disease.

I walk away from this meeting with a wealth of information and new ideas about emerging receptor and protein targets, novel and improved diagnostic techniques, etc. These new results are especially evident in the molecular biology field where it seems like the molecular biology is way out in front of the pharmacology. So, it is my impression that over the next few years we will see the pharmacology and functional studies start to catch up to the wide net cast by molecular techniques. Also, the variety of new ways to assess the tear film, especially non-invasive ones, are crucial for successful drug development programs that are desperate to use something other than Schirmer strips.

I am impressed by the depth and quality of the basic science presented here, and I am especially interested in how all this information can be *transformed* into new clinical approaches to *treat the disease*. Let's face it, there is not going to be a single cure or magic bullet for dry eye disease–a multi-faceted disease like dry eye requires a variety of treatment options for the patients. In my opinion, we can do better than artificial tears, punctal occlusion and steroids–*we have to do better* for the patients.

The paradigm shift towards inflammatory dry eye disease is the first of what I believe will be many new trails blazed, both in the clinic and in the lab.

My overall message is to congratulate everyone who contributed to this meeting from academia and industry for producing the new leads for drug discovery and development, and a plea to get these new ideas, compounds, treatments, techniques, etc. into the clinical setting as fast as possible.

I think a complex disease like dry eye is ultimately only truly understood at the level of the patient. As we heard earlier, with millions of patients in the U.S. alone, this presents a tremendous opportunity for pharmaceutical companies to meet a large unmet medical need.

At our next meeting in four years, I sincerely hope we will be talking about some new FDA approved pharmacologically active drugs for treating dry eye.

Addendum

Additional Conference Presentations

During the course of the Third International Conference on the Lacrimal Gland, Tear Film and Dry Eye Syndromes: Basic Science and Clinical Relevance, additional keynote, oral and poster presentations were made that are not summarized in this book. Abstracts of these presentations, which are titled below, may be found in a special supplement of the journal Cornea (volume 19, number 6, November 2000).

The structural basis of tissue specificity in normal and malignant breast: the role of signalling from surface receptors.

Evaluating the role of the main and accessory lacrimal glands.

Pharmacology of capacitative calcium entry channels in lacrimal acinar cells and other non-excitable cells.

Mucins of the ocular surface.

Regulators of mucin gene expression in the ocular surface epithelia.

The place of meibometry in the study of meibomian gland disfunction.

Clinical diagnosis, classification and management of meibomian gland dysfunction.

The proteinaceous composition of human tear fluid as determined by mass spectrometry.

Overexpression of biglycan induces ocular surface disorders in *Ktcnpr-Bgln* transgenic mice.

Effects of lasik on tear production, clearance and the ocular surface.

P2Y2 agonist, INS365: a potential dry eye therapeutic agent.

Secretory leukocyte protease inhibitor (SLPI) in conjunctiva of human.

Protein kinase C mediates UTP-induced rabbit corneal epithelial migration.

Alterations of the corneal epithelium and tear film after small - incision cataract surgery - first results.

Development of an instrument for the measurement of practical visual acuity and its clinical application in lasik patients.

Effects of inhibitor of adhesion molecule (K-7174) in Sjögrens syndrome model mice.

HLA DR expression in different locations of bulbar conjunctiva of early KCS eyes.

A novel non-invasive, in vivo technique for the quantitation of leukocyte rolling and extravasation at sites of inflammation in human conjunctiva.

The cytokines studies of organotypic cultures of ocular surface epithelium on amniotic membrane.

Ocular surface inflammation in conjunctival biopsies of patients with moderate to severe keratoconjunctivitis sicca (KCS) before and after treatment with topical cyclosporine A.

Novel autoantibody to salivary gland-specific 45-kDa antigen in patients with Sjögren's syndrome (SS).

Frictional characteristics of contact lenses and ophthalmic solutions.

Radioactive contamination of contact lenses during radioiodine therapy.

Immunophenotype of lymphocytic cells in cases of benign lymphoepithelial lesion with
    Sjögren-like dry eye.

Herpes zoster acute dacryoadenitis: a case report.

Superior limbic keratoconjunctivitis, graves disease and Sjögren's syndrome in japanese
    patients.

Images and analysis to help non invasive tear film diagnosis with the tearscope plus.

Table for ambiental evaluation, prevention and treatment of dry eye.

About the effect of an estrogen-ointment on the ocular conjunctiva in postmenopausal women
    with keratoconjunctivitis sicca.

Transplantation of cultivated corneal epithelial cells on amniotic membrane; preliminary
    clinical outcomes.

The long-term effect of punctal occlusion for the treatment of dry eye syndrome.

Electrodessication with punctal ring resection and epithelial debridement is very effective
    for permanent punctal occulusion.

AUTHOR INDEX

INDEX

Eukaryotic initiation factors
  eIF-2, 65—69
  eIF-4, 65—69
Evaporative dry eye (EDE), 389—94, 705, 1087*See also* dry eye syndrome
  androgen deficiency and, 390—93
  causes of, 390—94
  estrogen hormone replacement therapy and, 393—94
  lipid mixture concentrations and, 416
  meibomian gland dysfunction and, 390—93
  oil-water emulsion effects, 419—23
  sex steroid hormones and, 389—94
  tear supplements for, 419
Evaporative stress test, 1153—57
Evaporimeter, 1154—57
Excimer laser surgery. *See also* laser in situ keratomiluesis (LASIK); photorefractive keratectomy (PRK)
  corneal haze following, 1316
Exocrine cells
  CA^{2+} signaling, 175—81
  functions of, 175—76
  insulin-signaling, 27—30
Exotropia, 73
Extended wear contact lenses
  albumin deposit effects, 951—54
  dry eye symptoms and, 1011
  mucin balls and, 917—23
  surface protein profile, 957—59
Eye-associated lymphoid tissue (EALT), 842—43, 865
Eye discomfort rating scale, 503. *See also* ocular discomfort
Eyelashes, gaze direction and, 1210
Eyelids
  blinking motion, 593—94, 677—83
  cleansing, 704, 705, 706
  downward gaze and, 1208—10
  heaviness, 1234
  inflammation and, 1073
  lid margin disease, 733
  lid margin metaplasia, 1082
  lipid measurement, 517—20
  ocular dryness and, 1107
  sodium hyaluronate and, 1234
  surface temperature, 1227—31
  surgical reconstruction of tear meniscus at margin, 1263—68
  sweeping effect of, 524
  third, in dogs, 1278, 1279
Eye movements
  downgaze, 1088, 1205—10
  horizontal near focusing, 1205—10
  superior epithelial arcuate lesions (SEALs) and, 975

Eye movements (*cont d*)
  tear distribution and, 1088
  tear osmolarity and, 1091
  upgaze, 1088

Factor B, 952
Familial amyloidosis type II/Meretoja disease, 657, 659
Familial dysautonomia (FD)
  β-NGF, 74
  carrier status, 75
  corneal disease associated with, 631—32
  current therapies, 76
  defined, 71
  diagnosis, 75
  enzymatic mechanisms in corneal ulceration, 629—36
  genetic basis of, 75—76, 632—33
  information resources, 77—78
  investigational therapies, 76—77
  ocular manifestations of, 71—78, 632
  prenatal diagnosis, 75—76
  prognosis, 77
  symptoms, 73—74
  systemic features of, 73, 631—32
Farnesyl pyrophosphate synthetase, 466, 470, 471
FAS/FAS ligand-mediated apoptosis
  in cornea, 828
  in lacrimal gland, 773—75
FA synthase, 466, 470, 471
Fatty acids
  age and, 937
  dry eye symptoms and, 898
  Sjgren's syndrome and, 443—46, 449
  in tear film, 374, 375, 376
FBUT. *See* fluorescein breakup time (FBUT)
Ferning test of tears, 611, 612
Fiberscopic observation, of punctal plug extrusion, 1285—88
Fibronectin, contact lens wear and, 880, 881, 883
Fibrosis, corneal ulcers and, 631, 636
FK506, 1042
Flagellin, 269—72
Flow cytometry, 761—67, 778
Fluid secretion model, 191—97
Fluid transport
Fluorescein breakup time (FBUT). *See also* breakup time (BUT)
  accuracy of, 509—11, 1124
  Akorn Dry Eye test of, 1135—39
  evaluation of, 1128—29
  fluorescein instillation methods, 507—11
  hormone replacement therapy and, 1030, 1032
  LASIK treatment and, 712
  methods, 508—9

Glx, 343, 344
Glycerophospholipid (GPL)-rich microdomains, 208—10
Glycine, 618, 619
Glycocalyx, 350
Glycocojugates, 308, 341—44
Glycolipids, 207—11, 270
Glycoproteins, 347—50, 353—56
Glycosphingolipid (GSL), 208—11
Glycosylated lipophilin, 574
Glycosylation, 309
GM1, 270
Goblet cells
    characterization of, 301—5
    conjunctiva epithelial cells and, 651, 652, 653
    functions of, 301, 1071
    gefarnate-stimulated mucin production, 353—56
    growth and morphology of, 303
    inflammation and, 1073
    LASIK treatment and, 714—15
    mitogen-activated protein kinases (MAPK) and, 297—300, 303
    ocular surface shrinkage and, 363—68
    physiology of, 1068
    in Sj gren's syndrome, 368
    TTF-peptide synthesis in, 313—15
    in vitro systems, 301—5
Goggles, for environmental stress test, 1154—57
Gold particles, for particle-mediate gene transfer, 1299—1302
Golgi complex
    endomembrane compartments and, 54, 207
    [^{125}I]-EGF and, 215—16
    proteins associated with, 51
G protein, 193
G protein-coupled receptors (GPCRs), 175, 261
    host responses to flagellin and, 270—71
    MAPK activation and, 185, 188
    signaling, 42, 178—79
G protein-independent receptors, 175
G protein-linked nucleotide receptors, 271
GR-1, 855—57
Graft-*versus*-host disease (GVHD)
    chronic (cGVHD), 1041—44
    corneal blindness from, 1335—40
Gram-negative bacteria, 525, 551
Gram-positive bacteria, 525, 551
Granulation tissue, punctal plug extrusion and, 1285—88
Greater superficial petrosal nerve (GSPN), 225—29
GTPase accelerating (GAP) functions, of RGS proteins, 180

Hagman factor, 962
Harderian glands

aldosterone and, 433
Harderian glands (*cont d*)
    cyclosporine and, 433, 436
    functions of, 115, 431
    gangliosides and, 433
    lipase mRNAs expressed in, 115—18
    lipid production, 431—36, 437, 439
    pharmacological effects, 431—39
    phospholipids and, 433
    somatostatin-like immunoreactivity (SLIR) in, 81—83, 85
    tacrolimus and, 433
Headache, 1047—49
Head position. *See also* eye movements
    in horizontal near vision, 1205—10
Hematopoietic stem cell transplantation (SCT), 1041—43
Hemifacial spasm, 1241, 1242
Hereditary sensory and autonomic neuropathies (HSAN), 631
Herpes keratitis, 657
Herpes simplex virus, 746, 828
Herpetic corneal disease, 662, 663
15(S)-HETE
    conjunctival mucin secretion and, 323—26
    corneal epithelium mucin production and, 317—20, 329—32
    defined, 317
    for dry eye treatment, 317—20, 326, 329—32, 335—39
    functions of, 324, 326
    mucin production and, 335—39
Heterodimers, 573, 574
Heterotrimeric G proteins, 179—80
B-hexosaminidase
    carbachol and, 201, 207, 232—34
    glycosphingolipid (GSL)-rich microdomains and, 207, 208
    microtubules and, 201
    rabbit lacrimal gland, 33—38
High-density trans-Golgi network (hd-tgns), 207, 209—10, 215—16
High endothelial venules (HEV), 838, 863, 864, 870
High molecular weight kininogen (HMWK), 961—66
High performance liquid chromatography (HPLC), 935—36
Hilafilcon A, 968
Histidine, 618, 619
HIV vectors, 1310—13
HMG-CoA synthase, 466, 470, 471, 473
Holo-tear lipocalins (holo-TL), 557, 562—63
Homer proteins, 179
Homodimers, 573
Horizontal near focusing, 1205—10
Hormone replacement therapy (HRT)